WILLA CATHER

Willa Cather as Managing Editor of *McClure's* (c. 1910).

WILLA CATHER

The Emerging Voice

SHARON O'BRIEN

New York Oxford
OXFORD UNIVERSITY PRESS
1987

Oxford University Press

Oxford New York Toronto
Delhi Bombay Calcutta Madras Karachi
Petaling Jaya Singapore Hong Kong Tokyo
Nairobi Dar es Salaam Cape Town
Melbourne Auckland

and associated companies in
Beirut Berlin Ibadan Nicosia

Published by Oxford University Press, Inc.,
200 Madison Avenue, New York, New York 10016

Library of Congress Cataloging-in-Publication Data

O'Brien, Sharon.
Willa Cather : The emerging voice.

Bibliography: p.
Includes index.
1. Cather, Willa, 1873–1947. 2. Novelists, American—20th century—Biography. I. Title.
PS3505.A87Z746 1987 813'.52 [B] 86-12710
ISBN 0-19-504132-1 (alk. paper)

9 8 7 6 5 4 3

Printed in the United States of America
on acid-free paper

To my family,
especially to my mother,
and in memory of my father

Acknowledgments

While I was working on the book I was particularly struck by Willa Cather's description of the creative process as a "gift from heart to heart" because it captured my own experience. In one sense this book is my own, but in another sense it has been given to me by the friends and colleagues who have offered me support, encouragement, and intellectual companionship over the last fifteen years.

I trace the book's origins back to my years in graduate school when three people helped me learn how to think and to write. Daniel Aaron, my mentor and dissertation adviser, first suggested that I combine biographical with literary analysis; he gave me the foundations of this book along with hours of his time, and our ongoing discussions have contributed to its evolution. Barbara Solomon urged me to ask a historian's questions of my material, and she can see how much I owe to her advice and example. And my good friend and fellow dissertation-writer Glenda Hobbs first showed me how the pleasures of friendship and creative work could be united; I was aided in writing because I knew she was my sympathetic and demanding reader, asking then—as now—that I be my "best self."

Three other friends and readers were particularly important to me during the years when I was finding my way as a writer and a scholar; their enthusiasm for the project sustained me many times when I doubted its worth. With Mary Kelley I first began to explore the conflicts many nineteenth-century women writers experienced between gender and creativity, and our early collaborations gave me a sense of the shape my book should take. I have benefitted both from her work on the women writers who preceded Cather and from the sense of history she has helped to give me. In the course of our many conversations about politics, literature, culture, feminism, and friendship, Janice Radway helped me to trust my sense that the mother–daughter bond was important in Cather's development; her intellectual clarity and courage have aided me countless times, and I thank her in particular for stimulating my thinking about the theoretical and political implications of my work. Having team-taught two courses on women writers with Ellen Rosenman, I can only say that her influence and support have been profound; our years of collaboration at Dickinson College were a happy time when professional and personal lives seemed effortlessly integrated, and I owe much of my productivity during that period to our friendship.

Lonna Malmsheimer has the distinction of being the only person not employed or remunerated by a publishing company to have read the whole manuscript, all 950 pages of it: I thank her for this heroism, as well as for her long-standing encouragement of my plunge into interdisciplinary scholarship. Clarke Garrett gave me useful and tactful suggestions for my first chapter; Peggy Garrett helped me with my Jewett chapter and, on very short notice, guided my revisions for the last three chapters. Both Garretts have provided me with civilized dining, sound—if occasionally contradictory—editorial advice, and a sustaining friendship.

Many others have helped by reading drafts of chapters along the way, and—perhaps more important—by expecting me to complete the book. I am grateful to Laurence Davies, Miles Orvell, Genessee Pomfret, Peter Rabinowitz, John Rogers, John Sears, Maggy Sears, and Jonathan Strong both for their attention to specific chapters and, more generally, for letting me know that they valued me and my work.

For providing me a supportive context in which I could write, I thank, first of all, Cary Nicholas for offering a desk at the Philadelphia Women's Law Project—where I wrote many chapters before I discovered the word processor—and for helping me through my year of full-time writing. And I must mention my colleagues at Dickinson College—primarily but not exclusively located in the English Department—whose friendly harassment may actually have been useful. Ken Rosen and Bob Winston deserve particular notice: the former for his generosity in word and deed, the latter for his numerous inquiries after Willa—it was good to know that someone cared.

Turning a manuscript over to a publisher can be an ambivalent experience. Publishing of course signifies the successful end of writing, and yet one can fear what will happen when one's vulnerable and defenseless pages are given over to strangers. So I want to thank Bill Sisler and Joan Bossert, my editors at Oxford University Press, for their considerate, careful, and gratifying work with the manuscript: my book has found a good home.

I am grateful to various agencies and institutions that offered me financial support while I was traveling to collections, doing research, and writing. The American Philosophical Society gave me two travel grants, and the Dickinson College Research and Development Committee has generously funded travel and research on several occasions. A fellowship from the National Endowment for the Humanities allowed me to devote a year to writing, and the Harvard University English Department has helped to defray the costs of preparing the manuscript for publication.

I could not have written this book without drawing on the insights and achievements of a community of scholars. I describe the importance of feminist scholarship to my work in my Introduction, but here I must mention a special debt: to two Cather scholars whose ground-breaking editorial and interpretive work has made my book possible. If Bernice Slote and William Curtin had not devoted so much time and intelligence to reclaiming and publishing Willa Cather's early journalism, I would not have been able to understand the importance of her apprenticeship years. And I particularly want to thank William Curtin for his early and generous encouragement of my unorthodox approach to Willa Cather: at a time when feminist criticism had yet to make an impact on the profession and when Cather scholarship was largely preoccupied with formalist issues, he gave me confidence that I was on the right track in addressing questions of gender and sexuality.

All scholars depend heavily on the helpfulness and resources of libraries and historical societies, and this is particularly so in my case. Because Willa Cather forbade publication of her letters in her will, her critics and biographers have paraphrased her letters rather than quoting them directly, and I honor this tradition here. So I am grateful to the following libraries and historical societies for granting me permission both to consult their holdings and to refer to Cather's letters in my book: the Bailey/Howe Library, University of Vermont; the Collection of American Literature, Beineke Rare Book and Manuscript Library, Yale University; Bentley Historical Library, University of Michigan; Special Collections, Colby College Library; the Enoch Pratt Free Library, Baltimore, Maryland; the Holy Cross College Library; the Houghton Library, Harvard University; the Huntington Library; the Nebraska State Historical Society; the Newberry Library; the Pierpont Morgan Library; the Willa Cather Collection (#6494), Manuscripts Department, University of Virginia Library; and the William R. Perkins Library, Duke University.

And finally I owe an incalculable debt to my family, whose loving support and healthy irreverence have proved an effective combination. Thank you for taking me seriously and for not taking me too seriously.

S.O.

July 1986
Carlisle, Pennsylvania

Contents

Abbreviations

Editions of books by Willa Cather cited in the text (in chronological order):

April Twilights (1903), edited and with an introduction by Bernice Slote (Lincoln: University of Nebraska Press [Bison ed.], 1976): *AT*

O Pioneers! (Boston: Houghton Mifflin, 1913): *OP*

The Song of the Lark (Boston: Houghton Mifflin, 1915; rpt. University of Nebraska Press [Bison ed.], 1978): *SL*

My Ántonia (Boston: Houghton Mifflin, 1918): *MA*

Youth and the Bright Medusa (New York: Alfred A. Knopf, 1920; rpt. Vintage, 1975): *YBM*

One of Ours (New York: Alfred A. Knopf, 1922; rpt. Vintage, 1971): *OO*

Alexander's Bridge (Boston: Houghton Mifflin, 1912; rpt. Lincoln: University of Nebraska Press [Bison ed.], 1977): *AB; Alexander's Bridge* (Boston: Houghton Mifflin, 1922): *AB* (1922)

April Twilights and Other Poems (New York: Alfred A. Knopf, 1923): *ATOP*

A Lost Lady (New York: Alfred A. Knopf, 1923; rpt. Vintage, 1972): *LL*

The Professor's House (New York: Alfred A. Knopf, 1925; rpt. Vintage, 1973): *PH*

My Mortal Enemy (New York: Alfred A. Knopf, 1926; rpt. Vintage, 1961): *MME*

Death Comes for the Archbishop (New York: Alfred A. Knopf, 1927; rpt. Vintage, 1971): *DCA*

Shadows on the Rock (New York: Alfred A. Knopf, 1931; rpt. Vintage, 1971): *SR*

Obscure Destinies (New York: Alfred A. Knopf, 1932; rpt. Vintage, 1974): *OD*

Lucy Gayheart (New York: Alfred A. Knopf, 1935; rpt. Vintage, 1976): *LG*

Not Under Forty (New York: Alfred A. Knopf, 1936): *NUF*

Sapphira and the Slave Girl (New York: Alfred A. Knopf, 1940; rpt. Vintage, 1975): *SSG*

The Old Beauty and Others (New York: Alfred A. Knopf, 1948; rpt. Vintage, 1976): *OB*

Willa Cather on Writing (New York: Alfred A. Knopf, 1949): *OW*

Edited collections of Willa Cather's works:

The Kingdom of Art: Willa Cather's First Principles and Critical Statements 1893–1896, ed. Bernice Slote (Lincoln: University of Nebraska Press, 1966): *KA*

Willa Cather's Collected Short Fiction 1892–1912, ed. Virginia Faulkner (Lincoln: University of Nebraska Press, 1970): *CSF*

The World and the Parish: Willa Cather's Articles and Reviews, 1893–1902, ed. William M. Curtin (Lincoln: University of Nebraska Press, 1970), 2 volumes: *WP*

Uncle Valentine and Other Stories: Willa Cather's Uncollected Short Fiction, 1915–1929, ed. Bernice Slote (Lincoln: University of Nebraska Press, 1973): *UV*

Books about Willa Cather:

Edith Lewis, *Willa Cather Living* (New York: Knopf, 1953): *WCL*

Mildred Bennett, *The World of Willa Cather* (Lincoln: University of Nebraska Press [rev. ed.], 1961): *WWC*

Elizabeth Sergeant, *Willa Cather: A Memoir* (Lincoln: University of Nebraska Press [Bison ed.], 1963): *WC:AM*

Edited by Willa Cather:

Sarah Orne Jewett, *The Country of the Pointed Firs and Other Stories*, selected and arranged with a preface by Willa Cather (New York: Doubleday [Anchor ed.], 1956): *CPF*

WILLA CATHER

Introduction

Willa Cather did not make it easy for her biographers. Unlike Edith Wharton, who carefully left a packet of papers labeled "For My Biographer," in the last years of her life Cather set about destroying letters in her possession and urging friends to do the same. In her will she tried to regulate the public's access to her literary and personal texts, directing her executors to refuse radio, film, and dramatic adaptations of her fiction and forbidding publication of her correspondence. Having been exposed to severe critical attacks during the 1930s—when reviewers accused her of softness, romanticism, and escapism—Cather did not trust future generations of biographers and scholars to address her work or her life sympathetically. She sought to control interpretations of her life and her fiction in the only way that seemed certain—by reducing or eliminating the evidence on which interpretation could be based.

In her eagerness to manage her literary reputation, Cather particularly sought to deemphasize her early years of trial and experimentation. Designing the Library Edition of her fiction in the 1930s, she chose to begin with *O Pioneers!*, the first novel she liked to think of as her own. Thus she eliminated from official consideration the dozens of stories she wrote during her lengthy apprenticeship as journalist, high school teacher, and managing editor of *McClure's Magazine*. And yet if we take Cather's autobiographical portrait of the artist—*The Song of the Lark*—as a guide to her artistic development, then we are directed to consider precisely the early years that she later sought to underplay in the official version of her literary self. "Success is never so interesting as struggle," Cather wrote in her Preface to the 1932 English edition of *The Song of the Lark*, and it is her own years of struggle that concern me here: the years in which she was striving to define herself as woman and writer.[1] Although when I began this biography I intended to trace the connections between Cather's life and work over the span of her entire career, I became convinced that the story of her artistic emergence was important enough to merit a volume of its own.

"That is happiness," wrote Willa Cather in *My Ántonia*, "to be dissolved into something complete and great" (*MA*, p. 18). Most of the heroic protagonists in Cather's novels find happiness in this fashion, devoting their lives and energies to something larger than the self: the land, the family, art, religion. But before she could imagine such characters or "be dissolved" into the creative process, Cather would create a self and a voice. This was a difficult task for a young woman who found no acceptable models for identity and vocation in the late-Victorian culture. Much of Cather's energy during her apprenticeship years was absorbed by her attempt to fashion a female self that could be compatible with the artist's role.

Since the young Willa Cather equated art with masculinity, the story of her apprenticeship and artistic emergence contains two separate but related mysteries. How did she manage to move from male to female identification? The adolescent girl first expressed her rejection of Victorian definitions of femininity by posing as William Cather; later, the aspiring journalist and writer adopted masculine values in both life and art, seeking self-definition by denigrating and devaluing women. Yet by the time she wrote *Alexander's Bridge* (1912) and *O Pioneers!* (1913), Cather had dispensed with conventional categories of gender and in *O Pioneers!* portrayed a strong female hero who was not masculine. Reimagining and accepting womanhood, the novelist finally reestablished the connection she had broken with the women in her own past, a far more radical act than the young woman's male identification. What social, psychological, and literary contexts helped her to do so?

The second mystery is the literary one Cather shares with other writers: How did she become a creative artist? Like Walt Whitman, Cather had a long apprenticeship: twenty years separate her first short story from her first novel. Moreover, in most of the fiction she wrote during these early years, Cather was not, as she later observed, writing from her "deepest experience" (*AB* [1922], p. vi). Retrospectively she described her literary birth in *O Pioneers!* as the writer's inexplicable, intuitive discovery of her own voice and material, but the literary and personal advances Cather had made during the preceding two decades allowed her to create what she thought she had found: "a cadence, a quality of voice that is exclusively the writer's own" (*CPF*, p. 7). Unraveling this mystery will lead us back to the first, for as long as Cather denied her womanhood she was unable to speak authentically and powerfully as a writer.

Focusing on these issues, I have structured this biography both chronologically and thematically. The first part ("Backgrounds") explores Cather's childhood and adolescence, looking in particular at the impact of the mother–daughter bond on her search for identity; the second part ("Apprenticeship") examines what Cather later called the artist's years of "awakening and struggle," beginning with her college writings and ending with the first stories she published after her brief but important friendship with Sarah Orne Jewett;[2] and the third part ("Emergence") looks closely at the important transitional period in Cather's life as a writer: the years 1911 and 1912, when she wrote *Alexander's Bridge*, visited the Southwest for the first time, and returned to find *O Pioneers!* spontaneously writing itself.

As I trace the interconnections between Willa Cather's life and her fiction throughout these years, I offer the reader a shaped narrative, selecting details and incidents that illuminate her literary emergence. All biographers employ principles of selection, of course, and all biographies are interpretations as well as accounts of the life under consideration. All biographers write at a particular moment in history and read their subjects through the conceptual frameworks available to them at the time. And yet a biographer must try, as far as possible, not to impose the present upon the past, but to capture the meaning experience held for someone who lived in a different historical and social context. So I have tried to combine the outsider's and the insider's perspective, integrating detachment and sympathy in describing and interpreting Willa Cather's life and work.

Since I locate my subject in a specific historical and social background, I must also place my own book in context, making clear the interpretive frameworks through which I view Willa Cather's literary development. Although I employ a range of methods—biographical, historical, psychoanalytic, literary—in seeking to connect Cather's life and fiction, I have been particularly influenced by recent developments in feminist criticism. While I am indebted to an entire community of scholars, I must single out two books published in the late 1970s: Nancy Chodorow's *The Reproduction of Mothering: Psychoanalysis and the Sociology of Gender* and Sandra M. Gilbert and Susan Gubar's *The Madwoman in the Attic: The Woman Writer and the Nineteenth-Century Literary Imagination.*

When I was writing my dissertation in the early 1970s, historian Barbara Solomon asked me whether Cather's relationship with her mother might have been a significant influence. I thought not. A few years later, having read Chodorow's book, I began rethinking this question and rewriting the early chapters of the manuscript. Chodorow's analysis of the role of the mother in the daughter's psychological development allowed me to see important connections between Cather's personal and her literary life—to see, for example, why her friendship with Sarah Orne Jewett had been such an important catalyst. Gilbert and Gubar's provocative study of nineteenth-century women writers encouraged me to think more deeply about Cather's relationship to a female literary tradition and to apply Chodorow's model of female development to the dynamics of literary influence. And other advances in feminist literary and psychoanalytic theory reinforced my conviction that a new biography of Cather was needed that would look fully and seriously at the interconnections between gender and creativity.

So my biography differs from other biographies of Willa Cather because I regard her not simply as a writer who happened to be female but as a woman writer, one whose apprenticeship was distinguished by her struggle to resolve the culturally imposed contradictions between femininity and creativity. Once Cather resolved these contradictions, her work and her life became less marked by issues of gender. This does not mean that the novelist who published *My Ántonia* (1918) or *Death Comes for the Archbishop* (1927) was a genderless or androgynous being, writing in some rarified imaginative realm beyond culture. On the contrary: once Cather had reconciled the woman and the artist, she could write from her own necessarily female experience without feeling that she was limited to telling a woman's story. In her major novels she was both writing as a woman and telling a human story; by the 1920s, "female" and "universal" no longer seemed like mutually exclusive terms to her.

Cather's correspondence shows how strongly she viewed herself as a woman writer even though she realized that this category could lessen her literary reputation in the eyes of critics and readers who associated art with masculinity. Her first inclusion in a literary history—Grant Overton's *The Women Who Make Our Novels* (1918)—struck her as a dubious honor: she had been grouped among the ''authorines,'' she told her editor at Houghton Mifflin, a term suggesting the diminished stature women writers have traditionally been given in American literary culture.[3] And throughout the 1920s and 1930s Cather remained aware of the liabilities of being classified as a woman writer: she once told Sinclair Lewis that the critics cursed her because she did not write like a man; and she admitted to one of her readers that it was a great disadvantage to be a lady author—anybody who thought it wasn't was a fool and should read Virginia Woolf's *A Room of One's Own*.[4]

But despite these cultural pressures to separate the identities of ''woman'' and ''writer'' so laboriously integrated during her apprenticeship years, Cather never insisted that she be considered a writer rather than a woman writer. In fact, she once told a close friend that the critic who best understood her was Margaret Lawrence, whose *The School of Femininity* captured the real merit in her books.[5] Lawrence's 1936 study of British and American women writers is an unusual critical work for its time. Like Gilbert and Gubar, Lawrence attempts to define the ''artist quality peculiar to women'' by examining the imaginative patterns connecting seemingly unique and individual writers, among them Virginia Woolf, Edith Wharton, Katherine Mansfield, and Willa Cather. In Cather's writing, Lawrence found the ''freedom to be feminine.'' ''It is free from a depressing sense of its limitation as feminine writing,'' she observed of Cather's fiction, ''and is also free from a subconscious effort to be masculine. It is the work of a woman sure of herself as a woman.''[6] Given Cather's appreciation of *The School of Femininity,* to regard her as a woman writer rather than as a genderless artist is not to impose an uncongenial category upon her life and art, but to articulate her own experience of gender and creativity in the language of the present.

When I first began work on this book in the mid-1970s I was convinced that viewing Cather as a woman writer captured her perception of her own life and work, but I was not then prepared to portray her as a lesbian writer. I knew that friendships with women constituted the emotional center of her life, but I had no way of knowing whether or not these bonds were sexual, and I was not certain that she viewed herself as lesbian. Wanting, as far as possible, to express Cather's self-definition—not to impose a definition that might distort her experience of herself and her love for women—I did not refer to her as lesbian in the early drafts of the manuscript. But when I came across Cather's love letters to Louise Pound, written during her college years, I changed my mind. This important correspondence persuaded me that ''lesbian'' did in fact capture Cather's self-definition and that my biography should consider the impact on the creative process of her need both to conceal and to reveal her experience of desire.

In discussing Cather's lesbianism in my book, I know that I am disregarding one of her wishes. Judging from Cather's letters and her fiction, she wanted to be viewed as a woman but not as a lesbian. Her love for women was a source of great strength and imaginative power to her, but she feared misunderstanding and repudiation if this love were to be publicly named, quite a legitimate fear in her time. But if we are to understand not only this writer's life but also the complex interconnections between her life and her work, we must now take account of her lesbianism, which has long been assumed by most readers although excluded from official biographical consideration. Willa Cather's need to conceal and camouflage her sexual identity in both life and art arose from her society's condemnation of such love as deviant and unnatural. But since I am writing in a more enlightened era than hers, I can assume that I am addressing the sympathetic readers that Cather, given her different historical context, could not have imagined.

While some strains in feminist theory have helped me to see and to trace patterns in Cather's life and work, other recent developments in feminist criticism, as well as in psychoanalytic and literary theory, make biography itself a problematic genre. For those who reject the liberal humanist view of a unified, coherent self and envision in its place an unstable, shifting configuration of forces, biography—insofar as it portrays a stable, knowable self—seems to offer the reader a deceptive certainty. In this book I do not intend to represent my subject's core, essential self, a futile project since the self is always changing, always in the process of self-creation. But some patterns can be discerned linking an individual's many selves, and in this biography I trace these patterns by describing the fictions Willa Cather used to imagine and create a self. Imbued with the nineteenth-century faith in individualism that many people now question, Cather believed that she had discovered her authentic, essential identity in the Southwest in 1912 and then expressed that identity honestly and openly in *O Pioneers!* From my perspective, Willa Cather was creating rather than discovering a self by drawing on the cultural fictions available to her: the romantic story of self-discovery and the American story of self-transformation. So I use the romantic notion of the unique, essential self to explain Cather's literary and personal development because, in her social and historical environment, she used that belief to fashion the first half of her life.

In the "Preface" to her 1925 edition of Sarah Orne Jewett's *The Country of the Pointed Firs and Other Stories,* Cather praises Jewett's possession of an "individual" literary voice, which she attributes to Jewett's ability to convey the distinctive speech of her Maine countrywomen and sea captains. "She had not only the eye," Cather observes, "she had the ear." Because she paid attention to the voices of the people around her, Jewett was given "the finest language any writer can have," the cadences that composed her own "beautiful voice" (*CPF,* pp. 10, 7). Cather also created her literary identity when she became more concerned with capturing others' voices than with shaping her own. Having in turn listened to Willa Cather over the past several years, I now tell her story in my own language, hoping that her voice will be heard by my reader.

NOTES

1. Willa Cather, *The Song of the Lark* (Boston: Houghton Mifflin, 1915; rpt. 1938), p. v.
2. Cather, *The Song of the Lark,* p. vi.
3. Willa Cather to Ferris Greenslet, May 12, 1918, Houghton Library, Harvard University, Cambridge, Mass.

4. Willa Cather to Sinclair Lewis, n.d., Beineke Library, Yale University, New Haven, Conn., and Willa Cather to Mr. Bain, January 14, 1931, Bentley Historical Library, University of Michigan, Ann Arbor, Mich.

5. Willa Cather to Carrie Miner Sherwood, June 28, 1939, Willa Cather Pioneer Memorial Collection, Nebraska State Historical Society, Lincoln, Nebr.

6. Margaret Lawrence, *The School of Femininity* (New York: Frederick Stokes, 1936), pp. 356–57.

I

BACKGROUNDS

1

The Strains of Blood

> As we grow old we become more and more the stuff our forebears put into us. . . . We think we are so individual and so misunderstood when we are young; but the nature our strain of blood carries is inside there, waiting, like our skeleton.
>
> *My Mortal Enemy*

> One realizes that even in harmonious families there is this double life: the group life, which is the one we can observe in our neighbour's household, and, underneath, another—secret and passionate and intense—which is the real life that stamps the faces and gives character to the voices of our friends. Always in his mind each member of these social units is escaping, running away, trying to break the net which circumstances and his own affections have woven about him.
>
> "Katherine Mansfield"

Even when Willa Cather sought to imagine her "real life" by masquerading as a young man, she wanted to be rooted in the family's group life. Baptizing herself "William Cather, Jr." in adolescence, the defiant daughter signified her determination both to escape the private sphere Victorian America assigned to women and to sever the family "net" by taking a male name. Yet simultaneously she wanted to be the descendant of forebears who had occupied the public world before her. Cather's new name linked her to grandfathers William Boak and William Cather and to her mother's brother, William Seibert Boak, who was killed fighting for his Virginia regiment in the Civil War. Desperate to break free both from a provincial Midwestern town and the confinement of the female role, the young Cather acknowledged no similarities between herself and the older women in her family. What ties could link a brilliant, ambitious girl who wanted to be a doctor with pious farm women? Surely she resembled her heroic uncle who fought gloriously on the battlefield, not her self-effacing grandmothers who labored silently in the home.

In imagining her family's masculine blood flowing in her veins, the adolescent girl allied herself with Cather and Boak ancestors who had made their mark in the public world of male enterprise, leaving behind the written records that had become part of a community's official history: wills, deeds, paragraphs in local histories.[1] But this male lineage was only one strain in Willa Cather's family inheritance. In constructing her "real" self as a woman and a writer, the daughter would eventually re-weave the ties she had broken with the female branch of her family.

•

Although we associate Willa Cather with the Nebraska landscape, she was born in 1873 in Back Creek, Virginia, and spent the first nine years of her life in this small farming community a few miles west of Winchester, within sight of the Blue Ridge Mountains. In the "Biographical Sketch" Cather wrote for her publisher in the 1930s she connected the hierarchies in Virginia's "old conservative society" with the patrilineal inheritance of land. She had been born in an "ordered and settled" world, she wrote, where "the original land grants made in the reigns of George II and George III had been going down from father to son ever since."[2] Although the Cathers did not partake in the original royal land grants, they also passed down their land from father to son. Like many of the settlers in Virginia's Shenandoah Valley, the Cathers were descended from Scotch–Irish pioneers and farmers who had migrated down the valley from Pennsylvania in the late eighteenth century. Jasper Cather came from Ireland to western Pennsylvania in the 1750s, fought in the Revolution, and by 1786 had established a farm in Frederick County, a few miles west of Winchester, Virginia. He left his property on Flint Ridge to his sons, James and David.

Although the Back Creek region was not wealthy, the Cathers were among its most prominent citizens. Willa Cather's great-grandfather, James Cather, built on his father's heritage and became a successful farmer and stock-raiser. As the local magistrate—the representative and executor of law in his small community—James was widely respected for his compassion, judiciousness, and intelligence. Twice elected to the state legislature, he opposed slavery and secession but supported the rebel cause during the Civil War, loyal to what J. E. Norris' local history calls his "Southern blood and chivalry." A staunch Presbyterian, James acted on Christian beliefs by opening his home to "the orphan and deserving poor." He valued community service over individual profits—refusing to sell his grain to speculators if "he had reason to think the poor in his neighborhood had need of his supplies." But James frowned on charity without responsibility. He would attend sheriff's sales, remembers his grandson James Gore, buy up household goods, and hold them until the widows could repay him.[3] Reportedly gifted with a lively mind, James Cather made his household the neighborhood center for conversation as well as charity: "nowhere else in the district was there such talk, amusing, thoughtful, informed, alive."[4]

According to Norris' history, few men in Frederick County ever "bequeathed to posterity" as inspiring a legacy as did James Cather (p. 610). To his son William, James certainly left his Protestant virtues. Willa Cather's grandfather inherited his father's industry, sobriety, and piety without, perhaps, his conversational abilities. A taciturn, strong-willed patriarch, William Cather wanted his home protected from unseemly levity and recreation, forbidding cards and music on Sundays. A successful farmer like his father, William established a thriving enterprise at Willow Shade, where he settled in 1851; raising sheep for the Baltimore market, by 1863 he had more than doubled his holdings—increasing his 130 acres to 304. But the son rejected his father's religious and political beliefs by becoming a Baptist and supporting the Union during the Civil War, when Willow Shade became the headquarters for Union Army officers (*WWC*, p. 17).

Because of his Northern sympathies, William Cather played a more controversial role in the community than had his father. Appointed a U.S. marshal after the war, he continued to prosper while other Virginians suffered financial reverses. He was shunned by some family members and neighbors (he had sent his sons to West Virginia to escape conscription) who remained hostile until the 1870s when Willa Cather's mother invited them to a party and healed the rift. Others responded sooner to William's gestures toward conciliation; since he prospered during and after the war, he used his "Northern money" to send some young people to school, including Virginia Boak, Willa Cather's mother.[5] His most enduring record is architectural rather than written: the three-story brick farmhouse he built at Willow Shade is still standing today. With a fireplace in every room and a column-lined portico, the house was large and comfortable. Six willow trees shaded the front lawn, and a box hedge extended to a creek which was spanned by a small bridge. Willa Cather spent her childhood years in this pastoral setting. Willow Shade and its surrounding countryside stayed etched in her memory throughout her life; we can see how fresh her memories remained in the "Epilogue" to *Sapphira and the Slave Girl* (1940), written almost sixty years after she left her first home, where she describes both house and landscape in loving detail.

The Boaks left Willa Cather a more romantic Southern legacy than did the Cathers. Her maternal grandfather William Lee Boak served as a justice of the Berkeley County Court, a member of the Virginia House of Delegates, and an official in the Department of the Interior. After his death in 1854 his wife Rachel returned from Washington to her Back Creek birthplace with her children. The Boaks had stronger Southern sympathies than did the Cathers—they had relatives in Louisiana—and displayed a secessionist fervor more common in the Deep South than in a border region like Back Creek. William's three sons fought for the Confederacy and his namesake died in battle long before Willa Cather was born. But the younger William Boak's sister Virginia, Willa's mother, kept his memory alive for her children. Sharing William Boak's commitment to the Southern cause, "to the end of her life" she "cherished her brother's sword, and the Confederate flag."[6]

When the adolescent Willa Cather was surveying the family tree for a suitable ancestor, she chose to claim her military heritage and decided she had been named for her soldier uncle. In reality she was baptized "Wilella" after her father's sister, who had died of diphtheria in childhood, a fact Cather eventually tried to disguise when she replaced "Wilella" with "Willa" in the family Bible (*WWC*, p. 234). In the early 1900s, when Cather was hoping to be as heroic a writer as she imagined her uncle had been a warrior, she wrote a poem and a story entitled "The Namesake." Both works imagine resemblance and continuity between herself and William Boak, a more heroic forebear than the stolid, successful farmers in the Cather family or an aunt who died of diphtheria. Willa Cather wanted to partake in his bloodline: "Proud it is I am to know/ In my veins there still must flow,/ There to burn and bite alway,/ That proud blood you threw away" (*AT*, pp. 25–26).

Her own father had not fought in the Civil War and did not offer such a stirring legend, but Willa Cather was nevertheless proud to be Charles Cather's daughter, as the numerous close father–daughter relationships in her fiction attest. A gentle, well-read man, Charles Cather differed both from his father William and his more aggressive brother George. In recalling Willa Cather's early years in Nebraska, friends recount a typical scene: the nine-year-old girl sitting in Red Cloud's general store and "discoursing, with some prompting from her father, on Shakespeare, English history, and life in Virginia" (*WWC*, p. 1). In encouraging his daughter to speak in a public setting, Charles Cather was, doubtless unconsciously, challenging the cultural assumption that a woman's place was to be domestic and her voice to be silent. A father's pride in a gifted daughter was an important resource for a girl growing up in Victorian America; as the representative of male authority in the extradomestic world, Charles sanctioned his daughter's desire to succeed there.[7]

Charles Cather wanted his eldest daughter to speak in public, but he did not want her to be male—unlike the father of novelist Mary Virginia Terhune who once exclaimed "Ah, my daughter! if you had been born a boy you would be invaluable to me!"[8] In fact, as Cather's younger sister Elsie explains, Charles' acceptance of Willa's desire to attend college revealed his chivalrous deference toward women, not his desire that his eldest daughter achieve masculine success:

> None of the family has ever regretted that Father found it his duty to send his charming and brilliant daughter to college, even though the times were hard. It *was* his duty. Both he and Mother were from the south and had southern ideas about how women should be treated. It did make things pretty hard on the two older boys, but they too had been brought up in the southern tradition, and I never heard either one of them speak a word against it. . . . When Father saw how firmly Willa's heart was set on going to school *right now,* he did the best he could. (*WWC*, p. 233)

A "Southern gentleman refined almost to the point of delicacy," Charles Cather did not inherit the same economic drive the other Cather men possessed.[9] Charles neither established his own farm nor built his own house. When his father left for the Nebraska Divide in 1877, Charles took over the sheep farm at Willow Shade, and when he emigrated to Nebraska in 1883 to join his parents and to manage temporarily his father's farm, he soon decided he did not want to emulate his pioneer brother George and settle his own land. After eighteen months on the Divide he moved into Red Cloud and found more congenial work in a real estate and loan office. He supported his family but never prospered, lacking the entrepreneurial zeal his brother George possessed in more abundance. Unlike his brother, father, and grandfather, Charles did not seek public roles, and when he was forced to play one—as member of a committee investigating political corruption in Red Cloud—he disliked the publicity and controversy that resulted (*WWC*, pp. 24–26). As a result, Charles offered his daughter encouragement without coercion:

not driven to succeed in the public world himself, he did not burden her with his own thwarted ambitions.

In a sketch of a "Southern gentleman" written during her college years and based on her father, Cather suggests that a man's delicate sensibilities could prevent him from flourishing in a commercial world where one must be hard and aggressive to thrive: "He was a Virginian and a gentleman and for that reason he was fleeced on every side and taken in on every hand," she writes, concluding "That man had better go back to the South; it does not pay to be a Southern gentleman in the hustling Northwest" (*WP* 1, pp. 20–21). In "Old Mrs. Harris" Cather sketches another portrait of her father in Hillary Templeton, the "delicate," "boyish" man too "soft" to outwit the town's "hard old money grubbers" (*OD,* p. 112). Charles Cather resembled Mr. Templeton, thought Edith Lewis, who lived with Cather from 1908 until her friend's death in 1947. Lewis remembers Cather's father as a "kindly" man with a "hopeful, friendly disposition" who might have been "too trustful in his dealings with other men" (*WCL,* pp. 5–6).

Willa Cather was close to her father. As the oldest child she was his special companion in Virginia, spending time with him in the fields around Willow Shade, riding the mower, playing with their shepherd dog, accompanying him on his rounds. She had a "vivid memory of how her father would take her with him, carrying her on his shoulder, when he went to drive the sheep into the fold at evening," recalls Lewis (*WCL,* p. 7). Traveling in Provence on her first trip to Europe in 1902, Cather glimpsed a little girl riding on a cultivator with her father, and "the sight filled her eyes with tears" because it released such poignant memories (*WCL,* p. 57). She drew on this idyllic vision of the past in "The Swedish Mother," a sentimental poem in which the father is a maternal figure, tender as well as strong:

> Then come grandpa, in his arm
> Li'l' sick lamb dat somet'ing harm.
> He so young then, big and strong,
> Pick li'l' girl up, take her 'long,—
> Poor li'l' tired girl, yust like you,—
> Lift her up an' take her too.
> Hold her tight an' carry her far,—
> Ain't no light but yust one star.
> Sheep go 'bah-h,' an' road so steep;
> Li'l' girl she go fast asleep.
> (*ATOP,* p. 53)

Charles Cather may have been "fleeced" by Red Cloud's predatory moneychangers, but in his daughter's imagination he was, literally and metaphorically, the good shepherd: after his death in 1928 she contributed a stained glass window to Red Cloud's Episcopal church in his memory, which depicts Christ with his lambs and shepherd's crook.

A self-educated man who enjoyed reading poetry aloud to his children, Charles supported his daughter's desire to attend college, although he did not fully understand the urgency of her ambitions. After Willa's high school graduation, her father thought she might teach school while her mother "could see immediately that the girl should go on to college at once" (*WWC*, p. 30). Nor did Charles comprehend the extent of his daughter's literary reputation until 1920, when Sinclair Lewis told an Omaha audience that Willa Cather was "Nebraska's foremost citizen." Surprised to learn that the daughter he had once carried in his arms was a nationally acclaimed writer, he wrote her "a loving letter in which he paid tribute to her genius and her success." Cather returned the tribute when she reissued *April Twilights* in 1923, including "The Swedish Mother" in the volume and dedicating it "to my Father for a Valentine." After her father's death, she inscribed a friend's copy of *April Twilights* in memory of "my one winter in Red Cloud, after so many years, and of my gentle father to whom this book is dedicated" (*WWC*, pp. 26–27).

Although he had no land to pass on, Willa Cather's gentle father left his daughter a valuable legacy. Implicitly challenging the Victorian definition of male identity and role, Charles Cather gave her a personal, familial resource for questioning the social construction of gender. He is one source for a recurrent figure in his daughter's fiction—the sensitive man of integrity who places aesthetic, intellectual, or spiritual values above commercial ones. Such male characters include Carl Linstrum, Oswald Henshawe, and Tom Outland as well as those based more directly on Cather's father—Euclide Auclair, Hillary Templeton, and Henry Colbert. The daughter's fiction also features the Southern gentleman's antagonists—Nebraska's "hard old money grubbers" reappear as Wick Cutter, Bayliss Wheeler, and Ivy Peters, Cather's hated symbols of American materialism and male aggression.[10]

•

As Virginia Woolf observes in *A Room of One's Own*, the woman writer is both an inheritor and an originator who receives a literary legacy from female predecessors before she can develop her own voice. If she is fortunate she also receives an inheritance from female kin who leave, if not written texts, signs of strength, autonomy, or courage that she will need in daring to aspire to a literary vocation. In *Room* Woolf symbolizes that inheritance by the legacy the narrator receives from her aunt, Mary Beton, who wills her the money necessary for financial and literary independence.

In imagining herself William Boak's namesake nephew, Willa Cather did not recognize the inheritance she had received from her female forebears. The adolescent girl who claimed a male ancestor might have made the same accusation to the women in her family that the disappointed narrator in "On the Gull's Road" (1907) makes to the maternal woman he loves: "You [gave] me nothing" (*CSF*, p. 91). The women in Cather's family did not make their mark in the public realm: they did not cultivate and pass on the land, serve in the state legislature, or fight in the Revolutionary or the Civil War. Yet the adult writer eventually understood and recovered the connection between herself and the women she had known in her childhood—women like her great-aunt Sidney Cather Gore, her paternal grandmother Caroline Cather, and her maternal grandmother Rachel Seibert Boak. As she came to see, these women gave her more than a heritage of female strength and dignity. They were also unselfconscious artists who offered her a domestic tradition of creativity in their cooking and gardening skills. Women's private realms were regions in what Cather later called the "kingdom of art" (*KA*, p. 449): in her novels, kitchens and gardens are the spaces where nature is transformed into culture, rituals created, and order established. Cather's female relatives were also the first mediators between the child Willa and her cultural and literary inheritance. Even if their voices were not heard in public spaces, women were the teachers, readers, and writers in her family, the custodians of language.

Cather's female relatives believed in the importance and dignity of woman's work. Her great-aunt Sidney Cather Gore and grandmothers Caroline Smith Cather and Rachel Seibert Boak were self-reliant, hardy farm women who lived in an agrarian region where woman's productive economic role in the family had not been transformed by industrialization. Willa as a child was surrounded by older women whose arduous tasks—tending the vegetable garden, raising animals, sewing and quilting, preparing and preserving food, managing the household, childbearing and rearing—were as vital to the family's survival as their husbands' farming duties. Her forebears practiced the sexual division of labor common in rural societies, however, with men taking the farm, women the home and garden as their province. By extension, the public and civic world was men's domain, the private world women's.[11]

In this agrarian economy mothers were their daughters' educators and mentors, training them to fulfill the farm wife's role by transmitting essential household skills. Thus Willa's grandmother Caroline Cather inherited her domestic expertise and role from her mother, Ann Ellis. Here is a portrait of Ann Ellis sketched by George P. Cather, Willa Cather's uncle:

> Her great and wonderful physique was fortified with perfect health, great strength, powerful endurance and unsurpassed industry, which carried with them a practical knowledge of the work in her line of duty. She would take the wool as it came from the sheep's back, card it, spin it, weave it into cloth, and make it into clothes; and weave it into nice bed clothes and blankets. She would take flax and hackle it, spin it, weave it into linen, and make it into clothes, tablecloths, towels, bed-ticks and threads. She was a good tidy housekeeper, and a grand good cook.[12]

In the year of Willa Cather's birth—1873—these strong mother–daughter bonds still existed: Caroline Cather wrote several letters of advice to her newly married daughter Jennie, in which she assumes the roles of mother and mentor, passing on both household and marital wisdom.

Historian Carroll Smith-Rosenberg suggests that in the closely knit female world of nineteenth-century America, where domestic wisdom was transmitted from one generation of women to another, the "bonds that held mothers and daughters together in a world that permitted few alternatives to domesticity might well have created a source of mutuality and trust absent in societies where greater options were available for daughters than for mothers."[13] Learning methods of household management from their mothers, inhabiting a sequestered rural world where nondomestic roles for women were practically nonexistent, Willa Cather's female relatives escaped the conflict she experienced when she found her own mother neither a model nor a teacher for an ambitious, college-bound daughter.

Although rural women like Cather's female relatives performed essential social and economic functions, we should not exaggerate or romanticize the power they enjoyed in their families and communities. Some historians praise the relative egalitarianism of agrarian and frontier culture where female weakness and delicacy were not prized virtues, but even in agricultural societies power was divided unequally between men and women. Ownership of the land sanctioned male privilege. A woman might take charge of domestic activities within her sphere, but her husband was head of the family and of the household. In nineteenth-century Virginia, where women lived under English common law, a wife was not an independent economic or political agent: she could not make a contract, conduct a suit, write a will, or free a slave.[14]

Even though Willa Cather's female relatives accepted women's subordination within family and society and remained within the domestic sphere, these were not passive women. And so they left her a complex, contradictory legacy: in principle they accepted conventional definitions of woman's role and nature, but in practice they managed to find some room for self-expression and self-assertion within the constricted ideological and social worlds they inhabited.

One of Willa Cather's female relatives made her mark in the public sphere supposedly reserved for men. The residents of Back Creek renamed their village "Gore" in honor of her great-aunt Sidney Cather Gore, placing her name on the map as well as in the history books. A daughter could inherit personality traits from a strong-willed father even if she could not take over his land, and Sidney received "the most liberal share" of James Cather's "intellectual energy, his power of speech, his happy unresented dominion over others."[15] Creating more written records than did Cather's other female relatives—she left a diary, letters, and a passage in the local history—Sidney Gore was a forceful woman who managed to get her own way without ever striking others as unfeminine.

Born in 1823, Sidney married Mahlon Gore when she was nineteen and settled into the strenuous life of a Back Creek farmer's wife. But her husband died in 1860, and she was left with three young sons and a heavy burden of debts. Refusing her father's offer of shelter and financial assistance, she decided to maintain her own home and to support herself and her children; as Sidney explained in her diary, she intended to raise her children in her own way and did not want to incur any obligations. Like many other nineteenth-century women, Sidney found the widow's independence agreeable and declined to remarry. No second husband would interfere with her child-rearing values and practices.

Sidney may also have feared that a second husband might forbid or regulate the many projects she began after Mahlon Gore's early death. A woman of formidable energy, her motto was "A change of work is rest."[16] At first she managed the farm with hired help. Later she taught in the local school and became Back Creek's postmistress, finally regaining the financial security lost when her husband died. "By the exercise of rigid economy, unwavering industry, unceasing toil, adding advantage to advantage and losing no opportunity for acting wisely and well, prosperity was forced to come and each year brought a welcome increase," reports Norris' local history (p. 595). Practicing Benjamin Franklin's Protestant virtues throughout her life, Sidney eventually turned her home into a boardinghouse, added on several rooms, named it "Valley Home," and attracted vacationers who came to the Winchester area because of its reputation as a health resort. The years brought more "welcome increase," and by the end of the century, Sidney was a major landowner. She attributed her good fortune to God's will, but Norris makes no mention of divine intervention in characterizing her as a "preeminently successful" businesswoman and the town's first citizen (p. 595).

Sidney Gore's industriousness stemmed not from individual ambitions but from maternal concern, she explains in her diary, for the "widowed mother's first duty is to her orphaned children" (p. 118). Sharing this view of her endeavors, her neighbors never thought her unwomanly. As Norris admiringly notes, Sidney Gore's two "guiding principles" were her unselfish motherly hope to "advance her sons to useful positions in life" and an equally feminine "heaven-born" desire "to do good" (p. 595). Evidently Sidney's economic successes never rankled because, like James Cather, she returned much of her income to the community in charitable activities. In addition to helping the needy and tending the sick, Sidney made part of her home into an unofficial community center and social service agency. There she provided care and housing for young men studying for the ministry as well as for orphaned children and reforming alcoholics—she welcomed anyone requiring food, shelter, and a healthy Christian atmosphere. The bulwark of Back Creek's small Baptist congregation, she also singlehandedly raised the money to build a parsonage.

Deeply felt piety motivated Sidney's charitable activities and supported self-assertion. Like many other Victorian women she could justify even defiance of male authority by appealing to the higher court of God's will. Her conversion to the Baptist faith was her first independent act, which she defended to her Presbyterian father by evoking the need of each separate soul to work out its salvation "with fear and trembling," even if this process required a daughter to repudiate her parents' faith (p. 19). Later she sought to control unruly male behavior when she crusaded against card-playing and intemperance; in common with other evangelical reformers, Sidney thought the latter vice "the acme of human degradation and misery" and "the curse . . . of our land" (p. 44).

Although religion buttressed the inequality of the sexes, it was also the source and legitimation of Sidney's strength, as a representative anecdote from her diary illustrates. During the Civil War, twenty Union soldiers commandeered her home for the evening. They behaved quite decently, she confided in her diary, until to her disgust "They commenced *playing cards*." Distraught, Sidney "sought guidance from above"; receiving it, she decided that she "*could* not, and *would* not tolerate it." She asked them to cease. The soldiers ignored her. Her horror increased when they began betting, for she considered this abominable male practice a "God-dishonoring, soul-destroying" stepping-stone to other vices. A defenseless woman with twenty enemy soldiers, Sidney suffered momentary fear and indecision, but she determined to renew the struggle. She knew she had a powerful ally:

> My heart told me that God was stronger than they, and the sweet promise, "I will be the widow's God," gave me courage, and I returned again to express again my disapprobation of the proceedings in *stronger terms,* and to the circle *collectively,* thinking that some of them might be able to understand, at least a portion, of what I said.

Unable to ignore such militant feminine righteousness, the Yankees graciously surrendered. "One of them looked up inquiringly, threw down the cards, and said, ' 'Tis right, I stop, I stop.' They all very respectfully followed his example" (p. 79).

If Sidney Gore's victory over the soldiers reflects the informal power women could sometimes exert in nineteenth-century America, her role as Back Creek's postmistress gave her a more socially recognized position of authority within her small community. When Willa Cather based Mrs. Bywaters in *Sapphira and the Slave Girl* on Sidney Gore, she associated the postmistress' role with reading, writing, language, and power. As the dispenser and collector of mail within the community, Mrs. Bywaters is both the overseer of communication and a woman who combines words with actions. An opponent of slavery, the liberally minded postmistress rejects the values of her slave-holding community because she reads of alternative viewpoints in the *New York Tribune,* and she arranges for the slave girl Nancy's escape by writing an "important letter" to an abolitionist friend (*SSG,* p. 222).

The postmistress Sidney Gore also exercised a certain amount of power through reading and writing, private activities that nonetheless could connect a woman to the public sphere. Reading was her "favorite pursuit" as a child, and whenever she had a free moment she picked up a book (p. 4). She must have read and reread the poems of Moore, Byron, and Burns, since she frequently quoted their works to her children, and her diary shows her familiarity with Milton and the Bible. Desiring to make her life correspond to the shape of the Scriptures, every New Year's Day Sidney would begin with a reading from Genesis, hoping to complete the New Testament by Christmas. She was also well versed in the religious and sentimental poetry of the day, and the language she employs in her diary reveals the impact of this reading. Whether she is speaking of family ties, the beauties of nature, or God's will, her voice is submerged in conventional rhetoric.

Yet the flowery, formulaic language is not inconsistent with her self-expression (though at times it may cloak it), as when Sidney recalls using a biblical reference to express her wish that her Presbyterian father be reconciled to the religion she defiantly chose for herself: "I fervently prayed that he might speedily become reconciled to my course and fall out with his own. And that the scales would fall from his eyes so that he might see that unless he too repented, he would likewise perish" (p. 20). Although a reader of male-authored texts that circumscribed and defined femininity, Sidney was a selective reader who could use her sources to support her own views. To be sure, her prayers are expressed in a language inherited from patriarchal sources—the Bible, religious tracts, sermons—but at times she uses their rhetoric to assert herself and to influence others. As she writes, "Woman's sphere is limited. But thank God, her prayers are not" (p. 68).

Sidney did not rely solely on prayers for self-expression: she also wrote letters, hundreds and hundreds of letters. Like many nineteenth-century women she corresponded to maintain family ties, a form of writing connected with woman's role as custodian of the family's moral and emotional life. But she wrote letters for public purposes as well. During the Civil War she was the "scribe" of the neighborhood, her son James remembers, writing letters to soldiers for their illiterate parents. "I distinctly remember these dictations. The awkward broken sentences sounded quite differently when she read the finished letter, and the grateful father or mother showed their pride" (p. 95). Even if the written sentences differed from the speakers' words, Sidney tried as assiduously as any novelist to capture the flavor of another speaker's voice and language. "To have written, as I would have written for myself," she reflects in her diary, "would have been so foreign to their every-day intercourse, that *neither* could have understood it" (p. 97). Perhaps because she wrote so many letters for others, in one letter she imagined herself the spokewoman for her whole community. Writing Captain R. B. Muse and his men to warn them of the dangers of card-playing and drinking ("Satan's auxiliaries"), Sidney adopts a collective female voice: "Will you accept the feeble testimonials of tender regard and anxious concern of Mother, wives, sisters, friends" she asks, and signs her letter "The Loving Ones at Home"(pp. 104–9).

Sidney's letters were not merely "feeble" testimonials. Like Mrs. Bywaters, she used the letter as an instrument of power. Hearing that an acquaintance had been unjustly imprisoned, she promptly wrote to the Governor for a pardon "with such efficacy that he was liberated" (p. 147). To raise money for the Back Creek parsonage she wrote over four hundred letters. Although not a supporter of women's suffrage, Sidney was interested in political matters and thought that important people would be glad to hear her views. She read the newspapers every day and scanned the Congressional Record for "the most interesting speeches"; invariably reading led to letter-writing. She would inform Congressmen of her agreement or displeasure with their policies, and "occasionally her views were incorporated in subsequent speeches" (p. 149). Doubtless the Congressmen did not cite their sources, but nevertheless—even though she was not named—Sidney achieved an indirect form of public expression.

While such letters represent Sidney's desire to exert influence in the public world, her diary reflects her need to tell her own story. Some of its subjects are the conventional chapters of a woman's life (childhood, marriage, religious conversion, motherhood), yet the fact that this autobiographical text exists at all reveals a desire to see her life shaped in language other women in her family did not share. And there are unconventional chapters as well. From time to time Sidney includes some of her letters to public figures, and she writes extensively on her experiences during the Civil War. The document is thus a curious genre, intermingling the letter, the diary, and the autobiography. Some of the entries are made at the time of the events they describe, while some others—roughly beginning with the Civil War episodes—are retrospective. It may be that as time passed Sidney recognized that she was telling her own story and tried to shape it; the Civil War may also have caused her to see that she was recording her community's and her nation's story as well as her own, making her a more conscious author. By the Civil War years she was imagining her sons and possibly their descendants as readers ("When I am gone, I think my orphaned ones will cling to all I may have written, whether possessed of any merit or not" [p. 103]). In this sense her diary, like the copies of correspondence she included within it, was also a letter to be read by family members after her death.[17]

Assuming that women's voices should be heard even when they disagreed with male authority, even though she expressed herself through the private genres of letters and diary, Sidney Gore was still a woman writer who—however circumspectly, however cautiously—associated language with power. Hers was a useful legacy for a great-niece who sought to forge the same alliance. Sidney expanded the domestic sphere instead of leaving it, but Willa Cather knew her great-aunt during the years when she was widening its circumference and gaining a position of prestige and influence in their small community. Living a mile down the road from Willow Shade, Sidney was a frequent visitor in the Cather household, part of Willa's daily life in Virginia. Even more important than the days she visited may have been the one day when she did not: Sidney reserved Sundays for writing as well as worship, staying home to write letters and make entries in her diary—a sign to family members that she took her writing seriously (p. 16).

Willa Cather's affection and admiration for Sidney Gore endured after she left Back Creek for Nebraska. Although we cannot be sure whether she was conscious of the connection, when the adolescent girl named herself "William" she was aligning herself not only with the Williams in the family but also with the great-aunt who possessed a male name. When she graduated from the University of Nebraska, Cather returned to Virginia to visit with Sidney. Several years later, in Back Creek again, she found it depressing because those she loved, Sidney Gore among them, were dead.[18] But she had not departed from Cather's memory, as we can see from her contribution to *Sapphira and the Slave Girl*—written over forty years after their last meeting.

•

Caroline Cather, Willa's paternal grandmother, displayed female strength within more conventional limits: endurance and fortitude. Born Emily Anne Caroline Smith, Caroline married James Cather's son William in 1846. They had six children, four daughters and two sons, one of whom was Willa's father Charles. Only Charles and his brother George survived to adulthood; Caroline's four daughters died young, one from diphtheria (Cather's namesake aunt Wilella), the others from tuberculosis. As a wife Caroline incurred obligations that the widow Sidney Gore escaped. In 1877 William Cather decided to resettle on the Nebraska frontier, and like countless dutiful wives before her Caroline followed him to Webster County. It must have pained her to abandon family, friends, and the comforts of Willow Shade, particularly since the decision to emigrate most likely was not hers. In his study of the diaries and reminiscences of men and women who emigrated to California and Oregon, John Faragher finds that "Not one wife initiated the idea; it was always the husband. Less than a quarter of the women writers recorded agreeing with their restless husbands; most of them accepted it as a husband-made decision to which they could only acquiesce."[19]

Less educated and well-read than Sidney Gore and too self-effacing to think of translating her life into prose, Caroline nevertheless left records of herself in the letters she wrote to her newly married daughter Jennie Cather Ayre. Filled with loving, concerned maternal advice, these letters tell us about the author, something the modest Caroline never intended. (Afraid that her lack of education was revealed in her writing, Caroline told Jennie to burn her letters as soon as she read them because "I write so badly I dont want any Strangers to see them.") The writer of these fortunately preserved letters was a tireless worker who devoted herself to the farm wife's traditional tasks. Believing in the importance of woman's work, Caroline took biblical injunctions against idleness very seriously. "The Master never intended their should be any idlers in his vinyard," she told Jennie, her beautiful, pleasure-loving favorite whom Caroline feared might easily leave the grapes rotting on the vine.[20] In the mother's view, woman's onerous chores left no time for frivolity, repose, or even love-making. Describing an ideal daily schedule, she told Jennie to turn to her sewing as soon as she had made breakfast and cleaned her house. She shouldn't interrupt work even to kiss her husband, her mother counseled; if John wanted a kiss, "he will do it without your quiting your work."[21] After the day's tasks were completed, Caroline recommended the same useful pastime she enjoyed herself—immersion in the Scriptures. Like Sidney Gore and William Cather, she was a fervent Baptist.

Preaching and practice coincided in her life. Caroline's letters are filled with references to the task just completed, the task now awaiting. Nor did pioneer hardships dismay her: "I git along with the work real nicely," she wrote soon after her arrival in Nebraska, referring as always to the subject that absorbed her.[22] Completing her housekeeping duties brought Caroline satisfaction but not joy. Her life was hard, and she viewed it as a pilgrim's progress through temptation, trouble, and grief. Writing her daughter Jennie shortly after her marriage, Caroline spoke not of married happiness and fulfillment, but of the problems and setbacks the young couple should expect. Her daughter should be a "source of comfort" to her husband in "all the trials of life" Caroline counseled, for they would "meet with many trials and disapointments and need each others sympathies and encouragement."[23]

Her own life was filled with such trials—the sacrifice of Willow Shade, the resettlement in Nebraska, the illnesses and deaths of her daughters. After two weeks in her new home she had to endure the death of her "darling" Jennie whose vanities and frivolities made Caroline worry that she had "made an idle" of her favorite daughter.[24] When Caroline's other surviving daughter died shortly afterward, the bereaved mother donned mourning and wore black silk for the rest of her life. But she did not shirk her new duties: left with three orphaned grandchildren in her early fifties, Caroline gamely set out to raise another family.

Like her sister-in-law Sidney, Caroline was a deeply religious woman, but her piety did not similarly result in public manifestations of reforming zeal or organized charity. Her religious beliefs had private uses, giving her the strength to endure losses and sorrows without succumbing to bitterness or despair. God's ways might be inscrutable, but she knew they were just. A letter of consolation she wrote to Jennie after her daughter's first child died reveals how Caroline's religion helped her cope with life's tragedies. Jennie and her husband should not grieve too deeply for the "little Darling," for with his death they would have "one more tie" to bind them to heaven and "one less to earth." At least the child would never feel the grief his parents suffered: "it is the greatest comfort I have to feel that so many of our children are done with the sorrows and disappointments of this world."[25] In another letter, written as a brother-in-law was dying, Caroline grappled with an age-old question: Why, if God was just, did He allow good men to suffer? As a "poor short sighted mortal" she could not discover the reason, but her Saviour would make it "all plain" some day, she trusted.[26]

While it helped her to accept sorrow and loss, Caroline's faith also reinforced her belief in woman's subordination to male authority in marriage. The husband was God's representative within the family, and a wife should not question his wisdom. Her daughter must never command her husband or use the word "must" to him, Caroline advised ("it is not lady like"), and if they ever had a difference of opinion, it was the wife's place to yield. Throughout their married life Jennie should "make it [her] studdy to please him."[27] Although in her own marriage authority resided with her husband, Caroline could command male family members if her work required it. "I got my garden made last Tuseday," she wrote Jennie proudly. "Uncle John don the work and I don the bossing."[28]

Within her spheres—the kitchen and the garden—Caroline Cather did the bossing. These realms at first seemed limited and oppressive to Willa Cather, who rejected the posture of female subordination her grandmother in some ways embodied and advocated. But by the time she created the portrait of Grandmother Burden in *My Ántonia* she regarded woman's traditional tasks with respect. Drawing on her memories of Caroline Cather, Cather describes Grandmother Burden as a "strong woman, of unusual endurance" who survives uprooting by quickly reestablishing woman's domains in the wilderness: she plants a garden and fashions a homey, warm kitchen, decorated with curtains, flowers, and plants (*MA*, p. 10). Her grandson Jim's transition from his sheltered Virginia homeland to a vast, seemingly empty prairie "outside man's jurisdiction" is in turn aided by the realms within woman's jurisdiction—the kitchen and the garden, protected and ordered spaces where the boy feels at peace (*MA*, p. 7).

•

Rachel Seibert Boak, Willa Cather's maternal grandmother, left no written texts that libraries have preserved. Yet Cather gave this grandmother the fullest, most affectionate portrait of a family member in her fiction: she was the model for the self-denying grandmother in "Old Mrs. Harris," the short story Cather based on the tensions, misunderstandings, and affection linking three generations of women—barely disguised versions of her grandmother, her mother, and herself. Along with the "Epilogue" to *Sapphira and the Slave Girl,* this is the most autobiographical of her works; "Old Mrs. Harris" might have been called "Family Portraits," thought Edith Lewis (*WCL,* p. 6). Rachel Boak was also the model for Rachel Blake in *Sapphira,* the strong-willed, compassionate woman who defies her tyrannical mother to save the slave girl Nancy from rape.

Rachel Boak contributed more to Cather's fiction than Sidney or Caroline simply because she was more important in her granddaughter's life. She lived with the family in both Virginia and Nebraska and was a continuing emotional presence until her death in 1893. She may even have done more mothering than Willa's own mother (particularly as more children came along); she tended the little girl when her parents were away, read aloud to her, nursed her during childhood illnesses. Rachel Boak was also Cather's first teacher, introducing the child to the world of language, stories, and books she was to inhabit.

Born in Back Creek Valley, Rachel Seibert briefly attended boarding school and married William Boak in 1830 when she was only fourteen. She followed her husband to Washington, and when he died in 1854, she returned to Back Creek an impoverished widow with five children. Unlike Sidney Gore she accepted her father's offer of a house in the village. Charles and Virginia Cather lived with her after their marriage, and Willa Cather was born there in 1873. The next year the Cather family moved into William Cather's Willow Shade, Rachel accompanying them. There she became the household manager, relieving her daughter of cooking, cleaning, and other domestic tasks. After the family moved to Nebraska, she ran the kitchen, cooking for nine family members and two paying guests who came in for lunch (*WWC*, p. 21).

In "Old Mrs. Harris" Cather explains why Grandma Harris strives to spare her elegant daughter Victoria unnecessary toil. It was traditional in Mrs. Harris' native Tennessee, Cather observes, for a widow with a married daughter to consider herself an old woman, wear black dresses, and become a housekeeper. Mrs. Harris accepts this change in status "unprotestingly, almost gratefully." Believing that "somebody ought to be in the parlor, and somebody in the kitchen," she gladly leaves the parlor for her pampered daughter: "She wouldn't for the world have had Victoria go about every morning in a short gingham dress, with bare arms, and a dustcap on her head to hide the curling-kids, as these brisk housekeepers did" (*OD*, pp. 134–35).

In an 1897 article "Nursing as Profession for Women," Cather praises her grandmother's selfless concern for others. "My own grandmother was one of those unprofessional nurses who served without recompense, from the mere love of it," even though she had "a host of little children" to care for. "But you have had grandmothers of your own," she reminds her readers, "you know how it went. You remember the old woman who nursed you when you had scarlet fever, and walked the floor with you when you had whooping cough. Money will never buy such attendance for you again." Cather portrays her grandmother as the unofficial nurse for Back Creek residents as well as for her own family, risking her own life to help the sick and comfort the dying. Although she had "cares enough of her own, poor woman," when a "child was burned, when some overworked woman was in her death agony, when a man had been crushed under the falling timber, or when a boy had cut his leg by a slip of the knife in the sumach field, the man who went to town for the doctor always stopped for her on the way." During the diphtheria epidemics that swept Virginia in the 1850s, Rachel risked her life to help out: "I have often heard the old folks tell how during those dreadful diphtheria scourges . . . she would go into a house where eight or ten children were all down with the disease, nurse and cook for the living and 'lay out' the dead"

(*WP* 1, p. 320). Like many nineteenth-century Americans Willa Cather associated male power with the doctor's role, female powerlessness with the patient's; as the community's and family's unofficial nurse, however, her grandmother escaped this dichotomy, suggesting the possibility of female strength by her efficacy as a healer (as did Sidney Gore in founding a health resort for ailing city folk).

While Cather's male relatives were known for economic or political status, Rachel Boak was a community figure who attained public recognition through exercising woman's traditional role as tender of the sick and dying. But her fame was more fleeting than that of the Cather and Boak men, recalled only in conversation and memory; James Cather's achievements were memorialized in written words, whereas Rachel Boak's were transmitted by word of mouth ("I have often heard the old folks tell . . ."). But because they were received by Willa Cather as an eager listener, these stories were given permanence in her fiction, where she passes them on to future generations of readers of "Old Mrs. Harris" and *Sapphira.* Cather's memories of her grandmother's courageous nursing inform her portrayals of Grandma Harris and Rachel Blake, women who sacrifice their own safety and comfort to help people in distress or need.

Because Rachel Boak taught the young Willa Cather to read and write—in a sense ensuring her own written portrait years later—from the beginning Cather had reason to associate the acquisition and mastery of language with a maternal presence. Yet the texts Rachel Boak chose as schoolbooks were male-authored as well as irreproachably moral; she favored the Bible, *Pilgrim's Progress,* and Peter Parley's *Universal History.* (Bunyan's allegory remained one of Cather's favorite books; any child who had not read it, she said later, had "missed a part of his or her childhood" [*WP* 1, p. 336].) This early reading offered Cather models of heroism she never forgot, although she transformed them when she envisioned the artist as a questing pilgrim. Cather found it more difficult, however, to revise the gender roles she first encountered in these narratives. As in most of the literature Cather read in childhood and adolescence, male characters are the primary actors in this early reading. Women either take subordinate parts (generally defined by their relationships to men) or do not appear at all.

Possibly because these first stories featured male heroes, they stimulated the child's fascination with assuming another identity, which can be seen again both in the male impersonation of her Nebraska adolescence and the fiction writing of adulthood. Willa could be kept quiet for hours, her parents recalled, if she were allowed to enact a scene from Parley's *History.* She would place one chair upside down on another to make a chariot and become Cato, accompanied by an imaginary slave who ran beside her (*WCL,* p. 10). The child Willa was then exhibiting the same desire Cather said later drove her to write—to achieve the "bliss of entering into the very skin of another human being" by forgetting the self and identifying imaginatively with another (*WC:AM,* p. 111). And yet this early scene also anticipates the conflict Cather at first confronted in becoming a writer: to assume a powerful identity—master accompanied by slave—the female child must imagine herself as male.

In "Old Mrs. Harris," the story where she pays tribute to her grandmother Boak, Cather grants Grandma Harris an empathic version of the writer's ability to enter another self. The old woman expresses her power of identification and sympathy in anticipating others' needs. Blessed with the gift of self-forgetfulness her daughter and granddaughter lack, Grandma Harris is the only one of the three women who escapes the boundaries of her own ego and connects to others:

> Sometimes, in the morning, if her feet ached more than usual, Mrs. Harris felt a little low. . . . She would hang up her towel with a sigh and go into the kitchen, feeling it was hard to make a start. But the moment she heard the children running down the uncarpeted back stairs, she forgot to be low. Indeed, she ceased to be an individual, an old woman with aching feet; she became part of a group, became a relationship. (*OD*, pp. 136–37)

In writing, Willa Cather also "ceased to be an individual" and "became a relationship" as she entered imaginatively into the lives of her characters. In "Old Mrs. Harris" she portrays this literary connection between the grandmother and Vicki Templeton, a portrait of her adolescent self. Both grandmother and granddaughter love to read aloud to the Templeton children, and in one scene Cather shows us how this gift has been passed on when Vicki gladly takes over from her grandmother the task of reading *Tom Sawyer* to the rapt children. The line of inheritance is delicately drawn from the grandmother who "loved to read, anything at all, the Bible or the continued story in the Chicago weekly paper," to the granddaughter who passionately wants to attend university (*OD*, pp. 89–90). Cather makes this matrilineal inheritance even more direct by altering her own family history. In reality Charles Cather borrowed the money to send Willa to the University of Nebraska, but in the story Grandma Harris arranges for Vicki to be given the money she needs for her tuition. The gift of education is thus a female bestowal in "Old Mrs. Harris," similar to the inheritance the narrator receives from her aunt in *A Room of One's Own*.

Willa Cather's female relatives left her an ambiguous legacy. On the one hand, they could not offer her proof that female identity and the writer's vocation were compatible. Circumscribed by subordinate roles and domestic ideology, for the most part they expended their energies and imagination within the private sphere, not seeking to alter the script written for women by culture and society. In most ways they deferred to male authority: the stories of Caroline Cather and Rachel Boak, required to emigrate to Nebraska because the Cather men decided to settle there, epitomize women's traditional powerlessness to determine the course of their lives. Even Sidney Gore, who ranged the most widely within the limits of the female sphere, never openly challenged conventional wisdom regarding woman's role and nature.

On the other hand, these women provided her with varying examples of female strength and competence, demonstrating that women need not be passive and submissive angels in the house, but could be the cohesive forces binding family, church, and community. In creating a self and imagining a literary vocation, Cather managed to make use of the piety and selflessness in her female strain of blood—seemingly the two qualities most contradictory to the artist's role—by transforming them to suit her needs. Her many proclamations in the 1890s that art was her religion reveal continuity as well as contrast with her female relatives' ardent faith, as does her view in *The Professor's House* that art and religion "are the same thing, in the end" (*PH*, p. 69). And Cather's view that the artist must transcend self in the creative process revises the self-denial Rachel Boak exhibited in placing others' needs before her own.

But the most important gift of all was the discrepancy her female relatives displayed between the Victorian "cult of true womanhood" and real women's capabilities.[29] Women were supposed to be silent, and yet Sidney Gore found ample room for self-expression; women were supposed to be dependent, and yet Caroline Cather was the mainstay of her family; women were supposed to be prone to illness, and yet Rachel Boak was a healer. And so, even though these women could not grant Willa Cather the public sphere she wanted to occupy, by implictly exposing the contradictions between ideology and lived experience they gave her an imaginative space within which to redefine womanhood.

•

Speaking of black women writers, novelist Alice Walker observes that "our mothers and grandmothers have, more often than not anonymously, handed on the creative spark, the seed of the flower they themselves never hoped to see: or like a sealed letter they could not plainly read."[30] Although it took Cather several years to read the letter her female forebears had sent, by the time she wrote *O Pioneers!* (1913) she was beginning to trace the links between grandmothers and granddaughters, between farm wives and artists, between gardeners and writers. Just as Walker finds the seeds of her literary art in her mother's garden, so in *O Pioneers!* Cather suggests that Nebraska's immigrant farm women left their daughters an aesthetic inheritance. Mrs. Bergson is a tireless worker who makes a garden in the wilderness and preserves her family's cultural heritage, along with her jams and jellies, during their adjustment to a new land. The mother thus performs on a small scale what her daughter Alexandra—the novel's artist—will do on an epic one when she helps the wild land to release its powers of fertility and growth. In a 1921 interview Cather made her association between the artist's creativity and the farm wife's domestic tasks explicit: "The farmer's wife who raises a large family and cooks for them and makes their clothes and keeps house and on the side runs a truck garden and a chicken farm and a canning establishment, and thoroughly enjoys doing it all, and doing it well, contributes more to art than all the culture clubs" (*WWC*, p. 167).

As an ambitious young journalist who thought artists descended from the male heroes of epic and legend, however, Cather perceived no artistry in woman's domestic work, no links between herself and a female culture. Women's selfless concern with others' welfare and their private, communal activities seemed inconsistent with the self-expression, fame, and individual achievement Cather associated with literary greatness and wanted for herself. Imbued with Romantic ideology, convinced that the artist must be male, Cather saw a split between commitment to art and to human relationships, an incompatibility between creative work and love. Only men, she thought, possessed the singleminded devotion to an abstract cause artistic greatness required. But after meeting Sarah Orne Jewett in 1908, Cather recognized that even a literary art could be communal and selfless. Inspired by love for her Maine homeland and its residents, Jewett sought to pass on their stories, traditions, and lore in her work. Hers was creativity as preservation as well as self-expression, similar to the farm wife's work or the storyteller's art, more closely related to folk culture than was the fiction of Cather's first mentor, Henry James.

Precursors for Sarah Orne Jewett were also part of the female world of Willow Shade. There Cather learned to read male-authored stories, but there she also encountered women storytellers. As she later said, her first introduction to narrative was meeting the women who came down from the mountains to help out at Willow Shade—and to tell stories.[31] During busy seasons women from the remoter regions of Timber Ridge and North Mountain would come to Willow Shade to help with "spinning and quilting, butter-making and preserving and candlemaking." And when the "old women came from Timber Ridge to make quilts," Edith Lewis recounts, Willa would "creep under the quilting frames and sit there listening to their talk." Mary Ann Anderson, the model for Mrs. Ringer in *Sapphira,* was the "best of the story-tellers." The preserver and transmitter of local folklore, gossip, and legends, she "knew the family histories of all the countryside, and all the dramatic events that had become legends among the country people. Her talk was full of fire and wit, rich in the native idiom" (*WCL,* pp. 10–11).

Whenever Cather went back to Back Creek, as she did after her graduation from college, she would visit often with Mary Ann Anderson, catching up on local stories. Writing to Mrs. Anderson's granddaughter in 1941, Cather told her that her grandmother had had an unusual interest in following the story of people's lives; as a girl she talked with Mrs. Anderson for hours at a time, she recalled, curious to discover what had become of the people the storyteller remembered from her past. It was worth being sick as a child, she added, because an illness always produced a visit and a story from Mrs. Anderson.[32]

The Back Creek female storytellers' art closely resembles their quilting, a folk art that provided both the context and the excuse for socializing and storytelling. Both quilting and storytelling are communal practices in which a group of women share in the creative process, binding together pieces of fabric in their quilts, joining members of the community as well as past, present, and future in their tales. Frequently quilts became stories themselves as women recorded important family events among their patterns—births, marriages, deaths—and passed on these heirlooms to daughters, a kind of diary or family history crafted from their domestic skills. Listening to women's talk as she crouched under the quilting frames, the young girl heard the unwritten history of the community that never entered written records or public history. When Cather became a writer of fiction she likewise practiced an art of connection: she retold and reworked some of these community stories, passing them on to her readers, weaving together oral and written narratives, farm women and artists, past and future in her fiction.

Her most direct acknowledgment of the writer's debt to the women storytellers of her childhood is the "Epilogue" to *Sapphira,* in which Cather claimed to be writing fact rather than fiction; this was, she insisted, an unmediated record of her past. The narrator of the "Epilogue," we are asked to believe, is "Willa Cather"—

the author of *Sapphira and the Slave Girl*—who recounts the five-year-old Willa's life at Willow Shade. Prominent are the memories of the stories from which Cather derived her novel, stories the young child hears as she sits in the kitchen, listening to Mrs. Blake, Nancy, and Old Till talk. When Nancy—returned to Back Creek after twenty-five years—asks about the last days of Sapphira and her husband, the narrator confides: "That story I could almost have told her myself, I had heard about them so often" (*SSG*, p. 291). Then we realize that the child—who grew up to become the author Willa Cather—is indeed telling that story herself, both passing on the tales she once heard in the Willow Shade kitchen and proclaiming herself the inheritor of a tradition of female narrative. Historians, chroniclers, and entertainers, Mrs. Anderson and her cohorts gave Cather the gifts of an oral culture, stories she "remembered all her life" and used when she finally turned back to her Virginia past in *Sapphira* and discovered she could have written two or three novels from her material (*WCL*, pp. 10–11).

•

Excluded from the male-dominated public world, the women in Willa Cather's past nevertheless gave her an even more essential legacy for a writer than land: language. Words were webs that bound women together and at times connected them indirectly to the world of men.[33] Sidney Gore writing letters, Rachel Boak teaching the young Willa to read, Mrs. Anderson passing on folklore: this female legacy would ultimately contribute to the woman writer's attainment of literary authority.

And yet it is understandable that Willa Cather did not take pride in her personal and artistic inheritance from her female forebears until she had first declared her attachment to the male line, literary as well as familial. Traditional women were not the first models and mentors for a young writer desiring acceptance by the elite world of art and letters; female diary-keepers, letter-writers, and storytellers had no status in the world of high culture dominated by Henry James. Nor did these women think of themselves as public figures. As historian Mary Kelley has shown, even nineteenth-century women who wrote for publication—"literary domestics" like Catherine Sedgwick and Susan Warner—experienced great conflict in trying to reconcile their private roles as women with the author's emergence on the public stage.[34]

But just as she turned from Henry James to Sarah Orne Jewett and then to her Nebraska past for subject matter, so Cather eventually looked back to the women in her family and her Back Creek neighbors. The portraits she draws in *My Ántonia*, "Old Mrs. Harris," and *Sapphira and the Slave Girl* reveal how the daughter/writer finally drew on her female strain of blood. In this fiction she also gives something back: the public record most nineteenth-century women did not write but that the author Willa Cather could create in granting her maternal forebears a significance they doubtless would not have claimed for themselves.

Cather's imaginative shift in alignment from her male to her female forebears, which accompanied her transition from a male literary identity to a female one, was a complex, long, and gradual process, as we will see in ensuing chapters. But this development can be seen in two dramatically contrasting poems: "The

Namesake,'' published in *April Twilights* (1903), and ''Macon Prairie (Nebraska),'' included in *April Twilights and Other Poems* (1923). In the first the speaker imagines herself the spiritual and imaginative heir of her uncle; in the second, of her aunt.

Dedicating ''The Namesake'' to ''W.S.B., of the Thirty-third Virginia,'' Cather imagines herself the inheritor of William Boak's ''proud blood'' and the double of the lad ''Who once bore a name like mine.'' The likeness she asserts is one of power: the dead soldier exemplifies glory, fame, courage. Written in the early 1900s—when Willa Cather was hoping for a literary career but had not yet achieved one—the poem even asserts that the namesake niece/nephew will surpass the warrior's achievements: ''And I'll be winner at the game/ Enough for two who bore the name'' (*AT*, pp. 25–26). Despite the stress on inheritance, here the artist is imagined as a solitary figure. Cather connects heroism and achievement with a contest—she may win for both herself and her uncle, but she will win at a game he lost, thus surpassing and in effect replacing him.

But in the 1923 edition of *April Twilights*, published when her personal and literary identity were secure, Cather eliminated ''The Namesake'' and added ''Macon Prairie,'' a poem linking herself not with her uncle but her aunt Jennie Cather Ayre, Caroline Cather's much-advised daughter, who died two weeks after reaching Nebraska in 1877. Imagining a woman guiding the ''grave slow-moving millions'' to the West, in ''Macon Prairie'' Cather gives a female ancestor the public role that Cather and Boak women did not possess in reality. The patriarch William Cather moved his family to Nebraska, but in the poem his daughter is transformed into the female ''leader of the expedition'' who takes her family beyond culture and society (''They came, at last, to where the railway ended/ The strange troop captained by a dying woman'').

In this mythic version of family and pioneer history, a ''frail woman'' with an ''explorer's will'' possesses the courage, vision, and imagination to lead her family from the familiar to the unknown. Hence she can claim the future, becoming spiritual owner of an unmarked, unappropriated land which Cather gives a woman the right to pass on: ''And with her burning eyes she took possession/ Of the red waste,—for hers, and theirs, forever.'' The poem constructs a direct line of descent from aunt to niece by juxtaposing the older woman's death and the speaker's birth: ''She took me in her bed to sleep beside her,—A sturdy bunch of life, born on the ocean. . . . That night, within her bosom,/ I slept./ Before the morning/ I cried because the breast was cold behind me.'' Cather establishes a mother–daughter link between the dying woman and the speaker, who here receives a proud inheritance from her female strain of blood and symbolically acknowledges that she was an aunt's namesake.

Reversing gender roles and power relationships—both familial and social—Cather revises both personal and public history and creates a new American myth in ''Macon Prairie,'' making a place for women in the American Adam's story. The father is a man of ''silence'' who follows his dying daughter, whose illness is redefined as power. Cather thus grants power to her aunts who died of diphtheria and tuberculosis as well as to other nineteenth-century women who suffered from frequent illnesses: the ''burning eyes'' of the dying woman reflect the internal fires of the visionary as well as the fever of the consumptive. The only family member who is quoted in the poem, the aunt uses her power of speech to redefine Genesis, telling her brother that she wants an apple orchard planted around her grave "In memory of woman's first temptation,/ And man's first cowardice" (*ATOP*, pp. 57–59).

So, unlike the Cather women who brought their unquestioned religion with them, this pioneer envisions a new story for a new land, reflecting Cather's similar ability to rewrite traditional narratives—in "Macon Prairie" the biblical story of creation and the American narrative of Adamic exploration, naming, and regeneration. In her imagination and in her art, Cather eventually could invest her female ancestors with the power and authority she wanted to receive from them.

•

In her essay "Katherine Mansfield" Cather mentions three family relationships that threaten the self's "secret and passionate and intense" inner life: "loving husband and wife, affectionate sisters, children and grandmother" (*NUF*, p. 136). One relationship is revealingly missing from this list, the final, crucial female attachment we have yet to consider. Like many daughters, Cather had a more conflict-ridden bond with her mother than with other family members. Because mother and daughter were woven together so tightly in the family net, Virginia Cather seemed to pose a graver threat to the "real self" Willa Cather wanted to possess—a powerful, autonomous, expressive self—than did Caroline Cather, Rachel Boak, or Charles Cather. And yet the mother also consciously and unconsciously collaborated with her daughter in fashioning that self. Fraught with aesthetic as well as emotional dangers and possibilities, Willa Cather's strong and ambivalent bond with her mother deeply marked her evolution as woman and writer.

NOTES

1. I am indebted here, as throughout my book, to previous biographers: E. K. Brown, *Willa Cather: A Critical Biography*, completed by Leon Edel (New York: Knopf, 1953) and James Woodress, *Willa Cather: Her Life and Art* (New York: Pegasus, 1970; rpt. Lincoln: University of Nebraska Press [Bison ed.], 1975). I have also relied heavily upon Mildred R. Bennett, *The World of Willa Cather* (New York: Dodd, Mead, 1951; rpt. Lincoln: University of Nebraska Press, 1961), for information about Cather's childhood and adolescence. Bennett has done an invaluable service in interviewing Red Cloud neighbors and friends as well as Cather's relatives, preserving what amounts to an oral history of her Nebraska childhood and adolescence.

2. [Willa Cather], *Willa Cather: A Biographical Sketch/ An English Opinion/ and an Abridged Bibliography* (New York: Knopf, n.d.), pp. 1–2.

3. J. E. Norris, ed., *History of the Lower Shenandoah Valley* (Chicago: H. Warner, 1890; rpt. Berryville, Va.: Virginia Book Co., 1971), p. 611. (Subsequent page references included in the text.)) James Howard Gore, ed., *My Mother's Story* (Philadelphia: The Judson Press, 1923), p. 3.

4. Brown, p. 6.

5. For an account of William's postwar generosity, see Mildred R. Bennett, "The Childhood Worlds of Willa Cather," *Great Plains Quarterly* 2, no. 4 (Fall 1982): 204–9.

6. Brown, p. 15.

7. A close father–daughter relationship, historian Barbara Welter observes, was shared by several nineteenth-century suffragists. In their families the girl (usually the firstborn) enjoys an "extraordinarily close" tie with her father, who supports or guides her education. She later wants to compensate her beloved father for "the son she is sure he wished her to be" by achieving some kind of public distinction. See *Dimity Convictions: The American Woman in the Nineteenth Century* (Athens: Ohio University Press, 1976), esp. p. 6.

8. Marion Harland (Mary Virginia Terhune), *Eve's Daughters, or Common Sense for Maid, Wife and Mother* (New York: 1882), p. 59. Terhune evidently never recovered from her father's rejection, referring to his comment as a "calamity past human remedy" for which she still grieved (p. 59).

9. Mildred Bennett contrasts Charles to his "aggressive older brother George, who was intent on making money and participating in public affairs" (*WWC*, pp. 23–24). But Helen Cather Southwick, in an effort to show what "an enterprising and successful man my grandfather was," points out that Charles made $5,000 when he sold his livestock and farm equipment before moving into Red Cloud. See Southwick's "Memories of Willa Cather in Red Cloud," *Willa Cather Pioneer Memorial Newsletter* 29, no. 3 (Summer 1985): 11.

10. David Stouck briefly discusses the contribution Charles Cather made to his daughter's literary imagination in *Willa Cather's Imagination* (Lincoln: University of Nebraska Press, 1975), pp. 150–51, 228.

11. For a description on the farm wife's essential and time-consuming work, see Mary P. Ryan, *Womanhood in America: From Colonial Times to the Present* (New York: New Viewpoints, 1975), pp. 19–82. In her study of nineteenth-century women in Petersburg, Virginia (a city not unlike Winchester), Suzanne Lebsock finds that "It was productive labor, the growing of food and the making of clothing, that gave [women] the purest sense of accomplishment" (*The Free Women of Petersburg* [New York: Norton, 1984], p. 153).

12. Brown, pp. 10–11.

13. Carroll Smith-Rosenberg, "The Female World of Love and Ritual: Relations between Women in Nineteenth-Century America," *Signs: Journal of Women in Culture and Society* 1, no. 1 (Autumn 1975): 17.

14. Visions of preindustrial, Western, and frontier societies as havens of sexual egalitarianism can be found in Ann Oakley, *Woman's Work: The Housewife, Past and Present* (New York: Vintage, 1976) and Caroline Bird, *Born Female* (New York: David McKay, 1968). For an analysis of unequal power relationships in rural culture, see John Mack Faragher, "History From the Inside-out," *American Quarterly* 33, no. 5 (Winter 1981): 537–57. As Faragher argues, "The question of status is not primarily a question of what people do, but rather of the recognition they are granted for what they do and the authority that recognition confers. Despite the essential work done by Euro-American rural women there is little evidence to suggest that their husbands and sons granted equal power for equal work" (p. 541). As Lebsock observes, however, women found more legal protection for property rights, in equity than in common law (see pp. 23–25).

15. Brown, p. 7.

16. James Howard Gore, p. 113. Further references to be included in the text.

17. Because the diary was edited by her son, who provides his own commentary and narrative, we have no way of knowing whether we have the original text. Gore may have rearranged entries or, more likely, have omitted some.

18. Willa Cather to Mrs. Ackroyd, May 16, 1941, and to Elizabeth Sergeant, Septem-

ber 12, 1913, Barrett-Cather Collection, University of Virginia Library, Charlottesville, Va.

19. John Mack Faragher, *Women and Men on the Overland Trail* (New Haven: Yale University Press, 1979), p. 163. In her study of women's diaries from the Westward Movement, Lillian Schlissel notes a similar pattern. ''When women wrote of the decision to leave their homes, it was almost always with anguish, a note conspicuously absent from the diaries of men'' (*Women's Diaries of the Westward Journey* [New York: Schocken, 1982], p. 14).

20. Caroline Cather to Jennie Cather Ayre, September 10, 1873 and April 17, 1873, Nebraska State Historical Society, Lincoln, Nebr.

21. Caroline Cather to Jennie Cather Ayre, August 30, 1873. In another letter Caroline reminds Jennie of the dangers as well as the demands of woman's work, telling her to ''be careful and not scald yourself or get your clothes afire when you are canning or preserving'' (August 18, 1873).

22. Caroline Cather to Jennie Cather Ayre, January 22, 1875.

23. Caroline Cather to Jennie Cather Ayre, April 17, 1873.

24. Caroline Cather to Jennie Cather Ayre, August 30, 1873.

25. Caroline Cather to Jennie Cather Ayre, May 4, 1876.

26. Caroline Cather to Jennie Cather Ayre, January 22, 1875.

27. Caroline Cather to Jennie Cather Ayre, May 31, 1873.

28. Caroline Cather to Jennie Cather Ayre, April 17, 1873.

29. The phrase is from Barbara Welter's classic article, ''The Cult of True Womanhood: 1800–1860,'' in *Dimity Convictions,* pp. 21–41.

30. Alice Walker, ''In Search of Our Mother's Gardens,'' from *In Search of Our Mother's Gardens* (New York: Harcourt, Brace. 1983), p. 240.

31. Cather to Mrs. Ackroyd, May 16, 1941.

32. Cather to Mrs. Ackroyd, May 16, 1941.

33. As John Faragher observes, ''articulation was at the very core'' of rural women's female culture (''History from the Inside-out,'' p. 553).

34. Mary Kelley, *Private Woman, Public Stage: Literary Domesticity in Nineteenth-Century America* (New York: Oxford University Press, 1984). See in particular Chapter 5, ''Secret Writers'' (pp. 111–37).

2

The Great Unwritten Story

She [Mrs. Templeton] wasn't in the least willowy or languishing, as Mrs. Rosen had usually found Southern ladies to be. She was high-spirited and direct; a trifle imperious, but with a shade of diffidence, too, as if she were trying to adjust herself to a new group of people and to do the right thing.

"Old Mrs. Harris"

Ever since I could remember anything, I had heard about Nancy. My mother used to sing me to sleep with: *Down by de cane-brake, close by de mill, Dar lived a yaller gal, her name was Nancy Till.*

Sapphira and the Slave Girl

The mother–daughter bond, Adrienne Rich observes, is patriarchal society's "great unwritten story."[1] Yet Willa Cather wrote versions of that story over and over again in her fiction, sometimes concealed as a subtext in stories ostensibly concerned with other matters, sometimes disguised as a mother–son story *(My Ántonia* and *A Lost Lady),* sometimes transformed into images of space and landscape (*The Song of the Lark* and *The Professor's House*), and sometimes openly described (in later fiction like *My Mortal Enemy,* "Old Mrs. Harris," "The Best Years," and *Sapphira and the Slave Girl*). Even in the seemingly all-male worlds of *Death Comes for the Archbishop* and *Shadows on the Rock,* the mother is there in the mythic presence of the Virgin, Cather's and Catholicism's version of the archaic mother-goddess. Although Willa Cather did not find her mother a model or a mentor in her struggle to redefine woman's role and identity, the mother–daughter bond—with all its richness and resentments, its support and strife—ultimately was a source of great imaginative power to her.[2]

Mary Virginia Boak Cather ("Jennie" to her family) challenged the "cult of true womanhood," although in a different way from Rachel Boak or Sidney Gore. Attractive and imperious, Virginia was more the charming and vain Southern lady than the pious and submissive wife praised by Victorian advice-givers.[3] Born in 1850, she returned to Back Creek Valley after her father's death and grew up in the house her grandfather had purchased for his daughter Rachel and her five children. After briefly attending school in Baltimore (with William Cather's financial support), Virginia married Charles Cather in 1872.

She gave birth to Willa a year later, on December 7, 1873. Shortly afterward she moved, along with mother, husband, and new baby, to William Cather's Wil-

CATHER FAMILY

James Cather–Ann Howard

Perry | Sidney Gore | Amanda | Adelaide | William– Cardine Smith (1823–1887) (1827–1900) | John | Howard | Clark

George–Frances Smith | Charles– Mary Virginia Boak 1848–1928 1850–1931 | Alverna | Alfretta | Virginia | Wilella

Willa 1873–1947 | Roscoe 1877–1945 | Douglass 1880–1938 | Jessica 1881–1964 | James 1886–1966 | Elsie 1890–1964 | John 1892–1959

BOAK FAMILY

William Lee Boak–Rachael Seibert

Sarah Ellen | William (killed in Civil War) | Mary Virginia | Betty | Jacob | Clarence

Note: This genealogy identifies people mentioned in this book and is not complete.

low Shade. Her daughter was the first of seven children: Roscoe (1877), Douglass (1880), and Jessica (1881) were born in Virginia; James (1886), Elsie (1890), and John (1892) in Nebraska. Virginia was twenty-three when she had her first child, forty-two when she had her last. This childbearing pattern—spanning almost twenty years—deviates somewhat from the declining birthrate that historians now see as an important feature of women's history in the nineteenth century. At the beginning of the nineteenth century a white woman could expect to have seven or eight children, at the beginning of the twentieth century, three or four.

We cannot be sure whether Virginia Cather intended to have such a large family. She and her husband may have wanted to limit family size but not have succeeded, given that nineteenth-century birth-control methods were unreliable. In any case, she lived most of her adult life with a farm-sized family in a town-sized house, and if we can believe her daughter's autobiographical "Old Mrs. Harris," these numerous births may not have been initially welcomed. Depression, not sickness, causes Victoria Templeton to take to her bed in a darkened room when she learns she is to have another baby—evidently a common pattern in the Templeton family, for the servant Mandy exclaims "Miz' Harris, I 'spect Miss Victoria's done found out she's goin' to have another baby! It looks that way. She's gone back to bed" (*OD*, p. 177). But Mrs. Templeton is an attentive mother—particularly with her babies—and evidently Virginia Cather was as well. She "enjoyed nursing her babies" and never "turned them over to former slave wet nurses," as other Southern women did.[4]

All descriptions of Virginia Cather stress her concern with appearance, beauty, and dress, aspects of her life where she could exercise some control. A "handsome woman" who maintained the "strictest standards of poise" and took pride in her familiarity with high fashion, she carefully cultivated her appearance, at times changing her outfits several times a day. Virginia made her family the first audience for the ideal of femininity she sought to enact: "Always meticulous in appearance, never stepping out of her bedroom without first being perfectly groomed, Virginia Boak Cather allowed no one to see her until her lovely hair had been pinned up" (*WWC*, p. 29). Presenting herself to the public required even more preparation. She dressed carefully when she went out, carrying a small parasol to match her costume and frequently adorning herself with fresh violets.

In "Old Mrs. Harris" Willa Cather attributes the same beauty and poise to Victoria Templeton, an "attractive" woman blessed with a "tall figure and good carriage" (*OD*, p. 149). Even after bearing several children, Victoria takes pride in her youthful figure; she can still wear her "flowered organdie afternoon dress" without a corset and appear "smooth and straight" (*OD*, p. 125). The Templeton children only see Victoria after she has carefully stage-managed her appearance, but in *Sapphira and the Slave Girl* Cather describes the daughter's unwelcome entry into the maternal bedroom before the mother has completed transforming herself into an object ready for public viewing. Rachel Blake intrudes on Sapphira while she is "sitting at a dressing-table before a gilt mirror." Annoyed, Sapphira reflects that "Rachel had come in to disturb her at her dressing hour, when it was understood she did not welcome visits from anyone." Even though the mother's hair seems "in perfect order" to the daughter, "combed up high from the neck and braided in a flat oval on the crown," Sapphira resents being seen while she thinks her head is still "frowsy" (*SSG*, p. 13).

As these literary portraits suggest, from her mother Willa Cather learned that being a lady was a performance: the mother who does not want the family to see her until her hair is arranged is also the actress who does not want the audience backstage before the play begins. The similarity between the lady and the actress is directly conveyed in "Old Mrs. Harris" in which Victoria Templeton is "always ready to lend her dresses and hats and bits of jewellery for school theatricals" (*OD*, p. 112). Virginia Cather did the same when Willa and her Red Cloud playmates performed amateur dramas before an audience of family and friends. In later years, when Willa Cather "knew how to please her mother and with what gifts to do it," she chose jewelry, delicate lingerie, and imported perfumes, tributes to her mother's beauty as well as props for the lady's role. All the children "enjoyed buying their mother extravagant personal things" because they were "proud of her regal bearing, and they knew she would show them off to better advantage than anyone they knew" (*WWC*, p. 31).

Evidently Virginia's flair for the lady's role did not impress her mother-in-law: she did not conform to the sober, industrious, and pious pattern Caroline Cather wanted her son's wife to follow. Virginia was indulged by her mother and her husband, Caroline felt, and enjoyed having others cater to her needs. Not surprisingly, the hard-working Caroline disapproved of her daughter-in-law's pampering: "her mother and Charley has a happy time waiting on her," she commented dryly to her daughter Jennie. Virginia may have exaggerated physical complaints, Caroline hinted in the same letter, because she enjoyed the drama and attention an illness aroused. Shortly after her marriage, Virginia thought she was sick and "called the Dr in twice." Caroline and Sidney Gore—a formidable team of nurses—inspected the patient, but Virginia was just pregnant, they decided: "I think we understand her case as well as the Dr and think he was not needed as much now as he may be after [a] while," Caroline told Jennie discreetly. Sensing that Virginia enjoyed thinking her illness serious, Caroline decided not to risk an unpleasant scene by informing her she was only suffering from morning sickness: "I did not tel her so for she is so easy insulted I knew she would fly right up for she thinks She is awfully sick."[5] Veterans of several pregnancies, Caroline and Sidney had the right diagnosis: Virginia was several weeks pregnant with Willa.

As Caroline's letters to her daughter suggest, Virginia Cather differed from the industrious, uncomplaining housekeepers of her mother-in-law's generation. One could attribute this difference simply to variations in individual temperament, but broader changes in women's roles and opportunities also contributed to the contrast. Daughters like Jennie Cather Ayre or Virginia Boak, born into relatively prosperous families, did not face the same responsibilities their mothers had twenty or thirty years before. By the 1850s men like William Cather had built comfortable farmhouses and managed successful farms, having profitted from the shift from subsistence to commercial farming. Because the daughter's role as her mother's assistant was no longer as economically crucial in such privileged households, and since new educational opportunities for women had arisen, some fathers could afford to send their daughters to one of the female academies that had sprung up in Richmond, Baltimore, or nearby Winchester, as William Cather did for his daughter Jennie and future daughter-in-law Virginia. By contrast, neither Caroline Cather nor Sidney Gore had received formal education. Only Rachel Boak had briefly attended school before her early marriage at fourteen.

This generational difference might not have been apparent to a mother who scanned one of the catalogues for a female academy in the Winchester area, which reassured parents that the school's major purpose was to prepare young women for their proper sphere in life—the home. Stressing their commitment to inculcate moral virtues and teach gracious "accomplishments"—the curricula feature music, sketching, elocution, and French—the academies portrayed themselves as pious Christian households where parental teachers prepared young girls to become cultured, graceful partners for Virginia gentlemen. But the gracious homes the female academies envisioned their pupils enhancing differed from the rural household where Caroline Cather had gardened, preserved, sewed, and cooked. The seminaries intended their graduates to occupy the parlor, not the kitchen. And yet this was still a private realm. As historian Barbara Solomon observes, "It was assumed in the South, unlike the North in this period, that the well-bred female would not teach school; rather, her education should fit her to be a lady—polished, competent, and subservient."[6]

Back Creek girls like Jennie Cather or Virginia Boak who attended the seminaries may have separated from their mothers and their roles in ways that Caroline Cather or Rachel Boak might not have anticipated. A few semesters at boarding school constituted a period of freedom from maternal authority when a daughter would be exposed to new ideas and interests. At least two mothers in the Cather family—Caroline Cather and Sidney Gore—feared that boarding school might also be introducing corrupting temptations. Caroline's daughter Jennie attended a Winchester boarding school, and both women were disturbed by her pride, vanity, and self-absorption, dramatized by an interest in fashion and dress. "Silk was not made for traveling dresses," Sidney cautioned her niece; "wise ladies never use them for such purposes. You, like all inexperienced folks, have much to learn. No doubt you will learn fast if you will forever banish all superfluous *pride*." Jennie should seek to "adorn the mind" with quiet virtues, Sidney advised, not to "bedeck the body" with extravagant clothing and jewelry. Then she might one day become what Sidney considered a "*real* ornament," a "happy, useful wife." Boarding school was encouraging her niece to develop false values and shallow vanities, Sidney thought. "Many questions have arisen in my mind since you entered school," she wrote. "Such, as how much real profit will she realize? Will she . . . earnestly strive to fill her unfolding mind with *useful* knowledge?"[7] Caroline's letters to her "thoughtless vain little daughter" whose "extravagance and pride and neglect" troubled her reveal similar fears; she frequently scolds Jennie for caring too much for "fashion and dress" and for neglecting household tasks.[8] In addition to deflecting a daughter's interests from the Scriptures to silk dresses, fashionable schooling might thus cause a more serious problem, encouraging her to ignore her domestic duties.[9]

If her stay at boarding school first sparked Virginia Cather's interest in clothes, jewelry, and perfume, as it did Jennie Cather's, this was a maternal legacy Willa

Cather in part rejected during her adolescent male impersonation (although this was likewise a means of forming identity through the art of performance) and in part accepted when eventually she modified the lady's persona and costume to suit herself. But the mother's schooling may have allowed her to give her eldest daughter a more direct legacy. Female academies may have stressed the lady's accomplishments, but they also introduced their pupils to mathematics, history, geography, grammar, and rhetoric. Although Virginia did not redefine woman's role herself, her experience of formal education may have contributed to her support of Willa's desire to attend the University of Nebraska.

Virginia's boarding school years may also have introduced her to a pastime she passed on to her daughter: novel-reading. Whereas Rachel Boak and Caroline Cather concentrated on religious and didactic texts and Sidney Gore thought fiction frivolous, Virginia Cather enjoyed the genre some early nineteenth-century Americans viewed as questionable or even immoral because of its supposed power to divert young women from sober, pious, and useful pastimes.[10] Like many Southerners Virginia Cather loved Scott's novels, and one of her daughter's first memories was of her mother riding on horseback "to a distant neighbor's to borrow the 'Waverley novels' as they slowly found their way across the Atlantic and down into rural Virginia" (*WCL,* p. 194). She also enjoyed the popular romantic novels of the day by writers like Ouida, Marie Corelli, and Sarah Grand.[11]

Willa Cather savagely attacked Ouida and Corelli when she was struggling to find herself as a writer in the 1890s. Their breathless, melodramatic effusions convinced her that women did not have the aesthetic or emotional discipline to create great literature. Yet she had loved this fiction in adolescence—one of Ouida's heroes was then her favorite character in fiction—when she learned an important lesson from her mother's books: writing as well as reading could be a female activity (*WWC,* pp. 112–13).

Although in some ways Virginia Cather followed the traditional plot of nineteenth-century women's fiction—courtship, marriage, motherhood—she was not a self-effacing, deferential heroine. A strong-willed, imperious woman who overshadowed her gentle, more easygoing husband, Willa Cather's mother was "always the dominating figure in the family." The family disciplinarian, Virginia punished her children firmly, even tyrannically, Edith Lewis recalls, administering vigorous beatings with a rawhide whip: "her will was law, to show her disrespect was an unthinkable offense, and her displeasure was more dreaded than any other catastrophe that could happen" (*WCL,* p. 6). And yet she was also generous, thoughtful, and considerate, inspiring "great devotion and great deference" in her children. In Lewis' opinion it took Willa Cather years to understand her mother's volatile, paradoxical nature—"quick to resent, quick to sympathize, headstrong, passionate, and yet capable of great kindness and understanding" (*WCL,* p. 6, 156).

Yet this woman whose will was "law" to her children had restricted power in a society where law and custom were patriarchal. Although strong-willed in personality, Virginia Cather did not exercise much control over her own life. If she did not plan to bear seven children, as seems likely, she could not determine the size of the family she supposedly ruled. Nor could she decide where she would make her home: Virginia Cather may have dominated her husband in daily interaction, but the momentous decision to emigrate from Virginia to Nebraska was his. Charles Cather had visited the West in 1870, prospecting in Colorado, Wyoming, and New Mexico, and had been captured by the landscape. "The Prairies here are beautiful beyond description," he wrote his sister Jennie, "whilst in full view are the Snow capped peaks of the Rocky Mts. On the prairies the Antelope feed in abundance; they are the most beautiful thing you can imagine." Entranced by the Edenic possibilities of this untouched yet fertile land, Charles thought of resettling immediately: "It is a splendid country," he concluded, and "[I] think of making my home here next Spring."[12] When he decided to make his home in the West thirteen years later, joining his relatives in Nebraska, his wife had no choice but to follow.

Virginia's children saw her social powerlessness physically expressed when she succumbed to one of her frequent illnesses and the powerful, strong-willed family leader became a dependent invalid.[13] It is not possible to determine the exact nature of her many illnesses from the evidence we have. She may have suffered from physical ailments associated with the strains of pregnancy and childbirth, but it is also possible that her illnesses had a psychological component. At least two occurred at times of difficult transition: when she was pregnant with her first child and when she moved from the comfort of Willow Shade to the Nebraska frontier. As Carroll Smith-Rosenberg has argued, the hysterical illness not uncommon among nineteenth-century women frequently erupted when the indulged daughter encountered the heavy responsibilities of marriage and motherhood for which she was ill-prepared.[14] Although Rachel Boak managed household tasks, Virginia must have found the bearing and raising of seven children a jarring contrast to her free girlhood. The likelihood that Virginia Cather's illnesses were not only physical is also suggested by the miraculous cures she experienced: twice she was saved from what seemed a fatal illness when Dr. G. E. McKeeby, a Red Cloud physician who became the model for Dr. Archie in *The Song of the Lark,* paid a special visit to her bedside. In both cases Virginia began to improve as soon as she was assured of the doctor's attention (*WWC*, pp. 110–11).

Illness allowed Virginia refuge from family responsibilities, granted her momentary power as family members and doctors catered to her needs, and made her the star of a domestic drama. In this sense we can consider illness a performance as well as a kind of speech in which the body expresses what cannot be said openly: perhaps in her case a desire for the power and autonomy the social circumstances of her life did not provide. Whatever the source and meaning of Virginia's illnesses, in her daughter's fiction the figure of the crippled or ill matriarch represents the limitations of women's power in patriarchy. *My Mortal Enemy*'s Myra Henshawe, having sacrificed her own fortune to marry for love, spends her last years confined in a wheelchair plagued by noisy, disrespectful neighbors, chained to a husband who is not financially successful; while the crippled Sapphira Colbert, who vainly tries to transform her "crude invalid's chair" into a "seat of privilege," is constricted by social and economic forces she cannot control (*SSG*, p. 15). And in "Old Mrs. Harris" Cather sympathetically shows how Victoria Templeton's pregnancy-related illness (as well as her childlike selfishness) arises from her inability to exert much control over her life.

Virginia Cather was not an intrusive mother who sought to compensate for whatever power she lacked by attempting to control and dominate her children. On the contrary, she tried to promote their independence and autonomy, possessing, observes Edith Lewis, a "most unusual sympathy and understanding of her children's individuality" and giving them "almost complete freedom, except where the rules of the household were concerned. . . . She had her own absorbing life and she let her children have theirs" (*WCL,* p. 6). In later years Willa Cather recognized the importance of this gift. Her mother had really done a good job in raising seven children, Cather confided to one of her nieces; in addition to keeping them warm, clothed, and fed she had given her and her brothers and sisters the freedom to be themselves.[15]

•

In her essay "Katherine Mansfield," Cather describes human relationships as the "tragic necessity of human life." Everyone is "half the time greedily seeking them," she observes, and "half the time pulling away from them" (*NUF,* p. 136). The profound ambivalence in this statement—in which the self vacillates between desires for connection and separation—characterizes Willa Cather's relationship with her mother, which swirled with powerful and conflicting emotional currents. The daughter loved her mother and hated her, rebelled against her and sought her approval, enjoyed thinking her self completely different from her when she was young and recognized their similarities when she was older.

Although Cather experienced her bond with her mother as personal and private, it was affected by its social and historical context, as Cather later came to see: Virginia Cather embodied the socially defined role of the lady, attractive and strong-willed in the home yet powerless in the public realm. As Adrienne Rich observes, motherhood is both "experience" and "institution," composed both of personal, intimate feelings and of socially constructed and transmitted roles and relationships.[16] Some of the conflict between Cather and her mother arose from the fact that nineteenth-century women were expected to socialize their daughters in a subordinate role.

In later years, Edith Lewis recalls, Willa Cather "always said she was more like her mother than like any other member of her family" (*WCL,* p. 7). The author acknowledges this similarity in "Old Mrs. Harris," in which Mrs. Rosen's observation that "Dat Vickie is her mother over again" is supported by the women's shared names (*OD,* p. 84). One day Vickie will be Victoria. Yet before Cather recognized this resemblance she strove to establish difference. The daughter's most serious conflicts with her mother occurred during adolescence when she adopted her William Cather persona, but this defiance began during her Virginia childhood when she rejected the role of the Southern lady her mother wished her to play.

Reviewing *The Mill on the Floss* in 1897, Cather described her young self as a rambunctious tomboy and declared her identification with Eliot's defiant heroine Maggie Tulliver:

> haven't we all had pretty prim little cousins like Lucy Deane, whose hair curled naturally, and who were always neat when we were dirty, and mannerly when

> we were rude, who never tore their frocks nor dropped their fork at the table and whose China-blue eyes grew wide with astonishment at our tomboyish proceedings? And haven't we all just ached to push these immaculate cherubs into the mud—just as Maggie did? And haven't we all felt bitterly that our mothers secretly suffered from our plain brown faces and stubby noses and wished we were pretty like other children? (*WP* 1, p. 363)

Whether or not Virginia Cather "secretly suffered" from her eldest daughter's plain face, she did strive to make her look "pretty like other children." A childhood studio photograph reveals this maternal desire: Willa is wearing a delicate white dress with a lace collar, a cross around her neck, and a carefully placed ribbon holding her long ringlets neatly in place. *(See photograph section.)* Momentarily the tomboy daughter seems an "immaculate cherub," her hair naturally curly, her dress neat and clean, her face "pretty like other children" despite her "stubby nose" and stubborn mouth. As this photograph suggests, the lady's role is a social construct mothers impose upon daughters: Willa is transformed into a "little woman" by the proper costume and hairdo.

If this photograph reveals Virginia Cather's ideal vision of her daughter, what of the daughter's vision of her mother? Our only resources are the daughter's early memories, which feature the mother in contrasting roles and show the daughter both "pulling away" from and "seeking" her.

An early memory, which became a story Cather recounted to Edith Lewis, records what must have been a typical skirmish between rebellious daughter and imperious mother. One day an elderly judge was visiting Willow Shade. Adopting a courtly, paternalistic manner toward Willa, he "began stroking her curls and talking to her in the playful platitudes one addressed to little girls." But Willa refused to employ the playful platitudes with which little girls were supposed to flatter Southern gentlemen. Throwing away the script, the daughter "horrified" her mother, Edith Lewis recounts, by exclaiming, "I'se a dang'ous nigger, I is!" Willa Cather later attributed this outburst to the child's need to defy the "smooth, unreal conventions about little girls" prevailing in Virginia. Such conventions were part of Southern codes of etiquette that falsified communication and masked conflict, Cather later felt: "Even as a little girl she felt something smothering in the polite, rigid social conventions of that Southern society—something factitious and unreal. If one fell in with those sentimental attitudes, those euphuisms that went with good manners, one lost all touch with reality, with truth of experience. If one resisted them one became a social rebel" (*WCL,* p. 13).

We might interpret this story as Willa Cather's first rebellion against patriarchal authority, represented here by the condescending judge. In a sense it is. Yet the person "horrified" in her memory, revealingly, is not the judge, the person to whom the remark was ostensibly directed, but her mother—the real object of the daughter's attack. It was from Virginia Cather, not the visiting judge, that Willa learned "smothering" social conventions defining the role of the Southern lady, not realizing that her mother was in turn enforcing the judge's definition of femininity. Rejecting the mother's standards of propriety and hospitality by misbehaving in her realm of apparent power, the parlor, the daughter assumes the identity most opposed to the one her mother had fashioned for her. A dangerous "nigger" is a disruptive, unruly, lower-caste outsider—most likely male—not a proper (and powerless) little lady.

Since acting "like a lady" meant playing a part and assuming a role, in pulling away from her mother it is not surprising, then, that Willa Cather evolved other roles, wrote new scripts, and constructed different personas. A picture taken a few years later shows her as Hiawatha. (See photograph section.) Although still appearing in a white dress with hair carefully curled, she is gripping a bow and arrow, props for her recitation of Longfellow's poem but also symbols of masculine power. Later, in adolescence, she transferred the signs of maleness from her hand to her body, cropping her hair and adopting male dress. "Dang'ous nigger," Indian warrior, "William Cather, Jr."—her first dramatic roles and her first fictional characters—these personas anticipate Cather's love for the theater as well as her career as a novelist. But they also served the child's immediate need to reject the social role her mother wanted her to accept. She was the disappointment of her mother's life, Cather told a college friend, because she had never lived up to her mother's definition of a lady. Only her little sister Jessica promised to fulfill Virginia Cather's desire for "one lady in the family." [17]

Willa Cather's other enduring childhood memory reveals how the maternal "strain of blood" brought compassion as well as conflict. The mother–daughter reunion described in the autobiographical "Epilogue" to *Sapphira and the Slave Girl* (1940) later seemed the dramatic center of her childhood. Although Cather always insisted that her novels did not contain real-life portraits despite their roots in past experiences and remembered friendships, she made an important exception in this case. The "Epilogue" to *Sapphira* was literally true, every word of it, Cather insisted to friends, and it described the greatest event of her Virginia years: when Nancy Till, the slave girl who had fled to Canada before the Civil War, returned to Willow Shade to be reunited with her aging mother. [18]

In telling the story of this reunion in *Sapphira*'s "Epilogue," which is narrated from the point of view of the five-year-old Willa, Cather gives her own mother the central role in nourishing her daughter's creativity. Virginia had first whetted Willa's interest by telling her stories about Nancy Till and singing a lullaby about the escaped slave. "Ever since I could remember anything, I had heard about Nancy," she writes. "My mother used to sing me to sleep with: *'Down by de cane brake, close by de mill,/ Dar lived a yaller gal, her name was Nancy Till.'* " The mother continues to minister to her child's imaginative needs as the narrative proceeds. On the day Nancy is to return to Back Creek Willa is sick with a cold, but the mother persuades the two women to stage their reunion at Willow Shade so the child can be included in the unfolding drama. She then puts her daughter in her own bed so that she can watch the road from a good vantage point. When the stagecoach arrives, the mother quickly arranges a front-row seat for the daughter:

> Suddenly my mother hurried into the room. Without a word she wrapped me in a blanket, carried me to the curved lounge by the window, and put me down on the high head-rest, where I could look out. There it came, the stage, with a trunk on top, and sixteen hoofs trotting briskly round the curve where the milestone was.

Then we see the most memorable event of Willa Cather's childhood, as she describes it sixty years afterward:

> Till had already risen; when the stranger followed my mother into the room, she took a few uncertain steps forward. She fell meekly into the arms of a tall, gold-skinned woman who drew the little old darky to her breast and held her there, bending her face down over the head scantily covered with grey wool. Neither spoke a word. There was something Scriptural in that meeting, like the pictures in our old Bible. (*SSG,* p. 281–83).

In contrast to the scene of parlor rebellion, this memory evokes the harmonious bond between mothers and daughters in an exclusively female world. The setting for the reunion—the mother's bedroom—is the private space of warmth, informality, and harmony. The powerful image of mother–daughter unity expressed in the embrace is apparent also in the bond between Willa Cather and her mother: here they share a common purpose. Virginia Cather is the caring mother concerned with pleasing her daughter, not the repressive mother concerned with her daughter's pleasing Southern men. She is also the strong mother who tends her sick child, not the family invalid who needs attention herself.

The silent embrace conveys the women's deeply felt, inexpressible emotion and provides a contrast to the "platitudes" expected of the daughter in the parlor's social space. The silence in which authentic female speech and understanding take place also links the two pairs of mothers and daughters: neither Nancy nor her mother "spoke a word"; Virginia Cather carries her daughter to the window "without a word."[19] In finding "something Scriptural" in this mother–daughter reunion, Cather imputes a sacred, archaic quality to her childhood recollection. She suggests that this meeting had occurred before, that the two women were acting parts other women had already played. But the antecedent lies in Greek mythology, not Christian Scripture: the women re-enact the reunion of Demeter with her daughter Korê, or Persephone, celebrated in the Eleusinian mysteries.[20]

These two early memories dramatize the conflicting strains in Cather's bond with her mother, who both sought to confine and to expand her daughter's imagination. In the first memory, the parlor is the domestic space where the mother's power is only apparent, since the female roles possible there are those sanctioned in a male-dominated society. Like the powerful but crippled women rooted to wheelchairs in her daughter's fiction, on the parlor's public stage the imperious Virginia Cather, concerned with pleasing the judge, occupies a fixed, limiting space. In the second memory, the bedroom is a utopian space and a private stage where women can create and enact roles that are not defined by their relationships to men. And these roles are flexible: even the mother–daughter bond can be inverted. The daughter is tall, powerful, and self-assured; the mother little, meek, and uncertain. Their embrace visually symbolizes the possibility of role reversal, as Nancy holds her mother against her breast like a child. The continuity between mothers and daughters implied by this scene may be one reason why the memory endured in Willa Cather's imagination for so many years, an unopened letter, the seed for her last novel. There is a link between the mother who creates a drama for her daughter and the woman who writes the novel; the child for whom the scene was staged eventually became the creator who stages it in fiction.

It is understandable why this memory, rather than the scene of defiance, became the center of Willa Cather's last novel. In becoming a writer she would leave the parlor (the space where the mother assigns limiting female roles and socially acceptable words) and construct an imaginative space where she could speak in her own voice. In *Sapphira* Cather acknowledges the contribution Virginia Cather had made to this creative emergence, conveyed in the "Epilogue" by the link between the mother's voice and the writer's voice, the lullaby and the novel.

•

An adult's memories of childhood experiences do not, of course, constitute an objective record of the past. Memory distorts as well as records; the images retained over the years are selective and fragmentary, fictions the adult creates as well as retains, stories he or she employs to imagine a continuous self and to describe that self to others. Yet precisely because they *are* subjective, early memories constitute important psychological and biographical evidence. Willa Cather's stories and memories of her mother do not tell us what Virginia Cather was actually like, but they do tell us how she appeared in her daughter's imagination and memory. Featured in opposing roles as the repressive and the attentive mother, they dramatize an ambivalence other daughters experience.

Recent developments in feminist psychoanalytic theory suggest why Willa Cather retained these opposed memories from childhood. Nancy Chodorow's analysis of female personality development offers a way of reading Willa Cather's early memories and the struggle to define a female self they reflect.[21] Although Chodorow bases her theories on the twentieth-century patriarchal family in which the work of mothering is assigned to women, the many parallels between her account of female development and Cather's life and fiction suggest that her psychoanalytic story of the female self's emergence can help us to understand Willa Cather's.[22] Arguing that classic Freudian theory is inadequate in explaining female psychological development both because of its male bias and its stress on the oedipal period as the crucible in which adult personality is formed, Chodorow draws on object-relations theory in formulating her thesis, which sees the roots of adult personality in the mother-dominated preoedipal stage.[23] Chodorow contends that mothering by women results in asymmetrical personality development in women and men: women tend to define themselves in relationship to others, whereas men do not.

According to object-relations theory, the infant's earliest relational experience with its mother structures its later development. The infant's original state is one of fusion with the mother who satisfies its needs for warmth, food, and protection. In this period of extreme dependence the infant does not distinguish itself from the external world, which is coextensive with the mother: there are no ego boundaries separating self and other and no reflective ego. The dawning awareness of the self as separate is thus connected with the child's recognition that she is distinct from the mother, who aids in the separation process by mirroring back the child, confirming its existence, providing recognition as well as nurturance. Object-relations theorists thus stress that the self is born in relationship; separateness can only be established by a sense of relatedness to and difference from another. This ambivalent process of separation and individuation involves loss as well as gain: the child wants to be independent, yet increasing autonomy means losing the original union with the mother and abandoning the original state of oneness.

Separation and individuation are particularly complex psychological tasks for daughters, Chodorow argues, and never as fully completed as for sons. Because the mother identifies more closely with the female child (whom she is more likely to see as an extension of herself), she does not create as clear boundaries separating herself from her daughter as she does from her son. The sex differences make the male child clearly a "differentiated other" while the daughter is more easily considered part of the self. The daughter, in turn, may find it difficult to escape what Chodorow terms her "primary" sense of oneness with the being with whom she was once merged; unlike her brother, she cannot use the barrier of gender difference to establish her autonomy.[24]

The result of these mutual identifications between the mother and daughter, according to Chodorow, is that the girl retains more "permeable ego boundaries" than the boy and tends to define herself in relation to others rather than in opposition to them. The boy, on the other hand, more easily severs the infant's preoedipal attachment to the mother, and so maintains more rigid boundaries between self and other. "The basic feminine sense of self is connected to the world," Chodorow concludes, while "the basic masculine sense of self is separate."[25]

If we now return to Cather's early memories of her mother, we can see how they reflect psychological as well as social and historical patterns. In the childhood picture of the neatly dressed Willa as "immaculate cherub," we see the mother's vision of the daughter as an extension of herself. Even though Virginia Cather may have consciously promoted her daughter's separation and autonomy, she could not escape identifying with her. In the scene of parlor rebellion, we see the daughter struggling to break the symbiotic union between mother and daughter by assuming another identity, arbitrarily forming boundaries. In the memory of mother–daughter reunion, by contrast, we see continuity between mothers and daughters expressed not only in the dramatic meeting of Nancy and her mother but also in Virginia Cather's sympathetic, intuitive understanding of what would please her daughter.

The child's fusion with the mother during the preoedipal phase may lead to conflict in adult life for sons as well as daughters. As Dorothy Dinnerstein observes, the mother has fearful power. She confirms the separateness of the growing self, mirroring its worth and integrity. Yet she can also be experienced as a "menace to that self" and may even "encourage the child's lapses from selfhood, for she as well as the child has mixed feelings about its increasing separateness from her." So the mother has a dual role as both supporter and enemy of the self. She is the one who has the power to confirm the existence of the child, and yet she is also experienced as "the one who beckons her loved ones back from selfhood, who wants to engulf, dissolve, drown, suffocate them as autonomous persons."[26]

States of fusion in which the self dissolves its boundaries—religious ecstasy, sexual union, or the creative process—can be fearful to us since they recall the ambivalences of the mother–infant bond, raising the possibility of engulfment, incorporation, drowning. Yet such moments of merging with something larger than the self are also attractive because, in recalling the earliest union with the mother, they momentarily erase the separation between self and other that can be alienating and isolating.

Like many artists who abandon firmly defined selves in the creative process, Cather was drawn to undifferentiated states in which the boundaries between self and other dissipate. Her characters who find fulfillment most often do so by connecting to "something complete and great" that enriches, extends, and defines the self even as it dissolves it—religion, land, the family, art (*MA*, p. 18). Yet simultaneously she dreaded self-annihilation and obliteration, variously expressed in her fear of Nebraska's expenses of limitless prairie, her terror of drowning (a motif in much of her fiction), her suspicion that she might be no more than what she called a "cell in the family blood stream" (*WC:AM*, p. 143).[27]

Hence the psychological issues Nancy Chodorow and other theorists see in the mother–daughter bond are also enduring threads in Cather's life and fiction, possibly more intense for her as a lesbian than for heterosexual women since her lovers directly reinvoked the dynamics of the mother–daughter bond. Her desire simultaneously to possess a demarcated ego and to lose herself in something larger than the self took various forms in her life, marking the tensions she felt between the demands of art and of human relationships, between creative and sexual passion, between sheltered and exposed landscapes, between her search for an original literary voice and her wish to be part of a literary tradition. She strove for a precarious balance between self and other, desiring to be an inheritor of her family's strain of blood who retained individual identity instead of an indistinguishable "cell" in its bloodstream. As we will see in later chapters, Cather reconciled the seeming polarities of separation and connection, self-assertion and self-dissolution, most fully in the creative process where she could melt into characters or subject matter without losing adult separateness.

Returning again to Cather's memory/story of mother–daughter reunion in *Sapphira,* we can see dramatized a moment in which the daughter's opposing needs for separation and connection are satisfied. The passage from the "Epilogue" suggests the enduring power of mother–daughter attachments as well as the possibility of a female bonding that does not obliterate the individual self. This is a vision of reunion but not fusion; daughter and mother embrace but are not indistinguishable from each other. Power passes from mothers to daughters without harm to the self: Nancy cradles her mother in her arms, Willa Cather transforms the memory into art.

•

In her apprenticeship fiction, however, Cather was not able to transform mother–daughter conflicts into fictions that existed apart from herself. Not in control of this volatile psychological and emotional material, she projected it without mediation into many of the short stories she wrote in the 1890s and early 1900s. These stories do not constitute biographical evidence of the child's psychological and emotional world, but they reveal how strongly the psychodynamics of the mother–daughter bond underlay Cather's literary imagination and suggest the psychological as well as the aesthetic difficulties she would confront in becoming a writer.

Several of her early stories reveal the child's ambivalence toward a mother who is alternately rejecting and tender. Cather characteristically projects these two aspects of the maternal presence into different characters. "The Prodigies" (1897), "Flavia and Her Artists" (1905), and "The Profile" (1907) present beautiful, cold, narcissistic mothers who ignore or exploit their children because they are incapable of unselfish maternal love, while "The Burglar's Christmas" (1896), "A Resurrection" (1897), and "On the Gull's Road" (1908) feature selfless, loving mothers who devote their "youth and strength and blood" to their children rather than their own needs (*CSF*, p. 429). Like fairy tales, which oppose the benevolent fairy godmother to her malevolent opposite and double, the cruel stepmother or vindictive witch, such stories reflect the child's splitting of the mother into "good" and "bad" selves: unable to comprehend how the same person can be both caring and punishing, loving and rejecting, she dissociates loving feelings from angry ones by constructing two opposed maternal figures.[28]

The children in the first group of stories are weak and sickly, their frailness symbolizing their emotional starvation, while in the second group they are healthy and strong, since the "good" mother gives her child essential sustenance. The two stories presenting the fullest portraits of the rejecting mother and her nurturant opposite/double—"The Profile" and "The Burglar's Christmas"—are also the most directly autobiographical.

The brief prologue to "The Profile" provides clues to its genre and interpretation: it is a fairy tale requiring psychological as well as literary exegesis. Aaron Dunlap, an American portrait painter working in Paris, is attending a meeting of the Impressionists' Club. The painting under discussion, "Circe's Swine," anticipates the story's central character, an evil, seductive woman who casts an enchanting spell over her lover and daughter. The foreshadowing is complex, however, for she will be both the Circe-like witch and a grotesque, half-human monster resembling the "hideous" enchanted creatures in the painting.

After this introduction, Cather begins her fairy tale. Commissioned to paint the portrait of a young American girl named Virginia Gilbert, Aaron Dunlap is shocked to discover that a dreadful scar mars one side of the young woman's face, making a "grinning distortion" of her smile: "It was as if some grotesque mask, worn for disport, were just slipping sidewise from her face." The awful blemish seems all the more cruel because without it, he can see, the girl would have been unusually beautiful. Dunlap finds the encounter bearable only because Virginia acts as if she were the beautiful woman behind the grotesque mask, laughing and flattering him "almost with coquetry." Only when the two discuss the pose do her actions hint at her awareness of the disfigurement when she casually suggests a profile. As Virginia turns the dreadful scar away from him, Dunlap views her full, unspoiled loveliness, which he transfers to the canvas.

As the sittings continue, Dunlap becomes increasingly moved by Virginia's supposed courage. He falls in love and marries her, seeing himself as a fairy-tale prince who will, in time, make "her hurt . . . disappear, like the deformities imposed by enchantment to test the hardihood of lovers." But Virginia is not the enchanted princess whose ugly surface hides a pure, loving, and humble nature, but the evil enchantress; Dunlap has been casting himself as a character in the wrong fairy tale. The scar is an accurate sign of her character after all, for she hides a cruel, selfish nature.

Soon the husband discovers Virginia's two "all-absorbing interests" which intensify over time: a "passion for dress" and a "feverish admiration of physical beauty." These preoccupations eventually become obsessions, for despite her disfiguring scar Virginia is narcissistic. Her characteristic pose is before the mirror, either "surveying the effect of a new gown" or "carefully arranging her beautiful hair, which she always dressed herself." Cather progressively connects Virginia's "mania for lavish display" and devotion to the "pageantries of her toilet" with shallowness, self-centeredness, and cold-heartedness. The enchantress' worst sin is her inability to love anyone but herself, which Cather represents by her "marked" indifference to her firstborn daughter, a sickly child who is not as attractive as the mother would have liked. The daughter's physical weakness signifies her emotional deprivation; the frail little Eleanor does not receive enough love from the mother who prefers to spend her time with "Madame de Montebelloe and her handsome children," meanwhile neglecting the daughter whose unattractive face is not the mirror in which she wants to see herself.

Several parallels exist between Virginia Gilbert and Virginia Cather. The name is the most obvious tie, but they share personal traits as well: the spirited, strong-willed natures, the interest in dress, the concern with appearance. The mother's disappointment with the daughter who is not pretty enough to please her may be another link ("And haven't we all felt bitterly that our mothers secretly suffered from our plain brown faces and stubby noses and wished we were pretty like other children?").[29] In addition to giving the witchlike Virginia Gilbert the bad mother's evil traits—coldness, selfishness, narcissism—Cather inflicts her with a hideous physical deformity. Why was the scar an essential part of this fairy tale?

In one sense, Cather uses the scar as Hawthorne does physical signs—a black veil, a birthmark, a hunchback—to symbolize a hidden psychological or emotional reality. Yet Cather could have represented Virginia's hidden coldness in many ways; she could also have dispensed with physical symbolism and allowed the character's inner nature to be revealed solely through her actions. This authorial choice suggests not only the inexperienced writer's use of fictional clichés but also the daughter's revenge on the beautiful mother who made her feel she was not pretty or ladylike enough to win her approval—or to win her. In destroying the character's physical beauty Cather removes the sign of feminine success and power in her mother's social world, which is also the sign of the mother's unavailability to the daughter, since female beauty signifies the mother's sexual availability to men. As the fairy-tale plot demonstrates, only a prince can hope to rescue and possess a princess.

So the scar is a complex sign, reflecting the daughter's search for closeness as well as her feelings of abandonment, her love as well as her anger. In destroying Virginia's beauty Cather creates a bond between mother and daughter, a possibility suggested by the story's bizarre, melodramatic conclusion. Another character joins the Dunlap household—Eleanor Vane, Virginia's cousin, who befriends the lonely, sickly daughter (also Eleanor). "From the first," Cather comments, "the two Eleanors seemed drawn to each other." As the names suggest, there is mutual identification between this mother–daughter pair; Eleanor Vane is also the kindly, nurturant good mother. Yet by the story's end she has become externally the evil Virginia's twin, for just before she abandons her family Virginia arranges for an alcohol lamp to explode in her rival's face—leaving Eleanor with a disfiguring scar. Cather thus makes the good mother the bad mother's double as well as her opposite. The second scar suggests that these women are two halves of a divided self, as does Eleanor's last name (Vane/vain). The scar also links the disfigured benevolent mother with the physically imperfect child, so the final bonding between the two Eleanors is physical as well as psychological and emotional (*CSF*, pp. 125–35).

•

Discovering the all-providing good mother may not, however, be a blessing after all, "The Burglar's Christmas" suggests, for she has the power to deny the child separateness and autonomy. In this early story the protagonist's goal is the most intimate domestic space of all, the mother's bedroom; here it is the stage for fusion rather than reunion. At first this crudely melodramatic tale, which Cather published in the 1896 Christmas issue of the *Home Monthly,* the Pittsburgh woman's magazine whose editorship she had just assumed, seems to reflect only the inexperienced writer's adherence to the formulas of Victorian magazine fiction.

The plot is contrived and sentimental. Starving, obsessed with the desire for food, a destitute young man is driven by hunger and desperation to rob a fashionable Chicago mansion on Christmas Eve—which is, coincidentally, his birthday. He stealthily enters the strange house, enters a richly furnished bedroom, opens a dressing-case, and begins pocketing the rings, watches, and bracelets he finds there. Suddenly the thief discovers a familiar object—the "silver mug he used to drink from when he was a little boy." Before he or the reader can puzzle further, his mother enters the room. An angelic being, she embraces her long-lost son, tells him all is forgiven, and welcomes him back to the comfort of her inexhaustible love. In contrast to "The Profile," this mother's beauty is not scarred, for her affectionate nature matches her external loveliness; she showers the thief with unconditional love. "How could you rob your own house," she asks when the ashamed William, his pockets stuffed with her jewels, calls himself a common thief. "How could you take what is your own? They are all yours, my son, as wholly yours as my great love—and you can't doubt that, Will, do you?"

Eventually we discover the reason for this unlikely meeting. After the son had severed his ties with home, ceasing all communication with his parents, unknown to him they had moved from his original childhood home somewhere "down East" to Chicago. Hence his unwitting entry into the maternal bedroom. The story ends with the errant son literally and figuratively returned to the bosom of his family, renewed on Christmas Eve and reborn on his birthday.

It is tempting to dismiss this story simply as crude apprenticeship work. But despite its aesthetic deficiencies, "The Burglar's Christmas" is a crucial story in the Cather canon because its psychological themes anticipate the major fiction, where the writer developed, controlled, and transformed them in novels like *My Ántonia, A Lost Lady,* and *The Professor's House,* which also feature the search for a maternal figure. In this story, however, the author is too closely identified with the protagonist and his emotional drives to create fiction. "The Burglar's Christmas" is a dreamlike fairy tale of wish-fulfillment, not a consciously shaped narrative.

Some overt biographical clues link the author with her protagonist. His name is William, the name Cather took as her own in adolescence; his nickname is "Willie," also hers. William has used other unmentioned names during his travels just as the young journalist adopted various pseudonyms for her stories and articles, continuing the role-playing she began in childhood. Close to the author's age, William has followed a similar life pattern. A restless and ambitious child, he left home to make his way in the public world where he also attempted journalism. But as he stands cold and hungry in front of the Chicago mansion, the failure of his childhood dreams of personal success is clear. Defeats have reduced him to a level where he cannot satisfy even the basic human needs for warmth, food, and shelter, much less attain the independence and prestige for which he once seemed destined.

His deprived present stands in contrast to a bountiful past. Before William enters the strange house, he is overwhelmed by a memory linking maternal love with abundant food. He thinks of the "splendid birthday parties" his mother used to give him "and how he would eat and eat and then go to bed and dream of Santa Claus. And in the morning he would awaken and eat again. . . ." In his present despair, what he most wants to do is to retrace his steps and become a little child again, a goal Cather grants him as he finds "refuge and protection" in his mother's arms just as he had "in the old days when he was afraid of the dark." Loved and fed by his mother once again on his birthday, William manages to recapture the childhood bliss of complete gratification. As he sits in his old chair, like the male protagonists in Willa Cather's later fiction who reserve their deepest love for their childhood selves—Bartley Alexander, Jim Burden, Godfrey St. Peter—he feels a "sudden yearning tenderness for the happy little boy who had sat there and dreamt of the big world so long ago." As the story ends he has traveled back even futher into the past; the "rich content" he experiences sitting by a warming fire with hunger satisfied and sleep descending suggests the infant's satiated pleasure more than the little boy's happiness.

Such preoccupation with the mother–child bond was not uncommon in late nineteenth-century popular fiction. Given the Victorian exaltation of motherhood, Cather's use of the prodigal son motif, in which she replaces father with mother, was an effective way to exploit popular sentiment and to appeal to the female readership of the *Home Monthly.* But the story is not merely a conscious manipulation of literary conventions, as its intense evocation of a regressive fantasy that mingles the erotic and the maternal suggests. The sexual imagery of the theft is obvious, and the behavior of both parents suggests the wished-for resolution to the child's fantasy of sexual possession: the father is absent and permissive, the mother loving and generous, desiring exclusivity as much as her son ("I want you all to myself tonight").

But we do not even have to consider the covert meaning of passageways, darkened chambers, boxes, and jewels to see the eroticism in this mother–child reunion. Not only does the mother seem beautiful and seductive to the son, with her "superb white throat and shoulders" and her "impetuous and wayward mouth," but their passionate embraces are described in the language of romantic fiction. At first his mother's lips on his cheek "burnt him like a fire"; later, as the son ceases to struggle in her arms, he becomes a swooning lover overwhelmed by an irresistible sensual spell or a knight captured by *la belle dame sans merci:* "That soft voice, the warmth and fragrance of her person stole through his chill, empty veins like a gentle stimulant. He felt as though all his strength were leaving him and even consciousness."[30]

The mother is both a nurturing and an erotic power, and the child desires both preoedipal union and oedipal possession. Of the two, the preoedipal concerns are the most intense. The starving William's real goal is not the jewel box he intends to steal but the silver mug he discovers—the Holy Grail he did not consciously realize he was seeking. Although the author does not intend to condemn William's regressive urges—no authorial clues tell us to take his return to childhood as anything but a happy ending—the story implicitly reveals the costs as well as the benefits of returning to the mother. William may have gained food, contentment, and maternal love, but he has lost adult separateness and autonomy. Overpowered by his mother's embrace, drugged by her perfume, by the story's conclusion he is sliding into a state of preverbal infantile bliss which is remarkably similar to death.

Along with strength and consciousness, William loses a separate identity in his mother's arms: she regards him as an extension of herself. She and her son were always unusually similar, she tells him; even when he was a baby she experienced a "likeness" between them, and even after his years away, when he has had experiences of which she can know nothing, the mother still can say, "I know your every possibility and limitation, as a composer knows his instrument." Her denial of his separateness is complete and terrifying. "We are more alike than other people," she explains. "I am a woman, and circumstances were different with me, but we are of one blood. I have lived all your life before you. You have never had an impulse that I have not known, you have never touched a brink that my feet have not trod" (*CSF*, pp. 557–66).

Unlike the mother–son relationship Nancy Chodorow describes, this bond is characterized by fusion rather than by separation. As this anomaly suggests, the story's real concern is the mother–daughter relationship, a likelihood reinforced by the biographical overtones. If we take William as a stand-in for "Willie" Cather, the mother's over-identification with the child becomes more understandable; similarly, the intensity of William's rebellion as well as his urge to merge again with his mother can be connected with the daughter's struggle to carve out a separate identity against a figure doubly overwhelming as her first love-object and her same-sex parent.

Of course the biographical parallels are suggestive but not complete, for in writing the story Cather fought against the regressive impulses to which her character succumbed, succeeding in the adult world of journalism instead of returning home in defeat. Yet the unresolved tensions aroused by the daughter's ambivalent process of separation, also displayed in Cather's early memories, are the source of this seemingly formulaic and sentimental short story.

Reflecting the difficulties mothers and daughters have in establishing boundaries, "The Profile" and "The Burglar's Christmas" are important stories because they reveal many of the conflicts Cather would experience in becoming a writer. How could she connect with something that "dissolves" the self without being obliterated? How could she draw strength from her family's strain of blood without fearing that she and her mother shared "one blood" and one identity? How could she shape the emotional patterns of her life in art? The adolescent Willa Cather tried to find answers to these questions in her male impersonation, the adult writer in the creative process.

•

The loss of the infant's oneness with the mother's body, Dorothy Dinnerstein observes, is an "original and basic human grief" shared by men and women alike, and many of our adult strivings are efforts to regain "some part of the lost delight." But since the daughter who becomes heterosexual does not regain the mother in adult love relationships, in Dinnerstein's view she must renounce her first love even more completely than the son. This wrenching dislocation can give the daughter a continuing and "ancient, primal fear of loss."[31]

Like other daughters, Willa Cather experienced this primal loss when she realized that her mother's cultivation of dress and beauty signified her erotic allegiance to men. She also experienced the loss of her mother on a more symbolic level when she understood that the strong-willed, whip-wielding woman whose word was law within the family did not have real power in the larger society. And as the first of seven children, she experienced the loss of her mother in a less symbolic way: born into a world without rivals, she saw her mother's attention increasingly consumed by the demands of younger children.

Although Willa Cather remained particularly susceptible to change and dislocation throughout her life, she found many compensations for both the original loss of her mother and life's subsequent, inevitable deprivations and discontinuities. As a lesbian, she retained the love of women as the primary emotional and erotic force in her life; as an independent professional woman, she reversed the pattern of female dependency and powerlessness symbolized in her fiction by the crippled matriarch; as an artist whose imagination was often sparked by memories of lost ladies and distant landscapes, Cather could possess what she called "precious, the incommunicable past" in the act of writing (*MA*, p. 372). The ways in which the adult writer would transform loss into creativity are anticipated by the nine-year-old girl's response to the most traumatic event of her childhood, her separation from her Virginia homeland.

NOTES

1. Adrienne Rich, *Of Woman Born: Motherhood as Experience and Institution* (New York: Norton, 1976), p. 225.

2. Observing that Cather's relationship with her mother provided "a source of conflict integral to Willa Cather's imaginative life," David Stouck briefly discusses the mother's impact on the daughter's fiction (*Willa Cather's Imagination* [Lincoln: University of Nebraska Press, 1975], p. 208). See also Jane Lilienfeld, "Reentering Paradise: Cather, Colette, Woolf and Their Mothers," in *The Lost Tradition: Mothers and Daughters in Literature,* ed. Cathy N. Davidson and E. M. Broner (New York: Ungar, 1980), pp. 160–75, and Sharon O'Brien, "Mothers, Daughters, and the 'Art Necessity': Willa Cather and the Creative Process," in *American Novelists Revisited: Essays in Feminist Criticism,* ed. Fritz Fleischmann (Boston: G. K. Hall, 1982), pp. 265–98.

3. In *The Southern Lady: From Pedestal to Politics 1830–1930,* Anne Firor Scott demonstrates the contradictions between the popular image of the antebellum lady and the realities of women's lives (Chicago: University of Chicago Press, 1970), pp. 3–44.

4. Mildred R. Bennett, "The Childhood Worlds of Willa Cather," *Great Plains Quarterly* 2, no. 4 (Fall 1982): 204–9, esp. 204.

5. Caroline Cather to Jennie Cather Ayre, April 17, 1873, Nebraska State Historical Society, Lincoln, Nebr.

6. The goal of the Valley Female Institute was to inculcate the "quiet modesty of manner" that marked the "true and cultured woman," while the Virginia Female Institute professed to provide a "Christian household" where the daughter would acquire "the graces of refined manners" along with the "solid attainments of knowledge" and "Christian principles." The Episcopal Female Institute promised to encourage those "relations and enjoyments that do not violate any of the proprieties of female character, and are not inconsistent with a Christian profession" (undated catalogues, Winchester Public Library, Winchester, Va.). See Barbara Solomon, *In the Company of Educated Women: A History of Women and Higher Education in America* (New Haven: Yale University Press, 1985), p. 21. In her study of the plantation mistress, Catherine Clinton supports Solomon's findings: "Female academies and seminaries fitted women strictly for their preordained rôle in plantation culture: that of a well-read elite serving as wives and mothers to the master class. Unlike their northern counterparts, these women could not use their education to explore new avenues of experience." *The Plantation Mistress: Woman's World in the Old South* (New York: Pantheon, 1982), p. 137.

7. Sidney Gore to Jennie Cather, June 10, 1872, Nebraska State Historical Society, Lincoln, Nebr.

8. Caroline Cather to Jennie Cather Ayre, May 31, 1873.

9. Although Lois Banner does not connect women's increased interest in fashion, dress, and beauty with the development of boarding schools, she notes a similar pattern in observing that women who "wanted to be fashionably dressed" were no longer completely satisfied to "follow their mothers' roles." Responding to the economic, social, and ideological changes connected with modernization, women who followed the fashions—a trend Banner sees beginning in the 1820s and 1830s—wanted to be "ladies" rather than housewives (*American Beauty* [New York: Knopf, 1984], p. 27).

10. For a discussion of American skepticism toward fiction in the early nineteenth century, see Mary Kelley, *Private Woman, Public Stage: Literary Domesticity in Nineteenth-Century America* (New York: Oxford University Press, 1984), pp. 114–22. Even though some Americans condemned novel-reading, "from all indications the practice was

widespread'' (Clinton, p. 131). Most boarding schools did not offer courses in literature, let alone in novels, but girls still read them.

11. See Bernice Slote's description of the Cather family library in *KA*, pp. 38–40.

12. Charles Cather to ''My Dear Sister,'' October 30, 1870, Nebraska State Historical Society, Lincoln, Nebr.

13. For a discussion of women's health in the antebellum South, see Clinton, pp. 138–59.

14. Carroll Smith-Rosenberg, ''The Hysterical Woman: Sex Roles and Role Conflict in Nineteenth-Century America.'' *Social Research* 39 (1972), pp. 652–78.

15. Willa Cather to Helen Cather Southwick, September 17, 1946, Beineke Library, Yale University, New Haven, Conn.

16. As Rich also observes, the private realm of feeling and experience is of course not separate from the ways in which the institution of mothering has been defined and perpetuated in patriarchy (p. 13).

17. Interview in the *Omaha World-Herald,* February 1, 1920, quoted in Mona Pers, *Willa Cather's Children* (Uppsala, 1975), p. 85.

18. Writing to Dorothy Canfield, Cather asserted that the meeting between Nancy and Old Till had been one of the most moving events of her childhood (Willa Cather to Dorothy Canfield, October 14, 1940, Bailey Library, University of Vermont, Burlington, Vt.). Cather also told Viola Roseboro' that the ''Epilogue'' was literally true (Willa Cather to Viola Roseboro', November 9, 1940, Barrett-Cather Collection, University of Virginia Library, Charlottesville, Va.).

19. The silent embrace also suggests the preverbal roots of the mother–daughter relationship in the mother–infant bond, the time of fusion and closeness before language and patriarchy separate them—a period associated in French feminist theory with the potentiality of authentic female speech. Because, as Helene Cixous argues, being born into language cannot be separated from construction and constriction by masculine discourse, the task for women is to ''get rid of the systems of censorship that bear down on every attempt to speak in the feminine'' (''Castration or Decapitation?'' [translated by Annette Kuhn], *Signs: Journal of Women in Culture and Society* 7, no. 1 [Autumn 1981]: 41–55). Luce Irigaray poetically associates ''speaking in the feminine'' with the preoedipal mother–daughter relationship, the time of female bonding when the daughter inhabits a feminine space not yet limited or defined by men (''When Our Lips Speak Together'' [translated by Carolyn Burke], *Signs: Journal of Women in Culture and Society* 6, no. 1 [Autumn 1980]: 69–79; ''And the One Doesn't Stir without the Other'' [translated by Helene Vivienne Wenzel], *Signs: Journal of Women in Culture and Society* 7, no. 1 [Autumn 1981]: 60–67). For a discussion of the connections between French feminist theory and mother–daughter relationship, see Marianne Hirsch, ''Mothers and Daughter'' in *Signs: Journal of Women in Culture and Society* 7, no. 1 (Autumn 1981): 200–222.

20. Adrienne Rich discusses the Eleusinian mysteries in *Of Woman Born* (pp. 237–40). See also C. Kerenyi, *Eleusis: Archetypal Image of Mother and Daughters* (New York: Pantheon, 1967), pp. 13–94. These earlier studies of the Demeter–Persephone archetype anticipate recent work on the mother–daughter relationship in psychoanalytic theory; the stress on continuity between mother and daughter enacted in the ancient rituals parallels the theory of female identity developed by Nancy Chodorow, Dorothy Dinnerstein, and Jane Flax, who stress the mutual identifications characterizing the mother–daughter relationship. (See next note.)

21. Nancy Chodorow, *The Reproduction of Mothering: Psychoanalysis and the Sociology of Gender* (Berkeley and Los Angeles: University of California Press, 1978). See also Jane Flax, ''The Conflict between Nurturance and Autonomy in Mother/Daughter Re-

lationships and Within Feminism,'' *Feminist Studies* 4 (February 1978): 171–89, and Dorothy Dinnerstein, *The Mermaid and the Minotaur: Sexual Arrangements and the Human Malaise* (New York: Harper & Row, 1976). Like Chodorow, Flax and Dinnerstein draw on object-relations theory and stress the importance of the preoedipal stage in female development. The similarities among their theories and their contributions to feminist inquiry are reviewed in Hirsch, pp. 204–12. Chodorow's work has received criticism from a sociological perspective that argues the author places too much stress on interpsychic forces as determinants of female personality, implicitly takes the white, middle- or upper-middle-class family as a model, and does not take sufficient account of class, race, and historical context. (See Judith Lorber et al., ''On *The Reproduction of Mothering:* A Methodological Debate,'' *Signs: Journal of Women in Culture and Society* 6, no. 3 [Spring 1981]: 482–514.) In my view, however, Chodorow's work is the most useful theory we have for understanding female identity, and it is important to note that she integrates a psychoanalytic with a cultural and sociological perspective: the fact that women mother is not a biological or psychological inevitability, she argues, but the product of a family structure determined by historically contingent social, political, and economic systems (see ''The Sexual Sociology of Adult Life'' in *Reproduction*). Although Chodorow assumes a twentieth-century family with mothering assigned to one woman, Cather's family approximates this pattern closely enough: she received some care and nurturing from her grandmother, to be sure, but her own mother was the center of the family drama, seemingly for her siblings as well as for herself.

22. Of course, Chodorow's depiction of the typical female psyche formed in the patriarchal family does not perfectly describe Willa Cather (as it would not any individual woman). As a daughter seeking a different role from her mother's, Cather had a greater impulse than the generic female self described by Chodorow to break away from the mother. In addition, Chodorow's theory does not specifically account for women who become writers and lesbians instead of heterosexual mothers. Yet the conflicts Chodorow sees characterizing the female personality were central to Willa Cather, and so I will use her theories, modifying or qualifying them when specific historical or biographical contexts seem to demand an alternative reading. For a critique of Chodorow's failure to consider lesbianism, see Hester Eisenstein, *Contemporary Feminist Thought* (Boston: G. K. Hall, 1984), p. 94.

23. Feminist psychoanalytic theories that draw on male models of the female self constructed by Freud or Lacan—however revised—have been questioned by those feminists who argue that such interpretive frameworks still have their source in ''phallogocentric'' language (see Hirsch, ''Mothers and Daughters'' and Pauline Bart, ''Review of Chodorow's *The Reproduction of Mothering* in *Mothering: Essays in Feminist Theory,''* ed. Joyce Trebilcot [Totawa, N.J.: Rowman and Allanheld, 1984], pp. 147–52). I agree that there are difficulties in using terms like ''preoedipal'' to speak about female development. Not only does the word refer back to Freud and Oedipus, thus symbolically defining this important period of mother–daughter bonding in relationship to men, but it also suggests that this stage will, and should, lead to an ''oedipal'' stage—an ideological model of psychological growth and ''progress'' that I do not accept. I am not using ''preoedipal'' in either of these ways—I mean the term to be descriptive, not evaluative. And although the feminist criticism of object-relations theory and its terminology has merit, since this is the only terminology we have at the moment with which to talk about these concepts, I will be using it with the above qualifications.

24. Chodorow, pp. 166–67.

25. Chodorow, p. 169. Chodorow's argument in many ways parallels Carol Gilligan's theory of female moral development as presented in *In a Different Voice: Psychological Theory and Women's Development* (Cambridge: Harvard University Press, 1982). As Gil-

ligan points out, women base moral decisions on "a feeling of connection, a primary bond between other and self," and realize maturity through "interdependence and taking care" (pp. 46, 172).

26. Dinnerstein, pp. 111, 112.

27. Death by drowning either occurs or is feared in "Wee Winkie's Wanderings," "The Strategy of the Were-wolf Dog," "A Resurrection," "The Clemency of the Court," *Alexander's Bridge, The Song of the Lark,* "The Diamond Mine," "Coming, Aphrodite!" and *Lucy Gayheart.* In other stories Cather links abandonment to sexual passion with a metaphoric drowning: "Eric Hermannson's Soul," "The Bohemian Girl," and "On the Gull's Road."

28. See Dinnerstein's account of the child's ambivalence toward the mother and her discussion of "splitting" in *Mermaid,* pp. 95–105, and Bruno Bettelheim's analysis of the good mother/bad mother opposition in fairy tales in *The Uses of Enchantment: The Meaning and Importance of Fairy Tales* (New York: Knopf, 1976; rpt. Vintage Books, 1977), pp. 66–73.

29. Virginia Gilbert is not an exclusive (and perhaps not a fully conscious) portrait of Virginia Cather; Cather was drawing on the psychological and emotional dynamics of the mother–daughter relationship, but not self-consciously attempting to portray her mother in fiction as she did in "Old Mrs. Harris." In addition, by the time she wrote "The Profile," Cather had experienced two important love affairs with Louise Pound and Isabelle McClung. Both relationships doubtless contributed to the emotional undercurrents in "The Profile."

30. The blatant mingling of the maternal and the erotic in this story recalls Leon Edel's reflections on *The Professor's House,* in which the protagonist stubbornly clings to his womblike attic room and its maternal sewing dummies: "He wants his mother to be a mother and an erotic stimulus," Edel comments, and "above all he wants to possess her exclusively." *Literary Biography* (Toronto: University of Toronto Press, 1957), pp. 61–80.

31. Dinnerstein, pp. 60–64. As Nancy Chodorow and Susan Contratto argue, however, infantile yearnings for perfect maternal gratification are fantasies, not all of which can or should be gratified. See "The Fantasy of the Perfect Mother" in *Rethinking the Family: Some Feminist Questions,* ed. Barrie Thorne, with Marilyn Yalom (New York: Longman, 1982), pp. 54–75.

3

Transplanting

> For the first time, perhaps, since that land emerged from the waters of geologic ages, a human face was set toward it with love and yearning. It seemed beautiful to her, rich and strong and glorious. Her eyes drank in the breadth of it, until her tears blinded her. . . . The history of every country begins in the heart of a man or a woman. *O Pioneers!*

> You have not seen those miles of fields. There is no place to hide in Nebraska. You can't hide under a windmill. Willa Cather to Elizabeth Sergeant

Although we think of Willa Cather as the loving painter of the prairie landscape, her attachment to her new land coexisted with a fear and alienation she never fully transcended. "It has been the happiness and the curse of my life," she told an interviewer in 1921 (*WWC*, p. 140). Cather's discomfort was reflected in her decision to live and write in the East. She returned West periodically to renew ties with family, friends, and land, yet she never lingered too long. During the years of her major novels Cather doubled restlessly back and forth across the continent, spending part of the summer in Nebraska and then escaping to Pittsburgh, New York, New Hampshire, or her island retreat on Grand Manan, New Brunswick, to translate reawakened emotions and remembered impressions into literary form. Needing distance from the landscape that both inspired and frightened her, Cather rarely wrote in Nebraska.

When Willa Cather was proclaimed America's foremost regional writer in the 1920s, critics like H. L. Mencken liked to think of her as a daughter of the prairies whose fiction had been nourished by her Nebraska childhood. And at times Cather reinforced this image in interviews when she celebrated her ties to the Midwest. Yet as her friend and fellow novelist Dorothy Canfield rightly observes, "an imaginative and emotional response to the great shift from Virginia to Nebraska" is at the heart of Cather's fiction. Not her rootedness in Nebraska, but her transplanting from one radically different landscape and community to another at a formative, impressionable age powerfully affected her writing.[1]

Although Cather later emphasized her Nebraska background in interviews with Lincoln journalists, in the "Biographical Sketch" she wrote for Alfred Knopf (intended for a more neutral and national audience) she portrayed the contrast between Virginia and Nebraska as the catalyst for the child's imagination. Virginia was an "old conservative society," she wrote, where "life was ordered and settled, where the people in good families were born good, and the poor mountain people were not expected to amount to much." But in Nebraska the girl escaped this "definitely arranged background" where social class was inherited along with the family farm or mountain cabin, finding herself "dropped down among struggling immigrants from all over the world" for whom the future could be different from the past. Their story of transplantation and resettlement was far more compelling than Back Creek's stability: "Struggle appeals to a child more than comfort and picturesqueness, because it is dramatic. No child with a spark of generosity could have kept from throwing herself heart and soul into the fight these people were making to master the language, to master the soil, to hold their land and to get ahead in the world."[2]

Looking back on her childhood dislocation, the adult writer whose imagination kept returning to Nebraska and the Southwest considered the move fortunate as well as tragic. To see how this contradicatory experience helped to shape her literary imagination, we need to consider both what Willa Cather left behind in Virginia and what she discovered in Nebraska as she became a participant in the great American story: moving West.

•

Although Charles Cather had planned to stay in Back Creek, he had to leave Virginia when the sheep barn at Willow Shade burned down. The patriarch William Cather wanted all of his children in Nebraska and refused to rebuild, "virtually forcing Charles to join him." While Charles may have been reluctant to emigrate because his wife did not want to leave her friends and familiar landscape, he decided to join the rest of his family in a small farming community in Webster County, Nebraska, sixteen miles northwest of Red Cloud in the south-central portion of the state.[3] His brother George and sister-in-law Franc had been farming there since 1873; William and Caroline Cather had followed in 1877. As other Virginians arrived the region became known as "Catherton" in honor of George, the first settler.

Motives for emigrating varied. Some wanted to escape the defeated South's social turmoil, while other Virginians who left during the 1870s and 1880s were enticed by tales of Nebraska's rich soil and healthy crops, alluring visions for small farmers whose overused soil was becoming exhausted. William and Caroline Cather also had an urgent personal reason for leaving. One of their daughters had died of tuberculosis and another was in the grip of the disease, and her parents hoped she might improve in Nebraska's dry climate. Nebraska winters "are just the thing for people who have a tendency to have lung troubles," Franc Cather had written encouragingly to her relatives, "and every body who has come here for that, of whom I have heard, has been cured."[4] But the move came too late for the dying Jennie Cather Ayre (the aunt memorialized in "Macon Prairie").

The family group moving West in 1883 included the parents and their four children (Willa, Roscoe, Douglass, and Jessica), Rachel Boak, Virginia Cather's niece Bess Seymour, and Marjorie Anderson, daughter of Back Creek storyteller Mary Ann Anderson. (The family's domestic servant, Marjorie would later be the model for Mahailey in *One of Ours* and Mandy in "Old Mrs. Harris.") The travelers were spared the hardships suffered by earlier generations of emigrants: in 1883 they could travel all the way from Back Creek to Red Cloud by train. Once they left Red Cloud for William and Caroline Cather's farmhouse, however, traces of civilization diminished. Their destination was a small settlement in the midst of the flat, sparsely populated Nebraska Divide, the high, rolling plains between the Republican and the Little Blue Rivers, so the young girl confronted a wild land homesteaders were still struggling to subdue. "The roads were but faint tracks in the grass, and the fields were scarcely noticeable," she wrote in *O Pioneers!* "The record of the plow was insignificant, like the feeble scratches on stone left by prehistoric races, so indeterminate that they may, after all, be only the markings of glaciers, and not a record of human strivings" (*OP*, pp. 19–20). William Cather had built a frame house which, although less roomy than Willow Shade, still was a more comfortable dwelling than the small, dark sod huts in which many newly arrived Nebraskans lived. There Willa Cather spent her first eighteen months in Nebraska.

In moving from East to West Willa Cather was not merely exchanging one landscape for another: she was moving from an inscribed to an unwritten land. Edith Lewis mentions the "tragic" dislocation her friend felt when she had to leave Back Creek, where she loved "passionately . . . every tree and rock, every landmark of the countryside" (*WCL*, p. 8). This writing on the land composed what Cather termed "familiar earth" in "Macon Prairie." Even if a daughter could not legally inherit the land, she was its emotional and spiritual heir. The landmarks that made Virginia "familiar" were not merely well-known topographical features but also signs of her family's presence in the world before she was born: previous generations of Cathers had settled and farmed the land, building houses and fences, barns and graveyards.

When Cather reconstructed the five-year-old child's dramatic witnessing of Nancy and Old Till's reunion in the "Epilogue" to *Sapphira*, she included such signs in her narrative. The house that shelters the child on a "brilliant, windy March day" is one landmark created by a Cather forebear, while outside the window is a "macadam road with a blue limestone facing" and a "flint milestone with deep-cut letters" proclaiming "ROMNEY—35 MILES" (*SSG*, pp. 280–81). The Cather farmhouse thus faced a well-traveled thoroughfare—the main road between Winchester and Romney, West Virginia—whose milestones signified the directions people had chosen to take in the past and would, presumably, take in the future.

Whether constructed by nature or by people, landmarks are products of the human imagination, the visual equivalent of folktales and stories. What makes landscape meaningful in Cather's fiction are inhabitants who possess their world imaginatively and emotionally rather than economically, marking the land the way a writer marks a blank page. She was stirred not by the sublime, untouched American landscapes that inspired much nineteenth-century painting—large canvases in which people are either absent or dwarfed by the grandeur of Niagara Falls, the Great Plains, or the Rocky Mountains—but by signs of human endeavor and creativity. "To people off alone, as we were, there is something stirring about finding evidences of human labour and care in the soil of an empty country,'' remarks Tom Outland in *The Professor's House* when he comes upon some shards of pottery and an ancient irrigation ditch left by the Indians. ''It comes to you as a sort of message, makes you feel differently about the ground you walk over every day'' (*PH*, p. 194).

Willa Cather first found the links between landmark and narrative in Back Creek. In *Sapphira and the Slave Girl* she describes the child Willa's visit to the Colbert graveyard with Old Till ''to put flowers on the graves.'' Their pilgrimage is to a place where the ''record of human strivings'' on the soil is sharp and poignant, not ''feeble'' or ''indeterminate'': the gravestones are signs of human beings who are lying beneath the final mark they have made on the land. Yet gravestones are also signs that human stories continue, that the past flows into the present and the future. These graveside visits inspire Till to recount the histories and stories the epitaphs do not reveal: ''Each time she talked to me about the people buried there, she was sure to remember something she had not happened to tell me before. Her stories about the Master and Mistress were never mere repetitions, but grew more and more into a complete picture of those two persons'' (*SSG*, p. 292). Back Creek was thus ''familiar earth'' to Willa Cather not merely because she viewed its hills and valleys every day; it was a land with a human history, a marked land, a land with stories she could receive.

Losing this familiar landscape when she left for Nebraska, Cather also left behind the intimate, domestic space of Willow Shade. The comfortable house was composed of many rooms, but Cather's ''Epilogue'' to *Sapphira* suggests that the kitchen was the most memorable because of its associations with food and storytelling:

> Our kitchen was almost as large as a modern music-room, and to me it was the pleasantest room in the house,—the most interesting. The parlour was a bit stiff when it was not full of company, but here everything was easy. Besides the eight-hole range, there was a great fireplace with a crane. In winter a roaring fire was kept up in it at night, after the range fire went out. All the indoor and outdoor servants sat round the kitchen fireplace and cracked nuts and told stories until they went to bed.
>
> The tall cupboards stored sugar and spices and groceries; our farm wagons brought supplies out from Winchester in large quantities. Behind the doors of a very special corner cupboard stood all the jars of brandied fruit, and glass jars of ginger and orange peel soaking in whisky. Canned vegetables, and the preserved fruits not put down in alcohol, were kept in a very cold cellar. . . .
>
> Till and Nancy usually came for dinner, and after the dishes were washed they sat down with Mrs. Blake in the wooden rocking chairs by the west window where the sunlight poured in. They took out their sewing or knitting from the carpet-bag, and while the pound cake or the marble cake was baking in a slow oven, they talked about old times. I was allowed to sit with them and sew patchwork. Sometimes their talk was puzzling, but I soon learned that it was best never to interrupt with questions,—it seemed to break the spell. Nancy wanted to know what happened during the war, and what had become of everybody,—and so did I. (*SSG*, pp. 286–88)

Unlike the parlour, the "stiff" room where little girls were expected to utter polite platitudes, the kitchen is a warm, informal, "easy" space for storytelling where female culture—expressed in cooking, sewing, knitting—is not oppressive or constricting.

Cather found the topography and meaning of the kitchen echoed in Back Creek's external landscape. Not only were there stories in the land, but Back Creek Valley was a comforting space like the kitchen, sheltered and surrounded by the low hills of the Blue Ridge. Like Jim Burden, until she saw the Nebraska prairies Willa Cather "had never before looked up at the sky when there was not a familiar mountain ridge against it" (*MA*, p. 7).

In Nebraska, however, Cather found no protective mountain ridges and few marks on the land. In a 1913 newspaper interview she describes the child's response to this desolate landscape:

> I shall never forget my introduction to it. We drove out from Red Cloud to my grandfather's homestead one day in April. I was sitting on the hay in the bottom of a Studebaker wagon, holding on to the side of the wagon box to steady myself—the roads were mostly faint trails over the bunch grass in those days. The land was open range and there was almost no fencing. As we drove further and further out into the country, I felt a good deal as if we had come to the end of everything—it was a kind of erasure of personality.
>
> I would not know how much a child's life is bound up in the woods and hills and meadows around it, if I had not been jerked away from all these and thrown out into a country as bare as a piece of sheet iron. I had heard my father say you had to show grit in a new country . . . but . . . I thought I should go under.
>
> For the first week or two on the homestead I had that kind of contraction of the stomach which comes from homesickness. I didn't like canned things anyhow, and I made an agreement with myself that I would not eat much until I got back to Virginia and could get some fresh mutton. (*KA*, p. 448)

Cather struck a similar note in telling Witter Bynner, a colleague at *McClure's*, that she had almost died from homesickness during her first year in Nebraska. Having lived in the most beautiful part of the Shenandoah Valley, she wrote to Bynner in 1905, she had never realized that so much ugliness could exist. Someone who had never been to Nebraska simply could not imagine the bleakness and desolation a child confronted in the 1880s.[5]

Why was the Western landscape—which had engaged Charles Cather's imagination—in part a "curse" throughout his daughter's life? In Cather's fiction, structure, space, and landscape are always significant. To understand how her characters shape and are shaped by the worlds they inhabit, her readers must always consider the house, the kitchen, the parlor, the attic, the garden, the prairie, the rock, the mesa. Hence it is not surprising that Cather's memories of the child's encounter with the prairies are stories expressed in the language of space.

Her descriptions of Virginia and Nebraska, the contrasting landscapes of her childhood, reveal the psychological and emotional dynamics of the mother–daughter bond. In Virginia's pastoral world the child is cherished in a maternal embrace, "bound up in the woods and hills and meadows around it," merged effortlessly and safely with an Edenic world. The abrupt change in landscape signifies both the separation of self and other, and the loss of maternal protection and nurturance: the child is rudely "jerked away" and "thrown out" into a forbidding, sterile country "as bare as a piece of sheet iron"; the imagery suggests a rude, unwelcome expulsion from the womb to a wasteland where the transplanted self cannot grow. Loss and deprivation are embodied in starvation; deprived of her motherland, the abandoned child resolves not to eat much until she can return to Back Creek's consoling hills and Willow Shade's well-stocked kitchen.

Because the mother's body is the first landscape the child encounters, as several feminist theorists observe, we have the pervasive identification in Western culture of "woman" with "nature," an equation that dominates Cather's fiction as well as this early memory.[6] Like nature, the mother is, Dorothy Dinnerstein writes, a source "of ultimate distress as well as ultimate joy. Like nature she is both nourishing and disappointing, both alluring and threatening, both comforting and unreliable."[7] In Willa Cather's description of the Virginia and Nebraska landscapes, these qualities are split: Virginia is the good land/mother, Nebraska the bad. Her description further suggests that the move recapitulated the child's separation from the mother. This is an ambivalent process that brings independence as well as loss, just as leaving the womb signifies the beginning of the child's separate life as well as the end of the infant's supposedly blissful fusion with the mother. But Cather's early memory is more negative than ambivalent, stressing loss rather than birth or rebirth; indeed, the barren land both causes and reflects the death of the self ("it was a kind of erasure of personality").

Cather may have vividly recalled the threat of erasure because it never completely left her. As Elizabeth Sergeant reminds us, at times Cather was uneasy when she returned to the West because she did not want to "[die] in a cornfield" or be "swallowed by the distances between herself and anything else . . . as in childhood, again." Afraid that the wind might put her to sleep, she refused to "drowse and to dream" on the prairies because that sleep might be her last (*WC:AM*, p. 49, p. 79). Trying to explain the source of these unreasonable terrors to her Eastern friend, Cather could only say: "You could not understand. You have not seen those miles of fields. There is no place to hide in Nebraska. You can't hide under a windmill" (*WC:AM*, p. 49).

Throughout her life Cather feared the prairies' vastness. With no "familiar mountain ridge" against the sky to encircle the self, the self's boundaries might dissipate instead of expand. Hence on the prairies Cather was wary of states in which the ego relaxes its control; in a land without barriers or protection, dreaming, drowsing, and sleeping seemed perilously close to death. Her fear of erasure suggests the frightening aspect of the "permeable ego boundaries" Nancy Chodorow finds characteristic of the female personality.[8] If the boundaries of self dissolve completely, then there is no separation between subject and object, self and world: this state of dispersal and regression suggests not the infant's pleasurable merging with a nurturing mother but her frightening absorption by an obliterating one. In her fiction Cather often explores the terrifying aspects of such self-dissolution through both real and metaphoric drownings, evident here in the image of submersion ("I thought I should go under").

As Annette Kolodny has shown in her study of nineteenth-century women's imaginative responses to the Western landscape, apprehensions about the plains' vastness were not uncommon among the first settlers. But many Eastern women accommodated themselves to the new landscape, Kolodny finds, by projecting an "idealized domesticity" onto the Edenic prairie landscape, marked by cottages, groves of trees, and flowers, envisioning and creating "a garden that reflected back images of their own deepest dreams and aspirations." In doing so, such women revised the masculine fantasy of the American land-as-woman in which the male perceiver or adventurer is either a son seeking an innocent, blissful reunion with an all-providing mother or a sexual exploiter ready to ravage a virginal continent.[9]

This means of accommodation to a potentially forbidding landscape, however, was not Cather's maternal inheritance. One reason for the pervasive imagery of absence, barrenness, and death in her description of Nebraska may have been the fact that her own mother was not an actor in the drama of emigration and resettlement: not having initiated the decision to leave, she was uprooted like her daughter. Once she reached Nebraska, Virginia Cather could not begin to make an attachment to the new country. Pregnant with her fifth child, she was ill and bedridden for much of the time she spent on the Divide, eventually losing the baby.[10]

Unlike her husband, Virginia Cather was not prepared imaginatively or emotionally to appropriate and transform her new environment. Not only had Charles Cather made the decision to move to Nebraska—where his family had resettled—but since he had thrilled to the West's open spaces during his earlier trip to Colorado and New Mexico, he had, to use Kolodny's term, already constructed a "fantasy" that could help him to assimilate the experience of migration and resettling. Significantly, in Cather's memory the voice that urges accommodation to the "new country" is her father's, not her mother's: "I had heard my father say that you had to show grit."

But the mother who did not choose or desire this landscape could not give her daughter an imaginative means of understanding her new world. In contrast to the nineteenth-century women writers and diarists whom Kolodny studies, Virginia Cather did not busy herself fashioning a garden in the desert that would correspond to her "deepest dreams and aspirations." On the contrary, during the months the Cather family spent on the Divide, the mother's illness caused and signified her inability to master the new environment. Having lost her homeland and her baby, she too had suffered a devastating "erasure."

Doubtless Virginia Cather eventually found life in Red Cloud more comfortable than her sojourn on the prairies, but as a letter she wrote in 1895 reveals, after twelve years in Nebraska she still yearned for Virginia's "familiar earth." Writing to Caroline Cather, who had returned to Back Creek for a long visit, Virginia could not help thinking of the people and the landscape she might never see again. "How do you find everyone and every thing there in Va," she asked, "does it seem the same as it used to in the old times?" Even more than her daughter's, the mother's life had been "bound up" in the Shenandoah Valley, and Virginia went on to confess somewhat wistfully that she envied Caroline's seeing "the Mountains and Springs and trees and mossy brooks where as children

[we] had such good time[s].'' She concluded her letter by saying that she longed to see her Virginia friends and relatives ''so much.''[11]

Since Virginia Cather did not initiate the decision to leave Virginia's ''Mountains and Springs and trees and mossy brooks'' to move West, possibly the diminishment, annihilation, and loss her daughter later associated with her first response to the barren country (''as naked as the back of your hand'') were connected with a recognition of her mother's powerlessness (*WWC,* p. 140). Cather initially shared this powerlessness: the daughter also had not chosen to emigrate, as the passive verbs Cather uses demonstrate (she is ''jerked away'' and ''thrown out'' by incomprehensible, uncontrollable forces). If mother and daughter shared a legacy of passivity, they also momentarily shared the invalid's role in Nebraska: Willa Cather suffered a mild attack of what may have been infantile paralysis during this time and was ''supposed to use a crutch'' (*WWC,* p. 40).

And so the imagery of maternal loss, apparent in Cather's description of her move from Virginia to Nebraska, reflects more than the change from a pastoral to an unwelcoming landscape. During her months on the Divide the motherless daughter was ''thrown out'' in a world Virginia Cather did not understand and could not explore, a loss that reinforced the bleakness and emptiness she found in an unsettled land. Virginia's landscape features ''woods and hills and meadows,'' but Nebraska's is ''the end of everything.'' In this unmarked landscape the self is also erased.

When she created her alter ego Jim Burden, who also undergoes the wrenching move from Virginia to Nebraska in *My Ántonia,* Cather decided to make him an orphan. Depriving him of parents reflects the emotional, if not the literal, truth of her own uprooting; he is at first the abandoned, the erased, self. In describing Jim's introduction to Nebraska, Cather intensifies the journey into nothingness by having it take place at night. Surrounded by blackness, unable to distinguish landmarks or signs, Jim feels annihilated:

> Cautiously I stepped from under the buffalo hide, got up on my knees and peered over the side of the wagon. There seemed to be nothing to see; no fences, no creeks or trees, no hills or fields. If there was a road, I could not make it out in the faint starlight. There was nothing but land: not a country at all, but the material out of which countries are made. No, there was nothing but land. . . . I had the feeling that the world was left behind, that we had got over the edge of it, and were outside man's jurisdiction. I had never before looked up at the sky when there was not a familiar mountain ridge against it. But this was the complete dome of heaven, all there was of it. I did not believe that my dead father and mother were watching me from up there; they would still be looking for me at the sheep-fold down by the creek, or along the white road that led to the mountain pastures. I had left even their spirits behind me. The wagon jolted on, carrying me I knew not whither. I don't think I was homesick. If we never arrived anywhere, it did not matter. Between that earth and that sky I felt erased, blotted out. I did not say my prayers that night: here, I felt, what would be would be. (*MA,* pp. 7–8)

Having left behind the relationships that had defined his identity—with Virginia's land, with his parents, with his God—Jim is a blank slate. And with ''nothing to see'' in the enveloping darkness, there is no other against or with which he can fashion a self.

•

And yet the new land was a blessing as well as a curse, and Cather was later glad that her parents had taken her to Nebraska, cruel though the move seemed at the time. Why was this so?

Separation from the mother brings independence as well as loss; the orphan is the autonomous as well as the abandoned self. Even the mother's perceived powerlessness can be a blessing because her absence or diminishment may enhance the daughter's freedom. Growing up in a landscape and a society her mother did not choose and where she felt displaced, escaping the realm of the Southern lady, Cather had the chance to inherit and to redefine the power her mother had lost and to explore an unfamiliar, uncharted environment her mother did not understand or control. This radical break with the past held out the possibility for renewal as well as for annihilation.

In *My Ántonia* Cather suggests the positive, enriching aspects of transplanting when she hints that rebirth will follow Jim's mythic descent into the underworld of nothingness and nonbeing. Although he is ''erased, blotted out,'' Jim confronts a land of possibility seemingly beyond ''man's jurisdiction'' where a new self can be fashioned from the ''material out of which countries''—and selves—''are made.'' Freed even from his parents' spirits, the son can make fresh choices, and the next day the unmarked prairies' openness—roads are only ''faint tracks'' there—becomes an exhilarating metaphor of freedom. ''The new country lay open before me,'' Jim exults. ''I could choose my own way over the grass uplands'' (*MA*, p. 28).

Although Cather's choices were not limitless in Nebraska—the new land was not, in fact, ''beyond man's jurisdiction''—they were more varied than in Virginia, and the daughter had more freedom to find her ''own way'' in a less structured environment. Although she ''felt the break cruelly'' at the time, Edith Lewis writes, later Cather ''believed that for her the move was fortunate'' (*WCL*, pp. 12–13). With Virginia Cather ill, household order was relaxed and

> the older children were allowed to run about the country much as they pleased. Willa Cather spent a great deal of her time on horseback, riding about through the thinly-settled countryside, visiting the Bohemians and Danes and Norwegians who were their nearest neighbours, tasting the wild plum wine the old women made, eating watermelons with the little herd girls, who wore men's hats, and coolly killed rattlesnakes with clods of earth. (*WCL*, pp. 13–14)

This description of a young girl's adventures is somewhat romanticized. Lewis does not mention her friend's illness or her attendance at a prairie school, references that Cather also eliminates from her descriptions of her months on the Divide. Every moment was not a stirring adventure. Nevertheless, in contrast to her mother, the daughter actively formed an attachment to her new land, and retrospectively Cather contrasted her eager response to Nebraska with her mother's homesickness:

> This country was mostly wild pasture and as naked as the back of your hand. I was little and homesick and lonely and my mother was homesick and nobody paid any attention to us. So the country and I had it out together and by the end of the first autumn, that shaggy grass country had gripped me with a passion I have never been able to shake. (*WWC*, p. 140)

In this memory Cather first equates mother and daughter: both are homesick and ignored. But the recognition of sameness leads to the daughter's establishment of difference through her combat with the land. ''[Having] it out'' with the new country, she both differentiates herself from her homesick mother and surpasses her, mastering the land that left her mother incapacitated as she becomes mastered by it.

In Cather's description of her months on the Divide her verbs change from passive to active, reflecting the power she later attributed to her attachment to her new environment. The child at first is ''thrown out'' in a barren land, but after a while, like Jim Burden, she ''was all over the country . . . on foot, on horseback and in our farm wagons. My nose went poking into nearly everything'' (*WWC*, p. 77). Nebraska was not the ''end of everything''—an empty space—but a land that became ''familiar earth'' as the child created new landmarks: ''I knew every farm, every tree, every field in the region around my house and they all called out to me'' (*WWC*, p. 139).

Like Whitman's child who went forth every day, by investing herself in the physical and social world around her Cather simultaneously absorbed the world into the self; placing her mark on the new world, she was in turn marked by it. She did not long remain ''erased'':

> I go everywhere, I admire all kinds of country. I tried to live in France. But when I strike the open plains, something happens. I'm home. I breathe differently. That love of great spaces, of rolling open country like the sea,—it's the grand passion of my life. I tried for years to get over it. I've stopped trying. It's incurable.[12]

Writing to Dorothy Canfield in 1933, Cather likewise expressed her ''incurable'' passion for the West, comparing the wondrous feeling of escape she felt while writing to the happy expansiveness she knew as a prairie child. When she was taking a Santa Fe train to the West, Cather told her friend, she often awakened in her berth with the same exalted feeling she had known in childhood. Encircled by endless miles of empty country, enveloped by sky and wind and night, she regained the happiest feeling she had ever had. She could never find another place on earth she would love more than the Divide, Cather confided in another letter. Her country was Nebraska.[13]

Since the Victorian circumscription of female possibilities was reflected in spatial metaphors that contrasted woman's ''sphere'' or ''place'' with the extra-domestic world men claimed as their territory, it seems fitting that a woman writer who eventually challenged her society's narrow definition of femininity came of age in a landscape without clear barriers and limits. In Virginia the road from Winchester to Romney was clearly marked, as were the behaviors required of little girls; in Nebraska, where the signs of roads were "faint," there was more possibility for choice. In Virginia the daughter is "bound up" in the hills, an image suggesting constriction as well as connection; in Nebraska she escapes the parlor for an expansive space where the South's social codes no longer rule.[14]

The social meaning that Cather later attached to enclosed and open spaces can be seen in "Wee Winkie's Wanderings" (1896), the story of a little girl who wants to run away from a home that resembles Willow Shade. Here Cather sets in opposition domestic and nondomestic space; the house and its surrounding landscape are contrasting worlds coded "female" and "male." Winkie (one of Cather's childhood nicknames) is serving tea to her dolls, telling her surrogate daughters they cannot play along the creek because they must not "soil their white frocks." In her play the daughter imposes her mother's standards of deportment, staying within domestic space. Like the dolls to whom she attributes her rebellious urges, Winkie wants to escape the sphere defined by her mother and join her father in the fields, the male world of dirt and activity. But to do so is to sully the white dress she is wearing, and her mother refuses, telling her daughter she must either take a nap or do some sewing. Finding the choice between the bedroom and the parlor unacceptable, Winkie runs away into the male world outside the home. But her quest for freedom is eventually undermined by her fear of losing her mother's love. So she returns home defeated, enclosed once more in a space and an embrace that Cather defines as protective as well as oppressive.[15]

While Cather gained more freedom outside her new home in Nebraska than she would have enjoyed in her mother's territory in Virginia, the woman who loved Nebraska's expanses also commented that there was "no place to hide" on the prairies and was afraid to drowse and to dream on the Divide. Cather's terror of openness reveals that, whatever social freedoms she gained on the Divide, her ambivalent response to the new landscape never dissipated. But her contradictory feelings about Nebraska's vastness could be momentarily reconciled in her imagination, as an important passage in *My Ántonia* suggests. Although Jim feels "erased, blotted out" when he first arrives in Nebraska, he soon discovers a welcoming space in his grandmother's house: the kitchen, a "heavenly safe and warm" retreat like "a tight little boat in a winter sea." Venturing outside, he finds that the Nebraska plains are not uniformly barren and flat. Like the kitchen, Grandmother Burden's garden is a sheltered area Cather associates with maternal warmth. In this refuge Jim does not fear the prairies' vastness: "I wanted to walk straight on through the red grass and over the edge of the world, which could not be very far away." Although he claims not to be very interested in the garden as he imagines sailing off into the sky like the hawks floating over head, he makes these adventurous observations in a "sheltered draw bottom" where the wind is gentle and the earth warmed by the sun.

Only in this retreat does Cather allow Jim to "drowse and to dream." In one of the most beautiful passages in her fiction, he envisions death and sleep as natural processes to be welcomed rather than dreaded:

> I kept as still as I could. Nothing happened. I did not expect anything to happen. I was something that lay under the sun and felt it, like the pumpkins, and I did

> not want to be anything more. I was entirely happy. Perhaps we feel like that when we die and become a part of something entire, whether it is sun and air, or goodness and knowledge. At any rate, that is happiness; to be dissolved into something complete and great. When it comes to one, it comes as naturally as sleep. (*MA*, p. 18)

Separated from his mother, Jim regains the lost connection by discovering a maternal realm in the new landscape that paradoxically allows him to be independent while attaching him to something larger than the self. Fusion does not mean obliteration, for this ''erasure of personality'' offers transcendence rather than annihilation.

In this passage Cather creates a landscape offering what phenomenologist Gaston Bachelard calls ''intimate immensity.'' The spatial configuration in which a ''sheltered draw bottom'' connects to the vastness of land and sky satisfies what Bachelard sees as the contradictory human needs for security and adventure, boundedness and expansion. In contrast to the womblike, self-encapsulated bedroom where William was reabsorbed by his mother or the parlor where Willa Cather was bound tightly by social rules, Jim's partially enclosed space shelters him while exposing him to limitless expanses, moving from ''deep intimacy to infinite extent.''[16] When she wrote to Dorothy Canfield of her delight in Nebraska's openness, Cather mentioned another space of intimate immensity: her imagination was thrilled by the surrounding vastness of sky and land as she was crossing the plains on the Santa Fe railroad, sheltered in a berth. The landscapes to which Willa Cather was most drawn after leaving Nebraska—the Southwest, Provence, the island of Grand Manan—provided similar spatial configurations, in different ways offering both vastness and enclosure, mediating between Virginia's encirclement and Nebraska's openness.

Bachelard views the contradictory needs for adventure and security—satisfied by spaces that are simultaneously enclosed and open—as human desires. But these polarities are particularly intense for women, their reconciliation perhaps more difficult than for men. The daughter's preoedipal union with her mother may provide her a stronger sense of connection with the world than the son's, but states of fusion or exposure to vastness may be more threatening since they recall the mother's terrifying power to obliterate her separateness. Cather's description of space in her memories of Nebraska and in ''The Burglar's Christmas'' suggest the daughter's difficulty in shaping a separate self: both losing the mother and returning to her endanger autonomy. The child ''thrown out'' on the barren, trackless prairie is the deprived and the annihilated self; yet the seemingly fortunate William in ''The Burglar's Christmas'' who regains the nurturing, enclosed space of childhood is also erased by an overwhelming mother; he is ''swallowed'' by her just as Cather feared being swallowed by the prairies. As the author of *My Ántonia,* however, the daughter/artist could finally envision a paradoxical space in which the polarities of separation and connection were momentarily reconciled.

Willa Cather's struggle to establish an independent personal and literary self was also a search for a psychological, social, and geographical space where she could ''dissolve into something complete and great'' without being dissolved (*MA*,

p. 18). Her choice of a burial site reflects her discovery of such a place. Jaffrey, New Hampshire—where she wrote *My Ántonia*—is nestled in a small valley; like Willow Shade, it is protected from the sky by a mountain ridge, visible from Cather's grave. But this maternal world was associated not with confinement but with the daughter's adult creativity, the writing which allowed her to "[fade] away into the land and people of [her] heart," as she once defined the artist's mission (*CPF*, p. 7); yet in this artistic landscape she could also retain a secure, inviolate self. So she described Nebraska's open spaces in a landscape like Virginia's. Cather composed *My Ántonia* in a tent pitched in a Jaffrey meadow, describing the self's encounter with vastness as she wrote out-of-doors yet sheltered in a demarcated, enclosed space.

•

Willa Cather did not transform the Nebraska plains from alien to familiar earth solely by herself. Separation from home and mother did not mean abandonment; she made new attachments and formed new friendships. Within a short ride were neighbors Cather had not encountered in Virginia: the immigrants from Scandinavia, Bohemia, Germany, France, and Russia attracted to Nebraska in the 1870s and 1880s by promises of cheap land and fertile soil. Coming from the homogeneous Anglo-Saxon culture of white settlers in the Shenandoah Valley—where the sharpest divisions were those between Baptists and Presbyterians, supporters of secession and Union sympathizers—she was excited by the discovery of difference and volunteered to deliver mail to the immigrant settlements, an imaginative pretext for entering the homes of her neighbors.

In an essay Cather wrote for the *Nation* in 1923 she explained how these exotic neighbors had enlivened her world. The European settlers "spread across our bronze prairies like the daubs of color on a painter's palatte," she wrote, bringing vitality and shading to a "neutral new world." Back Creek was an isolated ghetto of native-born Americans in contrast to Nebraska, where foreign countries were only minutes away. "On Sunday we could drive to a Norwegian church and listen to a sermon in that language," Cather recalled, "or to a Danish or Swedish church. We could go to the French Catholic settlement in the next county and hear a sermon in French, or into the Bohemian township and hear one in Czech, or we could go to church with the German Lutherans." Whereas other Southerners were "provincial and utterly without curiosity," Cather remembered the excitement she felt in identifying with the immigrants' dramatic struggle to root themselves in a new soil and culture. They offered her "older traditions" than did her own family—the heritage of European language, custom, and culture from which Americans were severed and which Willa Cather would eventually want to claim as her own.[17]

The farmers' struggle to master the soil was a masculine enterprise in Nebraska, as it had been in Virginia. But mastering language—an even more potent inheritance for a writer—Cather associated with Nebraska's immigrant women, who had the time to tell her stories about the old country while their husbands were breaking the sod. She spent hours in the kitchens of the immigrant farm wives—thus regaining her favorite room at Willow Shade—while they prepared meals and gave her "the real feeling of an older world across the sea" (*KA*, p. 448). She would later record the stories of their daughters, the "hired girls" like Ántonia who worked in the kitchens of native-born Americans, but fittingly her first encounter with the immigrant women was in their own realm. They "made up for what I missed" in Nebraska, she said later, for "they understood my homesickness and were kind to me" (*KA*, p. 448). They also relieved her painful "contraction of the stomach"; there was one sure way to win their hearts, and she found it pleasing: "I must eat a great deal and enjoy it" (*WWC*, p. 167).

The immigrant farm women became part of Willa Cather's extended family. Her mother ill and absorbed with younger children, the daughter recruited a company of substitutes. In later years she always remembered these old friends at Christmas, sending off presents from Pittsburgh or New York, helping them out with gifts of money during the Depression. She wanted to maintain these ties, Cather told Dorothy Canfield in a 1929 letter, because they were her first circle of intimate friends; the farm women loved her because she had loved them first. She told Canfield that it always pleased her when a daughter wrote to her after a mother died and mentioned that the dying woman had spoken of Willa often—a comment suggesting that she may have felt herself in competition with her friends' real-life daughters. Perhaps with some cause, then, Virginia Cather feared that these were rival mothers. Cather once told Canfield that she couldn't stop off to see them when she returned to Nebraska because her mother was ill and jealous of the attention Cather might devote to the immigrant women. Even though she yearned to spend time with her beloved friends, she knew it would hurt her mother too much if she did.[18]

Language was not a barrier in these friendships. "Even when they spoke very little English," Cather recalled, "the old women somehow managed to tell me a great many stories about the old country. They talked more freely to a child than to grown people, and I always felt as if every word they said to me counted for twenty." After spending a morning with a pioneer woman who was baking or making butter, she would "ride home in the most unreasonable state of excitement," Cather remembered, feeling "as if they told me so much more than they said—as if I had got inside another person's skin. If one begins that early, it is the story of the man-eating tiger over again—no other adventure ever carries one quite so far" (*KA*, p. 449).

The child's excitement at entering "another person's skin" in the act of listening parallels the self-transcendence Cather experienced in the act of writing, for her a means of entering other worlds and other selves. Cather's description of the genesis of *The Song of the Lark* uses the same metaphor of entry: "Nothing, nothing . . . could equal the bliss of entering into the very skin of another human being," she told Elizabeth Sergeant in rapturously recounting her first meeting with opera singer Olive Fremstad, a partial model for Thea Kronborg (*WC:AM*, p. 111). Whereas the vast landscape threatened "erasure of personality," in the farm women's kitchens loss of self was both safe and exhilarating, just as it was for Jim Burden in his grandmother's protected garden and for Willa Cather at her work table. Although she feared being "swallowed up" by the trackless prairie, as the listening child or the "man-eating tiger" she reversed this pattern of incorporation; whether devouring the immigrant women's words or telling their stories in her own writing, she controlled the process of dissolving. In attaching herself to the barren land and its pioneer mothers, the uprooted child found emotional and imaginative nourishment, regaining Willow Shade's kitchen in her neighbors' homes, where food and stories were mingled.[19]

In an interview Cather once attributed her desire to write to the impressions she absorbed in the pioneer women's kitchens:

> Few of our neighbors were Americans—most of them were Danes, Swedes, Norwegians, and Bohemians. I grew fond of some of these immigrants—particularly the old women, who used to tell me of their home country. I used to think them underrated, and wanted to explain them to their neighbors. Their stories used to go round and round in my head at night. This was, with me, the initial impulse. I didn't know any writing people. I had an enthusiasm for a kind of country and a kind of people, rather than ambition.[20]

In fact, Cather was motivated to write by ambition as well as enthusiasm, and she knew a good many "writing people." But in saying that she was obsessed with the immigrant women's stories she made two complementary observations that accurately describe her narrative impulse. She was excited both by the stories the women told her about their "home country" and by the stories they lived. Telling both narratives in her fiction, Cather could both pass on their stories and record their lives, giving voices to women who did not fully command the language of their new country.

Although the Nebraska immigrant women continued the introduction to female narrative power begun by the Back Creek storytellers, like the women in Cather's Virginia past who accepted conventional definitions of womanhood, these women also gave her a contradictory legacy: one of creativity and circumscription, of strength and submissiveness. They told stories but did not publish them, and becoming a literary artist required challenging woman's traditional role as storytelling did not. Yet once Cather felt she had discovered her own voice and subject in *O Pioneers!*, she chose to see herself as the adopted daughter of these pioneer mothers— professionally as well as personally. Recognizing the creativity rural women channeled unobtrusively into the garden, the quilt, and the meal, she reestablished continuity not only with Nebraska's farm women but also with the storytellers of Timber Ridge and her female relatives.

•

The wrenching move from a sheltered, fertile landscape to a barren land was, Cather later thought, the greatest trauma of her childhood. But this dislocation also stimulated creativity and self-renewal. Suddenly plunging into separateness did not mean starvation, isolation, or annihilation in a hostile wasteland. A rich and varied world could be found in her new home, for Nebraska was, as she put it later, a "highway for dreamers and adventurers."[21] Once she was able to draw on it, Willa Cather's transplanting became a potent creative source. Recalled in memory and imagination, her Nebraska past became the subject matter for several novels after she reclaimed her Midwestern heritage in *O Pioneers!*

Cather's emigration from a settled to a frontier society also marked her fiction in more important, and perhaps less obvious, ways. This may have been a more fortunate move for a woman writer than even she realized because it included her in America's story. Cather had an experience no other writer of her generation, male or female, shared. No other writer moved West at the same crucial age: old enough to remember the past, young enough to adapt quickly to the new world. "Thrown out" on the prairies, she struggled to survive and flourish like all the emigrants and immigrants who settled this country. So when she turned to her Nebraska material in *O Pioneers!*—unlike many of her female literary predecessors and contemporaries— Cather did not limit herself to telling a conventional female narrative. Alexandra Bergson's taming of the wild land embodies the history of her culture as Cather makes a female character representative of the American experiment without seeming to be self-conscious about doing so. Having experienced dislocation, loss, and resettlement herself, unlike her contemporaries Edith Wharton and Ellen Glasgow, Cather could claim this dominant subject in American culture and literature as her own and envision the heroic leader of the pioneer venture as female.

A concern with uprooting, transplanting, and resettlement infuses much of her mature work. Cather was drawn again and again to individuals and groups forced to move from one home to another—a professor reluctant to leave an old house for a new one, the immigrants who settled the Divide, the French settlers in Quebec, the Spanish and French missionaries in the Southwest, the wandering Indian tribes who built their homes into the cliffs. Concerned with the human drive to create culture and civilization by making marks in a new landscape, Cather saw artists and pioneers as dreamers and adventurers on the same highway, driven equally by passion and imagination. She was particularly drawn to the rituals and traditions that provided continuity during the process of transplanting, those customs that allowed immigrants to integrate the past with the present and the future. Like Emerson and Whitman, Cather liked to celebrate the heroic individual, but she never advocated a radical break with the past or a separation from community. Her heroic protagonists invariably draw strength from their connection with people, places, and customs that sustain rather than oppress individual achievement. Having survived her own discontinuity by finding emotional sustenance and cultural continuity in the kitchens of her immigrant neighbors, Cather had personal grounds for praising woman's role in preserving life from one home to another. The rituals of domesticity—preserving, cooking, gardening, housekeeping—are the bearers of culture in her fiction, where establishing a home signifies the human ability to transform an empty world into an inhabited one.

Moving to the prairies may also have been beneficial for a writer since Nebraska was uninscribed in a literary as well as a topographical sense. Virginia had its writers and historians, but Nebraska was open land for the pen as well as the plow. "Except for some of the people who lived in it," Edith Lewis writes, "I think no one had ever found Nebraska beautiful until Willa Cather wrote about it" (*WCL*, p. 17). Inheriting her subject from her own past, Cather could be a literary originator in creating fiction from a land apparently as unpromising for fiction as it had once seemed for agriculture. "The history of every country begins in the heart of a man or a woman," Cather writes in *O Pioneers!* Nebraska's story, if not its chronological history, began in the uprooted child's emotional response to her new home and culminated in the mature writer's ability to discern "picture writing" in the sight of a plow encircled by the setting sun, an image of mythic enlargement that reflects her literary purposes in much of her fiction (*OP*, p. 65; *MA*, p. 245).

•

A seemingly less dramatic break awaited Willa Cather when her parents decided to move into the small prairie town of Red Cloud after a year and a half on the Divide. The distance was short, yet once again the move was from one world to another. The girl was taken from the vastness of the prairies to a structured physical and social space where right-angled streets reflected the townspeople's concern with order, propriety, and regimentation. In this new world spaces and roles for women were more clearly marked and less economically crucial than on the Divide. On the farm, woman's work contributed to the family's survival, symbolized by the centrality of the kitchen, the most important household space. There, as in Back Creek, Cather discovered women in charge of language and narrative as well as cooking, passing on cultural traditions in stories as well as recipes. But in Red Cloud, where men controlled the public world of education, law, government, and the professions, middle-class housewives—freed from the farm wife's productive role since the hired girls took over domestic duties—adjourned from the kitchen to the parlor. In that household space Cather saw women's language directed not to storytelling but to what she called the "tongue of gossip" in *My Ántonia* (*MA*, p. 210), where genteel women's speech enforces social control and repression.

Willa Cather's move from the prairies to the town was not simply an entrance into the petty "tyranny" Jim Burden describes in *My Ántonia* (*MA*, p. 219). In Red Cloud she once again found compensation for loss. Reading became the adolescent girl's passion. As she gradually replaced her storytelling women with Virgil, Ovid, Flaubert, and Kipling, she also found new mentors for her intellectual life and professional aspirations—a schoolteacher, a storekeeper trained in the classics, the town's two doctors. Yet because the literature she read, like the public world of power and authority in Red Cloud, was under "man's jurisdiction," becoming a legitimate inheritor of male-dominated culture seemed to require repudiation of her gender and discontinuity with the female worlds of Willow Shade and the Divide.

NOTES

1. E. K. Brown, *Willa Cather: A Critical Biography*, completed by Leon Edel (New York: Knopf, 1953), p. vi. Eudora Welty discusses the creative and emotional significance of the Virginia–Nebraska shift in "The House of Willa Cather" in *The Art of Willa Cather*, ed. Bernice Slote and Virginia Faulkner (Lincoln: University of Nebraska Press [Bison ed.], 1974), pp. 3–20.

2. [Willa Cather], *Willa Cather: A Biographical Sketch/ And English Opinion/ and an Abridged Bibliography* (New York: Knopf, n.d.), p. 2.

3. Mildred R. Bennett, "The Childhood Worlds of Willa Cather," *Great Plains Quarterly* 2, no. 4 (Fall 1982): 204–9, esp. 205.

4. Franc Cather to Jennie Cather Ayre, September 3, 1874, Nebraska State Historical Society, Lincoln, Nebr.

5. Willa Cather to Witter Bynner, June 7, 1905, Houghton Library, Harvard University, Cambridge, Mass.

6. See Simone de Beauvoir, *The Second Sex,* translated by H. M. Parshley (New York: Knopf, 1953); Dorothy Dinnerstein, *The Mermaid and the Minotaur: Sexual Arrangements and Human Malaise* (New York: Harper & Row, 1977); Annette Kolodny, *The Lay of the Land: Metaphor as Experience and History in American Life and Letters* (Chapel Hill: University of North Carolina Press, 1975); Sherry Ortner, "Is Female to Male as Nature Is to Culture?" in *Women, Culture, and Society,* ed. Michelle Zimbalist Rosaldo and Louise Lamphere (Stanford: Stanford University Press, 1974), pp. 67–87.

7. Dinnerstein, p. 95.

8. Nancy Chodorow, *The Reproduction of Mothering: Psychoanalysis and the Sociology of Gender* (Berkeley and Los Angeles: University of California Press, 1978), p. 169.

9. Annette Kolodny, *The Land Before Her: Fantasy and Experience of the American Frontiers, 1630–1860* (Chapel Hill: University of North Carolina Press, 1984), p. 8. Kolodny describes the male pattern in *The Lay of the Land,* pp. 3–9 and 148–60.

10. Bennett, p. 206.

11. "Cather Family Letters," ed. Paul D. Riley, *Nebraska History* (Winter 1973): 611.

12. *Lincoln State Journal,* November 2, 1921, p. 7.

13. Willa Cather to Dorothy Canfield, n.d. [1933], and Willa Cather to Dorothy Canfield, November 5, 1921, Dorothy Canfield Fisher Papers, Bailey Library, University of Vermont, Burlington, Vt. Dorothy Canfield (later Dorothy Canfield Fisher) retained "Dorothy Canfield" as her pen name, and so I will be referring to her by this name in the text.

14. Cather's joy in Nebraska's "open spaces," which coexisted with her terror, corresponds to symbolic use of space and landscape Sandra Gilbert and Susan Gubar find in nineteenth-century fiction by women, where enclosed, confining spaces like attics and parlors signify women's constriction within patriarchal society and language, and open, expansive spaces connote freedom and escape (*The Madwoman in the Attic: The Woman Writer and the Nineteenth-Century Literary Imagination* [New Haven: Yale University Press, 1979], pp. 83–93, 347–49). Cather, however, distinguished between the parlor and the attic, finding the second space congenial to creativity and self-expression.

15. Originally published in the November 26, 1896 issue of the *National Stockman and Farmer,* the story has been reprinted in the *Willa Cather Pioneer Memorial and Educational Foundation Newsletter* 17, no. 3 (Summer 1973): n.p. Bernice Slote, who describes the story as "almost certainly an autobiographical account from Willa Cather's Virginia childhood," makes the biographical links ("Willa Cather: Storyteller for Children" in the same issue, n.p.).

16. Gaston Bachelard, *The Poetics of Space,* translated by Maria Jolas (Boston: Beacon Press, 1969), p. 202.

17. Willa Cather, "Nebraska: The End of the First Cycle," *The Nation* 117 (1923), 236–38.

18. Willa Cather to Dorothy Canfield, n.d.

19. The young Edith Wharton, as Cynthia Griffin Wolff contends, found writing a "feast of words" that similarly compensated for a lonely childhood and a cold, rejecting mother. See *A Feast of Words: The Triumph of Edith Wharton* (New York: Oxford University Press, 1977), pp. 9–54.

20. Latrobe Carroll, *The Bookman* 53 (May 1921): 213.

21. Cather, "Nebraska: The End of the First Cycle," p. 236.

4

Shaped Imaginings

> The books we read when we were children shaped our lives, at least they shaped our imaginings, and it is with our imaginings that we live.
>
> *The World and the Parish*

> We have all dwelt once in that kingdom of lost delight, that fair domain where . . . we could be Caesar or Napoleon at will, and where all the splendid dramas of history were enacted again in one small brain.
>
> *The Kingdom of Art*

The Cathers moved to Red Cloud in 1884. According to local newspapers, the family wanted to be closer to schools and doctors, but Virginia Cather also found the isolation of pioneer life uncongenial. Their destination was a thriving prairie town. A division point on the Burlington Railroad where eight passenger trains a day stopped on the run between Kansas City and Denver, Red Cloud enjoyed a healthy economic and cultural life in the 1880s, attracting Easterners and European immigrants who believed in its future. Stores, hotels, banks, schools, churches, and newspapers prospered along with Nebraska's wheat and corn. The year after the Cathers arrived the opera house was built. Along with the school, it was the most important public offering the town made to Cather's artistic and intellectual development, hosting the traveling theatrical companies that entranced her.

Charles Cather opened a real estate and loan office, and the family moved into the frame house close to the center of town which Cather memorialized in *The Song of the Lark,* "Old Mrs. Harris," and "The Best Years." In contrast to Willow Shade the new house was small and cramped, particularly for a fast-growing family. Having spent her adolescent years living in a crowded house, Cather could speak with authority of the terror of losing an "individual soul" in the "general family flavour" and of the struggle "to have anything of one's own, to be one's self at all," a strain which "keeps every body almost at the breaking-point" (*NUF*, pp. 135–36).

During her Red Cloud adolescence, Cather was preoccupied with protecting and discovering what she termed her "real self" in "Katherine Mansfield" (*NUF*, p. 136). The immigrant women on the Divide who entranced her with their stories could not aid this quest for identity. These women played important roles in her life as substitute mothers, bearers of culture, and storytellers, but their lives followed the traditional female plot: marriage, motherhood, domesticity. Cather had to look elsewhere to find narratives that included ambition.

In bitter early short stories like "The Sculptor's Funeral" as well as in novels more flattering to Nebraska, Cather exploited Red Cloud as the model for the repressive, stultifying Midwestern small town hostile to creativity, spontaneity, and individualism. And yet while Cather may have been lashed with the "tongue of gossip" when she dressed as a boy and devoted herself to dissection and vivisection, in Red Cloud she also discovered alternative stories for a woman's life, the stories her neighbors on the Divide did not tell. As she reveled in adopting other selves like so many masks—Caesar or Napoleon, the amateur doctor, an alchemist in a town parade—she was both the adolescent seeking self-definition and the future creator of character and narrative. Cather blurred the distinctions between theater and everyday life most flagrantly in 1884 when she began her extended performance as William Cather. We must place that seemingly aberrant episode in the context of Cather's Red Cloud adolescence, seeing how her love of books featuring adventurous heroes, her delight in performance and play, and her identification with townspeople who commanded knowledge and power prepared for William Cather's flamboyant entrance.

•

In her "Biographical Sketch" for Alfred Knopf, Cather mentioned her childhood reading as a formative influence on her literary career: "All the while that she was racing about over the country by day, Willa Cather was reading at night."[1] Her juxtaposition of reading and "racing about" was apt: reading was a means of escaping the "family flavour" that permeated the Cather household; in books she could leave a constricted, boundaried world and create a self that momentarily existed beyond what Cather later termed the "rigid horizons of social life" (*KA*, p. 310). Cather understood how important reading was to her even at the time. Asked to name "an attachment exceeding all other endearments in intensity" in a friend's album book, the fourteen-year-old girl wrote "books" (*WWC*, p. 113).

In praising her Red Cloud schoolteachers and the storekeeper/classicist Will Ducker (who taught her Latin and Greek), Cather told Edith Lewis that these were the "first persons she had ever known with any intellectual background, any interest in ideas and the culture of the past." Important as these mentors were, however, they were not quite the first adults who "helped her to find her way in the world of imaginative thought" (*WCL*, p. 19). Cather's delight in books originated in her own family. Her parents were not intellectuals, but they were educated people who enjoyed reading, both a communal and a solitary activity in the Cather family. Although her mother had left much behind in Virginia, she brought her love for fiction to Nebraska, while her father participated in a Victorian family custom: reading poetry aloud to the assembled children. In contrast to Edith Wharton's parents, who esteemed literature but "stood in nervous dread of those who produced it," the Cathers valued writers as well as books and named their youngest son after Virginia novelist John Esten Cooke.[2]

The Cathers' substantial family library ranged from standard classics to contemporary popular literature:

> Among the Cather books which have been preserved are some complete editions of the standard nineteenth-century English and American classics—Dickens, Scott, Thackeray, Poe, Hawthorne, Ruskin, Emerson, Carlyle. There are volumes of Shakespeare and Bunyan, anthologies of poetry, the works of Campbell and Moore, a few translations of Greek and Roman classics, some general histories, a number of miscellaneous religious books (many from Virginia), books on the Civil War, home-bound volumes of the *Century* and ladies' magazines of the eighties. There were also a good many popular romances around the house . . . novels by Marie Corelli, Ouida, Sarah Grand, Richard Harding Davis, and dozens of now forgotten names. (*KA*, p. 39)[3]

Among the family resources were Cather's most-loved childhood books, the romantic adventure stories that the young journalist deemed the "highest form of fiction" in 1895—*Treasure Island, Huckleberry Finn* and *Tom Sawyer, Robinson Crusoe, The Swiss Family Robinson, The Count of Monte Cristo, Otto of the Silver Hand* (*KA*, p. 232). Cather also had a "Private Library" with most books numbered, labeled, and inscribed with the signature she affected after 1888, "Wm Cather Jr" (*KA*, p. 39). Her prized books included Pope's translation of the *Iliad,* Jacob Abbott's *Histories of Cyrus the Great* and *Alexander the Great,* George Eliot's *Spanish Gypsy,* Carlyle's *Sartor Resartus,* Shakespeare's plays, the *Odyssey, Prometheus Unbound,* and the book she read and reread, *Pilgrim's Progress.*[4]

Along with her parents, Cather's Aunt Franc (Frances Smith Cather) encouraged her niece's interest in books and reading. A graduate of Mount Holyoke Female Seminary, Franc met George Cather while teaching school in Winchester and left for Nebraska with her new husband in 1874. Her letters home reveal a strong-minded, resilient woman undaunted by the rigors of her new environment. Writing her sister-in-law Jennie Cather Ayre, she disclaimed praise for her successful adaptation: any "true woman," Franc thought, could "not only live unmurmuringly but happily, with the husband of her choice, under difficulties more insuperable than finding a home in a new country."[5] Franc brought her New England-bred interest in literature into her "new country," eventually drawing on her Mount Holyoke education when she arranged a series of literary discussions and introduced the farmers to Emerson and other transcendentalists.

Looking back, Cather felt that her Aunt Franc—her first representative of New England literary culture—had "distributed more manna in the wilderness than anyone else" (*WWC*, p. 14). In an article for the *Home Monthly,* she describes a childhood visit to this aunt who had been a "literary woman" in her day and who "still cared for books more than for anything else in the world." Although her literary tastes were far from avant-garde ("you could not convince her any poetry worth reading had been written since Byron died, or that anyone but Scott had ever written a respectable novel"), in introducing her niece to Byron she told stories of the Romantic artist's rebellion against bourgeois society, a narrative Cather later used to imagine her own life:

> She told me all about those days when people used to talk of the wild doings of Byron and Moore in Venice; and of how Byron used to write by the light of a hundred candles. . . . And then when Lord Byron went off to the wars and

> devoted all his fortune to the Greek cause, how the girls all bought his picture and chanted:
>
> The isles of Greece, the isles of Greece!
> Where burning Sappho loved and sang.

The aunt passed on other stories "of all the great men of her youth," including the poets Moore and Campbell, as well as legends of "the greatest man of all"—Napoleon, the adolescent girl's hero (*WWC,* p. 112). "And she would tell me how . . . the startled world just stood still and let him do as he would, as the Greeks waited dumbly for the thunderbolts of Jupiter" (*WP* 1, pp. 351–52).[6]

Red Cloud offered Cather other readers and libraries. The Wieners, who lived close by, spoke French and German (Mrs. Wiener was French) and owned a large collection of European literature in translation; they gave Cather access to their library, just as the Rosens do Vicki Templeton in "Old Mrs. Harris." In that story the Rosens' elegant parlor is a sanctuary of art and culture with Raphael engravings "in pale gold frames" on the walls and leather-bound books sheltered in bookcases with doors. The room's velvet-carpeted hush, which Cather contrasts with the noisy chaos of the Templetons' house where books are not treated reverently—their table is "piled high with old magazines and tattered books, and children's caps and coats" (*OD,* p. 80)—is the fitting environment for Vicki's first encounter with one of the "world's masterpieces," an illustrated edition of *Faust* (*OD,* p. 106).

Although in "Old Mrs. Harris" Vicki's pleasure in the Rosens' house "never made her dissatisfied with her own" (*OD,* p. 104), the Wieners' house was the first of many Cather did prefer to her own: gracious homes inhabited by cultured Europeans or Easterners where education and refinement were inherited along with the family silver, where leather-bound books were uncontaminated by ladies' magazines and children's caps. In Lincoln there were to be the houses of the Westermanns, the Pounds, and the Canfields; in Pittsburgh, those of the Siebels and the McClungs; in Boston, the cultural shrine of 148 Charles Street where Annie Fields and Sarah Orne Jewett reigned at the tea table. Once Cather had left Red Cloud and devoted herself to the art and literature she associated with these other houses, she at first was distressed by the crudeness of her Nebraska origins and the inadequacy of her prairie education. But the adolescent girl did not make distinctions between high- and middle-brow culture or between culturally defined "masterpieces" and books of lesser distinction but wider popular appeal. Indiscriminately devouring *Treasure Island,* the *Iliad,* the *Swiss Family Robinson,* Emerson's essays, and Ouida's romantic fiction, like Vicki Templeton, Cather "had not been taught to respect masterpieces; she had no scale of that sort in her mind. She cared about a book only because it took hold of her" (*OD,* p. 106).

Cather was introduced to the literature of the classical masters in an unlikely place: the Red Cloud store where Will Ducker, an educated Englishman with a passion for Latin and Greek literature, was working for his more conventionally successful brothers. Ducker recognized someone who shared his passion and took her on as an after-hours pupil; together they read Virgil, Homer, Ovid, and Anacreon. The two developed an affectionate relationship, Ducker regarding Cather as a daughter, she developing a "passionate hero-worship" of the fatherly scholar who was unveiling the mysteries of translation (*WCL,* p. 22). Even after she began formal study of the classics at the University of Nebraska, she continued her tutorials with Ducker during the summer.

Cather's grandmother Rachel Boak could teach her to read her own language, but only a man who had received the classical education denied most women in the nineteenth century could initiate her into the languages Cather, like many other women writers, came to associate with masculine power, knowledge, and privilege. As Elaine Showalter has observed, in the nineteenth century the "classical education was the intellectual dividing line between men and women; intelligent women aspired to study Greek and Latin with a touching faith that such knowledge would open the world of male power and wisdom to them."[7] This "touching faith" is reflected in George Eliot's fiction in which classical languages are the barriers surrounding provinces of masculine knowledge that ambitious heroines like Maggie Tulliver and Dorothea Brooke long to cross. "It was really interesting," reflects Maggie, the tomboy heroine with whom Cather identified in her review of *The Mill on the Floss,* "the Latin Grammar that Tom had said no girls could learn: and she was proud because she found it interesting."[8]

Determined to master this forbidden knowledge, women writers like Elizabeth Barrett Browning and George Eliot were self-taught, so Cather was fortunate in finding her own teacher and mentor in Will Ducker. But the fact that he was male foreshadowed the difficulty she would later confront: classical literature, like the nineteenth-century masterpieces on the Cather family bookshelves, was written by men. How could a woman be a legitimate heir of this exalted tradition? Classical influences are apparent in Cather's early writing, particularly in *April Twilights* where she slavishly and reverently copies ancient models. Not until *My Ántonia* did she feel herself equally the inheritor of the pioneer women's stories and classical poetry; by then she could proudly connect herself with Virgil with the epigraph *("Optima dies . . . prima fugit"),* and even more boldly appropriate his literary mission: "*Primus ego in patriam mecum . . . deducam Musas*"; "for I shall be the first, if I live, to bring the Muse into my country" (*MA,* p. 264).

Whatever deficiencies Willa Cather later found when she regarded her Nebraska education with a critical eye, she gained a rich and varied literary background by drawing on the resources family and friends offered. The importance of her parents' sharing and sanctioning her love for books becomes more apparent if we contrast her adolescence with that of her contemporary Edith Wharton. Although raised in a seemingly more privileged environment, as a member of New York's anti-intellectual aristocracy Wharton belonged to a group who thought writing a disreputable profession and reading a possibly subversive activity. Although the girl had access to her father's library where she spent her happiest hours communing with "the great voices that spoke to me from books," she found no support for reading from her parents.[9] Their "attitude toward the adolescent girl," Cynthia Griffin Wolff observes, "was precisely calculated to diminish her self-esteem." Wolff imagines Wharton as an "undernourished scavenger" desperately rummaging through "the family's literary leftovers." Wharton herself referred to her childhood and adolescence as an "intellectual desert."[10]

By contrast, Cather's family and adult friends offered her a literary feast. The image of the tattered books mingled with children's caps in "Old Mrs. Harris" at once suggests the casual intermingling of reading with family life Cather enjoyed and the presence of books Edith Wharton did not find in her father's library—popular children's classics and adventure stories. Her imaginings shaped by these narratives, at the same age when Edith Wharton was preparing for her debut, Cather was planning her entrance into Red Cloud society as William Cather (*WP* 2, p. 852).

•

The adventure stories and historical romances by Kipling, Stevenson, Dumas, Pyle, and others which Cather had in mind when she exalted romance over realism in her 1890s journalism resembled the literature the adolescent girl loved. Classical myths, Germanic and Norse folklore, epic poems, and biographies of Alexander and Napoleon: all were stories of heroes who exercised courage, physical prowess, and daring in overcoming obstacles. Although they represented various genres and national literatures, such narratives exhibited the self-reliance she found in Emerson, her favorite prose writer, who encouraged all readers of "Nature" to go forth and build their own worlds.

Some of the heroic stories she read supported the Victorian ideology of competitive individualism by turning the classical veneration of the heroic and the Romantic exaltation of the self to bourgeois purposes. Jacob Abbott's *History of Alexander the Great,* one of the treasured books in her private library, makes the biography of the classical hero conform to America's rags-to-riches plot. Abbott's Alexander is a fitting heir of Achilles: virile, courageous, ambitious; he was, Abbott asserts, "intent solely on enlarging his own personal power."[11] Simultaneously he is the self-made conqueror whose rise to kingly success parallels the ascent of Horatio Alger's orphan heroes to middle-class comfort. Like much of Cather's childhood reading, the book assumes a male readership, young boys who might someday profit from Alexander's example and become captains of industry.

Reading for Cather was not a process of passively absorbing a text that objectively existed outside herself, but an active, interactive process in which she partly created the books she read. Through identifying with characters she found another way of "entering another person's skin" and dissolving the boundaries between self and other.[12] Unlike the uprooted child who suffered a frightening "erasure of personality" when she was "thrown out" on the Nebraska prairies, the adolescent girl could control this process of self-dissolution (*KA*, p. 448). Hence it was pleasurable. When she picked up a book like *The Count of Monte Cristo* and became a powerful hero, her desire for control and efficacy, undermined when she was moved from Virginia to Nebraska without choice, was rewarded and strengthened.

Although literary critics warn us not to think of characters in fiction as real people, Willa Cather did not make such academic distinctions. Historical and literary heroes were her friends, companions, and other selves. "We have all dwelt in that country where Anna Karenina and the Levins were the only people who mattered much," she observed later, and known a time when Porthos, Athos, and d'Artagnan were "vastly more real and important" than the folks who lived next door (*WP* 1, p. 259). During her Red Cloud adolescence, the musketeers were her "three best friends" (*KA*, p. 323). So while characters were fictional constructs—creations of words on a page—they produced real responses in the reader who created them as they created her.

Since the development of the novel in the eighteenth century, reading has been an activity more associated with women than with men. Perhaps because their own lives lacked educational and social opportunities, girls and women found imaginative release in novels that compensated for restrictions in the domestic sphere. Paradoxically, however, romantic fiction, which was the staple of women's reading in the nineteenth century, led female readers back to domestic roles. The marriage plot in which the heroine's self-worth is demonstrated by the hero's love validated the social institutions that limited women's autonomy and thus prompted them to read fiction.

Feminist critics have begun to explore the ways in which reading reconfirms women's social world. As Rachel Brownstein observes, women readers are "inclined to understand one another, and men, and themselves, as characters in novels." Identifying with the heroine of a romantic novel and thus admiring her own "idealized image," the female reader interprets life through the frame of fiction and channels her fantasies into "becoming a heroine." And since the heroines of the eighteenth- and nineteenth-century fiction Brownstein examines are stars of the marriage plot, the female reader's imagination is often "shaped," to use Cather's term, by romantic—not Romantic—ideology.[13]

But Cather's imaginings were not shaped by the heroine's plot.[14] Most of the books she read told male stories about the world. The characters who offered her idealized images of the self possessed the freedom and self-determination Emerson exalted as an American right in "Self-Reliance" (while assuming, as did most Americans, that individualism was a male prerogative). Certainly the heroes Cather loved enjoyed the power to determine their destinies that both the heroines of fiction and her own mother lacked. During these years Virginia Cather was simultaneously finding her freedom increasingly circumscribed by the demands of young children and reading versions of the romantic plot in novels by Ouida and Marie Corelli, thus finding her social role reconfirmed by the supposedly alternative fictional worlds she entered. Meanwhile her daughter was becoming "Caesar or Napoleon at will" (*KA*, p. 339).

It was not unusual for a Victorian girl to read adventure stories or to identify with courageous heroes, but in adolescence most female readers turned their attention to the heroine's story. Although Willa Cather did not succumb to this literary and social rite of passage, she gradually realized that the stories she loved were not intended for women. Books like *Treasure Island* possess "an atmosphere of adventure and romance that gratifies the eternal boy in us," Cather commented in a 1895 review (*KA*, p. 323), and in her 1890s journalism she assumes a male reader when she recommends her childhood favorites to parents desiring "stirring, honest, and manly boys' books" (*KA*, p. 337).[15] But until she became more aware of the sexual polarization in adult Victorian society, the young book-lover did not know that she was reading the wrong plot. Identifying with the male heroes who possessed the power and autonomy she wanted for herself, Cather did not at first grasp what she later called the "hateful distinction" between boys' and girls' reading: the "fact that I was a girl never damaged my ambitions to be a pope or an emperor" (*WP* 1, pp. 337, 368). Later she would refer nostalgically to the time when boys and girls could share books as well as possible futures: "Well, of course there are 'boys' books' and 'girls' books,' but I prefer the books that are for both" (*WP* 1, p. 337).

•

The impact of Willa Cather's reading on her imagination is evident in her childhood play. Drawing on books, she delighted in acting out dramas with her brothers and with the Miner girls—Mary, Irene, Margie, and Carrie—her neighbors and closest Red Cloud friends. There were two central spaces for play and fantasy: an island sandbar in the Republican River and her attic bedroom, associated in her fiction with the child's and the artist's creative imagination.

Far Island was the long sandbar in the Republican River that Cather and her brothers discovered. Seeing this secluded realm in the context of *Treasure Island* as well as other books that featured island settings—*The Swiss Family Robinson, Robinson Crusoe, Huckleberry Finn*—the children made the island a stage for drama and invention. She and her brothers camped out, built fires, and enacted scenes from her adventure novels. The sandbar's landmarks were given names from *Treasure Island*—"Robber's Cave," "Pirate's Island" (*WWC*, p. 171).

Cather expressed the connection between the child's play and the adult's writing in her first published book, *April Twilights* (1903). In the dedicatory poem, addressed to her brothers Roscoe and Douglass, the speaker asks that the three regain their lost childhood selves and become once again dreamers who "lay and planned at moonrise,/ On an island in a western river,/ Of the conquest of the world together." Saluting the children's "vanished kingdom," the speaker evokes a time when the Republican River was the Aegean Sea and the playmates Greek warriors and bards who shared "Starry wonder-tales of nights in April" (*AT*, p. 3).

The importance of Far Island's "vanished kingdom" is apparent in the frequent appearance of island settings in her fiction, whose remoteness from the social world suggests the magical, transformative power of the imagination. In two stories she draws directly on memories of her childhood realm—"The Treasure of Far Island" and "The Enchanted Bluff"—and there are other islands, briefly glimpsed in *Alexander's Bridge, My Ántonia,* and *Lucy Gayheart,* which also reincarnate the dreams, aspirations, and creativity of children.

Cather did not have to travel as far as the Republican River to find a space for dreaming. There was an island in the house—her attic bedroom, an unfinished loft which she at first shared with her brothers Roscoe and Douglass. This was a children's domain where she and her brothers possessed the "freedom to be alone with their own thoughts and fancies and experiences" (*WCL*, p. 18). Sometime during her adolescence Virginia Cather decreed that her eldest daughter should have a room of her own. Perhaps motivated by notions of propriety as well as sympathy with her daughter's need for privacy, she had an ell-shaped wing partitioned from the main attic. At first Cather mourned the loss of the shared dormitory, but she soon grew to love her own space. The parents respected their daughter's claim to her territory and kept the room locked when she left for college, not assigning it to one of the younger children as might have been expected in such a crowded household.

Unlike the attic rooms confining women in *Jane Eyre* or "The Yellow Wallpaper," symbols of women's enclosure and oppression in patriarchy, Cather's childhood bedroom appears in her fiction as an architectural metaphor of the female protagonist's search for creativity, identity, and autonomy.[16] In the autobiographical first section of *The Song of the Lark,* Cather describes the acquisition of her own bedroom as "the beginning of a new era in Thea's life," one of the "most important things that ever happened to her." Until she acquires her own space, like the beleaguered family member Cather imagines in "Katherine Mansfield," Thea had lived in "constant turmoil" surrounded by relatives and friends who provided companionship and affection but "drowned the voice within" (*SL,* p. 58). In her attic retreat Thea begins to discover the voice or self that is her own—what Cather called the secret, passionate, and intense "real life" in her Mansfield essay, the life hidden beneath the domestic surface (*NUF,* p. 136).

In Cather's late story "The Best Years," published in the posthumous *The Old Beauty and Others,* the attic is both a communal space the adolescent girl shares with her brothers and a romantic retreat where she can have her own "dream-adventure":

> "Upstairs" was a story in itself, a secret romance. No caller or neighbour had ever been allowed to go up there. All the children loved it—it was their very own world where there were no older people poking about to spoil things. And it was unique—not at all like other people's upstairs chambers. In her stuffy little bedroom out in the country Lesley had more than once cried for it.

Although the most secluded realm in the house, the attic in "The Best Years" is the most connected to the outside, a space of "intimate immensity": "In this spacious, undivided loft were two brick chimneys, going up in neat little stair-steps from the plank floor to the shingle roof—and out of it to the stars!" Hints of the outside world's mysteries can enter the children's safe realm: they could never sleep until they had heard the ten o'clock whistle of the westbound passenger train, Number Seventeen, the locomotive connecting Red Cloud with Denver and beyond. As they charged "over the great Western land," Cather reflects, "towering locomotives" like Number Seventeen meant "power, conquest, triumph" to the listening children (*OB,* pp. 75–138).

Associated with childhood dreams in *The Song of the Lark* and "The Best Years," the attic room is the center of adult creativity in *The Professor's House.* Cather's description of the professor's attic study with its maternal sewing dummies evokes both her childhood retreat and her discovery in adult life of another attic room that fostered creativity: the sewing room in the McClung household which her friend Isabelle McClung converted into a study for Cather. In later years she lost Isabelle, but she found replacements for the attic workroom. Cather always returned to the same two small rooms on the top floor of the Shattuck Inn, her summer retreat in Jaffrey, New Hampshire, which "had sloping ceilings," Edith Lewis recalls, "like her attic room in the old days in Red Cloud" (*WCL*, p. 103). Cather finally recaptured both the island and the attic when she built her cottage on the Canadian island of Grand Manan, located a few miles off the Maine coast; it had "a large attic from which she could look out over the cliffs and the sea," her favorite place to work (*WCL*, p. 131). In her Grand Manan attic—as in Jaffrey, Pittsburgh, and Red Cloud—Cather found creative freedom in the part of the house Gaston Bachelard associates with dreaming and reverie.[17]

•

Cather's family also gave her the chance to be a storyteller as well as a reader. At times she felt her privacy threatened by the presence of too many people in too small a space, but her siblings were also the first audience for her own stories. So in momentarily adapting a maternal role—the older sister entertaining the younger children—Cather experimented with her future vocation.

Red Cloud offered other opportunities for storytelling, role-playing, and self-dramatization. Like other Victorian children she was encouraged to give recitations, and the Red Cloud papers praised her "elocutionary powers" and her "extraordinary self-control and talent." One of her "most popular selections" was "Hiawatha," recorded in an early photograph, "during the recitation of which she held a bow and arrow and fell to one knee at the proper moment to shoot into the imaginary frost." *(See photograph section.)* She participated in the 1889 Fourth of July parade by representing a local pharmacy, dressed as an alchemist in "black velvet with a mantle and knee pants," and with bottles and surgical instruments attached to her costume (*WWC*, pp. 170, 176–77).

There were collaborative stage ventures as well. Like Louisa May Alcott and her sisters, Cather and the Miner girls presented plays to an assembled audience of family and friends, Cather taking the male role as had Alcott. Her cousin Bess Seymour, who acted in Red Cloud community plays, improvised costumes from Virginia Cather's extensive wardrobe, and the children used the front hallway as a makeshift stage. By the time she was thirteen Cather was inventing and staging her own dramas (*WWC*, p. 171). Recruiting the Miners as fellow actors, she used their parlor as a stage—transforming the space associated with her mother's rules of polite conduct. The next year the players were ready for a larger audience. They took over the Opera House and invited the whole town to a performance of "Beauty and the Beast." Margie Miner played the Beauty, Mary Miner the Beast, and Willa Cather the merchant–father, dressed in "suit, hat, and waxed mustache" (*WWC*, p. 175).

Cather's fascination with amateur theatricals was intensified by the traveling stage companies that visited Red Cloud after the Opera House opened in 1885. She and her friends saw the staples of nineteenth-century theater and opera: *The Count of Monte Cristo, Uncle Tom's Cabin, The Bohemian Girl, The Mikado.* Reminiscing about the Opera House, Cather recalled how she and her friends would stand for hours studying every word on the advance posters, wondering whether their parents would let them attend every other night or just the opening and closing performance. If the company arrived on the night train, she and her friends would be waiting at the station, for it was so "delightful to watch a theatrical company alight." Then the children might invent some errand that would take them to the hotel after the actors. Some of the fortunate ones might later see the leading lady "breakfasting languidly at 9."

Even more entrancing than the train arrivals was the illusion of reality the players created on stage. Like reading, viewing drama allowed Cather to escape the boundaries of self and imaginatively enter another world. The theater could create this fusion between audience and actors more effectively than the cinema, she later thought, for

> only a living human being, in some sort of rapport with us, speaking the lines, can make us forget who we are and where we are, can make us (especially children) actually live in the story that is going on before us, can make the dangers of that heroine and the desperation of that hero much more important to us, for the time much dearer to us, than our own lives.

Because the actors, even in these second-rate traveling companies, could command "breathless, rapt attention" and evoke "deep feelings" in their audience, they plunged her and her friends into the "middle of lives so very different" and transported the children from Red Cloud to "Mme. Danglar's salon in Paris" or Simon Legree's cabin (*WP* 2, pp. 955–58).

The power Cather's play and play-acting gave her to imagine alternatives to the female role is perhaps best demonstrated by her most elaborate childhood scheme: the creation of a town from crates and packing-boxes of which she was mayor. "Sandy Point," as the children called it, boasted several packing-box stores, a post office, a fire department, a newspaper, and a graveled Main Street running the length of the Cathers' backyard (*WWC,* pp. 172–73). Cather created this miniature city shortly after her family doctor was elected mayor of Red Cloud. Hoping to emulate him, in the world of play—as in reading and drama—she could assume the powerful roles assigned to men in adult life. And so in Sandy Point she took all the male roles—mayor, banker, and newspaper editor—while the more conventional Miner girls ran a candy store and a millinery shop. In "The Way of the World," a short story Cather based on this childhood episode, she associates masculinity and power in casting her alter ego as male. Speckle Burnham both possesses the force of personality to "organize a community" and the imaginative energy to make others join in his play. He has the "right to be autocratic," Cather observes, because his "creative imagination" is the "real site" of the town (*CSF,* pp. 395–404).

•

Even before she assumed a male persona, Willa Cather was not shaping her imaginings through identification with her mother and her role. But Red Cloud offered her other adult "friends of childhood" who could fill the role that mothers played for other daughters (*SL,* p. 3). The professional people in Red Cloud she wanted to please or emulate were more accessible, if less exotic, models than Odysseus, D'Artagnan, or Napoleon. Will Ducker was her informal tutor, but she also formed close bonds with her teachers when she began formal schooling. Two were women, and along with Ducker they were guides who helped Cather "find her way in the world of imaginative thought" and implanted "the early ideals of scholarship and art that gave direction to her own life and work" (*WCL*, p. 19).

Mrs. Goudy, whose husband was principal of the high school, became a friend who maintained a correspondence with her former pupil for years after Cather's graduation. Edith Lewis found her friend's letters to her former teacher "filled with the new discoveries she was making about life and people, and about herself," the kind of letters "that are only written in the confidence of being infallibly understood" (*WCL*, p. 20). Evangeline King, Cather's teacher during her second year in Red Cloud, was even more helpful for a young girl seeking mentors and supporters. Cather later termed that year one of the happiest she had ever spent and described Miss King, a "stalwart young women" with mirthful eyes and a kind, sympathetic manner, as the first person she had ever cared for deeply outside her own family. "I had been in her class only a few weeks when I wanted more than anything else in the world to please her," she recalled. "During the rest of that year, when I succeeded in pleasing her I was quite happy; when I failed to please her there was only one thing I cared about and that was to try again and make her forget my mistakes." Success in the classroom brought her the "one thing [she] cared about," Miss King's approval (*WWC*, pp. 256–57). The teacher remained a friend and confidante throughout the Red Cloud years, becoming an important ally when the girl began to find the community parochial and confining. In "The Best Years," Cather drew on her childhood affection for Miss King in portraying Miss Knightly, the competent, kindly school inspector who ignores the rules to help her underage protégée get her first teaching job.

The first women Cather knew who possessed authority in the public sphere, these teachers provided a link with her first informal instructors—her grandmother and the storytellers of her childhood—as well as with Sarah Orne Jewett, her literary mentor. But the adolescent girl wanted to be a surgeon, not a teacher or a writer, so she identified more strongly with adult men. Her bonds with Will Ducker and the town's doctors anticipate her apprenticeship to her literary and professional mentors Henry James and S. S. McClure.

Cather shared Will Ducker's interest in scientific experiments as well as in the classics. He had a small laboratory in his house, and she often helped him with his experiments. The town's doctors, Dr. McKeeby and Dr. Damerell, were more direct professional models. "How I loved the long rambling buggy rides we used to take," Cather later reminisced. "We went over the same roads this summer. I could tell who lived at every place and about the ailments of his family. The old country doctor and I used to talk over his cases. I was determined then to be a surgeon" (*WWC*, p. 114). Dr. Damerell allowed the apprenticeship to extend beyond companionship and discussion, once letting her administer chloroform to a patient before he performed an amputation (*WCL*, p. 28). Elmer Thomas, unofficial historian of Webster County who was also one of the favored children, describes Damerell as a family man, childless himself, who "loved young folks" and was happy to have Cather's companionship on "his long drives into the country."[18]

Cather also pursued her medical studies independently. Setting up a makeshift laboratory in her basement, she began dissecting toads and frogs. Her practices appalled many of the neighbors and made her the subject of disapproving gossip, particularly when she reportedly moved on to cats and dogs. But she persisted. Vivisection was the only way the future surgeon could learn about anatomy and circulation, so she disdained the local busybodies and even began to explore embalming. Her unorthodox medical interests dominate the entries she made in a friend's album book in 1888. In recording her likes and dislikes, Cather termed her ideal of perfect happiness ''Amputating limbs,'' her favorite summer's pastime ''Slicing Toads,'' and her all-around favorite amusement ''Vivisection.'' Her chief ambition in life was consistent with her chief delights: "To be an M.D." (*WWC*, pp. 112–13).

A fascination with vivisection and amputation is an unusual trait in any adolescent, but it seems particularly odd when we consider the ''strange abhorrence for physical defect'' Willa Cather later displayed, a profound distaste for mutilation and infirmity that her friend Elizabeth Sergeant termed one of her ''characteristic temperamental twists.'' According to Sergeant, Cather felt that the human body ''should remain intact and as God made it''; as if supporting this observation, Cather wrote "Lop away so much as a finger and you have wounded the creature beyond reparation'' in ''The Profile'' (*WC:AM*, pp. 71, 164; *CSF*, p. 125). Yet minor characters with missing limbs or mutilated bodies keep appearing in her fiction, most dramatically in *Shadows on the Rock* where almost every background character has some sort of grotesque defect. And occasionally there is the spooky appearance of the amputated limbs themselves, the ''man's hand snapped off at the wrist'' in ''Behind the Singer Tower,'' the severed hand mysteriously surfacing in the soup in *Shadows on the Rock* (*CSF*, p. 45).[19]

Both the adult's fear of mutilation and the adolescent's fascination with amateur surgery may have had an unconscious source in a childhood trauma Cather recounted to Edith Lewis:

> One rather alarming incident happened when she was five years old. She was playing by herself in an upstairs room at Willow Shade, at some distance from the rest of the household, when a half-witted boy on the place, the son of one of the servants, slipped into the room, showed her an open clasp-knife, and told her he was going to cut off her hand. In recalling it, she said that she was very much frightened, but she knew instinctively that she must not show she was frightened. She began talking playfully to the lad, and coaxing him to the window, showed him a tall tree that grew outside, its branches almost touching the house. She suggested that it would be very amusing to climb out on one of the branches, and in this way get down to the ground. The strategem worked, for the boy forgot what he had come to do, and climbed down the tree. (*WCL*, p. 10)

Whether this is the memory of an actual event or a complete fiction, this childhood story conveys the threat of castration, mutilation, and powerlessness. Power is at first attributed to the male victimizer who threatens violence and bodily harm. The supposedly safe space—the upstairs room, a version of the attic where Cather later felt sheltered—is thus vulnerable to invasion, just as the female body can be entered or, in this case, mutilated. But then the power shifts from invader to potential victim as the child triumphs through her command of language and narrative. She creates a story and a game, adopts a persona ("she knew instinctively that she must not show she was frightened"), and invents a "strategem." If this subtle exercise of power can be considered a woman's strategy, it is also a writer's, a means of exerting control through language that Cather later enjoyed in her own "upstairs room."

In adolescence, however, Willa Cather was not yet a writer who had mastered the language. Just as the surgeon displayed the ability to enter another person's skin which Cather later metaphorically attributed to the writer, so the adolescent vivisectionist and toad-slicer literally reversed the power relationships with which this memory begins, inflicting on her animal subjects the mutilation she once feared for herself as she also prepared for an adult profession invested with power and status. But her anxieties did not dissipate, as the occasional appearance of severed limbs in her fiction suggests. As we will see, injuries to the hand—the writer's tool—were connected throughout her life with literary conflicts and insecurities; the "long stretches" in later years when "she could not use her right hand at all" had psychological as well as physical causes (*WCL,* p. 186).

Her novels demonstrate Cather's association of mutilation with the use—or misuse—of power that is most often portrayed as masculine. A recurrent fictional pattern links predatory, sexually aggressive, or exploitive men with violence, weapons, and sharp instruments; in *My Ántonia* there is the would-be rapist Wick Cutter with his appropriate name, and in *A Lost Lady* Ivy Peters carries a red leather case containing "tiny sharp knife blades, hooks, curved needles, a saw, a blow-pipe, and scissors" (*LL,* p. 23).[20]

In the gruesome opening scene of *A Lost Lady,* the phallic Ivy Peters—who walks with "unnatural erectness, as if he had a steel rod down his back"—slits a woodpecker's eyes, suggesting an element of self-condemnation in Cather's portrayal of the male mutilator (p. 20). As repulsive as Peters is, with his case of dissecting instruments partly derived from a "taxidermy outfit," he is a version of her adolescent self as well as of the knife-wielding boy in Virginia. The solution to female powerlessness, this scene implies, is not to adopt either the oppressor's or the male role, as she did in adolescence, but to recapture the five-year-old child's solution: to find power through language, power that would not threaten nature or women (the blinded woodpecker is addressed by the attacker as "Miss Female" [*LL,* p. 23]).

Medicine was an understandable professional choice for the adolescent girl, even if there were no unconscious motivations connected with the Willow Shade incident. There were a number of reasons why she might have found doctors appealing figures. Doctors were among the only professional people a rural child encountered (journalists and bankers did not make house calls on sick children) and so provided her with her most direct examples of public success and power. As healers of the sick, doctors might combine a man's authority with a woman's compassion, as did Dr. McKeeby who nursed the child through what "was probably a siege of poliomyelitis." Subsequently she "worshipped" him (*WWC,* p. 110). (When she drew on her memories of Dr. McKeeby in creating Dr. Archie,

Thea's friend and mentor in *The Song of the Lark,* Cather cast the doctor in such a nurturing role: Archie wraps Thea's chest in a hot plaster, dresses her in a clean nightgown he has warmed before the fire, and tucks in the blankets while thinking "he would cherish a little creature like this if she were his" [*SL,* p. 10].) And in curing disease through medicine and treatment, doctors were probably the only figures in Cather's childhood world who possessed the almost magical power over the environment also exerted by the heroic characters she admired.

In the Cather family, as in nineteenth-century America, illness and health were physical and psychological states invested with social issues of gender and power. Willa Cather's attraction to the medical profession in general and to Dr. McKeeby in particular should be connected with the daughter's relationship to her mother and the feminine role she represented. Like the first persona she adopted, the "dang'ous nigger" in the parlor, Cather's desire to be a doctor suggests her need to separate herself from her mother and the lady's role (*WCL,* p. 13). Yet the adolescent girl's planning a medical career was a more complex and ambiguous phenomenon than her childhood outburst, revealing the daughter's desire both for attachment and for separation. Overtly, of course, Cather's wish to "be an M.D.," her "chief ambition" registered in the album book, reflected a desire to reject her mother's example and to enter a vocation awarded prestige and authority in the public world. But Cather had reason to associate the doctor's role with her mother's approval as well as her helplessness. Virginia Cather relied on doctors during pregnancy and childbirth as well as in illness. The daughter must have been aware of her mother's respect for—as well as her dependence on—the physician who attended her.

In fact, a serious illness during which the mother became attached to Dr. McKeeby immediately preceded the daughter's medical apprenticeship. Early in 1884 Virginia became ill and the elderly doctor who examined her, evidently despairing of his ability to effect a cure, "just knelt by her bed and began to cry." Dr. McKeeby, however, came to the rescue, achieving his patient's confidence along with her recovery. "When Mrs. Cather opened her eyes and saw the capable-looking, dignified man," reports Mildred Bennett, "she relaxed, confident he could save her." A similar incident occurred several years later, suggesting an emotional and psychological dynamic at work in Virginia's recuperation. Dr. McKeeby was taking the train from Colorado Springs to Chicago. When it stopped in Red Cloud a few minutes, he stepped down and talked with some people in the station who told him that Virginia Cather was desperately ill: "They say she hasn't many more hours to live." This was the second time Dr. McKeeby would effect a miraculous cure ("Once again Mrs. Cather opened her eyes, relaxed at the sight of him, and began to improve" [*WWC,* pp. 110–11]).

The girl who witnessed the doctor's mysterious, mesmerizing power over her mother may have felt that, if she became a doctor herself, she could both reverse the power dynamics of the mother–daughter relationship and win her mother's love and approval. Cather's choice of specialty may also reflect the same mixture of love and anger that underlies her portrayals of "good" and "bad" mothers in her early stories. As a knife-wielding surgeon, she would have to inflict pain in order to cure, taking on the role her mother possessed when she was at her most powerful and punishing: the family disciplinarian who whipped the disobedient children. Simultaneously, of course, she would be trading places with her early persecutor, the boy who threatened to cut off her hand, as well as differentiating herself from her mother when she was at her most helpless.

In planning to be a doctor, Cather was hoping to assume the dominant role in a relationship that social historians view as paradigmatic of Victorian sex roles. The woman patient's dependence on the all-knowing doctor has struck many commentators as an embodiment of the Victorian drama of female submissiveness and male control. In her examination of nineteenth-century medical advice and treatment of women, Carroll Smith-Rosenberg has shown how the physician's "would-be scientific views reflected and helped shape social definitions of the appropriate bounds of woman's role and identity," an observation supported by Charlotte Perkins Gilman's "The Yellow Wallpaper" in which the doctor's medical advice to his ailing wife—quiet, bed-rest, and no writing—reflects the male power to define and circumscribe female identity in the larger society. Other Victorian girls besides Willa Cather wanted to be doctors, but as Regina Morantz observes, most female physicians accepted Victorian definitions of feminine identity and role. Medicine could be considered a "natural extension of woman's sphere and peculiarly suited to the female character, which was self-sacrificing and empathetic," the view taken by Sarah Orne Jewett in creating her physician–heroine in *A Country Doctor*. By contrast, medicine seems to have attracted Cather because she associated the profession with the male role.[21]

"Novelists, opera singers, even doctors," Cather later declared to Elizabeth Sergeant, "have in common the unique and marvelous experience of entering into the very skin of another human being. What can compare with it? she said, her eyes shining" (*WC:AM*, p. 111). That Cather associated the doctor's profession with creative power is suggested by this somewhat unlikely catalogue mingling the literal and metaphoric powers of entry and transformation. Equating the doctor with phallic potency and creativity, she links the patient with helplessness and emptiness. The extreme self-disgust Cather later experienced during periods of illness, when she assumed a dependent role and resembled her mother at her weakest, suggests that her adolescent desire to be a doctor also arose from her need to define herself in opposition to his subordinate partner, the female patient—a role that Cather shared with her mother after her arrival in Nebraska. During and after an illness, Elizabeth Sergeant recalls, Cather was characteristically gloomy and self-disparaging. She remembers once accompanying Cather from Boston to New York "in this typical down-hearted state of mind in which, through physical illness, she had, in a way that was baffling to me, lost her own self-respect" (*WC:AM*, p. 90). At another time—when Cather spent "many sick and feverish weeks" in the hospital with an infected scalp—her "comments on her illness sounded the note of 'The Profile,' " Sergeant remembers. "A physical blemish is so abnormal it creates a mental deformity. Willa went so far as to say she deserved derision, such as was given to lunatics in Dickens' time!" (*WC:AM*, p. 122). Writing her childhood friend Irene Miner in 1942, shortly after returning home from a long hospital stay, Cather was similarly self-condemning: I hate to be sick, she told Irene, I hate to be defective.[22]

In choosing to be a doctor, the adolescent girl was thus simultaneously choosing not to be a defective female patient, aspiring to the role she associated with power, health, and wholeness, shunning the one she associated with weakness, sickness, and mutilation. She was also rejecting the identity and role that seemed passed down by the women in her family: not only by her mother but also by the aunts who had died of tuberculosis and diphtheria, including the one for whom she had been named.

•

Although the young Cather found new stories and patterns for imagining a self in Red Cloud, Victorian ideology and social practice restricted the possible identities and roles a female child could assume. By the time she was fourteen, Cather recognized that being a girl did damage her ambitions to become heroic and powerful, and this mutilation she refused to endure. As in her childhood story, she overcame this threat with a "strategem" when she began a four-year performance as William Cather before an audience of family, friends, and disapproving townspeople. There was a history of twins in the Cather family; in transforming herself from Willa into William she was becoming, in a sense, her own twin brother. He would have a better chance of accomplishing childhood ambitions shaped by the hero's plot.

NOTES

1. [Willa Cather], *Willa Cather: A Biographical Sketch/ An English Opinion/ and An Abridged Bibliography* (New York: Knopf, n.d.) p. 2.
2. Edith Wharton, *A Backward Glance* (New York: Appleton, 1934), p. 68.
3. Bernice Slote, who reconstructed Cather's childhood reading, offers a full discussion of the family library and reading interests in *KA*, pp. 35–43.
4. For a fuller catalogue of her private library, see *KA*, pp. 39–40.
5. Franc Cather to Jennie Cather Ayre, April 28, 1874, Nebraska State Historical Society, Lincoln, Neb. According to Mildred Bennett, Cather "was not displeased when people told her she 'took after' " her literary aunt (*WWC*, p. 14).
6. William Curtin makes the connection between the nameless aunt Cather refers to in this 1897 column and her Aunt Franc (*WP* 1, p. 349).
7. Elaine Showalter, *A Literature of Their Own: British Women Novelists from Brontë to Lessing* (Princeton: Princeton University Press, 1977), p. 42.
8. George Eliot, *The Mill on the Floss,* ed. Gordon S. Haight (Oxford: Oxford University Press, 1981), p. 147. The strange language, Eliot writes, "gave boundless scope to her imagination," and the "mysterious sentences" were "all the more fascinating because they were in a peculiar tongue of their own, which she could learn to interpret" (p. 147).
9. Wharton, p. 68.
10. Cynthia Griffin Wolff, *A Feast of Words: The Triumph of Edith Wharton* (New York: Oxford University Press, 1977), pp. 43–45, esp. p. 45.
11. Jacob Abbott, *History of Alexander the Great* (Philadelphia: Henry Altemas Co., 1900), p. 74, p. 230.
12. For a psychoanalytic interpretation of the reading process, see Norman Holland, *The Dynamics of Literary Response* (New York: Norton, 1974). Holland's argument that

the reader's identification with characters involves a mixture of projection and introjection accords with Cather's description of her relationship to characters who became part of her as she became part of them. For other analyses of the role of the reader in constructing literary meaning, see Wolfgang Iser, "Interaction between Text and Reader" in *The Reader in the Text,* eds. Susan R. Suleiman and Inge Crosman (Princeton: Princeton University Press, 1980), pp. 106–119, and Stanley E. Fish, "Interpreting the *Variorum*" in *Reader-Response Criticism,* ed. Jane P. Tompkins (Baltimore: Johns Hopkins Press, 1980), pp. 164–84.

13. Rachel M. Brownstein, *Becoming a Heroine: Reading About Women in Novels* (New York: Viking, 1982), pp. xviii, xiv. For an analysis of the ways in which contemporary women's romances both reflect the woman reader's dissatisfactions with her social role and reaffirm patriarchal institutions and structures, see Janice Radway, *Reading the Romance: Women, Patriarchy, and Popular Literature* (Chapel Hill: University of North Carolina Press, 1984). As Radway observes, the fantasies of escape in which women readers indulge stave off "the need or desire to demand satisfaction in the real world" (p. 212).

14. Nina Baym interprets the heroine's plot of nineteenth-century women's fiction more positively than Radway does that of twentieth-century romances, seeing women's fiction more as a "protest against long-entrenched trivializing and contemptuous views of women" than as an acceptance of female subordination as the price of romantic love (*Woman's Fiction: A Guide to Novels by and about Women in America, 1820–1870* [Ithaca: Cornell University Press, 1978], p. 29). Cather, however, did not read women's fiction in this way (see Chapter 8).

15. For other examples of Cather's recommendations for children's fiction, see *WP* 1, pp. 339–448.

16. For a discussion of attics as architectural symbols of confinement, see Sandra M. Gilbert and Susan Gubar, *The Madwoman in the Attic: The Woman Writer and the Nineteenth-Century Literary Imagination* (New Haven: Yale University Press, 1979), pp. 89–90, 347–49, 425–26, 533–34. Cather's childhood experience with a delightful attic bedroom, a gift from her mother, may have allowed her to invest this traditional literary symbol of women's oppression and incarceration with a positive meaning.

17. See Gaston Bachelard's *The Poetics of Space,* translated by Maria Jolas (Boston: Beacon Press, 1969), pp. 3–37. Bachelard observes that "up near the roof all our thoughts are clear. In the attic it is a pleasure to see the bare rafters of the strong framework" (p. 18).

18. Elmer Alonzo Thomas, *80 Years in Webster County* (Hastings, Neb.: privately published, 1953), p. 112.

19. See Mildred Bennett's analysis of this pattern in the Introduction to *CSF,* pp. xxxvii–xxxviii. Bennett also associates Cather's preoccupation with mutilation with the childhood trauma Lewis recounts. Of course, it is possible that the childhood memory Cather related to Lewis was itself a fiction, a product of the obsession rather than the cause.

20. Consider also Emil Bergson and Frank Shabata in *O Pioneers!,* both associated with guns and death; the feral Frank Ellinger in *A Lost Lady;* the vicious, reptilian Buck Scales in *Death Comes for the Archbishop;* the casual rapist Martin Colbert in *Sapphira and the Slave Girl.*

21. Carroll Smith-Rosenberg, "Puberty to Menopause: The Cycle of Femininity in Nineteenth-Century America," in *Clio's Consciousness Raised: New Perspectives on the History of Women,* ed. Mary Hartman and Lois W. Banner (New York: Harper-Torchbooks, 1974), p. 24, and Regina Markell Morantz, "The Lady and Her Physician" in *Clio's Consciousness,* p. 50. For other studies of nineteenth-century medical opinions and the

male physician–female patient relationship, see Carroll Smith-Rosenberg, "The Hysterical Woman: Sex Roles and Role Conflict in Nineteenth Century America," *Social Research* 39 (1972): 552–78; G. J. Barker-Benfield, *Horrors of the Half-Known Life: Male Attitudes Toward Women and Sexuality in Victorian America* (New York: Harper & Row, 1976); and Barbara Ehrenreich and Deirdre English, *For Her Own Good: 150 Years of the Experts' Advice to Women* (New York: Doubleday, 1979). In *Read This Only to Yourself: The Private Writings of Midwestern Women, 1880–1910* (Bloomington: Indiana University Press, 1982), Elizabeth Hampsten, drawing on diaries and letters of women like Virginia Cather, finds that women "wrote about sexuality, disease, and death as though they expected to slide by degrees from one to the other" (p. 102).

22. Willa Cather to Irene Miner Weisz, April 18, 1942, Newberry Library, Chicago, Ill.

5

Enter William Cather

I like to be like a man. *My Ántonia*

Proud it is I am to know
In my veins there still must flow,
There to burn and bite alway,
That proud blood you threw away . . .
"The Namesake"

The "hateful distinction" that Willa Cather eventually discovered between boys' and girls' fiction reflects a larger distinction: that between male and female roles in Victorian society (*WP* 1, p. 337). Since those roles were imposed most strictly in adolescence, the phrase has a further connotation. If the dislocating move from Virginia to Nebraska was the first major discontinuity in her life, the "hateful distinction" between childhood and adolescence was the second. Although this was a more difficult break to heal, Cather found compensation for loss and disruption in storytelling, as she had after the move to Nebraska. But this time she told the story herself.

In 1888 the fourteen-year-old Cather decided to become the hero of her own life story when she created the masculine persona she sustained for the next four years. Employing the transforming power of dress and disguise, she distinguished herself from other Red Cloud girls by cropping her hair, donning boyish clothes, and naming herself "William Cather, Jr.," or, reflecting her career interests, "William Cather, M.D." Photographs from this period show hair at crew-cut length or slightly longer and clothes ranging from the masculine to the severely feminine: she wears a soldierly costume with Civil War cap (initialed "W.C."), a coat and tie with a jaunty derby, a light jacket with a dark blouse, a plain striped shirtwaist dress with boater hat and a man's gold watch, and a simple white blouse. Occasionally there are signs of the feminine—a scarf, a ruffle, a ribbon—but the visual impression is overwhelmingly masculine. Two of these photographs are snapshots (Willa Cather in soldier costume and with her two Red Cloud friends), but the rest are studio pictures taken by a professional photographer, which suggests her commitment to William Cather as a public self. *(See photograph section.)* Cather's real-life performance merged with her on-stage play-acting when she took the role of the merchant–father in the "Beauty and the Beast" at the

Red Cloud Opera House: she played the role so convincingly that one play-goer refused to believe that she was not a boy (*WWC,* p. 175).

Willa Cather's unorthodox dress and manner brought her "considerable notoriety" and made her the subject of "much talk around town," remembers Elmer Thomas in his history of Webster County, where he gives us our only eyewitness account. "I remember Willa Cather most for her masculine habits and dress," he writes. "This characteristic in those days was far more noticeable because it was very seldom that women appeared dressed other than in strict feminine attire." But she was impervious to criticism, he continues, and "even boasted that she preferred the masculine garb" as well as the "masculine sex." Thomas was evidently disconcerted by his cross-dressing neighbor: "To me, she was never attractive," he explains, "and I remember her mostly for her boyish makeup and the serious stare with which she met you. It was as if she said, 'stay your distance buddy, I have your number.' Enough, I did." Other young men seem to have kept a wary distance as well, for, Thomas reports, Cather did not have a "romantic love entanglement" with any of Red Cloud's eligible bachelors.[1] Even though Cather's male role-playing lasted only four years, her overturning of feminine norms was so unsettling that the talk endured and became legend. When I first visited Red Cloud in 1973, I was told that Cather had been a "hermapherdite" who "wore men's shoes—had 'em made special."[2]

In maintaining this rebellious performance for so long, Cather risked ostracism, disapproval, and social rejection in both Red Cloud and Lincoln, where she entered the University of Nebraska as William. She must have had important reasons for doing so. In a sense her male impersonation was her first major work of fiction, a text in which she was both author and character. Like any text, "William Cather" contains several meanings; like any recurrent private action, Cather's performance had diverse and even contradictory sources and resulted from a complex interplay of social and psychological factors. For purposes of analysis I will examine these sequentially, but we should keep in mind that social and psychological pressures were operating both interdependently and simultaneously. Let us consider the social causes first, placing Cather's defiant act in the context of the Victorian ideology of gender.

•

In her fiction and letters, Willa Cather frequently lamented the loss of childhood. In *The Professor's House* her alter ego/protagonist, Godfrey St. Peter, reflects that "adolescence grafted a new creature into the original one," forever altering the child's "original self." In childhood he possessed his "realest" and most authentic identity, the Professor reflects; afterward his life seemed "ordered from the outside," socially determined rather than individually chosen (p. 267). Just as the Professor nostalgically views his childhood self as original and authentic, free from social and sexual pressures, so Cather in a letter to a friend commented how pleasant she found her childhood memories, because young children were neither very male nor very female.[3] Childhood thus represented a time of freedom and wholeness to Cather when the self was not yet socially engendered.

Although Willa Cather did not, in fact, escape social and familial attitudes toward gender in childhood, like many Victorian girls she was somewhat freer to express traits socially defined as ''male'' in childhood than she was in adolescence. Victorians considered childhood a grace period when boys and girls could share interests and activities. Many authors of advice books even recommended active tomboy childhoods for girls and urged parents to let their daughters engage in active sports and games with their brothers. ''The peculiarities that characterize each sex rarely become pronounced before childhood passes into youth,'' one doctor commented reassuringly. Childhood was thought to be a time of relative equality as well as of similarity between the sexes. As domestic novelist and advice-giver Mary Virginia Terhune put it, ''Our sons and daughters start even.''[4]

But they did not finish even. Victorian advice-givers could recommend adventurous, independent childhoods because they viewed tomboyism as a common phase through which little girls passed on their way to the safe harbor of femininity. Consistently they integrated their endorsement of tomboyism with traditional definitions of woman's identity and role. The rowdy tomboy would make a better wife and mother than her prissy, house-bound sister, they agreed, for participation in boyish sports and games developed the health, strength, and competence she would later need in fulfilling a taxing domestic and maternal role. Daughters who led passive, inactive lives as children were likely to become sickly mothers whose enfeebled reproductive organs would in turn produce weak or defective children.

Although these prescriptive writers imagined a vital continuity between a relatively gender-free childhood and conventional womanhood, in fact their depiction of female adolescence introduced a crucial discontinuity between childhood and adolescence that Willa Cather first confronted in Red Cloud. Dr. John Kellogg summed up the change. In puberty, he explained, the tomboy naturally abandoned her romping and forsook the woods and fields where she once roamed with her brothers, unable to resist the lure of more compelling spaces: the ''kitchen, washroom and the garden,'' which he grandly proclaimed ''nature's gymnasia'' for adolescent girls. Another doctor concentrated on the gender differences adolescence naturally brought. ''Here they part as girl and boy, to meet as man and woman,'' he explained, ''and when they meet again, the change is so great that one scarcely recognizes the other.'' The emerging man then regarded his former playmate as ''the weaker being whom he is bound to protect,'' while the emerging woman acquiesced in her subordinate role with ''a sense of gratified vanity.''[5]

As such prescriptive texts suggest, adolescence was the crucial period of socialization when Victorian girls were taught the feminine virtues of dependence, submissiveness, selflessness, and passivity.[6] Of course, the prescriptive literature does not tell us what views of female roles Willa Cather confronted in Red Cloud, but other evidence suggests that the town's definition of femininity did not differ much from the Victorian norm. Articles appearing in Red Cloud papers during the mid-1880s reveal an ideology of girlhood consistent with these larger patterns. The frivolous, novel-reading, pleasure-seeking girl was criticized (''the girl of sixteen who will neither sew nor do housework has no business to be decked out in finery and rambling about in search of fun and frolic,'' sniffed the *Red Cloud Chief*) as was her opposite, the ambitious girl with her eye on a medical career (''Why, what man in his right mind would ever marry a woman doctor?'' wondered the *Webster County Argus*). The papers reserved their praise for the sober, industrious, submissive girl who devoted herself to domesticity. ''Lovable'' girls, explained the *Argus,* were ''girls without an undesirable love of liberty and craze for individualism; girls who will let themselves be guided; girls who have the filial sentiment well developed.''[7]

Cather was not persuaded by such editorial admonitions. Like Louisa May Alcott's Jo March, she scorned feminine confinements and refinements. The self-dramatizing entries she made in a friend's album book in 1888 show a contempt for proper girlish pastimes that recalls Jo's spirited rejection of ''niminy-piminy'' femininity.[8] In the category ''My idea of real misery,'' Cather entered ''Doing fancy work,'' and ''The greatest folly of the Nineteenth Century'' she defined as ''Dresses and Skirts.'' The items she most desired if ''Shipwrecked on a desolate island'' reflect William Cather's pragmatic preferences: ''Pants and Coat.'' Her entries praising amputation (her favorite ''Amusement'') and toad-slicing (''Occupation during a summer's vacation'') reveal William Cather's other unfeminine and gossip-rousing interests (*WWC,* pp. 112–13).

The ideology of femininity conveyed in the local papers was reinforced by Red Cloud society. Cather's journey from country to town brought her into a social world where women's possibilities were restricted. The town's leaders and professional people were male, and Red Cloud women desiring employment confined themselves to the serving roles traditionally assigned them: newspapers in the 1880s show women as teachers, dressmakers, milliners, receptionists, maids, and prostitutes. The town's respectable married women engaged in a few decorous activities that could be accommodated with the feminine role. In 1888, the year Cather adopted her male pose, the Red Cloud papers mention a Chatauqua Circle, an equestrienne club for ''prominent ladies,'' a Women's Relief Corps, a ladies' cooking club, a Baptist women's ice cream festival, and the Red Cloud chapter of the Women's Christian Temperance Union: hardly dramatic future activities for a reader of *Treasure Island* and *The Count of Monte Cristo.*[9]

Willa Cather's reaction to such confining social definitions of femininity can only be imagined from her rejection of them. She left no written record of her response to adolescence: her William Cather masquerade is the only diary entry we have. But we do know from more traditional diaries of the stress other Victorian tomboys experienced when they were forced to abandon childhood freedoms. The seventeen-year-old Frances (''Frank'') Willard described her mother's imposing a feminine dress code as the ''day of her martyrdom,'' and Louisa May Alcott, who later said she had been born with a ''boy's spirit under [her] bib and tucker,'' began a life of conflict when she tried, often unsuccessfully, to repress her ''boy's'' spirit and cultivate the womanly virtues of submissiveness and silence.[10] But whereas Willard and Alcott accepted, at least externally and temporarily, the Victorian definition of femininity, Cather's metamorphosis into William signified her attempt to fashion an independent, autonomous, and powerful self. Just as the child had proclaimed herself a ''nigger'' to avoid speaking the ''playful platitudes'' young girls were supposed to utter in the parlor (*WCL,* p. 13), so the adolescent girl became William Cather in order to avoid becoming a platitude, a conventionally assigned identity.[11]

Although William Cather's emergence seems a sharp break with her past, in a sense Cather's male impersonation was the logical extension of her active childhood. This continuity between childhood and adolescent selves is revealed in the album book where she listed Emerson as her favorite prose writer and Napoleon as her favorite "character in history." Signing her entry "William Cather, M.D.," Cather signaled her desire to achieve the heroism Napoleon embodied and Emerson urged. Earlier she had not found her gender a barrier to ambition, but after puberty's "hateful distinction" arrived, she was more likely to find the world "not fixed but fluid" as William Cather.[12]

In adopting a male persona Cather was being rebellious, theatrical, and bold, but she was not being particularly creative. To construct this alternate self she could only tap a cultural inventory of roles and selves; understandably, she was trapped by her contemporaries' polarization of gender traits and roles, the same dichotomy that Alcott evinced when she referred to her adventurous self as her "boy's spirit." Other nineteenth-century women who donned male dress similarly were unable to transcend Victorian sex roles. Desiring the autonomy and freedom that it seemed only men possessed, such women decided to cross rather than to blur gender boundaries and passed themselves off as men for several years. "This world is made by man—for man alone," commented one male imposter. "Is it any wonder that I determined to become a member of this privileged sex?"[13]

Male imposters who convinced friends and co-workers they were men were also performers, but Cather's audience, unlike theirs, knew she was acting. When "William Cather" walked down Red Cloud's main street in coat, pants, and a derby hat, a gaping Elmer Thomas could see both the female and the male selves, just as a nineteenth-century theater-goer could see both Sarah Bernhardt and Hamlet at the same time. In trying to cross gender boundaries Cather was blurring them for her audience, calling into question her community's polarization of masculine and feminine as the male imposters did not.

In changing her costume the young performer was altering the most prominent sign of gender. A form of communication as well as a symbolic representation of the self, dress is a kind of speech. What one "says" through dress can vary from the socially imposed speech of the uniform—in which the dresser is, in a sense, spoken by the social group—to more varied and personal forms of clothing which dramatize self-expression. In Victorian society the customs of dress mirrored the distinctions between the sexes; male and female clothing were sharply distinguished. Social historians point in particular to the connection between middle-class women's subordination and clothing that confined, weakened, and even endangered them—the stays, corsets, and cumbersome dresses that both caused and signified inactivity and passivity.[14]

In challenging nineteenth-century definitions of womanhood, women like Rosa Bonheur and George Sand—who often adopted male dress—went even further than did feminist dress reformers like Amelia Bloomer, but in both cases rejecting conventional female attire reflected their desire to reject the conventional female role. We see the same pattern reflected in the final scene of Kate Chopin's *The Awakening* (1899) in which the heroine discards her clothing and the social role it signifies. Once Edna Pontellier has cast off her false clothes and external self, however, she has nothing to wear and no one to be. Like language, dress mediates between the individual and the group. Without clothes—and without the identities and roles they signify—the naked individual has no place in a social world, and so Edna must choose the ocean and death. In contrast, when Cather put away her clothes and donned William's, she was not leaving the social world but demanding an alternative place in it. Although inspired by her fascination with the theatre and acting, unlike her amateur performances in the Miners' parlor this was a political act. She was using the transformative power of costuming and performance to reject Red Cloud's confining definition of femininity and to construct an alternative self—powerful, heroic, male.

But her photographs reveal an undeveloped subplot in the William Cather story, other transformative possibilities in her costume. The ribbon, the ruffled blouse, and the scarf suggest the girl's desire to redefine rather than to reject female identity, to find a way to express the human possibilities her society divided between male and female. As we will see in subsequent chapters, her attempts at redefining womanhood are mirrored by changes in her dress.

•

In the Cather family the dress-conscious Virginia Cather symbolized the connection between female identity and feminine clothing, an equation Cather stresses in those characters based on her mother. The "passion for dress" Virginia Gilbert displays in "The Profile" (*CSF*, p. 129) is particularized in "Old Mrs. Harris," where the stylish Victoria Templeton pleases her children with her "new dotted Swiss, with many ruffles, all edged with black ribbon, and wide ruffly sleeves" (*OD*, p. 124). Willa Cather's rebellious self-transformation may have had a personal source in her observation of the importance her mother attributed to dress. When Virginia Cather lent her dresses for use in the girls' amateur theatricals, thus equating the lady's attire with the actor's costume, she doubtless hoped that Willa would play a female part. But since daughter, like mother, was using dress to establish an identity, Cather's male impersonation was, in part, a maternal inheritance.[15]

For Edith Wharton dress was also a sign of maternal identity. Her mother Lucretia possessed an "inexhaustible" memory for the "details of dress," Wharton recalled, never forgetting her "martyrdom" when she had to wear a "home-made gown of white tarlatan" and "old white satin slippers" for her debut. But this humiliation was finally reversed by the triumphant "annual arrival of the 'trunk from Paris,' " when Lucretia and Edith shared "the enchantment of seeing one resplendent dress after another shaken out of its tissue-paper."[16] Unlike Cather, Wharton inherited her mother's enchantment with feminine dress as well as, temporarily, her definition of the female role. In Wharton's autobiography she remembers her own reaction to an invitation to dine with Henry James: "I could hardly believe that such a privilege could befall me, and I could think of only one way of deserving it—to put on my newest Doucet dress, and try to look my prettiest!" Decades after the dinner-party, the established author could "see the dress still—and it *was* pretty; a tea-rose pink, embroidered with iridescent beads" (p. 172).

That the adolescent Edith Wharton inherited her mother's fascination with female dress and that Willa Cather repudiated Virginia Cather's does not mean that

Cather's choice of dress was unconnected with her mother, however, or that her male performance arose only in response to social forces. Other rebellious girls faced the same social constrictions that Cather did, and many wished for male freedoms; as novelist Mary Virginia Terhune commented in 1882, "It is an uncommon [illegible]nt to meet a woman who . . . would not own that at some period of her life, she had wished she had been born a boy."[17] Why did Willa Cather go to extremes that others did not and turn wishes into action?

In answering this question we need to turn from the social to the psychic and the familial aspects of Cather's life, keeping in mind that cultural forms and expectations structure even the most private realm of all—the unconscious. William Cather's emergence satisfied multiple and sometimes contradictory desires central to the mother–daughter relationship. Trying both to sever and to maintain her bond with her mother, Cather created an "interior theater"—to use the phrase Elizabeth Fifer applies to Gertrude Stein—in which mother and daughter took the main roles as authors, actors, and audience.[18]

One specific event triggered Willa's transformation into William. Virginia Cather used to arrange her daughter's long hair as well as her own, thus casting her both as the ideal little woman and as an extension of herself. The photograph of Cather as a child with long ringlets demonstrates the mother's care in making her daughter appear as pretty and ladylike as possible. But one day in 1888 Virginia Cather could not perform this traditional task. Recuperating from the birth of her son James, Virginia was too ill to arrange her daughter's hair. So Cather went to the barber and asked for the crew-cut we see in one of her William Cather photographs (*WWC*, pp. 178–79). *(See photograph section.)*

The context for William Cather's emergence is significant. Since the mother's arrangement of the daughter's hair reflects her imposition of a social identity and role, cutting the hair is the daughter's declaration of independence and autonomy, similar to her defiance in the parlor. Yet Virginia Cather's failure to fulfill a maternal duty could also have seemed a sign of maternal coldness or abandonment, made more dramatic by her absorption with the new baby. By cropping her hair the rejected daughter could in turn reject the mother.

That Willa Cather's initial act of rebellion occurred during a maternal illness—a sign of weakness and dependence—is also significant. To the child the mother is an all-powerful being, but many adolescent daughters come to a painful realization: because mother is a woman, she fulfills a subordinate role in a society where fathers possess power in the public world. As Simone de Beauvoir observes, the adolescent girl faces a crisis when maternal weakness is unveiled:

> Up to this time [the daughter] has been an autonomous individual: now she must renounce her sovereignty. . . . A conflict breaks out between her original claim to be subject, active, free, and, on the other hand, her erotic urges and the social pressures to accept herself as passive object.

Although most girls adapt to the subordinate feminine role the mother represents, a few feel a "burning sense of deception," de Beauvoir notes, and reject their once-powerful mother and her now-restricted identity.[19] So Cather's "deception''—her costuming as William Cather—may have been the daughter's response to the mother's deception, exposed when Virginia Cather's social powerlessness was dramatized and represented by her illness.

In ''Katherine Mansfield'' Cather uses the metaphor of the net to describe intense family attachments that both confine and connect, restrict and enrich the self. Although she did not mention the mother–daughter bond in her essay, her reference to the ambivalent self that is ''half the time greedily seeking'' human bonds and ''half the time pulling away from them'' is an apt description of the drama the adolescent girl was playing out with her mother (*NUF,* p. 136). As Nancy Chodorow observes, the adolescent daughter engages in a ''replay'' of preoedipal and oedipal conflicts as she struggles to sever the ties of identification with her mother and to achieve a separate identity while simultaneously experiencing continuity and sameness: the adolescent girl ''alternates between total rejection of a mother who represents infantile dependence and attachment to her, between identification with anyone other than her mother and feeling herself her mother's double and extension.'' To break the strongest link in the net of family attachments, Cather tried a strategy that Chodorow sees as a common ploy daughters use to differentiate themselves from their mothers: creating ''arbitrary boundaries by negative identification (I am what she is not).''[20] Becoming a boy by altering the signs of femininity with which her mother was most identified—hair and dress—is perhaps the most extreme way Cather could have chosen to declare her difference. But the intensity of the rejection also suggests the daughter's fear of sameness.

Cather's male role-playing also represented a temporary identification with her father. The close relationship she enjoyed with Charles Cather in childhood continued during adolescence, when Red Cloud neighbors could see her ''hanging about her father's office.'' During the period when she was aspiring to a medical career, the daughter naturally gravitated toward the parent who occupied and symbolized the public sphere. At this time she began to treasure her physical resemblance to her father, developing an identification to counter her tie to her mother: the adolescent girl ''gloried in the fact that her own fine skin and dark blue eyes'' were like her father's (*WWC,* p. 26).

Yet Cather's turn to her father and male identification must be seen as more than a rejection of her bond with her mother. In fact, it was an act both of ''pulling away'' and of ''greedily seeking'' a maternal attachment. As Chodorow observes, girls turn toward their fathers because of their ambivalence toward their mothers, acting in a drama of longing as well as hostility, identification as well as separation. ''Every step of the way,'' Chodorow writes, the adolescent girl ''develops her relationship to her father while looking back at her mother—to see if her mother is envious, to make sure she is in fact separate, to see if she can this way win her mother, to see if she is really independent.'' The daughter's turn to her father, Chodorow concludes, ''is both an attack on her mother and an expression of love for her.''[21]

Among the motivations Chodorow mentions for the adolescent girl's turn to her father is one we have still to consider: the possibility that Cather's male impersonation was an ''expression of love'' for her mother. As Chorodow argues, daughters never abandon the intense preoedipal attachment to their mothers. If the daughter becomes heterosexual she transfers her erotic drive from mother to father and then to other men, but does not make "final and absolute commitments to heterosexual *love*." The daughter maintains the primary emotional tie to the mother, which she may then enact through intimacies with children and with female friends.[22] And the daughter who becomes a lesbian, as did Cather, does not abandon the child's erotic bond with the mother but replays it in her sexual and romantic relationships (*WWC*, p. 26).

Cather's transformation into William can also be seen, then, as a statement of her desire to continue the search for her mother's love. As William she could rival the males her mother favored, competing with her father, her brothers, and the new baby. Moreover, as William she could join the competition for maternal affection without fearing the "erasure of personality" she found terrifying, protecting herself from merging and sameness by the same action that declared her attachment (*KA*, p. 448). As her mother's son, the daughter could both receive and seek her mother's love without being crippled by dependency and powerlessness.

•

It is curious that Virginia Cather continued to encourage her daughter's ambitions during this troublesome phase. Edith Lewis mentions the mother's "unusual sympathy and understanding of her children's individuality," and it takes unusual sympathy indeed for a genteel, fashion-conscious Southern lady to tolerate a teenage vivisectionist and cross-dresser (*WWC*, p. 6). And yet Virginia Cather strongly supported her adolescent daughter's eager curiosity, as a family anecdote reveals. One day Professor Shindelmeisser, the town's talented but erratic piano teacher, complained to Virginia about her daughter. Willa wasn't applying herself, he said. Instead of playing her assigned piece, she bombarded him with questions about his native Germany. If that was so, Virginia replied, he should come twice a week instead of once (*WWC*, p. 154). In this case the mother joined the daughter in redefining female behavior in the parlor; she was willing to employ the piano teacher as a storyteller rather than as an instructor in feminine accomplishments.

Perhaps inspired by memories of her own schooling, Virginia Cather encouraged her daughter's intellectual and cultural aspirations in other ways, endorsing her "strenuous, untiring, hungry efforts to extract the essential facts and implications of knowledge and culture" from her Red Cloud mentors (*WC:AM*, p. 18). She also gave her adolescent daughter the private space where she could construct her "real self"—her beloved attic bedroom—and later supported Cather's desire to attend college when her husband thought his daughter might be content to teach at a local school.

It is possible that Virginia Cather tolerated her daughter's rebellion because in some ways she identified with it. She too had suffered displacement and constriction, but as a wife and mother, Virginia did not have her daughter's power of transformation. In moving from Willow Shade to the cramped Red Cloud house she had lost status, servants, financial security, and perhaps most important, the chivalric Southern world in which she glowed. And Virginia had to endure these blows as her family increased in size, further limiting her freedom. But she retained her Southern pride, which evidently did not win her many friends in a judgmental Midwestern town. According to Mildred Bennett, Cather sensed that "some of the neighbors thought her mother haughty" (*WWC,* p. 29), an impression corroborated by Elmer Thomas. The Cather children "boasted of their mother being 'the only lady in town,' " he reports. And "sure enough, her toilet and dress along with a haughty air, substantiated in our minds this alleged superiority complex."[23]

Since mother and daughter were both outsiders in Red Cloud, where neither woman fit the expected feminine mold, Virginia Cather may have identified with her rebellious daughter who was enacting the assertiveness and anger she could not express herself. The Southern lady may not have been as distressed by William Cather's emergence as one might have expected because the daughter was openly defying the confining aspects of the female role that the mother avoided passively in her frequent illnesses. Cather may also have sensed this alliance and enjoyed challenging her mother's enemies; William Cather was doubtless provoking the same town gossips who thought her mother haughty and superior. The daughter was thus expressing the "masculine" self that her strong-willed mother partly repressed and partly channeled into the family relationships she dominated.

•

Virginia Cather was not the only family member who stimulated Cather's adolescent role-playing. The daughter was overtly aligning herself with the men in her family, choosing in particular her maternal uncle William Boak as her "namesake" and symbolic progenitor when she named herself "William Cather, Jr." As Cather's sister Elsie commented in a letter to Mildred Bennett, it was not unusual for people in her sister's generation to change their names at will. Few records were kept, and name-changing was a simple matter of making an addition to the family Bible or of informing friends and family (*WWC,* p. 234). Several Cather children made slight alterations to their given names: Roscoe Boak Cather changed his middle name to "Clark"; Douglas Cather added another "s" to his first name; James McDonald Cather dropped the "Mc" from his middle name.

But Willa Cather's name changes were more numerous, elaborate, and creative than those of her siblings. Intimately connected with her evolving self-definition as woman and writer, her various namings are the overt signs of a developing myth of self-creation. Although male writers like Twain and Whitman also renamed themselves in the process of evolving an artistic identity and role, the symbolic connection between naming and the search for identity is particularly strong for women in a patriarchal society. Women's inability to trace a matrilineal inheritance through their last names signifies their impotence to name, or define, the self. In controlling her own naming, then, Cather was appropriating the traditionally male power of self-definition.[24]

As we know from Willa Cather's fiction, names held almost magical significance for her. Like Dickens, Melville, and Hawthorne, she frequently chose names for her characters as signs of identity: Lyon Hartwell ("The Namesake"), Bartley Alexander, Alexandra Bergson, Thea Kronborg, Jim Burden, Tom Outland, Godfrey St. Peter, Captain Forrester, Lucy Gayheart, among others, are symbolic and allusive. And in an endnote to *Sapphira and the Slave Girl,* she confessed that some names fascinated her "merely *as* names":

> In this story I have called several of the characters by Frederick County surnames, but in no case have I used the name of a person whom I ever knew or saw. My father and mother, when they came home from Winchester or Capon Springs, often talked about acquaintances whom they had met. The names of those unknown persons sometimes had a lively fascination for me, merely *as* names: Mr. Haymaker, Mr. Bywaters, Mr. Householder, Mr. Tidball, Miss Snap. For some reason I found the name of Mr. Pertleball especially delightful, though I never saw the man who bore it, and to this day I don't know how to spell it. (p. 295)

Even the names that were "merely" names to Cather are strongly imbued with their functional or descriptive origins: we can see in them the haymakers, householders, and dwellers-by-waters. Certainly the names she gave to herself had symbolic resonance.

Because Cather later created her own myth of origins when she claimed that she was named for William Boak, some confusion has surrounded her original naming. She was, as we now know, called "Wilella" after her father's sister, the aunt who died in childhood; this name is on the birth certificate and in the family Bible, although Cather later altered "Wilella" to "Willa." But Wilella was never used, perhaps because her parents thought it too formal. In an 1874 letter, Charles Cather wrote "We call her Willie after our little sister," so evidently the aunt had also been known by that nickname. Her parents also shortened the name to "Willa." "I am sure that my sister was never again referred to as anything but Willa or Willie," Elsie Cather observed, who also recalled that her sister's silver baby mug was engraved "Willa" (*WWC,* p. 234). Thus Cather's later fondness for her childhood nickname "Willie" (which she mentions in letters to Red Cloud friends) and her preference of Willa to Willela were not renamings. These were the names her parents had informally given her, despite the "Wilella" on the birth certificate.

But Willa Cather's choice of middle names *was* a personal statement. Here was space for creativity and self-definition: not given a middle name at birth, she could name the "real" self that belonged to her and not to her family. In childhood she became "Willa Love Cather," naming herself after Dr. Love, the family physician who attended at her birth. Taking his role as well as his name, she gave birth to a part of herself; in doing so, she anticipated her later identification with doctors, the powerful men who rescued her mother from illness. During her university days when she was struggling to escape provincial ignorance, Cather latinized her middle name, claiming the classical scholar's role with "Lova" (*WWC,* p. 235).

Cather's substitution of "Sibert" for Love/Lova was a more significant alteration. Her grandmother Boak's maiden name, it was also the middle name of her soldier–uncle William Seibert Boak. Cather altered the spelling when she adopted it, making it her own. Exactly when she made the transformation is unclear. Edith Lewis asserts that she took on "Sibert" when she had to provide her Red Cloud schoolteacher with a middle name for class records, but Mildred Bennett claims that the change occurred sometime during the 1890s (*WCL*, p. 19; *WWC*, p. 235). In any case, Cather was using the name in print by 1896. Several reviews in the Pittsburgh *Leader* were published with the by-line "Sibert" and by 1900 she was signing stories published in *The Library*, a Pittsburgh magazine, "Willa Sibert Cather." When Knopf became her publisher in 1920, Cather dropped the middle name, only keeping the initial in the monogram on her stationery, "W S C."

Both Cather's adoption and rejection of "Sibert" reflect her development as a writer. At first this link with her uncle gave her name the weightiness a novice writer needed, but once her professional status was assured and she transferred from Houghton Mifflin to Knopf after *My Ántonia* was published, a move reflecting her growing artistic and vocational self-confidence, she could become simply "Willa Cather" on the title page. It is likely, too, that Cather eventually recognized an irony: that her use of her uncle's middle name made her full name sound too feminine. Three-named writers were generally women, and women Willa Cather did not want to associate with—popular authors like Harriet Beecher Stowe, Mary Roberts Rinehart, Louise Chandler Moulton, Clara Louise Burnham, Ella Wheeler Wilcox. When Willa Sibert Cather became Willa Cather, she symbolically left their company.

The adolescent girl's becoming "William" displays an even stronger need to define the self than her middle-name variations, however, because the first name is more intimately connected with personal identity. Since her childhood nickname was "Willie," her choice of William reveals the adolescent's desire to leave childhood and its diminutives behind. One variant of her new name ("William Cather, M.D.") signifies her identification with mentors Will Ducker and the town's two doctors while the second ("William Cather, Jr.") reveals her desire to associate herself with the several Williams in the family.

Elizabeth Sergeant tells us that Cather "considered her name a link to her two grandfathers, William Cather and William Boak" (*WC:AM*, p. 13). But Cather's insistence that she had been named for William Seibert Boak points to a more important identification with her uncle. Cather so staunchly and adamantly promoted this myth of origin that Edith Lewis, confronted with the spectre of Wilella by Leon Edel (who was threatening to include the name in the official biography), could only reinvoke it. "She hated the name Wilella," Lewis told Edel in some distress. "Miss Cather always said she was named after her uncle who was killed in the Civil War."[25] Lewis loyally sustained the fiction in her own memoir, telling her readers that her friend had been named "after a son of Mrs. Boak" (*WCL*, p. 4).

A myth one creates about one's naming and maintains to an intimate friend for over thirty years must have had great significance. The gallant warrior who died fighting for his Southern homeland, William Boak was her most heroic ancestor. The young girl who worshipped Napoleon, considered Emerson her favorite prose writer, and devoured male adventure fiction must have been stirred by his legend. Imagining herself his namesake, she at once created a link between her family past and her individual future and placed herself in a heroic tradition. As the photograph of the adolescent Willa dressed in military garb and soldier's cap suggests, by assuming the warrior's persona she took on his power. *(See photograph section.)*

In a way this ancestor whom she had never met satisfied her need to form identifications with admired adults better than one whom she knew. Like a character in a novel, William Boak did not limit her with his actual presence, so she could create the perfect relationship in fantasy, defining herself as well as William Boak in the process. Even more useful may have been the fact that her uncle was a family hero who died in youth. A completed life of male accomplishment might have been intimidating, or at least have left little room for the girl's imaginative insertion of herself into his narrative. But since William Boak's life was incomplete, Cather could even envision herself finishing the heroic story her uncle had begun to write, as she did in her 1903 poem "The Namesake" ("I'll be winner at the game/ Enough for two who bore the name").

In the poem (1903) and the story of the same name (1907) which Cather derived from this drama of identity and inheritance, the uncle is both the other self and the ancestor, the twin brother and the progenitor. In transforming the uncle into her masculine double, her opposite-sex twin, Cather was in part drawing on her vision of the brother–sister relationship, which in turn may have been influenced by the history of twins in the Cather family. This bond, she later thought, was the most mysterious and intimate one possible between two people. Of the same blood yet a different gender, the brother was the sister's spiritual twin and her masculine self. Cather found such a bond in Wagner's *Die Walküre,* where she was fascinated by the story of the twins Siegmund and Sieglinde, separated at birth and rejoined as siblings and lovers, two halves of one whole.[26] She found another literary depiction of a close brother–sister bond in George Eliot's *The Mill on the Floss:* "I wonder why it is that no one else has ever been so successful in painting the strongest and most satisfactory relation of human life, the love that sometimes exists between a brother and sister. . . . It is more than a tie of blood; much more" (*WP* 1, p. 363).

Willa Cather enjoyed close relationships with her brothers and childhood playmates Roscoe and Douglass throughout her life. She visited them frequently in the summers, traveling to the Southwest, Wyoming, and the Black Hills for hiking and camping trips, welcomed Roscoe's daughters into her Eastern retreats at Bank Street and Grand Manan, and mourned their deaths. These were her most intimate, conflict-free family ties. Yet these real-life relationships do not sufficiently account for her view that the brother–sister bond was "much more" than a tie of kinship.

Although reversing the Romantics' incestuous obsession with the sister, Cather similarly saw the opposite-sex sibling as embodying a part of the self. We can see this in *The Song of the Lark,* where Thea's discovery of self as artist is dramatized by her brilliant performance as Sieglinde. To show her triumph, Cather describes the opera's recognition scene: Sieglinde discovers her identity when she recognizes that the man she loves is Siegmund, the twin brother from whom she has been separated at birth. Hence her discovery of self is carried out in relation, rather than in opposition, to another person who is both the brother and the hidden or repressed part of the self, which can then be incorporated into the whole.

Seeming to signify the difference between the sexes, the brother–sister relationship more profoundly symbolizes, as it eventually did to Willa Cather, men's and women's similarities. The opposite-sex twin or sibling exposed the artificiality of social categories of gender which construct men and women as opposite. Although the brother is male, he shares a common inheritance with the sister, and that bond may symbolically outweigh the greater power and privilege that accrue to men's supposed difference from women.

In the poem and the story she derived from her imagined bond with her uncle, Cather portrays the uncle more as a sibling, the speaker's other self, than as an inhibiting father. In her poem "The Namesake" Cather anticipates Thea's climactic recognition scene by describing the speaker's discovery of self in relationship to an ancestor who is also a twin. Dedicating the poem to "W.S.B., of the thirty-third Virginia," Cather creates a persona who imagines a physical and spiritual correspondence with the fallen uncle. Uncle and nephew/niece are twins who never met, yet the oneness the poem evokes is exhilarating rather than threatening because it leads to the speaker's commitment to adult achievement:

> Proud it is I am to know
> In my veins there still must flow,
> There to burn and bite alway,
> That proud blood you threw away;
> And I'll be winner at the game
> Enough for two who bore the name.
> (*AT*, p. 26)

Although at times Willa Cather feared being an anonymous "cell" in the family bloodstream, here the poem's imagery of a blood tie is sustaining and invigorating. She extends the process of identification in two directions; here the daughter views the progenitor as an extension of herself, as well as the reverse. But when *she* controls this process of identification with someone other than the mother—as the adolescent Willa did when she became William, as the adult writer did when she created fictional characters—she gains autonomy and power, not the defeated return to childhood she grants William in "The Burglar's Christmas."

Cather's short story "The Namesake," published four years after the poem, also celebrates her male inheritance and develops parallels between her imagined bond with William Boak and the creative process. Whereas she leaves the speaker's future achievements vague in the poem, in the story the warrior–uncle is the progenitor of an artist. The story's artist is a sculptor who incarnates the "teeming force" of the American pioneer spirit in his work. He creates several sculptures drawing on Western themes as well as monuments to Civil War heroes. His name, Lyon Hartwell, suggests the artist's natural power and emotional depth and recalls Willa Cather's depiction of herself as the "man-eating tiger" who devoured the pioneer women as she absorbed their stories (*KA*, p. 449).

The story centers on Hartwell's description of the source of his finest sculpture, "Color Sergeant," a "handsome lad in uniform, standing beside a charger." The work had its genesis in the mysterious "feeling of union" he experienced with an uncle for whom he was named, a soldier killed in the Civil War. Although Hartwell never knew him, one day, he explains to a group of friends, he "came to know him as we sometimes do living persons—intimately, in a single moment.'' Having received his inheritance from this return to origins—the gift of self and art through his sympathetic identification with his uncle—the nephew creates his most brilliant piece, the sculpture modeled on his symbolic parent. Simultaneously he becomes the parent who gives birth to himself as artist.

At the time she wrote both ''Namesake'' s, Cather envisioned creativity as a male privilege. Similarly, the adolescent girl's male impersonation signified her search to inherit authority and power from the male line. Yet just as her William Cather performance was in part a drama in which she and her mother were the primary actors, so the bond Cather created with William Seibert Boak reveals a maternal subtext. The Boaks, not the Cathers, were the heroic warriors who fought in the Civil War. Hence the power and glory the adolescent girl wanted for herself were possessions invested in the mother's family. Willa Cather saw this association dramatized in her mother's treasuring the signs of power, her inheritance from her brother: the Confederate flag and the sword (which is also the sign of male power and of brother/sister recognition and union in *Die Walküre*). The mother thus preserved the hero's memory and told his story.

Her mother's bond with her dead, heroic brother was the most dramatic family version of the brother/sister story Cather later found so potent. Since she thought of the brother as the sister's other self, in identifying with her uncle Cather was also connecting to part of her mother, to the powerful ''masculine'' self buried with William Boak. But the daughter was imagining a different version of the brother/sister story from the one her mother experienced. Unlike her mother, Cather would have the power to incorporate the autonomous self, to become both brother and sister, and to develop her mother's undeveloped powers.

In becoming William in adolescence and Willa Sibert Cather a few years later, Cather was thus creating a covert and doubtless unconscious bond with her mother (and perhaps also with her grandmother Boak, since ''Seibert'' was her maiden name). Overtly, however, the adolescent daughter was scorning matrilineal connections. At this point she could only invest a male relative with the identity and role she wanted to receive. Although she abandoned her male impersonation after two years at the University of Nebraska, her William Cather masquerade foreshadows the early stages of her literary apprenticeship when she tried to be the stylistic son of Henry James.[27] Only after she had received the gift of female literary inheritance from Sarah Orne Jewett did Cather rewrite her family mythology, transforming herself from her uncle's son to her aunt's daughter in ''Macon Prairie'' *(ATOP)*.

Cather's male impersonation was an imaginative, daring means of resolving the social and psychological conflicts that adolescence posed. By naming the desires she then believed were male, she preserved them. Yet this was only a partial baptism. In accepting the prevailing definition of male and female identity, Cather was also accepting the culture's polarization of ''masculine'' and ''feminine.'' Hence this was a period of fragmentation rather than integration, of conflict rather than acceptance. She had yet to learn all the lessons she would derive from the brother–sister bond: that the sister had a ''masculine'' self, that she and her brother enjoyed a common heritage. The girl became William not because she thought she was male but because she did not want to be female, and that choice suggests self-contempt as well as self-expression, a potentially debilitating self-denial. As long as Cather devalued women, she devalued herself; as long as she devalued herself, she could not commit herself fully to writing. In reconciling the seemingly contradictory identities "woman" and "writer," she would ultimately challenge and revise social definitions of gender, the process we trace in ensuing chapters.

•

In later comments and interviews the adult writer always stressed childhood and adolescence as crucial stages in her artistic formation. In her most frequently quoted interview she declared that the artist acquires subject and theme "under fifteen years of age," while at other times she extended the writer's formative period to twenty.[28] Although such observations highlight the importance of her emigration to Nebraska, the conflict she experienced in moving from childhood to adolescence marked her literary imagination as strongly as did the transition from Virginia to the Divide.

Cather's playing with a male identity in adolescence illuminates her fictional preoccupation with childhood. Linked with her veneration of what she termed the "precious, the incommunicable past" in *My Ántonia* (p. 372), childhood in her work is frequently portrayed as an idyllic state to which adult protagonists long to return. Like many other American writers, Cather was influenced by the romantic and pastoral vision of Edenic childhood as a metaphor for the American past, an association she develops most fully in *My Ántonia* and *A Lost Lady* where childhood is equated with the nation's lost innocence. But she also viewed childhood nostalgically as a period of freedom and unselfconscious wholeness when sisters were both equal to and indistinguishable from their brothers. In her fiction, this Edenic state is lost when the "hateful distinction" between the sexes arrives and the adolescent must enter the fallen world of adult sex roles and repress the part of the self then assigned to the other gender. The Fall—her metaphor for the loss of childhood—is thus a descent from unity to fragmentation as well as from freedom to restriction.

Cather portrays this downward movement most strongly in "The Treasure of Far Island" (1902), an early story portraying the loss of childhood androgyny. As children, Douglass Burnham and his tomboy playmate Margie possess traits supposedly inherent in the other sex: Douglass has a feminine love of beauty and adornment, while Margie is a fiery "wild Indian" who singlehandedly fights the boys who taunt her. But this idyllic, playful period ends when they enter the fallen world in adolescence and accept male dominance and female submission. The adult Douglass reflects on their loss:

> That night, after our boat had drifted away from us, when we had to wade down the river hand in hand, we two, and the noises and the coldness of the water frightened us, and there were quicksands and sharp rocks and deep holes to shun, and terrible things lurking in the woods on the shore, you cried in a different way from the way you sometimes cried when you hurt yourself, and I found that I loved you afraid better than I had ever loved you fearless, and in that moment

> we grew up, and shut the gates of Eden behind us, and our empire was at an end.

The treacherous landscape of "sharp rocks" and "deep holes" and "terrible things lurking in the woods" suggests a connection between puberty and sexual fears, but the passage even more explicitly equates the loss of innocence with the children's acceptance of traditional sex roles; the tomboy Margie, who never "stood in need of masculine protection," becomes timid and fearful and Douglass discovers that he finds her dependence attractive (*CSF,* pp. 265–82). Yet the Fall can be partly redeemed. As an artist, Douglass (whose name Cather borrowed from her playmate brother) is a case of "arrested development" who continues the child's unified self and imaginative power into adult life, retaining access to childhood's androgynous Eden.

As a novelist who could enter imaginatively into both male and female selves and create characters who combine traits supposedly divided between the sexes, Cather also eventually recaptured the unity she seemed to have lost in adolescence. Her frequent comparison of her work to play suggests how, in the creative process, the novelist managed to evade both gender distinctions and oppressive social rules. In writing she regained what she termed childhood's "kingdom of lost delight . . . that kingdom where we could be Caesar or Napoleon at will," a realm interchangeable for the adult writer with her "kingdom of art" (*KA,* pp. 337, 417).

•

Willa Cather concluded her Red Cloud years with a fittingly symbolic event when she delivered her high school graduation speech at the Red Cloud Opera House. Her William Cather masquerade culminated in this farewell performance as she faced her largest audience in the space set aside for drama, entertainment, and community rituals. The Opera House is the last and most public of the stages where she dramatized herself in childhood and adolescence—the Virginia parlor where she became a "dang'ous nigger," the Red Cloud parlor where she presented amateur theatricals with the Miner girls, and the town's streets and public spaces where she presented herself to her audience as her twin brother (*WCL,* p. 13). A spirited defense of her medical experimentation and a searing attack on her narrowminded critics, the speech was still remembered in the 1950s when E. K. Brown, researching his biography of Cather, found it described as a "startling dissonant note in a conventional program."[29] In fact the speech was both dissonant and conventional, for the young orator was trapped by the social categories of gender she seemed to be rejecting.

The high school commencement address—then as now—is a form whose content is structured by the social context in which it is delivered. Facing an audience of parents, neighbors, and teachers, the young orator generally reconfirms the values the listeners expect to hear. Cather spoke after her fellow graduates, both young men, had followed the rules of the genre. The message of Willa Cather's fellow-graduates must have been warmly received in their business-oriented community. The first extolled the virtues of "Self-Advertising," assuring his audience that self-promotion was a "great boon to those men who think that advertising doesn't pay." The second took as his subject "New Times Demand New Measures and New Men." In this seemingly forward-looking address he articulated American values dating back to Franklin's *Autobiography:* aspiring entrepreneurs should devote themselves to their studies, he proposed, because a good education would enable the businessman to "improve the opportunity" and increase his fortunes. Seemingly individualistic, these aspiring self-made men were simply following the scripts their society had written for such ambitious young men.[30]

In her speech, however, Willa Cather seemed to throw away the scripts and to reject Red Cloud's definition of male as well as female roles. She was dressed as a boy, yet she scorned the pragmatic values the preceding speakers were promoting. Unlike her colleagues she seemed to be making her own rules, using the stage—the realm of the actors she venerated—to present a challenge to male as well as female members of the community.

Cather chose to defend her controversial experiments with animals. Entitled "Superstition vs. Investigation," the speech supported the investigator's right to explore natural phenomena freely. Contending that the forces of superstition and empiricism had been in conflict throughout history, Cather endorsed science. Local critics could scarcely have missed the self-reference when she termed scientific experimentation the "hope of our age . . . and yet upon every hand we hear the objections to its pursuit. The boy who spends his time among the stones and flowers is a trifler, and if he tries with bumbling attempt to pierce the mystery of animal life he is cruel." Restrictions on amateur experimentation would derail scientific progress, she contended, for "If we bar our novices from advancement, whence shall come our experts?"[31]

Although Cather was defending her iconoclastic behavior and defying her neighbors' petty values, she was not being as individualistic as she might have seemed to her audience. The young orator placed herself in a tradition of experimenters who became respected authorities: Bacon, Newton, and Harvey. Claiming authority by declaring the speaker the descendant of powerful forefathers, Cather's speech suggests why her male impersonation would continue, although in subtler forms, for several years. Some men, perhaps most men in her immediate social world, might be soulless businessman and inarticulate conformists. But the great scientists with whom she aligned herself were, like her fallen and heroic uncle, male. To inherit this tradition, it then seemed, she had to be a son rather than a daughter.

NOTES

1. Elmer Alonzo Thomas, *80 Years in Webster County* (Hastings, Neb.: privately published, 1953), pp. 112–14.

2. Conversation with proprietor of the Royal Hotel, Red Cloud, July 1974.

3. Willa Cather to Mr. Bain, January 14, 1931, Beineke Library, Yale University, New Haven, Conn.

4. Tullio Verdi, M.D., *Mothers and Daughters: Practical Studies* (New York: 1877), p. 31; Mary Virginia Terhune, *Eve's Daughters: Or, Common Sense for Maid, Wife, and*

Mother (New York: 1882), p. 46. For a fuller discussion of the views Victorian advice-givers held of tomboyism, see Sharon O'Brien, "Tomboyism and Adolescent Conflict: Three Nineteenth-Century Case Studies," in *Woman's Being, Woman's Place: Female Identity and Vocation in American History,* ed. Mary Kelley (Boston: G. K. Hall, 1979), pp. 351–72.

5. J. H. Kellogg, M.D., *Ladies' Guide in Health and Disease* (Des Moines: 1883), p. 188; Verdi, p. 74.

6. See Carroll Smith-Rosenberg, "Puberty to Menopause: The Cycle of Femininity in Nineteenth-Century America," in *Clio's Consciousness Raised,* ed. Mary Hartman and Lois Banner (New York: Harper-Torchbooks, 1974), pp. 23–29. The medical literature, Smith-Rosenberg concludes, implies that female adolescence was a traumatic period that required an "often painful restructuring of intrafamilial and social identities" (p. 28). Her argument is based on medical literature, however, not on case studies of adolescent girls.

7. "Girls," *Red Cloud Chief,* February 13, 1885, p. 1; "Female Physicians," *Webster County Argus,* September 29, 1887, p. 5; "Girls That Are Lovable," *Webster County Argus,* March 24, 1887, p. 3.

8. Louisa May Alcott, *Little Women* (Boston: 1868; rpt. New York: Airmont, 1966), p. 17.

9. *Webster County Argus,* April 13, 1888, p. 3. For representative listings of the social activities of Red Cloud's genteel women, see the "Local Department" of the *Argus,* especially August 6 and 20, 1885, p. 3.

10. Frances Willard, *Glimpses of Fifty Years: The Autobiography of an American Woman* (Chicago: 1889; rpt. Source Book Press, 1970), p. 69; *Louisa May Alcott: Her Life, Letters and Journals,* ed. Ednah D. Cheney (Boston: Little Brown, 1928), p. 20. Referring to her tomboy childhood, Alcott went on to say: "No boy could be my friend till I had beaten him in a race, and no girl if she refused to climb trees, leap fences, and be a tomboy" (p. 20).

11. In her 1898 article "The Woman Who Wants to Be a Man," Anna Mearkle could have been speaking for the adolescent Cather: "The woman who wants to be a man—what is it that she really wants? . . . She wants to be what she may be and ought to be, a fully developed human being . . . not to be a male. It is man who keeps insisting on the distinction of sex,—woman would willingly forget it" (*Midland Monthly* 9 [1898]: 176).

12. "Nature," *Selections from Ralph Waldo Emerson,* ed. Stephen E. Whicher (Boston: Houghton Mifflin [Riverside ed.], 1957), p. 56.

13. Quoted in Jonathan Katz, *Gay American History: Lesbians and Gay Men in the U.S.A.* (New York: Crowell, 1976), p. 256. For a fuller account of female transvestism and male impersonation in American history, see Katz' chapter "Passing Women: 1782–1920" (pp. 209–79). Susan Gubar analyzes the connections between female cross-dressing and modernist women writers' subversion of patriarchal categories of gender in "Blessings in Disguise: Cross-Dressing as Re-Dressing for Female Modernists," *Massachusetts Review* 22, no. 3 (Autumn 1981): 477–508.

14. Cather also suggests that the appearance of demure femininity is a calculated role in "The Profile," where the duplicitous Virginia Gilbert acts the part of the fairy-tale heroine; Alcott does the same in *Behind the Mask.* In American literature, as in our society, performance, clothing, and identity are interrelated for men as well as women. In Franklin's *Autobiography* and its popular literary descendants, the Horatio Alger stories, dress is the all-important sign of class, respectability, and status in a supposedly egalitarian and classless society. Consequently the crucial moment in these stories occurs when the ragged street urchin is given a new suit, the first step toward upward mobility.

15. On the social and symbolic significance of Victorian women's dress, see Helene E. Roberts, "The Exquisite Slave: The Role of Clothes in the Making of the Victorian Woman," *Signs: Journal of Women in Culture and Society* 2, no. 3 (Spring 1977): 554–69, and Lois Banner, *American Beauty* (New York: Knopf, 1983).

16. Edith Wharton, *A Backward Glance* (New York: Appleton, 1934), p. 20. Future references in the text will be to this edition.

17. *Eve's Daughters,* p. 76.

18. Elizabeth Fifer, "Is Flesh Advisable? The Interior Theatre of Gertrude Stein," *Signs: Journal of Women in Culture and Society* 4, no. 3 (Spring 1979): 472–83.

19. Simone de Beauvoir, *The Second Sex,* translated by H. M. Parshley (New York: Knopf, 1952; Vintage ed., 1974), pp. 376–77.

20. Nancy Chodorow, *The Reproduction of Mothering: Psychoanalysis and the Sociology of Gender* (Berkeley: University of California Press, 1978). See Chapter 8, "Oedipal Resolution and Adolescent Reply," esp. pp. 138, 137.

21. Nancy Chodorow, "Mothering, Object-Relations, and the Female Oedipal Configuration," *Feminist Studies* 4, no. 1 (February 1978): 137–58, esp. 151.

22. *The Reproduction of Mothering,* p. 140.

23. Thomas, p. 112. In "Old Mrs. Harris," Victoria Templeton is similarly criticized by disapproving neighbors who view her as prideful and selfish because her mother performs most of the household tasks. A "belle" in Tennessee, Victoria, now in the Midwest, is "not very popular, no matter how many dresses she [wears]"; the neighbors seem to "take sides against her," and she suffers subtle "thrusts from the outside which she [cannot] understand" (*OD,* pp. 129, 128).

24. For an insightful discussion of the connection between Twain's and Whitman's name-changing and their emerging artistic identities, see Justin Kaplan, "The Naked Self and Other Problems" in *Telling Lives: The Biographer's Art,* ed. Marc Pachter (Washington, D.C.: New Republic Books/ National Portrait Gallery, 1980), pp. 36–55.

25. Leon Edel, "Homage to Willa Cather," *The Art of Willa Cather,* ed. Bernice Slote and Virginia Faulkner (Lincoln: University of Nebraska Press, 1974), p. 193. Cather also disliked the name "Willa," preferring her Nebraska nickname "Willie" throughout her life. She would have changed her name when she first began to write, she said later, if she had known she was going to write very much. (Willa Cather to Annie Pavelka, May 19, 1936, Willa Cather Pioneer Memorial, Red Cloud, Neb.; Willa Cather to Mr. Johnson, n.d. [1928], Beineke Library, Yale University, New Haven, Conn.) In her letter to Johnson she admits, however, that a name change would have hurt her mother's feelings.

26. In reviewing a performance of *Die Walküre* Cather associates the sword, power, brother/sister recognition, and incestuous love: "Sieglinde enters, and seeing the sword in his hand knows that her deliverer has come," while Siegmund "knows that this woman is his sister and his bride. The scene which follows is probably the most exalted love scene ever set to music" (*WP* 2, pp. 623–26). In "The Garden Lodge," Caroline Noble's love for Raymond D'Esquerré, the man who both awakens and represents her repressed passions, momentarily surfaces when she plays Sieglinde to his Seigmund and he sings "Thou art the Spring for which I sighed in Winter's cold embraces," the "most exalted" lover's aria from *Die Walküre* (*CSF,* p. 195). In at least one instance Cather reworked a real-life experience with one of her brothers into a heterosexual love story. Her descent from a windmill with her brother Roscoe during a summer storm was transformed into one of the most erotic episodes in her fiction in "Eric Hermannson's Soul" (1900), where the male lover, as in "The Garden Lodge" and *Die Walküre,* both signifies and awakens the woman's passionate self. For a description of this episode, see *WWC,* p. 33.

27. Sandra M. Gilbert analyzes the literal and metaphorical significance of gender-coded clothing in "Costumes of the Mind: Transvestism as Metaphor in Modern Literature," *Critical Inquiry* 7, no. 2 (Winter 1980): 391–417.

28. E. K. Brown, *Willa Cather: A Critical Biography,* completed by Leon Edel (New York: Knopf, 1953), p. 3.

29. Brown, p. 46. Willa Cather's graduating class consisted of three students, and all three gave speeches. She may have been fortunate in attending a new public school, for as Barbara Solomon points out "the basic contradiction between women's education and their probable futures became obvious at their school commencements. . . . If ever a girl forgot her place, at graduation she was reminded, for usually young women were not allowed to read their parts at this public event; rather, their speeches were read by adult substitutes or were delivered personally at a private session" (*In the Company of Educated Women: A History of Women and Higher Education in America* [New Haven: Yale University Press, 1985], p. 28). As Solomon also observes, "being an orator was an important expression of masculinity" (p. 29). Although distressed by Cather's male impersonation, by allowing her to speak in public the Red Cloud community was momentarily granting her the recognition accorded men in Victorian society.

30. Brown, p. 46.

31. The text of Cather's graduation speech can be found in Brown, pp. 43–46.

6

Divine Femininity and Unnatural Love

> What a girl she is anyway! What simplicity and elegance. How the slenderness of the waist, the pose of the head, the carriage, the firmness of the flesh, even the severe elegance of the dress with its little cloud of wonderfully painted illusion about the breast combine to suggest that innate aristocratic refinement which should be the outcome of wealth and culture.
>
> *The World and the Parish*

> What a shame that feminine friendship should be unnatural.
>
> Willa Cather to Louise Pound, June 15, 1892

Unlike the nine-year-old child uprooted from Virginia's pastoral landscape, the young woman who left Red Cloud in September 1890 for Lincoln and the University of Nebraska chose to transplant herself. Reversing the direction her parents had taken, Cather made the first stop in her pilgrimage to the Eastern centers of art and culture where she later settled: Pittsburgh, Boston, New York.

In describing Thea Kronborg's exhilarating departure from her provincial Colorado home for Chicago in *The Song of the Lark* (1915), Cather gives her eastward-bound heroine a sense of completeness and self-sufficiency: ''Everything that was essential seemed to be right there in the car with her. She lacked nothing'' (*SL*, p. 157). But at sixteen, Cather was not as confident as she later imagined her fictional alter ego to be. Although she struck her college classmates as aloof and arrogant, Cather's forceful manner masked insecurities and conflicts. As a country-educated student entering the university, Cather sensed, as Edith Lewis later observed, a ''whole continent of ignorance surrounding her in every direction, like the flat land itself; separating her from everything she admired, everything she longed for and wanted to become'' (*WCL*, p. 28). Years later, when Cather was traveling in sophisticated Eastern cultural circles, she still felt insecure because of her prairie background, embarrassed by her ''Boeotian ignorance'' when Mrs. Fields discovered that she did not know John Donne's poetry, and intimidated by Dorothy Canfield's cosmopolitan self-assurance and fluent French when the two traveled in France (*NUF*, p. 65).[1]

During her years at the University of Nebraska Cather continued to devour American, English, Continental, and classical literature. Yet as she became increasingly familiar with the Western literary tradition, she felt simultaneously initiated and excluded. She had many reasons for sensing herself an outsider: as a woman, she was separated by gender from the authors of the works she fervently admired; as an American, she suffered from the same sense of cultural inferiority that distressed such male compatriots as Henry James and T. S. Eliot; and as a middle-class Nebraskan she lacked the aristocratic Bostonian's sense of cultural privilege which she later associated with Mrs. Fields and her Beacon Hill drawing room. The girl who needed to change her middle name from "Love" to "Lova" in order to signify her inclusion in a classical heritage was not yet the self-assured daughter of Virgil she had become by 1918 when she chose her Latin epigraph to *My Ántonia*.

Even though Cather later felt her Lincoln years had not fully dispelled her prairie crudeness, the town and the university offered her intellectual and aesthetic enrichment. Only twenty years old in 1890, Lincoln was settled by Easterners committed to preserving social and cultural amenities in their new environment. Bernice Slote describes Lincoln's "transplanted culture" in the late nineteenth century:

> And among the firstcomers were families of education or money or both, bringing along their tapestries and oriental rugs, their fine linen and china, pianos and libraries. . . . And along with smaller boxes were great houses of brick or wood in chaste and classical lines, or spun wonders of Victorian Gothic. They moved in, took out their white kid gloves, subscribed to *Century*, shipped in oysters frozen in blocks of ice, and tried to keep life very much as it had been in Ohio, New York, Illinois, or Virginia. (*KA*, p. 7)

The emigrants brought their love for professional theater along with their dress clothes and opera glasses, and because Lincoln was a regular stop on East/West rail lines it became a major center for touring companies, with two bustling theaters that hosted a hundred dramatic companies a year (*KA*, p. 7). Drawn to the stage as she had been in Red Cloud, during her college years Cather could see some of the best-known actors of the day: "Bernhardt in Sardou's *La Sorcière*, Modjeska in *Antony and Cleopatra*, Julia Marlowe in *The Love Chase*, Mary Shaw in *Ghosts*, Richard Mansfield and Mrs. Patrick Campbell, all played in Lincoln in the 1890's," remembers Edith Lewis, a Lincoln native (*WCL*, p. 37).

Like other Midwestern land-grant universities founded during the 1870s and 1880s, the University of Nebraska was an egalitarian, co-educational, venture. The faculty was a heterogeneous group, drawn from Eastern and European backgrounds, while the students were talented and serious, many of them sons and daughters of farmers and shopkeepers who had to make personal and financial sacrifices to send their children to college. Some of Cather's Nebraska classmates and colleagues also went on to distinguished futures: Roscoe and Louise Pound, Dorothy Canfield, John Pershing, Alvin Johnson (later head of the New School for Social Research).[12] Cather enrolled in the preparatory school as a "second prep"; as the graduate of an untried small-town high school, she required a year of additional work before matriculating as a freshman in 1891.

Perhaps most important to Cather's literary development was the institution's most intangible asset: this was an intellectual community where people cared about, and fervently discussed, ideas. The central student organizations were the debating and literary societies where student orators presented papers and argued the merits of poets, artists, and writers. It's hard to imagine that at the University of Nebraska today a student could gain Willa Cather's "sudden elevation to fame" by following her route: impressed by her essay on Carlyle, one of her professors had it printed in the *Nebraska State Journal* and soon everyone was talking about it (*WCL,* p. 31). A vital interest in literary opinions was not restricted to faculty and students. According to Edith Lewis, during the early 1890s controversy "raged" in Lincoln over one of the English professor's teaching methods, and for a time literature was "as live an issue in the town as politics or social gossip" (*WCL,* p. 35).[3]

After arriving in Lincoln, Cather settled herself in a boardinghouse, reportedly the "best-eating place in town" (*WWC,* p. 215). She threw herself into the scholar's life, rising at five to begin studying. Still intending to be a surgeon, during her first year she took science courses and signed her letters home "William Cather, M.D." Pursuing her strong interest in the classics, she enrolled in Greek and Latin courses and decorated her room, as would Jim Burden, with a map of Rome. Later Cather viewed her youthful intellectual zeal with tolerant amusement. Writing Mariel Gere in 1896, she wondered how her friend could have put up with her while she was affecting the scholar's role and spouting Greek, and in the 1920s she humorously reminded Dorothy Canfield of her failure to understand why Canfield could have preferred such frivolous activities as party-going to translating Victor Hugo with her.[4] But at the time she took herself seriously. Her courses in the classics were particularly sacred, but she was also reading a wide range of contemporary literature. By the spring of 1891 she had begun to transfer her professional aspirations from medicine to writing.

Cather soon became involved with several campus publications. Along with Louise Pound she edited *The Lasso,* a short-lived literary magazine, was first associate and then managing editor of *The Hesperian,* a more enduring student venture, and finally became literary editor of the class yearbook. Eager to master the spoken as well as the written word and enter the masculine realm of oratory, Cather joined the university's literary and debating societies where she gained a reputation for articulate invective or "roasting" (*KA,* p. 21). College dramatic productions absorbed some of her time, and she continued to refine her William Cather act by taking male roles.

After establishing her reputation as a college journalist, in her last two years at the University she gradually took more and more time away from her studies to write for the Lincoln papers. She contributed a column on the arts called "The Passing Show" to the *Journal,* reviewed the plays brought by the touring companies, and placed articles in the Lincoln *Courier,* a weekly paper. In all, Cather churned out over three hundred columns and reviews and became a respected professional writer and critic while still in college. One observer of the local media placed her at the "head of the Lincoln writers," and in 1895, Elia W. Peattie, a nationally known author and columnist for the *Omaha World-Herald,* prophesied that Cather was "destined to win a reputation for herself" as a journalist (*KA,* pp. 26–27).

As in Red Cloud, Cather was helped by male mentors who recognized her talent. Ebenezer Hunt and Herbert Bates, members of the English department, encouraged her to write essays and short stories; Charles Gere, publisher of the *Journal* and father of her good friend Mariel Gere, supervised her newspaper career, aided by the *Journal*'s managing editor Will Owen Jones; and Dr. Julius Tyndale, drama critic for the *Evening News,* helped her to learn the reviewer's craft. The support Cather received for her intellectual and professional ambitions from the university and professional community sharply contrasts with the experience of most women writers born in the nineteenth century. As a journalist Cather could develop her writing skills and professional ambitions in the same training ground that fostered male writers of her own and later literary generations—Harold Frederic, Stephen Crane, Theodore Dreiser, Ernest Hemingway. At the same age when Edith Wharton was selecting a pretty dress in hopes of attracting Henry James' attention at a dinner party, Cather was working with Charles Gere and Will Owen Jones as a colleague. Even though Lincoln later seemed to her an unsophisticated backwater compared to New York or London, the city and the university were kind to her ambitions.

During her college years Cather also formed close friendships with other women which provided a sustaining network after her second uprooting. Mrs. Westermann, the matriarch of a large family whose home was a refuge of European civility and warmth, took Cather in as another child; later she would be the model for Mrs. Ehrlich in *One of Ours,* a hearty, generous woman who creates a nurturing and gracious atmosphere in her German household. Mrs. Gere, another "woman of great charm and warmth of hospitality," treated Cather "like one of her own daughters" (*WCL,* p. 38). This attachment also endured; after Cather left Lincoln for Pittsburgh, she frequently wrote Mrs. Gere when she was homesick or lonely. Other close friends were young women: Mariel Gere, a classmate and trusted, sisterly confidante; Louise Pound, another Nebraska student and Cather's romantic love for her first years in Lincoln, who later gained a doctorate from the University of Heidelberg; Dorothy Canfield, the schoolgirl daughter of University Chancellor James Canfield, six years younger than Cather, who also earned a doctorate before becoming a Pulitzer Prize–winning novelist and well-known literary figure. Cather formed more peripheral ties with the actresses whose touring companies stopped in Lincoln, reveling even in momentary friendships with the women who embodied the theater's magical power.

•

Since Cather had escaped from Red Cloud's ladies' clubs and sewing circles to an emancipated milieu where women like Dorothy Canfield and Louise Pound were interested in more than "babies and salads," as she described limited feminine preoccupations in "Tommy, the Unsentimental" (1896), we might have expected that she would abandon male identification shortly after arriving in Lincoln (*CSF,* p. 474). But Cather maintained her William Cather pose for her first two years at the university before she modified her male persona by letting her hair grow. Even then she continued to dress mannishly in suits and ties, shirtwaists, and starched shirts with cuffs. Cather used to leave her coat in the men's cloakroom, recalled Mariel Gere, and people meeting her for the first time often found her freakish. Another classmate described other students' initial response to this disturbingly androgynous figure:

> While the students were sitting in the classroom waiting for the instructor to arrive, the door opened and a head appeared with short hair and a straw hat. A masculine voice inquired if this were the beginning Greek class, and when someone said it was, the body attached to the head and hat opened the door wider and came in. The masculine head and voice were attached to a girl's body and skirts. The entire class laughed, but Willa Cather, apparently unperturbed, took her seat and joined the waiting students.[5]

In 1948 her Nebraska classmates were asked to record their memories of their famous colleague. Almost every respondent mentioned her rejection of feminine dress and manners. "She was the first girl that I ever saw in suspenders," wrote one, still marveling after fifty years, while another mentioned her "dark man-tailored suits" with stiff collars, string ties, and four-in-hand cuffs. "I never remember her wearing a dress," one commented, although another remembered her with a "sober, mannish-cut dress" and a third recalled her fondness for a "starched shirtwaist" and a "comparatively short skirt, both of which were somewhat uncommon in those days."

Cather's classmates did not view her as mannish solely because of her unorthodox dress. Their references to her fondness for ties and suspenders are mingled with mentions of personality traits socially defined as masculine: "assertive," "energetic," "outspoken," "individualistic," "superior," "independent," "forceful," "strong," "self-confident," "brilliant," and "egotistical" as well as "mannish" and "boy-like." Cather simply did not like "girls' ways or manners," remembered one classmate. In conversation, she "affected the slang expressions used by the boys" and claimed that her name "was the feminine of William." The nickname she then preferred was "Billy": "Willa was just plain Billy to all of us."[6]

That Cather's desire to be "just plain Billy" arose from her worship of the power and force considered masculine in her society can best be seen in her undergraduate views of football. The sport's violence and aggressiveness were delighting some Americans and shocking others during the 1890s, and Cather joined those who applauded football's hard-hitting drive. "It makes one exceedingly weary to hear people object to football because it is brutal," she wrote impatiently in 1893. "Of course it is brutal" (*KA*, p. 212). Indeed its brutality constituted its appeal for Cather, who associated the sport's "manly" virtue with its violence:

> Athletics are the one resisting force that curbs the growing tendencies toward effeminacy so prevalent in the eastern colleges. Football is the deadliest foe that chappieism has. It is a game of blood and muscle and fresh air. It renders distasteful the maudlin, trivial dissipations that sap the energies of the youth of the wealthier classes. . . . It doesn't do Cholly or Fweddy any harm to have his collar bone smashed occasionally. . . . The field is the only place that some young men ever know anything of the rough and tumble of life. Like the fagging system at Eton it is good because it lays the mighty low and brings down them

> which were exalted. Neither his bank book nor his visiting list can help a man on the eleven, he has nothing to back him but his arm and his head, and his life is no better than any other man's.(*KA*, p. 213)

In addition to revealing the middle-class, Midwesterner's contempt for decadent Eastern aristocrats and a democratic preference for red blood to blue, this passage shows Cather's association of male power with primitive force, combat, and violence. To her, the gridiron was another battlefield where warriors like her namesake uncle William Boak attained manhood by inflicting and withstanding physical pain and injury. Seeing "manliness" established in a sport where there were winners and losers, vanquishers and vanquished, Cather was at this point accepting a model of masculine identity based on dominance and submission. Hence her undergraduate love of football can be seen as a logical outgrowth of her childhood worship of the conquerers, generals, and heroes of adventure stories who mastered their worlds. Many of her contemporaries thought football would prepare young men to compete in a Darwinian capitalist economy. As historian Jackson Lears has argued, the late nineteenth-century "cult of strenuosity" and worship of force manifested, in part, in the popularity of football and other sports "reinforced bourgeois values of discipline and productivity" and instilled traits "which would later prove useful in a business career."[7] But Cather saw the game as a link with an epic past, not a commercial future. "It is one of the few survivals of the heroic," she thought, of "the old stubborn strength that goes clear back to the days of the Norman conquest" (*KA*, p. 213).[8]

Later Cather preferred a more androgynous man. The admirable male characters in her fiction—Carl Linstrum, Godfrey St. Peter, Bishop Latour—possess refined intellects and aesthetic sensibilities rather than red-blooded manly vigor and athletic prowess, and the muscular male character who could have smashed collar bones with ease—*A Lost Lady*'s feral Frank Ellinger—is a virile-seeming coward. By the time she wrote this fiction, Cather had freed herself from prevailing definitions of masculinity as well as femininity, necessarily related processes since these social constructs were related through their opposition. Cather's questioning of gender categories begins in the late 1890s and early 1900s; the author of "Paul's Case" (1905), which sympathetically explores the plight of a sensitive, aesthetic young man, who reveals "effeminacy" if not homosexuality, had obviously begun to shed the values held by the football-loving college student. But during her undergraduate years, men like these later protagonists would have struck her as effeminate aesthetes in need of a good workout with the Nebraska eleven.

•

Failing to question the culture's polarization of gender, "Billy Cather" could reject the female role she found limiting only by continuing to repudiate her sex. Her need to differentiate herself from the mass of ordinary females is evident in her college journalism, where she frequently expressed her contempt for women in tones ranging from amused dismissal to bitter condemnation.

One classmate later referred to Cather's "cruel, cynical, unjust and prejudiced criticism of everyone on campus" as reflected in her college writings.[9] Perhaps he had in mind her "Daily Dialogues, Or, Cloak Room Conversations as Overheard by the Tired Listener," an acerbic satire on sorority girls. In her sketch the sorority "sisters" foster not female community but competitiveness for male attention as two seductive recruiters from "Gamma Gamma Lambda" (named "Calypso" and "Circe") try to ensnare a desirable new girl to gain higher status with the local fraternity than their rival sorority. The sketch suggests an additional motive for Cather's fondness for plain, tailored, "mannish" dress: her perception that elaborate female dress was ornamental plumage designed to increase a woman's sexual appeal and marketability. As expert in parodying fancy female clothing as her mother was in wearing it, Cather has one of her recruiters create a verbal portrait of the dress a Gamma girl wore to a fraternity party as bait for the unaligned candidate: "Pale green surah, with canary colored trimmings, and old gold sash, made Empire, with Medici collar, cut V neck, and, *en train,* Watteau pleat in the back, and *jabot* of real lace shirred down the front in cascades." The superficiality Cather attributes to the sorority girls extends to their supposedly intellectual "literary" meetings where members consume "dainty refreshments" and read papers on topics like "Incident in My Summer Vacation."[10]

The literary clubs that were proliferating in Lincoln as elsewhere in America during the 1890s impressed Cather only as sororities for older women. She frequently ridiculed the Lincoln clubs' pretentious claims to foster female friendships and intellectual enrichment. "In the first place women have no particular talent for good fellowship," she charged, ignoring the evidence of her own friendships in her urge to castigate ordinary women. Slaves to domesticity, women could not converse intelligently when they congregated because they could not "leave their family affairs behind them" or "resist declaiming upon the faults of their last maid or the high marks their daughters get at the high school or university." The "intellectual female" was described with particular derision: she could be glimpsed "haunting the public libraries," Cather informed her readers, "stretching the seams of her best black silk, handling massive volumes and writing unreadable notes with her kid gloves on." And once a week she and her literary friends would meet and "mingle the 'glories that were Greece and the grandeur that was Rome' with tea and muffins and Saratoga chips." A recent initiate into the classics' high seriousness, Cather resented the ladies' assumption that knowledge could be graciously nibbled along with light refreshments: that notion, like the clubs, she found "just a little ludicrous" (*KA,* pp. 179–80).

The *Ladies' Home Journal* was Cather's symbol of the ladies' club mentality. Trivializing high culture to suit the feminine mind, the magazine mixed palatable samples of art and literature with articles directed toward women's narrow domestic interests, Cather complained, cheerfully juxtaposing breathless praise of Shakespeare and Dante with revelations that "bonnets and elaborately trimmed toaks [toques] will be worn for the theatre this winter," unabashedly following an article on Beethoven's sonatas with one telling readers "whether it is proper to kiss a young man good night after returning from a party." Such debasement of literary and artistic subjects was a feminine trait, Cather assumed: "There have been foolish men like Stevenson and Thackeray and George Meredith," she noted ironically, "who considered literature a craft and an art." But the *Journal* writers, their eye on the undiscriminating female reader, could just "blossom out into literature off-hand" (*KA*, pp. 187–88).

Feminine traits praised by upholders of the cult of true womanhood were Cather's favorite targets of abuse in her college journalism. If domesticity made women unfit for fellowship, piety and purity made them unfit for art. Women's upholding of middle-class morality was proof to Cather not of their superiority but of their Philistinism. Art could not be judged, she thought, by the narrow, repressive moral codes women enforced. In one article she raged at Lady Isabel Burton's destruction of her husband's translation of *The Scented Garden,* a collection of erotic Persian poetry and the "most cherished of his manuscripts." She took the occasion to generalize about male and female natures:

> The amusing womanishness of the action was almost enough to compensate for the very great loss that English literature sustained in the destruction of so valuable a work. A man would have felt in duty bound to publish the work which had cost his friend so many years of unremitting labor, and fit or unfit would have sooner cut off his own right hand than destroy so great a monument of man's scholarship.

Cather was sure that Lady Burton's act was representative of her sex. "How exceedingly like a wife and like a woman" it was, she reflected, to sacrifice the poetry of Persia for "home-and-fireside respectability" (*KA*, pp. 185–86).

Although Cather's denunciation of culture-hungry clubwomen and female prudes may have struck her Lincoln readers as boldly unorthodox, she was betraying her adherence to conventional wisdom when she considered women a sex with common, innate characteristics. Seemingly iconoclastic in her views as in her dress, Cather was being rebellious only in condemning what others praised. Unlike such feminist writers of the 1890s as Charlotte Perkins Gilman, Mary Wilkins Freeman, and Kate Chopin, she did not question whether social institutions and expectations might have defined and confined women. Just a few years after Cather ridiculed women's supposedly inherent liabilities, in "The Yellow Wallpaper" (1899) Charlotte Perkins Gilman gave a harrowing account of a woman's mental breakdown. Gilman attributes her unnamed protagonist's disintegration to a confining, male-imposed social role that forbade autonomy and stifled creativity, a cultural imprisonment symbolized by the nursery with barred windows in which she is incarcerated. Had Cather then possessed Gilman's insight into the social forces limiting women's avenues of expression and creativity, she might have found male impersonation unnecessary.[11] But since she accepted the Victorian ideology of gender the feminists were calling into question, at this point Cather could reject what she disliked about the feminine role and identity only by devaluing women and exalting men.

The distance between Cather's misogynist views and the contemporary feminist position can best be seen in her review of Elizabeth Cady Stanton's *Woman's Bible.* Stanton and her co-editors were beginning an important task. Attempting to expunge sexism from the Bible by calling their readers' attention to examples of female oppression as well as heroism, they tried to advance women's social emancipation by challenging religion's symbolic representations of the feminine.

Cather found their efforts a "valuable contribution to the humorous literature of the day." It was remarkable, wasn't it, that these "estimable ladies"—who lacked scholarship, linguistic attainments, and theological training— had the temerity to attempt a "task which has baffled the ripest scholarship and the most profound learning." Contemptuous of Stanton's feminist goals, she scorned her intention to recover the heroism demonstrated by biblical women: "Ruth, Deborah, Bath-sheba and Esther are each taken up and idealized and romanced about and fondled and wept over" (*WP* 2, pp. 538–40).

When it was Cather's turn to speak of women and the Bible she chose to perpetuate the stereotypes Stanton was trying to overturn. Contributing to a symposium on "Man and Woman" in the Lincoln *Courier,* Cather answered a reader's question—"Does not the Bible teach that God created woman subject to and subordinate to man and is it not a dangerous presumption in her to claim to be his equal?"—by citing women who manipulated or controlled men through allure or guile—Eve, Delilah, Jezebel. "Woman may be man's inferior but she makes him pay for it," Cather concluded, promoting a vision of female "power" that supported rather than challenged traditional definitions of the feminine (*WP* 1, p. 127).

Categorizing women in her college writing ("What a womanish thing to do," "How exceedingly like a wife and like a woman," "Women have no talent for good fellowship"), at this point Cather was unable to tell their stories. In creating a character like Alexandra Bergson or Thea Kronborg who had her own story to tell, Cather would later reject the ideologies and narratives her society used to define the feminine. Somewhere in her consciousness she knew that women could be strong, vibrant, and creative storytellers, but Cather's early journalism shows her retelling patriarchal myths and reinforcing the stereotypes that denied women individuality, complexity, and power.

•

The bonds Cather formed with other women during her college years would have surprised those classmates who viewed her as a loner who "had no friends and wanted none." Certainly she had no boyfriends, they thought: she was "so mannish herself" that she "scared off" possible suitors. One classmate summed up the popular impression: Willa Cather was "reserved and indifferent to ordinary people and consequently appeared to be rather lonely."[12] Elizabeth Sergeant later contributed to the image of Cather as a solitary, celibate priestess of art, which some of her critics also have perpetuated, when she suggests in her at times unreliable memoir that Cather had sacrificed a personal for an artistic life. According to Sergeant, Cather was forced to reject close human ties to devote herself to art. Didn't the artist necessarily become "entangled" with other people, Sergeant recalls asking her friend. Some women might marry and raise a family, which was "beautiful, if you had it in you," Cather reportedly answered, but for her "To be free, to work at her table—that *was* all in all. . . . There were fates and fates but one could not live them all. Some would call hers servitude but she called it liberation! Miss Jewett, too, had turned away from marriage" (*WC:AM*, pp. 115–16).

The misleading impression left with the reader of Sergeant's book is that Cather had to choose between two mutually exclusive alternatives: human relationships, as represented by heterosexual marriage, and the artist's lonely autonomy. But the dichotomy was a false one. At the time of their conversation Cather was living in New York with Edith Lewis, as she did until her death in 1947. Alone at her work table but not in her life, like Jewett Cather formed close ties with other women, including, for a time, with Sergeant herself. Her romantic affair with Louise Pound; her supportive friendships with fellow writers like Sergeant, Dorothy Canfield, and Zoë Akins; her abiding passion for Isabelle McClung; her enduring affection for childhood friends Carrie and Irene Miner; her dependence on loyal mate Edith Lewis: these were the threads from which Cather wove her emotional life. Unlike her mentor and friend Sarah Orne Jewett, a member of an earlier literary generation, however, Cather did not openly refer to same-sex friendships or explore women's intimacies in her fiction. As this reticence suggests, her love for women was problematic as well as sustaining.

Cather's ambivalence toward love between women surfaces in her 1895 review of three women poets—Sappho, Elizabeth Barrett Browning, and Christina Rossetti—in which she refers to the creative strength women writers could find in their "power of loving." But in commenting on Rossetti's "Goblin Market," a poem in which some twentieth-century readers discern a lesbian subtext, she refers to the "loathsomeness of our own folly in those we love" as the moral of Rossetti's enigmatic tale of sexual temptation, fall, and redemption (*WP* 1, pp. 142–45).[13]

Although Cather's juxtaposition of her "Goblin Market" discussion with praise of Sappho's love poetry places Rossetti's poem in a lesbian context and her reference to "loathsomeness" and "folly" suggests self-condemnation, the remark by itself is ambiguous. But in her often-quoted essay on the craft of fiction, "The Novel Démeublé," first published in 1922, Cather illuminates her experience of love for women during the 1890s as well as her reluctance to portray female friendship in her novels:

> Whatever is felt upon the page without being specifically named there—that, one might say, is created. It is the inexplicable presence of the thing not named, of the overtone divined by the ear but not heard by it, the verbal mood, the emotional aura of the fact or the thing or the deed, that gives high quality to the novel or the drama, as well as to poetry itself. (*NUF*, p. 50)

Here Willa Cather asks her readers to consider the importance of an unnamed, absent presence in the literary text. Whereas phrases like "overtone," "verbal mood," and "emotional aura" suggest ineffable realms of experience and feeling language cannot capture, Cather's startling phrase "the thing not named" has another connotation: an aspect of experience possessing a name that the writer does not, or cannot, employ.

By the time she wrote "The Novel Démeublé," Cather was a sophisticated novelist well-read in fin-de-siècle literature who must have been aware of the similarity between the phrase she made central to her literary aesthetic and the phrase used as evidence at Oscar Wilde's trial: the "Love that dared not speak its name.''[14] Certainly the most prominent absence and the most unspoken love in her work are the emotional bonds between women that were central to her life. From one perspective, Cather in ''The Novel Démeublé'' is the modernist writer endorsing allusive, suggestive art and inviting the reader's participation in the creation of literary meaning. But from another, she is the lesbian writer forced to disguise or to conceal the unnameable emotional source of her fiction, reassuring herself that the reader fills the absence in the text by intuiting the unwritten subtext. That Willa Cather was a lesbian should not be an unexamined assumption, however, but a conclusion reached after considering questions of definition, evidence, and interpretation.[15] Hence before returning to Cather's undergraduate friendships, I need to address a currently controversial question in feminist criticism. What is a lesbian?

•

In the last few years the importance to feminist criticism of defining both ''lesbian'' and ''lesbian writer'' has been addressed, although no conclusions have been reached. For good reason, genital sexual experience with women has been the least-used criterion. As several critics have observed, to adopt such a definition requires the unearthing of ''proof'' we do not think necessary in defining writers as heterosexual—proof, moreover, that is usually unavailable, as is the case with Willa Cather. And even if it were, to define lesbianism in narrowly sexual terms ignores the possibility that a woman who never consciously experienced or acted on sexual desire for another woman might possess a lesbian identity more broadly construed (for example, defined to include primary emotional bonding with women) and assumes a rigid separation between the sexual and the nonsexual, certainly an inadequate view of human experience.[16]

Two major approaches to this dilemma have extended the meaning of lesbianism beyond sexual, narrowly conceived definitions. The formulation of the first that has received the most attention is Adrienne Rich's frequently quoted definition of the ''lesbian continuum.'' Rich intends the term ''to include a range—through each woman's life and throughout history—of woman-identified experience; not simply the fact that a woman has had or consciously desired genital experience with another woman.'' Blanche Cook also makes genital sexuality secondary by defining lesbians as ''women who love women, who choose women to nurture and support and to create a living environment in which to work creatively and independently.''[17]

For critics who need more precise methodological tools, the problem with Rich's and Cook's definitions is their overinclusiveness. We might want to distinguish, for example, between women who defined themselves as lesbian and women who did not.[18] Such definitions also blur important distinctions we see in women's experiences in varying historical periods and cultures, such as the eighteenth and nineteenth centuries when romantic female friendships were normative, the modern period (beginning roughly in the last two decades of the nineteenth century) when lesbianism became a cultural category ideologically linked with deviance, and the period since the women's movement of the late 1960s when lesbian feminists defined their own experience and identity and increasingly found in marginality the source for a radical critique of patriarchal and heterosexual institutions.

In answer to these difficulties, Ann Ferguson offers a historically grounded statement in which self-identification as lesbian is an essential component of personal identity. If "lesbian" is to be a cognitive and emotional category for an individual woman—a prerequisite for self-identification—"lesbianism" must be a concept in her social environment: "A person cannot be said to have a sexual identity unless there is in his or her historical period and cultural environment a community of others who think of themselves as having the sexual identity in question." According to Ferguson, then, a lesbian is a "woman who has sexual and erotic-emotional ties primarily with women" or who "sees herself as centrally involved with a community" of such women and who is a "self-identified lesbian." This identity is possible only in the modern period, since, according to Ferguson, "there was no lesbian community in which to ground a sense of self before the twentieth century."[19]

Although Ferguson's dates need to be questioned—as we will see, Willa Cather possessed a lesbian "sense of self" in the late nineteenth century—I find her more limited and precise definition useful for understanding the complex relationships among self, culture, and text in the late nineteenth century when "lesbian" was emerging as a social category and a personal identity. Because the concept "lesbian" existed in the culture when Willa Cather began to write, the creative process meant something different for her than it did for an earlier writer like Sarah Orne Jewett—just as the act of writing for Jane Rule or Rita Mae Brown, who may refer directly to lesbian relationships, is different from what it was for either Cather or Jewett.[20]

In adopting this historically linked definition we must address the equation made between lesbianism and the culturally imposed concept of deviance. To do so is neither to perpetuate an oppression we want to reject nor to require an acceptance of deviance as a component of lesbian identity. But it is important to explore the interplay between the cultural definition of lesbianism and Cather's self-definition as a woman writer before we can understand the relationship between her sexual identity and creative expression.[21]

In deciding whether to define Willa Cather as a lesbian writer and then in determining her individual experience of lesbianism, we cannot use her silence—her failure to weave the emotional threads central to her life directly into her fiction—as a clear basis for deduction. The fact that Cather "could not, or did not, acknowledge her homosexuality" either in the public sphere or in her writing does not mean that she did not acknowledge it to herself; moreover, Cather's reticence in itself does not tell us how she experienced lesbian identity.[22] A problem for critics interested in these questions has been the scarcity of biographical data. Cather's destruction of her letters to Isabelle McClung, however suggestive this may be of a romantic attachment, removes the evidence of it. Fortunately an important collection of letters from Cather to Louise Pound written during her college years has recently become available for examination. Filled with unguarded emotional expression and schoolgirl self-dramatization—in sharp contrast to Cather's later reticence—these love letters are central documents in understanding her sexual identity.

•

Writing her college friend and fellow novelist Dorothy Canfield in 1921, Cather characterized her undergraduate years as a tempestuous, passionate era when she was continually being overwhelmed by emotional storms.[23] The storm center of the early 1890s was Louise Pound. The two young women were at first drawn together by their shared interest in the arts. Cather and Pound collaborated in the fall of 1891 as associate editors of *The Lasso,* the college literary magazine, and acted together in the drama society. In December 1892 Cather had a prominent role in "A Perjured Padulion," a five-act satire on university life written by Pound.

Three years ahead of Cather at the university, Pound was the fabled "New Woman" of the 1890s, the decade when the college woman, the sportswoman, and the professional woman symbolized female emancipation. Daughter of a prominent Lincoln family, a brilliant student, talented musician, outstanding athlete, and campus leader, Pound earned a master's degree from the University of Nebraska in 1895 and gained her doctorate from the University of Heidelberg, where she passed her exams in two instead of the usual seven semesters. Returning to Lincoln as professor of English, she went on to a distinguished career as a philologist and folklorist, managing to write several books, hundreds of articles, and to found and edit the journal *American Speech,* all the while teaching five courses a semester without benefit of sabbatical. Although her Nebraska colleagues never made her chair of the department, she eventually conquered another patriarchal academic bastion when, in 1955, she became the first woman president of the Modern Language Association at the age of eighty-two. That long-overdue triumph was only surpassed when a few months later she was the first woman elected to the Nebraska Sports Hall of Fame. "First woman again," she wrote to a friend after receiving this honor. "Life has its humors."[24] Although not embittered by the discrimination—and at times hostility—she incurred during her long career, Pound was acutely aware of the injustices professional women encountered. "When a man does well, it is taken for granted that he is typical. When a woman does well (So strong is the tradition), it is still thought to need explanation; and it is taken for granted that she is not typical but the product of special circumstances."[25]

When Willa Cather first met her, Pound had already begun to accumulate her athletic triumphs. An expert cyclist, skater, tennis player, and golfer, she won medals and titles in several sports, defeating male opponents to become the University of Nebraska tennis champion in 1891 and 1892. In a reminiscence Pound wrote for her fiftieth reunion, she mentioned other sports: "some experience in soft ball coaching, skiing on neighboring hills, riding and swimming—at one time had the high diving record."[26] Pound's brother Roscoe claimed his sister could "pitch a good curve ball, could bat, throw and field with the best of us." So while Cather was defining herself against the culturally prescribed feminine identity and role, Pound was encouraging other women to develop skills and interests the cult of domesticity suppressed. Pound even managed and captained the women's basketball team and helped to organize a women's military company which "drilled with the heavy Springfield rifles of the '80s and made a record at target practice."[27] Although Pound is wearing boyish garb in the photograph she and Cather had taken together (perhaps as a sign of their relationship), a later photograph reveals a more conventionally feminine appearance: a delicate-featured young woman with long hair gracefully swept up in a bun, wearing a dress with fashionable leg-of-mutton sleeves and a gold necklace. *(See photograph section.)* Proud of her red hair, Pound eventually formed an organization of red-headed women called "The Order of the Golden Fleece."[28]

Despite her loudly proclaimed contempt for the average woman, during her college years Cather was a rapt admirer of female beauty. In a column for the *Journal* she lavishly praised the portraits she discovered at a local art exhibit—aesthetic constructs of idealized femininity. At the same time that she was extolling the "manly" power of force and violence on the football field, she was captivated by "divine femininities" in the art gallery: "It is a privilege and a blessing to be alone with three such divine femininities for half an hour," she wrote, "for they are made beautiful and are not constantly doing and saying things to spoil it all." One portrait in particular entranced her: "What simplicity and elegance. How the slenderness of the waist, the pose of the head, the carriage, the firmness of the flesh . . . combine to suggest that innate aristocratic refinement which should be the outcome of wealth and culture" (*WP* 1, pp. 124–25). This "divine" femininity resembles the women to whom Cather was attracted in life and art: her dress-conscious, handsome Southern mother, the well-born Isabelle McClung, and her fiction's lost ladies—Marian Forrester, Myra Henshawe, Sapphira Colbert. There is also a hint of Louise Pound, beautiful daughter of an upper-class Lincoln family, and Cather may have been drawn to her, as to other women in her life, because she combined Virginia Cather's elegance and refinement with social, cultural, and intellectual sophistication. Moreover, as athlete and drill sergeant, Pound was associated with the male worlds of sport and military power to which Cather was attracted. So there were many reasons why Willa Cather fell in love with her.

Although Cather's Nebraska classmates recalled her as a rebellious loner who "had no friends and wanted none," in two intense, self-revealing letters to Pound she abandoned her public pose and revealed a private self (or perhaps a persona dramatized for her friend's benefit): the beseeching lover, infatuated, insecure, and melodramatic.[29]

Writing just before leaving Lincoln for Red Cloud in June 1892, Cather revealed both her infatuation and her insecurity. She had seen her at a party and had neglected to tell her friend how beautiful she looked in her new gown. Evidently finding feminine dress appealing in this instance, Cather went on to mention what she found most striking about Louise's costume—the neck, the train, the color—and to praise the wearer's beauty, adopting the same stance she assumed in the art gallery as worshipper of female perfection. Louise had looked stunning, Cather confided, and she enjoyed making a young man who was also admiring her feel envious. She wrote also of the care that she had devoted to choosing a parting gift: an edition of FitzGerald's *Rubáiyát of Omar Khayyám,* which, she confided, she loved as much as it was possible to love another person's work. Then Cather confessed the emotional conflicts plaguing her. She described how strange it felt that she wouldn't be seeing Louise for a while and admitted to being jealous of her other friends (presumably women). Cather hadn't realized her feelings were so strong; it was irrational that she should feel this way, she knew, when three years before she had never seen Louise and three years later . . . — Cather did not finish her sentence, evidently envisioning the diminishment or end of the friendship. It wasn't right that three years should make any difference, but she knew that they would; she supposed the two of them would laugh at their youthful intensity someday as other women did, but that thought made her feel terrible. It would be better to hate each other than to laugh, she thought. It was so unfair that feminine friendship should be unnatural, but she agreed with Miss De Pue (a classmate) that it was. Cather ended the letter apologizing for her silliness and hoping that Louise would pardon her. She signed it "William."[30]

Over a year later Cather was still totally preoccupied with Pound. She wrote Mariel Gere, a mutual friend and Cather's romantic confidante, promising to try to speak of Louise no more than once a day if Mariel came down to Red Cloud for a visit. At the same time she wrote a long letter to Pound, pleading for proof of affection and a visit to Red Cloud. At first she portrayed herself as plunged into depression without knowing why; the rest of the letter makes it clear that her low spirits were connected with her doubts about Pound's love and commitment. The promise of a visit from Mariel and her two sisters was her only relief from misery. Evidently trying to spark her friend's jealousy, Cather told her that she planned to have a wonderful time with the Geres, even though, she also acknowledged, their visit would be second best. Wasn't there a chance Louise could come? She might persuade the Geres to wait. She knew that Louise would either dispel or intensify her terrible sadness, and she risked rejection because she was so anxious to see her; only Louise could relieve her sense of worthlessness. Cather then injected a melodramatic note: if Louise did come to Red Cloud, she might as well bring a pistol and deliver a miserable soul from its earthly torment. If Louise did not come, it was goodbye.[31]

Perhaps Pound's decision not to visit signified the cooling of her feelings. In any case, this diminishment, and finally a break, occurred. Despite Cather's bravado, she was deeply hurt. In the spring of 1894 Cather published a "savage lampoon" of Pound's brother Roscoe in the student newspaper. Whether the attack resulted from or accelerated Pound's withdrawal is unclear, but after the sketch appeared there was a sharp rift between Cather and the Pound family.[32] Although Cather occasionally corresponded with Pound in later years, she never rebuilt the friendship, as she did with Dorothy Canfield after a similar break.

•

How are we to interpret these letters as psychological and biographical documents? Examining only Cather's revelation of feeling, we can term them love letters. Pound is the all-powerful being who controls Cather's emotions, making her blissful or miserable as she offers or withdraws affection. The range of emotion Cather expresses—jealousy, worshipful admiration, insecurity, self-condemnation, depression—further reveals a turbulent, passionate attachment. In contrast, the tone Cather adopts when writing Mariel Gere, while friendly, close, and intimate, lacks the instability, intensity, and melodrama of the Pound letters.[33]

The question remains whether these love letters reveal a lesbian attachment. There is no suggestion that Cather and Pound were lovers physically as well as emotionally, although Cather was enraptured with her friend's beauty and enjoyed besting male rivals. But a crucial reference signifies that this was a lesbian relationship according to the definition I am using here. When Cather told Pound that, while it was unfair that feminine friendship should be unnatural, she nonetheless agreed with Miss De Pue that it was, she betrayed a self-conscious awareness, shared by her community, that women's friendship constituted a special category not sanctioned by the dominant culture. Cather did not use the word "lesbian" to describe her feelings; the word was not yet in common usage in the 1890s and we would not expect her to employ such terminology in a love letter in any case.[34] But her interpretation and experience of her love for Pound—and hence her view of herself—was mediated by a cultural category that defined female friendships like theirs as deviant. Cather's response to that definition is complex, however; on the one hand she challenges the ideological yoking of same-sex friendships with deviance by terming it "unfair," while on the other, she reveals her acceptance of conventional wisdom by agreeing, however reluctantly, with its accuracy.

Cather's self-conscious awareness that her involvement with Louise Pound placed her in a suspect category sharply distinguishes her from women who enjoyed romantic same-sex friendships earlier in the century. As Carroll Smith-Rosenberg, Nancy Cott, and Lillian Faderman have shown, close, caring, sensual, and passionate attachments between women were commonplace and accepted in early and mid-nineteenth-century America when women created their own nurturing world of "love and ritual" in a sexually polarized society.[35] Victorian culture encouraged the development of these bonds without specifically naming them, so a woman who loved other women did not have to make her attraction to such intimacies central to her self-definition. Since the ideology of passionlessness, to use Nancy Cott's term, viewed women as nonsexual beings, they could participate in tender, even sensual, bonds without viewing their love as sexual.[36] Lesbianism was not a category for organizing and defining women's emotional and sexual experience in their world; hence, such women did not possess lesbian identities according to the definition used here.

As a member of this earlier generation, Sarah Orne Jewett—Cather's friend and mentor in the early 1900s—experienced her love for women in a social/historical context different from that of her younger colleague. Women were also the center of Jewett's emotional world. She frequently expressed sensual as well as emotional longings in letters and poems, unselfconsciously referring to the hugs and kisses she missed when a friend was absent and looking forward to physical closeness upon a reunion.[37] Her "Boston marriage" with Annie Fields, whose Charles Street home she shared for part of each year, was her most important bond. The two women supported each other emotionally and professionally, and when Jewett returned to her South Berwick home for her yearly visit she wrote to Annie almost daily. "Here I am at the desk again," Jewett begins a typical letter, "all as natural as can be and writing . . . to you with so much love, and remembering that this is the first morning in more than seven months that I haven't waked up to hear your dear voice and see your dear face.''[38] Jewett is assured of a mutual, enduring affection, and unlike Cather, she views this attachment, like her return to Maine, as ''all as natural as can be.'' Jewett's other friendships, which ranged from the adolescent girl's romantic infatuations to the mature writer's treasured companionships with other New England literary women, were similarly regarded as natural by Jewett's social community and hence by herself.

Because public definitions of this form of intimacy had begun to change in the twenty-five years separating Cather's young womanhood from Jewett's, the younger woman's most private realms of personal experience—love, sexuality, identity, self-concept—were altered. Cather's view of her college attachment to Louise Pound as unnatural correlates with the recent work of Nancy Sahli and Lillian Faderman, who chart an important and gradual shift in the social definition of female intimacies during the last decades of the century as the concept of lesbianism, linked with perversion and deviance, emerged in the medical literature and to some extent in individual consciousness. The innocent female world of loving friendship in which Jewett and Fields established their Boston marriage was crumbling in the 1890s, now viewed by historians as ''a crucial transitional period in the conceptualization and social experience of homosexual relations'' in both America and Great Britain.[39] Whereas earlier in the century women's friendships were consistent with their dependent status, the affection between women who were declaring their equality—or even more unsettling, their similarity—to men threatened the social, moral, and sexual order. The creation of a category of ''deviance'' then served as a means of social control as well as of boundary-setting.

Of course the medical literature did not, by itself, create and define homosexual identities that people then ''uncritically internalized''; the development of homosexual identities, as George Chauncey points out, involves a ''complex dialectic between social conditions, ideology, and consciousness.''[40] The medical literature itself responded to social changes already underway, and consciousness could be altered without exposure to Kraft-Ebing or Havelock Ellis. One advice-giver, seemingly unaware of medical opinion, shared the uneasiness about female attachments that Henry James betrayed in *The Bostonians* (1886), a ''study of one of those friendships between women which are so common in New England,'' without possessing James' insight into the perversities of heterosexual intimacies.[41] In 1895, Ruth Ashmore—a popular writer whose columns in the *Ladies' Home Journal* Willa Cather read—warned female readers against forming exclusive romantic bonds with other young women. ''I like a girl to have many girl-friends; I do not like her to have a girl-sweetheart,'' Ashmore cautioned. Such infatuations were ''silly'' and short-lived and might have disastrous long-term effects, she cautioned, for a girl who squandered her love on other girls might not have enough left for ''Prince Charming when he comes to claim his bride.''[42]

Placed in their social and historical context, Willa Cather's letters to Louise Pound reveal the effects that the public categorization of female friendship as lesbian, deviant, and unnatural had on her private experience. Unlike Jewett, she could not write unselfconsciously affectionate or passionate letters to a woman she loved. The self-condemnation suggested by her reference to the ''loathsomeness'' of love's folly in her commentary on "Goblin Market" does not typify Cather's perception of herself and her emotional nature, however; as her letters to Pound reveal, she also objected to the categorization of female friendship as unnatural. To understand Cather's individual experience of lesbianism, then, we cannot extrapolate from social ideology to consciousness. What did it mean for her to adopt an "unnatural" identity? To what extent did she internalize, reject, redefine, or enjoy the emerging identification of homosexuality with deviance?

•

As we have seen, Cather's childhood imaginings of herself and her future were shaped by her readings in epic, myth, and romantic adventure fiction like *Treasure Island*. Identifying with male heroes who possessed the power and autonomy she desired for herself, Cather's decision to become William Cather was only the most overt sign of her adolescent defiance, since she also shocked Red Cloud's bourgeoisie with her interest in vivisection and dissection. The adolescent girl's attraction to a rebellious posture thus anticipates her acceptance of an identity linked with "unnaturalness" a few years later; she was already a self-proclaimed rebel when she met Louise Pound. In addition, since lesbianism was frequently associated, as Chauncey notes, with "inversion," a young woman who had defiantly adopted male dress and name in adolescence might well have been more aware that her attraction to other women was "deviant" than one who strongly identified herself with women.

During her college years Cather passed through what she later called her "Bohemian" phase, a period of unorthodox behavior she ultimately repudiated when she became less interested in opposing convention. Scornful of piety and prudery—feminine vices in her opinion—when she heard that a friend had sent a copy of Alphonse Daudet's shocking novel *Sapho* to a religiously minded young woman, she told Mariel Gere that this pious maiden would doubtless soon join her in corruption.[43] As a drama reviewer Cather frequented the alternative world of the theater; she developed crushes on several actresses whom she treated to champagne and found herself frequently impoverished because she could not refuse their requests for loans. Whether her actress friends were self-identified lesbians or not, they challenged conventional sex roles by choosing a profession Victorians considered of doubtful respectability, and among them Cather could even see women who had theatrical license to flaunt sexual "inversion" and take male parts on the stage just as she was doing in life. She sent Pound an effusive poem dedicated to an actress which celebrates woman's erotic power ("For I dream of a smile with its shimmer/ Of silver and yellow of wine,/ And something that never has left me/ Had birth in your eyes and mine"), and in a letter to Mariel Gere's sister, written shortly after she arrived in Pittsburgh, Cather confessed she was going to the theater to see a favorite's glorious anatomy—she couldn't help it, the devil was in her.[44]

And there were other subgroups of friends: the small but select "bevy of admiring students" impressed by her atheism, the young women who also read questionable French fiction.[45] Participation in these alternative social worlds shows Cather's willingness to adopt a role of difference, if not "deviance"; and the members of what were—in Victorian America—relatively avant-garde groups would have been among the first to be aware of the emerging conceptualization of homosexuality.

Cather's reading was perhaps an even more important means for achieving a consciousness of "unnatural" sexuality. If her imaginings of powerful destinies were shaped by her childhood reading in epic and romance, her fantasies in the 1890s of love and sexuality came from non-American literature that addressed sexual matters far more frankly than was possible in William Dean Howell's America, where fiction was expected to be, as Cather bitingly phrased it, a "young ladies' illusion-preserver" (*KA*, p. 281). In her columns for the Lincoln, and later the Pittsburgh, papers Cather commented, often approvingly, on a wide range of fin-de-siècle writers, including Swinburne, Rossetti, Wilde, FitzGerald, Baudelaire, and Verlaine, demonstrating her familiarity with the decadence, ennui, and sometimes perverse eroticism characterizing this literature.

Cather's ability to read Greek and French admitted her to a literary world where homosexual attachments were portrayed with a freedom unknown even in Oscar Wilde's England. In Sappho she found a poet who celebrated the delights and agonies of love between women. Cather read Sappho during her college years, and in 1907 wrote "The Star Dial," a poem revealing her identification with this literary and sexual foremother as she assumes Sappho's voice: "All my pillows hot with turning,/ All my weary maids asleep;/ Every star in heaven was burning/ For the tryst you did not keep./ . . . Never fear that I shall chide thee/ For the wasted stars of night,/ So thine arms will come and hide me/ From the dawn's unwelcome light."[46] Evidently Sappho's poetry formed a bond between Cather and Pound, for Cather refers to her verse in one of the letters. Understandably the two young women were drawn to this poet of "love and maidens" where they found their own experience of romantic love mirrored.

French literature, particularly that of the southern regions of France, was Cather's passion during her college years and throughout her life.[47] For her a warm, sensuous land bathed in light, Provence was linked in her imagination with the southern Mediterranean world—North Africa, Italy, Greece, Turkey, and Persia. French writers like Gautier and Flaubert imbibed the "oriental feeling" of these exotic lands and conveyed it in fiction that "palpitate[d] with heat, like a line of sand hills in the South that dances and vibrates in the yellow glare of noon." A land where the "great passions" were "never wholly conventionalized," France to Cather was the symbolic location of the Other, the sign for everything repressed or feared in commercial, puritanical northern climes: it was the decadent, liberated realm of the senses (*KA*, p. 138). Cather's attraction to the fiction of Gautier, Flaubert (*Salammbô* in particular), and Daudet is connected with her love for other works where unconventional passions flourish in the Mediterranean world's "oriental" soil: *Antony and Cleopatra, The Arabian Nights, The Rubáiyát of Omar Khayyám*.[48]

In addition to palpitating sand hills and spice-laden breezes, nineteenth-century French fiction offered Cather a direct portrayal of sexual "inversion" and lesbian desire. As George Stambolian and Elaine Marks have observed, French writers have often exploited the "value of homosexuality as a transgression" in order to question sexual boundaries and social conventions.[49] Some of the French writers Cather read were both attracted and repelled by the figure of the lesbian or the androgynous woman, in part inspired by George Sand, whose portrait Cather eventually enshrined in her New York apartment.[50] Two novels that gripped Cather's imagination in the 1890s featured such heroines: Gautier's *Mademoiselle de Maupin* and Daudet's *Sapho*. In Gautier's novel Cather found a cross-dressing, bisexual heroine who seduces both a male and a female lover, a character who became the "prototype of the lesbian in literature for decades afterward," while in *Sapho*—a cult item among Cather's Lincoln friends—she encountered the more characteristic association of the lesbian with the evil, perverse temptress.[51] In commenting on the source of *Sapho's* power, Cather evidently found language an inadequate resource. The novel "involves shades and semitones and complex motives," she wrote, "the struggling birth of things and burnt-out ghosts of things that it baffles psychology to name" (*WP* 2, p. 688), anticipating the phrasing she used to articulate her literary aesthetic in "The Novel Démeublé."

Structuring as well as mirroring her experience of self and friendship, these literary corollaries to her emotions and desires bestowed an ambiguous gift. In Sappho's "broken fragments" she found an intense celebration of female sensuality, the power, joy, and despair of romantic love, and a goddess presiding over a female world of love and eroticism (in that "one wonderful hymn to Aphrodite") (*WP* 1, p. 147). But whereas the sexual love of women was "all as natural" as could be in Sappho's poetry, in French literature Cather encountered the lesbian viewed through male eyes. An enticing but threatening figure, associated with artificiality, perversity, and cruelty, she was the femme fatale's deadly sister. Such books presented the male construct of woman as Other, an embodiment of erotic power whose autonomy made her "unnatural" since, paradoxically, she did not conform to the socially defined feminine identity and role. That Cather could not help having her imagination shaped by such male-defined images is evident in her review of English contralto Clara Butt's "serpentine" performance of "The Enchantress," the song of a diabolical fatal woman who entraps and destroys her lovers. Butt's voice was "Circe-like," Cather thought, and her cheeks as unwholesome as a Burne-Jones' model. She made her think of "the verses of all the Degenerates," the "paintings of the Pre-Raphaelites," the "deadly verse of Baudelaire," and Verlaine's "feverish, overstrained, unnatural" poetry (*WP* 2, pp. 647–50).[52]

Cather's arrival at a lesbian identity by the 1890s, then, resulted from several psychological and social processes. Her adolescent male impersonation signified her attraction to a rebellious posture and may have predisposed her to respond to the emerging identification between lesbianism and sexual "inversion"; her membership in unconventional Lincoln subgroups gave her social communities that rejected or questioned dominant Victorian values; her reading in fin-de-siècle, Greek, and French literatures introduced her to the "oriental" world of hedonism and unconventional passions and gave her differing images of the lesbian. That Cather's lesbian identity was complex and even contradictory, however, is suggested by the letters. Attracted to a "deviant" identity because she delighted in opposing herself to bourgeois, Philistine culture, Cather did not fully or uncritically internalize the emotionally crippling definition of lesbianism as "sick" or "perverse" and even challenged the equation of female friendship with the unnatural. And yet simultaneously she could not help accepting it.

Cather's ambivalence did not fully dissipate in later years. After the affair with Louise Pound ended she found more enduring and supportive relationships, first with Isabelle McClung and then with Edith Lewis. Yet she could never declare her lesbianism publicly. Becoming more conservative and private as her fame increased, Cather shunned overt political activity and refused to align herself with feminism. And in her fiction she never wrote directly of the attachments between women that were the emotional center of her life. However "natural" they may finally have seemed to her, Cather knew she could not name them to a twentieth-century audience. By 1911, the year before Cather published her first novel *(Alexander's Bridge),* Annie Fields was being urged by her editor to eliminate the more intimate correspondence from her edition of Jewett's letters; he did not want people reading these passages "wrong."[53]

•

How did Cather's lesbian identity affect her experience of emotional attachments, both during her college years and throughout her life? In answering this question we need to proceed cautiously since important biographical evidence is lacking—the letters to Isabelle McClung. But Nancy Chodorow's theory of feminine development illuminates patterns in Cather's love affairs with women as well as recurrent constellations in the fiction. To employ psychoanalytic theory at this point is not to propose a theory of causation or to "explain" Cather's lesbianism by intrapsychic factors. We do not have enough biographical information to assess her early years in any detail, and in any case psychoanalysts offer such conflicting theories of the origins of homosexuality and lesbianism that it seems fruitless to search for an intrafamilial source when it seems that a rejecting and an overly attentive mother can cause the same phenomenon.[54] The ways in which psychological, cultural, social, and physiological factors may interact to produce lesbian desire may never be definitively explained. Our concern here is with the effect Cather's lesbianism had on her experience of intimacy—and with the effect her experience of intimacy had on her fiction—not with its causes.

As Chodorow argues, when the daughter becomes heterosexual and transfers her erotic drive from the mother to the father, she simply adds the father to the mother–daughter configuration, creating, as Chodorow puts it, a "relational triangle." But the daughter who becomes lesbian does not split the sexual from the emotional and transfer her erotic attachment from mother to father; unlike her heterosexual sister, she regains her mother more completely in the love relationship, perpetuating an emotional/erotic dyad rather than a triangle. Hence lesbian relationships "recreate mother–daughter emotions and connections" even more completely than do heterosexual ones.[55]

Cather's attraction to women as both maternal and erotic figures can be seen in her description of the "divine femininities" in the art gallery. She found two portraits particularly compelling. One was "Benson's girl," the slender-waisted, firm-fleshed beauty who suggested artistocratic refinement. The other was "Newman's woman":

> The first thing one notices in Benson's figure is the shoulders, in Newman's the bust. . . . Newman's woman is of a cruder stock than Benson's, her ancestors were not so good. She has warmer sympathies, but her intelligence is not so keen. She has more strength, perhaps, but not so much energy. . . . Both Newman's women [two of his portraits were displayed] are rather large, rather too generously built. Benson's girl is all lightness and pose. There is a century of refined ancestry between her and Newman's women. Newman's women are loosely built; their flesh is beautiful, but not especially firm, and their clothes fit them loosely. Benson's girl is all concentration. (*WP* 1, pp. 124–25)

This juxtaposition and contrast evoke the women to whom Cather was drawn in both life and art. Newman's strong, peasantlike figure of "warm sympathies" and crude, but strong, "stock" suggests the immigrant women who mothered her on the Divide as well as the maternal Ántonia Shimerda; in Benson's captivating aristocrat there is a glimpse of her elegant Southern mother, Louise Pound, and Isabelle McClung as well as her fiction's autocratic ladies. Cather's description of the two portraits remarkably anticipates the two sewing dummies to which Godfrey St. Peter is so attached in *The Professor's House* (1925), versions of the archetypal feminine underlying clothes and costumes. Newman's "generously built" maternal figure with her prominent bust becomes the figure the Professor straightforwardly terms "the bust," which seems (deceptively) to offer a mother's loving protection; the counterpart to Benson's "girl" is the elegant, alluring model with its "smart wire skirt" and "trim metal waist line," which seems to be "on tiptoe, waiting for the waltz to begin" (*PH*, pp. 17–19). At times—as in *The Professor's House* and *My Ántonia*—Cather divides nurturing and erotic power between two women; at times she combines them, as in early stories like "The Burglar's Christmas" (1896) or "On the Gull's Road" (1907). In either case, her fiction reveals that women retain the mother's primary sexual and emotional force.

If, as psychoanalyst Michael Balint argues, the "return to the experience of primary love" in which "all needs are satisfied" is a main goal of adult sexual relationships, then the lesbian's return to "primary love" and oneness with the mother is potentially even more complete—and therefore potentially both more satisfying and threatening—than that of heterosexual women, whose male lovers do not recreate the mother as intensely.[56] Thus lesbianism potentially offered Cather both the powerful attraction of union with a mother/lover and the possibly fearful dissolution of the self's boundaries.

As we have seen, Cather was both attracted to states of merging and fusion and terrified by them. She feared "erasure of personality," whether by dying in a cornfield or by losing the self in romantic love (*KA*, p. 448). The imagery Cather recurrently uses in her fiction to represent the dangers of sexual desire—imagery of dissolvement, annihilation, and diffusion—conveys this threat to the self; and the intertwining of passion and death in her fiction—frequently death by drowning—suggests the presence of the unnamed mother–daughter attachment, since death is the ultimate state of oneness and self-dissolution. Cather's frequent use of male characters and narrators to confront erotically compelling women thus should be connected with the daughter/writer's psychological need to place the barrier of gender between herself and erotically powerful maternal presences, as well as with the lesbian writer's socially imposed obligation to camouflage "unnatural" love.

If Cather's relationships with women raised some psychological and emotional difficulties, they also brought many benefits—both emotional and practical—which she could not have found in conventional heterosexual arrangements. While the bond with Louise Pound seems to have replayed the power dynamics of the mother–daughter configuration—Cather assuming the daughter's supplicant posture—with other women she could exchange roles, playing both mother and daughter. With women partners she was freer to reject the socially defined female role than she would have been with a man who accepted the Victorian definition of "wife" as submissive helpmate. As Simone de Beauvoir observes in seeking to explain why many women artists and writers are lesbians, "being absorbed in serious work, they do not propose to waste time in playing a feminine role or in struggling with men. Not admitting male superiority, they do not wish to make a pretense of recognizing it or to weary themselves in contesting it."[57] So although lesbian relationships posed the psychic dangers of absorption and incorporation by the mother/lover, if these dangers could be contained—as they were for Cather in the attachments that endured—they offered her equality (even, at times, dominance), emotional well-being, and the autonomy she needed to write.

Although we cannot say that Cather's social and historical milieu "caused" her lesbianism, it is possible that she found herself directed toward relationships with women by the same external forces that were labeling feminine friendship unnatural. Many New Women of the 1890s who challenged traditional sex roles and pursued professional careers chose women as their companions and lovers, whether or not they were self-identified lesbians. Straying too far from the conventional path, such women could not accommodate desires for independent achievement with the selflessness demanded by marriage and motherhood or accept a heterosexual relationship structured by male dominance and female subordination. Like Cather, they reconciled needs for autonomy and love with women who encouraged or shared their ambitions.[58]

•

The likelihood that Cather's continuing to choose female lovers was in part connected with these social forces—and thus a rejection of heterosexuality and marriage as social institutions that subordinated women—is suggested by the fact that she was not indifferent to men. To grant her a lesbian identity, then, is not to categorize or to circumscribe her experience of desire. Elizabeth Sergeant's discreet remark that men "did not pose the problem known to most unmarried women under forty" is misleading (*WC:AM,* p. 63). Although Cather never formed an attachment with a man that equalled the emotional depth of her bonds with women, it would be incorrect to say that men never roused her emotional, romantic, or sexual interest. Writing Mariel Gere in 1896, Cather reveals her attraction to a man and a woman in the same evening. Attending a local dance with her brother Douglass, she encountered a friend who had been co-editor of the university literary magazine; he was devoted to her all evening, she told Gere, and they danced every dance until two in the morning. She followed this pleased description of male attention with an account of meeting a handsome girl, a Miss Gayhardt (later Cather drew on the name and the meeting for *Lucy Gayheart* [1935]). After the dance was over the two went to bed and talked almost until dawn about books and theater, getting only a few hours of sleep.[59]

Red Cloud's inarticulate males were an unappealing group, but in Lincoln, Pittsburgh, and New York she met men of greater sensitivity and refinement who also valued art, literature, and music—"feminine" men if defined by Victorian stereotypes—and she found some of them attractive. During her Lincoln days frequent escorts were Dr. Tyndale and Charles Moore, whose father owned the real-estate firm for which Charles Cather worked.[60] In Pittsburgh Cather met several young men who displayed some interest in her, and she referred to their attentions in letters to Lincoln friends. She received one proposal of marriage from a young doctor and told Mariel Gere that, although she was considering it, she didn't really care for him; in another letter she mentioned a second man's romantic attentions, but said that she didn't want their friendship to develop in that way because she valued her liberty too much.[61] Flattered by the interest the composer Ethelbert Nevin took in her, she wrote Gere a rapturous letter describing an afternoon walk the two had taken when he had bought her a bunch of violets; later he dedicated one of his compositions to her.[62] And in 1912 Cather became infatuated with a young Mexican whom she met during her trip to the Southwest. The romance contributed to her creative emergence in *O Pioneers!,* as we will see in Chapter 18.

In considering the impact lesbianism had on Cather's fiction we must be careful to distinguish her love for women, which endured, from her male identification, which did not. Some of Cather's critics and biographers assume that her William Cather phase persisted as male identification, largely because of her use of male narrators and personas in her fiction.[63] But these were in part strategies necessary to conceal "unnatural" love that do not necessarily signify Cather's endorsement of masculine values and assumptions, which in fact she challenges in her fiction.

In fact, during the years when Cather's sexual and emotional interest in women was intensifying, she was also gradually deserting both male impersonation and identification. The college student who was praising football's "manly" violence was also beginning to discard her William Cather persona. Cather was persuaded by Mrs. Gere to grow her hair and adopt more feminine dress, she said later, but even though Mrs. Gere may have been the spark for this transformation, she was not the sole cause.[64] By the end of her Lincoln years Cather was ready to abandon overt signs of male identification since they had served their major function, aiding the adolescent girl's separation from her mother and her rejection of the feminine role.

Two photographs taken in 1895, her senior year, reflect this transition. *(See photograph section.)* They show a young woman with long hair carefully arranged and wearing a new costume, that of the elegant—even the overdressed—woman. For a trip to Chicago to enjoy the opera, she chose a cloak "trimmed with Persian lamb, a boa for her neck and an ostrich-plume hat, considered a fashionable necessity." Attending her Graduation Ball, she wore an "ivory net over [an] ivory satin" gown trimmed with "gold sequins" and posed for her picture on an ornate, brocaded couch wearing elbow-length gloves (*WWC,* n.p.). In later years Cather increasingly evinced her mother's interest in fashion, enjoying the silk dresses that Isabelle McClung helped her select at Liberty's of London; impressed by Elizabeth Sergeant's "Avignon dress with its lace fichu and sprigs of flowers," Cather wanted to "put it in a story" (*WC:AM,* p. 114). Cather often struck some people as mannish, but responses like Sergeant's, who found her personality "powerful, almost masculine," reveal not Cather's continuing male identification but the observer's identification of the powerful with the masculine (*WC:AM,* p. 34).

•

Although granting her more autonomy in life, initially Cather's love for women granted her less in art since she was, as we have seen, captive of the stereotypes male writers used to describe women's sexual and emotional power. In addition, because she needed to conceal lesbian love, in her apprenticeship writing she was frequently a slavish imitator of male-authored texts, using a predecessor's verbal or stylistic costume to cloak feelings she could not express directly.

So, while we might have thought that Cather's lesbianism would have promoted the woman writer's literary independence, at first the young writer's relationship to male predecessors was characterized by worshipful admiration. Seeking a literary inheritance from Shakespeare and Carlyle just as the adolescent girl had sought a family legacy from William Boak, Cather continued her male impersonation in a subtler form throughout the 1890s even as she modified her dress. Had she been content to remain a journalist, Cather might have found the reconciliation of female identity and vocation less problematic. But when she discovered that she wanted to devote herself to art—to be Henry James, not Henrietta Stackpole—Cather experienced the "anxiety of authorship" afflicting the nineteenth-century woman writer who faced a male-dominated literary tradition.[65] As her first literary reviews and opinions reveal, she firmly identified art with masculinity.

NOTES

1. Willa Cather to Dorothy Canfield, Dorothy Canfield Fisher Papers, Bailey Library, University of Vermont, Burlington, Vt.

2. For more information about the early years of the University of Nebraska and Willa Cather's classmates, see Bernice Slote, "Writer in Nebraska," *KA,* pp. 8–9, and *Willa Cather's Campus Years,* ed. James Shively (Lincoln: University of Nebraska Press, 1950), pp. 11–27.

3. Bernice Slote describes the controversy in more detail in *KA,* p. 18.

4. Willa Cather to Mariel Gere, May 2, 1896, Nebraska State Historical Society, Lincoln, Nebr.; Cather to Canfield, June 21, 1922.

5. Mariel Gere to E. K. Brown, December 13, 1948, E. K. Brown Papers, Beineke

Library, Yale University, New Haven, Conn. The classmate is quoted in James Woodress, *Willa Cather: Her Life and Art* (New York: Pegasus, 1970), p. 33.

6. Shively, pp. 120–42.

7. T. J. Jackson Lears, *No Place of Grace: Antimodernism and the Transformation of American Culture, 1880–1920* (New York: Pantheon, 1981), p. 108. Lears' analysis of the martial ideal, the romantic cult of self, the popularity of adventure fiction, and the worship of sport in late nineteenth-century America as interrelated cultural phenomena places Cather's worship of heroism, military power, and football in a broader context. See especially "The Destructive Element: Modern Commercial Society and the Martial Ideal," pp. 97–139.

8. As Lears points out, some Americans also linked their worship of sport with an antimodernist endorsement of the values represented by the knight and the chivalric code. Like the football player, the "Saxon warrior could provide a leaven of ferocity and suggest possibilities for physical revitalization" (p. 101). Antimodernist militarism served both to support and to condemn commercial capitalism, he argues: "Antimodern militarists, like their ideological forebears, applauded economic growth while they feared its cultural consequences" (p. 101). During her undergraduate years, however, Cather's sentiments were less contradictory since her fascination with bygone heroic times was not motivated by a commitment to "economic growth."

9. Shively, p. 135.

10. Shively, pp. 95–97.

11. For an account of the development of Gilman's feminism, see Mary A. Hill, *Charlotte Perkins Gilman: The Making of a Radical Feminist, 1860–96* (Philadelphia: Temple University Press, 1980), in particular pp. 187–209.

12. Shively, p. 139.

13. See Sandra M. Gilbert and Susan Gubar, *The Madwoman in the Attic: The Woman Writer and the Nineteenth-Century Literary Imagination* (New Haven: Yale University Press, 1979), pp. 564–75.

14. In a comment on Wilde's trial written in 1895, Cather described his "sins of the body" as "small" in comparison with his sin against art but nevertheless judged him "deservedly" imprisoned. While her opinion may show the influence of her Nebraska readership, Cather expresses the same difficulty with naming she later described in "The Novel Démeublé": "I did not know whether to give the name of the author of [Wilde's "Helas"] or not, for he has made even his name impossible." She ended her commentary with a quotation from Browning's "The Lost Leader" that begins "Blot out his name then" (*KA*, pp. 391–93).

15. Jane Rule included a chapter on Cather in *Lesbian Images* ([New York: Doubleday, 1975], pp. 74–87). More recently Deborah Lambert described Cather as a "lesbian who could not, or did not, acknowledge her homosexuality" in "The Defeat of a Hero: Autonomy and Sexuality in *My Ántonia*," *American Literature* 53, no. 4 (January 1982): 676–90, esp. 676. Phyllis Robinson is less forthright—in part continuing the genteel tradition of Cather's previous biographers by never applying the word "lesbian" directly to Cather—although she does discuss Cather's romantic attachments (*Willa: The Life of Willa Cather* [New York: Doubleday, 1983], esp. pp. 57–62). E. K. Brown wondered about Cather's sexual identity; there is a card among his papers at Yale with the heading "Homosexuality." But as the official biographer, working closely with Edith Lewis, Brown could not make his private speculations public (E. K. Brown Papers, Beineke Library, Yale University, New Haven, Conn.).

16. Cather's letters to close friends such as Elizabeth Sergeant, Zoë Akins, and Dorothy Canfield, located at the University of Virginia Library, the Huntington Library, and the

University of Vermont Library, respectively, do not refer explicitly to erotic or physical attraction. They do, however, reveal the emotional intensity Cather channeled into her relationships with women.

17. Adrienne Rich, "Compulsory Heterosexuality and Lesbian Existence," *Signs: Journal of Women in Culture and Society* 5, no. 4 (Summer 1980): 631–60, esp. 648; Blanche W. Cook, " 'Women Alone Stir My Imagination': Lesbianism and the Cultural Tradition," *Signs: Journal of Women in Culture and Society* 4, no. 4 (Summer 1979): 718–39, esp. 738.

18. For a fuller discussion of such problems, see Bonnie Zimmerman, "What Has Never Been: An Overview of Lesbian Feminist Literary Criticism," *Feminist Studies* 7, no. 3 (Fall 1981): 451–75, esp. 456.

19. Ann Ferguson, "Patriarchy, Sexual Identity, and the Sexual Revolution," *Signs: Journal of Women in Culture and Society* 7, no. 1 (Autumn 1981): 159–72, esp. 165.

20. For a different view, see Josephine Donovan's "The Unpublished Love Poems of Sarah Orne Jewett," *Frontiers* 4, no. 3 (1978): 26–31.

21. To explain, for example, why Cather left unnamed emotional and sexual realms for which Gertrude Stein devised a lesbian code, we need to question whether Cather experienced lesbian identity differently from Stein. On Stein's use of coding, see Catharine Stimpson, "The Mind, the Body, and Gertrude Stein," *Critical Inquiry* 3, no. 3 (Spring 1977): 489–506; Elizabeth Fifer, "Is Flesh Advisable? The Interior Theater of Gertrude Stein," *Signs: Journal of Women in Culture and Society* 4, no. 3 (Spring 1979): 472–83; Richard Bridgman, *Gertrude Stein in Pieces* (New York: Oxford University Press, 1970); and Linda Simon, *Biography of Alice B. Toklas* (Garden City, N.Y.: Doubleday, 1977).

22. Lambert, p. 676.

23. Cather to Canfield, April 8, 1921.

24. Louise Pound Papers, Nebraska State Historical Society, Lincoln, Nebr.

25. *Selected Writings of Louise Pound* (Lincoln: University of Nebraska Press, 1949), pp. 309–13.

26. Quoted in Evelyn Haller, "Louise Pound and the Taxonomic Rage to Order," paper presented at the Modern Language Association, 1983, p. 5. Haller provides an informative account of Pound's intellectual and professional achievements.

27. Roscoe Pound, "My Sister Louise," *Omaha World-Herald Magazine* (July 21, 1957), p. 2.

28. Pound's reluctance to associate women's emancipation with the emulation of men is suggested by a letter in which Cather mentions her friend's refusal to employ any of her masculine names; the fact that she lacked a decent first name was a great trial to Louise, Cather confided to Mariel Gere, and strange to say Louise wouldn't call her "Love" (then her middle name as well as the endearment Cather hoped for), at least in public. Cather to Gere, June 1, 1893.

29. To examine the Pound letters at Duke University Library, scholars must obtain special permission from the library and the university.

30. Willa Cather to Louise Pound, June 15, 1892, Pound Collection, Duke University Library, Durham, North Carolina. *The Rubáiyát* was linked in the Western imagination with homosexuality as well as with unconventional heterosexual passions, an association Cather may have made in selecting the gift.

31. Cather to Gere, June 30, 1893; Cather to Pound, June 29, 1893.

32. In a letter to Dorothy Canfield, E. K. Brown acknowledged the emotional trauma that terminating the relationship with Louise and the Pound family caused Cather: "I know a good deal about the quarrel with the Pounds, but so far have decided against including any mention of it. I have somehow to suggest the emotional disturbance it caused Miss

Cather, which was very intense, as some of the letters show. I think I can do that without going into the events'' (E. K. Brown Papers, Beineke Library, Yale University, New Haven, Conn.).

33. The contrast between romantic love and friendship is dramatically evident in two letters Cather wrote one day apart. On June 29, 1893, she wrote her long, histrionic letter to Pound begging for a visit and for relief from suicidal despair. The next day she wrote Gere and invited her to bring her sisters to Red Cloud without the anxious commentary with which she embroidered her letter to Pound. Cather to Gere, June 30, 1893.

34. As Catharine Stimpson observes, the ''first citation for lesbianism as a female passion in *The Shorter Oxford English Dictionary* is 1908, for 'sapphism' 1890'' (''Zero Degree Deviancy: The Lesbian Novel in English,'' *Critical Inquiry* 8, no. 2 [Winter 1981]: 363–79, esp. 365).

35. The phrase is from Carroll Smith-Rosenberg's now-classic article ''The Female World of Love and Ritual: Relations between Women in Nineteenth-Century America,'' *Signs: Journal of Women in Culture and Society* 1, no. 1 (Autumn 1975): 1–29, esp. 8. See also Nancy Cott, *The Bonds of Womanhood: ''Woman's Sphere'' in New England, 1780–1835* (New Haven: Yale University Press, 1977), pp. 160–96, and Lillian Faderman, *Surpassing the Love of Men: Romantic Friendships between Women from the Renaissance to the Present* (New York: Morrow, 1981), pp. 157–77.

36. See Nancy Cott, ''Passionlessness: An Interpretation of Victorian Sexual Ideology, 1790–1850,'' *Signs: Journal of Women in Culture and Society* 4, no. 2 (Winter 1978): 219–36, esp. 233.

37. See Glenda Hobbs, ''Pure and Passionate: Female Friendship in Sarah Orne Jewett's 'Martha's Lady,' '' *Studies in Short Fiction* 17, no. 1 (Winter 1980): 21–29.

38. Quoted in Faderman, p. 201. For a fuller discussion of Jewett's views and experience of friendship, see Hobbs and Donovan (nn. 37 and 20 above).

39. George Chauncey, Jr., ''From Sexual Inversion to Homosexuality: Medicine and the Changing Conceptualization of Female Deviance,'' *Salmagundi* nos. 58–59 (Fall 1982/Winter 1983): 114–46, esp. 114. Nancy Sahli also describes this transition in the conceptualization of female friendships in ''Smashing: Women's Relationships Before the Fall,'' *Chrysalis* 8 (Summer 1979): 17–27, esp. 25. See also Faderman; and Esther Newton, ''The Mythic Mannish Lesbian: Radclyffe Hall and the New Woman'' and Martha Vicinus, ''Distance and Desire: English Boarding-School Friendships,'' *Signs: Journal of Women in Culture and Society* 9, no. 4 (Summer 1984): 557–75 and 600–622, respectively.

40. Chauncey, p. 115.

41. *The Notebooks of Henry James,* ed. F. O. Matthiessen and Kenneth R. Murdock (New York: Oxford University Press, 1947), p. 47.

42. Ruth Ashmore, *Side Talks with Girls* (New York: 1895), pp. 122–23. For Cather's scathing critique of this ''shining literary light,'' see *KA,* pp. 187–89.

43. See Cather to Gere, July 11, 1891, where Cather ironically refers to her attempt to keep Mariel from reading *Sapho* as her one Christian effort.

44. Willa Cather to ''Neddius'' (her nickname for Ellen Gere), [1898?], Nebraska State Historical Society, Lincoln, Nebr. For Cather's response to an actress who took ''trouser roles,'' see her review of a performance by Johnstone ''Johnnie'' Bennett, comic actress, male impersonator, and ''hail-fellow-well-met'' (*WP* 2, p. 543).

45. Shively, p. 133.

46. ''The Star Dial'' appeared in *McClure's* 30 (December 1907): 202. Cather prefaced the poem with an epigraph from Sappho (in Greek).

47. ''There seemed to be a natural affinity between her mind and French forms of art,'' Dorothy Canfield recalled. ''During her undergraduate years she made it a loving duty to read every French literary masterpiece she could lay her hands on'' (*KA*, p. 37). See Michel Gervaud, ''Willa Cather and France: Elective Affinities,'' *The Art of Willa Cather*, ed. Bernice Slote and Virginia Faulkner (Lincoln: University of Nebraska Press [Bison ed.], 1984), pp. 65–83. Cather began to learn French only in her third year at Lincoln, but some of the literature, particularly the novels, was available to her earlier through translations.

48. Cather praised Sir Richard Burton's ''matchless translation of those glorious Arabian romances'' (*WP* 1, p. 186). Burton's sixteen-volume translation contained a ''Terminal Essay'' devoted to homosexuality where he associated this ''popular and endemic vice'' with a ''sotadic zone'' that geographically duplicates Cather's ''oriental'' regions (*The Book of the Thousand Nights and a Night*, translated by Sir Richard Burton [London: 1886], Vol. X, p. 208).

49. George Stambolian and Elaine Marks, ''Introduction,'' in *Homosexualities and French Literature*, ed. George Stambolian and Elaine Marks (Ithaca: Cornell University Press, 1979), p. 26.

50. See Isabelle de Courtivron, ''Weak Men and Fatal Women: The Sand Image,'' in Stambolian and Marks, pp. 210–27.

51. Faderman, p. 266. Mario Praz offers the classic study of the fatal woman in Romantic literature in *The Romantic Agony*, 2nd ed. (New York: Oxford University Press, 1970).

52. Since Cather had read Baudelaire during her college years, she may have encountered his poems ''Lesbos'' and ''Les Lesbiennes'' in which he associates the femme fatale and the lesbian.

53. For a discussion of Mark DeWolfe Howe's editorial suggestions and restrictions, see Donovan (n. 20 above).

54. In *Love Between Women* (London: Duckworth, 1971) Charlotte Wolff summarizes the conflicting interpretations on pp. 118–21. Wolff's perspective, like that of many commentators on lesbianism until the last few years, is Freudian and hence judgmental, connecting lesbianism with an ''immature'' desire for union with the mother (p. 60).

55. Nancy Chodorow, *The Reproduction of Mothering: Psychoanalysis and the Sociology of Gender* (Berkeley: University of California Press, 1978), p. 200.

56. Quoted in Chodorow, p. 194.

57. Simone de Beauvoir, *The Second Sex*, translated by H. M. Parshley (New York: Knopf, 1952; rpt. Vintage Books [1974]), pp. 450–73, esp. p. 459.

58. Faderman documents the rise of such same-sex ''marriages'' in the late nineteenth century (pp. 190–230). For an analysis of the career and marriage patterns of the first generations of college women, see Roberta Frankfort, *Domesticity and Career in Turn-of-the-Century America* (New York: New York University Press, 1977).

59. Cather to Gere, May 2, 1896.

60. See Woodress, p. 85. Charles Moore gave Cather a snake ring that she ''always wore on the little finger of her right hand'' (Helen Cather Southwick, ''Memories of Willa Cather in Red Cloud,'' *Willa Cather Pioneer Memorial Newsletter* 29, no. 3 [Summer 1985]: 12).

61. Cather mentions the proposal in her letter to Gere of April 25, 1897. On January 10, 1897, she had referred to the close friend (Mr. Farrar) with whom she did not want to have a romance. It is possible that Cather may have exaggerated the male attention she was receiving—as well as her interest in it—to please her audience. Mariel Gere was one of

her more conventional friends, and other letters suggest that Cather was eager to show Gere that she was becoming less Bohemian in her new environment.

62. Cather to Gere, April 25, 1897 and January 10, 1897.

63. Lambert, pp. 677, 680. See also Carolyn G. Heilbrun, *Reinventing Womanhood* (New York: Norton, 1979), pp. 79–81.

64. Cather to Gere, April 24, 1912.

65. See Gilbert and Gubar, pp. 45–92.

7

Manly Arts

> A good football game is an epic, it rouses the oldest part of us, the part that fought ages back down in the Troad with "Man-slaying Hector" and "Swift-footed Achilles." . . . The world gets all its great enthusiasms and emotions from pure strain of sinew. *The Kingdom of Art*

> The great secret of Shakespeare's power was supreme love, rather than supreme intellect, supreme love for the ideal in art. . . . He did not want to be studied; he just wanted to be loved. He did not write to make men think; he wrote to make men feel. "Shakespeare and Hamlet"

Cather's college writings introduce two important models of the artist which she developed in her journalism throughout the 1890s: the heroic warrior and the divine creator. The first model she derived from her reading of epics, sagas, romances, Romantic and Victorian prose and poetry, as well as from the ideology and iconography of the late nineteenth century. The second drew more exclusively on Romantic aesthetics. Both were male. Although these categories are not rigid (some writers Cather mentions—like Thomas Carlyle—combine traits of both), they are nevertheless useful rubrics for examining the conflicts between gender and artistic vocation she experienced in the 1890s.[1]

The first model—represented most prominently in Cather's writing by Rudyard Kipling—is the more aggressively, narrowly masculine, the more tied to accepted social definitions of gender. The second model, which offers a less conventionally masculine view of the creator, is both more complex and more paradoxical than the first. Although the divinely inspired artist is, from one perspective, the more exalted and patriarchal of the two, many of the traits he exhibits in Romantic theory as in Cather's journalism—like receptivity, sympathy, and selflessness—are culturally associated with the feminine. Shakespeare is Cather's exemplar of this second artist; she would later echo her undergraduate description of his creative power in portrayals of Sarah Orne Jewett and herself. As such continuity suggests, the second model of the artist provided more possibilities of transformation and metamorphosis than the first. In considering Cather's early views of art and the artist, we thus have to be careful not to consider the male literary tradition a monolithic entity. We need to separate those inherited assumptions and models that were oppressive to a woman writer from those that offered her a means of subverting the seemingly undeniable union of masculinity and creativity.

Let us begin with the first model, the product of what might be called Cather's muscular aesthetics. She found this version of an artistic self the more compelling one as a young woman trying to live out the hero's plot, but she would eventually discard it when she left William Cather behind.

•

Writing in the *Hesperian* in 1893, Willa Cather made what might seem an unlikely connection for a woman writer between art and athletics. She compared Homeric epic and football, the bone-crushing sport which, as we have seen, she praised as a rousing cure for foppishness, "chappieism," and Eastern effeminacy. After conceding that the sport was "brutal," she went on to say:

> So is Homer brutal, and Tolstoi; that is, they alike appeal to the crude savage instincts of men. We have not outgrown all our old animal instincts yet, heaven grant we never shall! The moment that, as a nation, we lose brute force, or an admiration for brute force, from that moment poetry and art are forever dead among us, and we will have nothing but grammar and mathematics left. The only way poetry can ever reach one is through one's brute instincts. "Charge of the Light Brigade," or "How They Brought Good News to Aix" [sic], move us in exactly the same way that one of Mr. Shue's runs or Mr. Yont's touchdowns do, only not half so intensely. A good football game is an epic. (*KA*, p. 212)

To the young Willa Cather, a good epic was also a football game. She gloried in the virility of art, combat, and athletics. Displays of aesthetic, military, or physical prowess often reminded her of the heroic era "when Greece was young" and "art and athletics went together" (*KA*, p. 214), although more recent heroic eras sometimes came to mind: football, she thought, embodied the "bulldog strength" of the Anglo-Saxon race which went back to the "days of the Norman conquest" (*KA*, p. 213).

Underlying Cather's association of sports, war, and epic poetry is a set of metaphoric equivalences—weapon/sword/pen/penis—that reveal her equation of creativity both with paternity and with an aggressive, phallic masculinity.[2] The links between her uncle's sword, the dissector's knife, the surgeon's scapel, and the writer's pen are literal as well as metaphoric. When she turned from medicine to writing, Cather was still embracing a profession she considered socially and symbolically masculine and grasping a tool of power.

Commenting on the difference between the French and the English languages in an 1898 article, Cather made the connections among words, weapons, and the masculine creative force explicit. Contrasting the precision of French with the inexactness of English, she imagined the clumsy Anglo-Saxon receiving a language as "unwieldy as his own giant battle-ax matched with the French rapier" (*WP* 2, p. 583). In a review of Robert Hichens' *The Green Carnation* (a satire of Oscar Wilde), Cather explained the distinction between French and English language/weaponry:

> The English were made for slower, heavier speech. The Frenchman cuts like a rapier, but the Englishman crushes like a sledge. If he studies French word-

> fencing he only makes a clown of himself. He can never apply his power through a slender blade, and he is at best only a poor imitation. The affectation poisons his style, his vigor and his whole personality. He loses not only his art but his manhood. (*KA*, p. 136)

Whereas Virginia Woolf thought the male sentence a ''clumsy weapon'' in a woman writer's hand, here Cather is concerned only with choosing the proper weapon, a perilous business since the writer who selects the wrong weapon/word castrates himself, losing his ''manhood'' along with his art.[3] Elsewhere equating sexual and artistic mutilation, Cather chided Kipling on his marriage to a woman who might tame him and his writing: ''Remember your own words, 'It cripples a man's sword arm' '' (*KA*, p. 318). As college editor, she used this martial rhetoric of masculine creativity to describe the aggressive posture she wanted the *Hesperian* to assume, assuring her readers that ''If there is any fighting to be done, we will be down in the line fighting on one side or the other.'' The young editor encouraged her staff to ''pummel'' their opponents and she declared her intention to enter the lists armed not ''with a lance or sword, but with a good stiff stub pen'' (*KA*, pp. 11–12).

Judging from the caustic reviews Cather published in Lincoln papers, her ''stiff stub pen'' was a replacement for a sword rather than an alternative. Will Owen Jones, her co-worker on the *Journal,* portrayed Cather during these years as a ''meatax young girl'' who terrified even actors of ''national reputation'' with her slashing, satiric reviews. Another journalist mentioned her entertaining way of ''crucifying'' actors (*KA*, pp. 16–17). Cather's targets ranged from actors who desecrated Shakespeare, to professors whose teaching methods she disliked, to fellow students who irked her. She soon developed a reputation for articulate invective on the campus, and the *Nebraskan* (the *Hesperian*'s rival) dedicated a sketch of a smoking pen to '' 'Billy' of journalist fame,/ Who writes her roasts in words of flame/ And gives it to everyone just the same'' (*KA*, p. 22).

Cather's association of football and Homeric epic, swords and pens, contributed to her vision of the artist as an aggressive warrior–hero. Even though she sometimes despaired for the survival of this mighty figure in modern times, she thought that Robert Louis Stevenson and Rudyard Kipling measured up to epic standards. Stevenson died ''with his spurs on,'' she noted approvingly, ''in the high tide of a great novel'' (*KA*, p. 315), while his successor Kipling ''had conquered an empire before he was out of his 'teens'' (*WP* 2, p. 556). This artist–hero was linked in Cather's imagination with the primitive and the barbarian. Creative power, she thought, was ''red lava torn up from the bowels of the earth where the primeval fires of creation are still smouldering,'' a force displaying the ''savagery of the stone age and arous[ing] in the individual the forgotten first instincts of the race'' (*KA*, p. 118). Gothic art, she speculated, must have been sustained for centuries ''just on the brute momentum it got when the old Goths used to throttle polar bears with their naked hands.''[4]

Cather found in Whitman and Carlyle contemporary inheritors of the Goths' stone-age savagery. She liked Whitman's ''primitive elemental force,'' thinking him ''sensual'' in the ''frank fashion of the old barbarians'' who ''smacked their lips over the mead horn'' (*KA*, pp. 351–32). Cather's Carlyle was a descendant of the Vikings, the ''hot blood of the old sea kings'' still raging in his veins. These passionate pagans tapped the potent and the untamed in their art: ''volcanoes, and earthquakes, and great unsystematized forces'' (*KA*, pp. 422–25). This heroic artist did not have to be literally as muscular as the warrior or athlete. His strength was imaginative, and as Cather observed ''despite his gaunt frame and sunken cheeks, Stevenson gave us some of the most strong and vigorous literature produced by any man of our time'' (*KA*, pp. 311–12).

Like combat and athletics, much of the literature Cather admired during the 1890s she straightforwardly defined as masculine. She termed the romances she loved ''noble and manly,'' praised Kipling for creating ''manly'' men and urged him to tell war stories, approvingly described poetry by Bliss Carman and Richard Hovey as ''thoroughly manly and abundantly virile,'' and recommended Stevenson, Kipling, and Anthony Hope to her readers because they allowed their heroes ''to love and work and fight and die like men'' (*KA*, pp. 313, 317, 354, 232). Occasionally a woman writer displayed the literary virility Cather admired. In a 1902 commentary on ''Poets of Our Younger Generation,'' she commended Boston poet Louise Imogen Guiney's commitment to will, action, and force, which Guiney frequently expressed in hymns to a chivalric past. Cather singled out ''The Vigil at Arms,'' a lyric describing a knight's fasting before departing on a quest (''manhood's way''), for special mention. ''There is no black flaunting of modern pessimism here,'' Cather noted approvingly, no ''bitterness affected, no rhapsodic exaltation of failure. The sentiment, calm and submissive to decree as it is, might be Emerson's own'' (*WP* 2, p. 884).[5]

In imagining the artist as a manly warrior, conqueror, or knight, Cather was far from being original or iconoclastic. During the late nineteenth century—the 1890s in particular—many Americans combated their fear of overcivilization, effeteness, and fin-de-siècle decadence by embracing a cult of manhood, strenuous living, and military vigor for which Teddy Roosevelt became the advocate and symbol, the West the symbolic American location, and ''manly'' the key adjective. Purveyed in editorials, speeches, advice books, sermons, and novels, this ideology had various manifestations: a fascination with the medieval and the chivalric inherited from Victorian Britain; the proliferation of best-selling romantic fiction like *When Knighthood Was in Flower;* the rise of muscular Christianity; the militarization of Protestant youth groups; the development of college athletics; and, after the turn of the century, the founding of the Boy Scouts.[6] Although the larger political function of these related social and symbolic systems was to rationalize such imperialistic ventures as the annexation of the Philippines and the Spanish–American War, a similar rhetoric characterized male attempts to reappropriate the profession of writing which struck some observers as having been feminized along with American culture during the 1870s and 1880s. (''The monthly [magazines] are getting so lady-like that naturally they will soon menstruate,'' observed novelist and doctor S. Weir Mitchell.)[7] The vision of the heroic artist–warrior can thus be seen, in part, as an expression of male fears of literary as well as of social and political impotence; Cather was probably not conscious of the irony that some of her male contemporaries thought writing a female-dominated occupation.

In addition to viewing the writer and the literary profession through the masculine ideology dominating the period, at times Cather described the writer's relationship to both reader and subject as a drama of dominance and submission like that played out on the gridiron and the battlefield. If the artist were a conqueror or warrior, then the reader was his colonized subject: art was "like the Roman army," she thought, for in overwhelming readers, viewers, or listeners "it subdues a world, a world that is proud to be conquered when it is by Rome" (*KA*, p. 54). At times Cather imagined the creative process similarly as a struggle to subdue intractable subject matter, as when she envisioned Carlyle "[wrestling] with his great ideas, finding them difficult to express in words, so great, so ungainly were they" (*KA*, p. 424). Although the sedate Emerson was not the mead-drinking, lip-smacking natural man she assumed Whitman to be, Cather was impressed by his association in "The Poet" of creative power with the artist's possession and domination of nature and subject matter. Addressing his Poet as "land-lord! sea-lord! air lord!," Emerson prophesies that he "shall have the whole land for thy park and manor, the sea for thy bath and navigation, without tax and without envy; the woods and the rivers thou shalt own, and thou shalt possess that wherein others are only tenants and boarders."[8] Cather later alluded to this passage in "The Treasure of Far Island" when she described the child/artist as a member of "that favored race whom a New England sage called the true land-lords and sea-lords of the world" (*CSF*, p. 265).

Cather's acceptance of an aesthetic theory infused with the late nineteenth-century ideology of masculinity and Emersonian self-reliance affected her assessment of writers, novels, and literary movements in the 1890s, contributing in particular to her dislike of aestheticism and decadence, her distaste for realism, and her infatuation with romance and adventure fiction. Although she was impressed by some fin-de-siècle poets—notably the French *symbolistes*—Cather found the work of Oscar Wilde and his American imitators marred by "drivelling effeminacy" as well as inauthenticity (*KA*, p. 135). The displays of dissipation she found in this poetry impressed her as signs of "maudlin and disgusting rot," not heroic rejection of bourgeois values. The poetry and prose produced by third-rate aesthetes and decadents—drunken young man who would be better advised to keep their confessions to themselves—she found contemptible, inane, and (worst of all) puny (*WP* 1, p. 155; *KA*, p. 135).

Cather was even more hostile to what she termed "encroaching realism and 'veritism' " and other forms of "literary unpleasantness" like naturalism than she was to the *Yellow Book* and the *Chap-Book* (*KA*, p. 312). Zola was her favorite target. She held him responsible for infecting American writers with the foreign contagion of pessimism and despair; without Zola's influence, social realist Hamlin Garland might not have been dragged down into the "primal slime" of political fiction (*WP* 1, p. 141). Although there were many reasons for Cather's dislike of realism and naturalism—including her belief that novelists should neither be reporters nor crusaders for social justice—the ideology of the realist/naturalist narrative was the most important. These writers told her the wrong story.[9] Their protagonists were either defeated by external forces or required to adapt to intractable social structures, and Cather was angered by these threats to self-reliance,

American individualism, and the decade's cult of virility. Even the crimes committed by Zola's characters, she complained, were not romantic acts of defiance or rebellion but the weak, unwilled reflexes of "joyless, besotted men who sin because they can do nothing else" (*WP* 1, p. 140). As a woman striving to escape the traditional female role, Cather wanted to read a different story about the relationship between the individual and society than the one she found in *Germinal, Main-Travelled Roads,* or even *The Rise of Silas Lapham.* She wanted to know that individual force, will, and passion could make a difference, that Emerson's "Self-Reliance" was a reliable guide to human endeavor.

Cather continued to find stories she liked to read in the genre that had dominated her childhood reading: romance, in her view the "highest form of fiction" where, as in Guiney's chivalric odes, she could enjoy the optimism she demanded from narrative. A faithful apologist for writers like Kipling, Stevenson, Dumas, and Hope throughout the 1890s, along with other fin-de-siècle American readers who made the revived genre popular, Cather found in the heroes of romance "something that is denied to most heroes in modern fiction"—the power to act. Whereas other fiction she found "full of the futileness of effort," in romance she found effort both glorified and efficacious. Because romance locates power in the individual rather than in the social world, its practitioners could tell Cather a story "whose possibilities are as high and limitless as beauty, as good as hope." The woman reader who identified with the genre's heroic protagonists could enter Emily Dickinson's house of possibility—a utopian realm beyond existing social arrangements—and hope that she could achieve their victories herself.[10] The protagonists of realistic and naturalistic fiction could not convey the sense of limitless possibilities Cather wanted confirmed in the stories she read.

Of course by the late nineteenth century the triumphant individualism of romantic adventure fiction was increasingly a fantasy for the male as well as the female reader. This pre-capitalist genre had its origins in epic and medieval romance, as Cather recognized when she called *The Prisoner of Zenda* a "romance [in] a dress suit, a real romance with war and blood and love and honor, like the romances of the Grail or the Holy Sepulchre" (*KA,* p. 319). Like the popular adventure fiction that currently attracts male readers—Westerns, detective stories, thrillers—late nineteenth-century romances were mass-marketed daydreams that attempted to satisfy and defuse longings for the power, efficacy, and recognition increasingly unattainable for men as well as for women in a society being transformed by industrial capitalism.[11] Hence most romance writers set their heroic tales in the historic past or in exotic, timeless realms—Anthony Hope's Ruritania, Owen Wister's mythologized American West—safely distant from the grubby or oppressive realities of the industrial and urban present.[12]

•

Cather's association of art with virility explains her fondness for Kipling, to her the representative manly writer and warrior–hero. She delighted particularly in his celebration of force and energy. Equating these desirable values with masculinity, she worried that Kipling's marriage might feminize and soften him. Writing in the Lincoln *Journal* in 1894, Cather found it discouraging that this rough-hewn writer was ''trying to live a respectable Puritan life in Vermont and be a full-fledged family man''; she would rather have heard that ''he had taken to opium or strong drink or that he had married a half-caste woman and was raising vermilion hades out in India'' (*KA*, p. 317). Afraid that Kipling's marriage might deflect him from his craft, she advised him to keep his priorities straight:

> If the climate is not good for Mrs. Kipling then remember that you were married to your works long before you ever met her. Alas! there were so many men who could have married Mrs. Kipling, and there was only you who could write *Soldiers Three*. (*KA*, p. 318)

Cather met Kipling shortly after moving to Pittsburgh in 1896. Evidently he did not disappoint her because in 1899 she praised his ''tremendously virile'' sensibility. He was much more of a ''man'' than his Romantic precursors, she contended, and she ''hate[d] to think of him'' in the effete and effeminate company of other British poets. ''He doesn't go well with Byron's impossible collar or Tennyson's curls and Shelley's lace ruffles. I'd rather think of him loafing back of the barracks with Mulvaney and Learoyd and Ortheris and the dog that had the mange. He is really too much of a—well, too much of a man to be called a British bard'' (*WP* 2, p. 555).

Finding Kipling a soldier ''to be reckoned with'' on the literary battlefield, a writer more at home with mangy dogs than ruffled collars, Cather liked the worship of force she found in his work. Kipling celebrated ''the energy of great machines,'' she wrote approvingly, ''of animals in their hunt for prey, of men in their hand-to-hand fight for a foothold in the world.'' Kipling and Zola were, she argued, the only contemporary writers who commanded the ''virility of the epic manner,'' and she quoted a catalogue of freight cars from Kipling's story ''007'' to prove her point. Zola could also have written such a powerful description, Cather acknowledged, but he would doubltess have proceeded, tediously and remorselessly, to extract every ''evil smell that is to be got out of a freight yard'' (*WP* 2, pp. 557–59). Kipling thus offered her epic sweep without an intruding nose for social injustice.

Because she thought that Kipling's fiction gave the reader the power and freedom lacking in daily life, Cather did not want him to be seduced by realism and naturalism, literary schools for the defeated and despairing. He should continue to write the kind of fiction she wanted to read: ''Don't tell us petty stories of our own pettiness,'' she pleaded. ''We have enough little Harvard men to do that. Tell us of things new and strange and novel as you used to do. Tell us of love and war and action that thrills us because we know it not, of boundless freedom that delights us because we have it not.'' To tell these stories, she advised, Kipling should leave his Vermont home—and symbolically the social world—to find an uncharted land beyond culture which Cather associated with the East: ''Go back where there are temples and jungles and all manner of unknown things, where there are mountains whose summits have never been scaled, rivers whose sources have never been reached, deserts whose sands have never been crossed'' (*KA*, p. 317).

The difference between the young Cather's evaluation of Kipling and Virginia Woolf's can help us see both the extent of Cather's male identification in the 1890s and the changes she would undergo in the next few decades as she separated creativity from masculinity. Woolf also considered Kipling the representative "manly" writer, but she was not similarly enchanted with his bristling virility. Kipling's preoccupation with manhood made his works inaccessible and uninteresting to a woman reader, she thought: "It is not only that they celebrate male virtues, enforce male values and describe the world of men; it is that the emotion with which these books are permeated is to a woman incomprehensible." Like Cather, Woolf linked such "unmitigated masculinity" with Kipling's worship of force; unlike Cather, she disliked this alliance, and thought that other women would as well. Because Kipling's phallic fiction lacked "suggestive power," Woolf wrote, it was paradoxically impotent, unable to "penetrate" the female mind:

> the emotion which is so deep, so subtle, so symbolic to a man moves a woman to wonder. So with Mr. Kipling's officers who turn their backs; and his Sowers who sow the Seed; and his Men who are alone with their Work; and the Flag—one blushes at all these capital letters as if one had been caught eavesdropping at some purely masculine orgy. . . . Thus all [his] qualities seem to a woman, if one may generalise, crude and immature. They lack suggestive power. And when a book lacks suggestive power, however hard it hits the surface of the mind it cannot penetrate within.[13]

In later years Cather came closer to Woolf's view, eventually finding both Kipling's manhood and his prose overdone. In her 1926 preface to Stephen Crane's *Wounds in the Rain* she compares the two writers and finds Kipling's excessively masculine pose—which Crane at times adopted but, she then thought, fortunately transcended—a bit embarrassing. One of Crane's battle sketches she thought "done in an outworn manner that was considered smart in the days when Richard Harding Davis was young, and the war correspondent and his 'kit' was a romantic figure. This sketch indulges in a curiously pompous kind of humour which seemed very swagger then." In poking fun at this "chesty manner" of writing which "came in with Kipling," Cather separated herself from the college student who had loved Kipling's "tremendously virile" tales. Looking back on her literary enthusiasms of the 1890s, she confessed that "When one re-reads the young Kipling it [the "war-correspondent idiom"] seems a little absurd . . ." (*OW*, p. 68).

Cather's changing opinion of Kipling's chesty manner is only one sign of the more profound development we are tracing: her gradual separation from male-identified views of the artist, of the relationship between the artist and subject matter, and of the creative process. In *O Pioneers!* we can see Cather's rejection of the ideology of masculinity that dominated her thinking in the 1890s. The novel's protagonist Alexandra Bergson—Cather's female Alexander, a creator and conqueror—tames the wild land not by exerting "brute force" but by giving herself up to its power and beauty with "love and yearning" (*OP*, p. 65). Possessed by the land more than possessing it, Alexandra rejects a model of ownership and authorship based on the rights of dominance and force, just as Cather was experiencing the creative process as an abandonment, rather than an aggrandizement, of the ego.

The aesthetic theory Cather later articulated in "The Novel Démeublé" even more closely resembles Virginia Woolf's. In that important essay she rejects her earlier view of the relationship between author and reader as a power struggle in which one is "subdued" by the other. In contending that literature must possess what Woolf called "suggestive power" and Cather defined as the "inexplicable presence of the thing not named," she imagines reading as an act of collaboration and empathy, not a contest (*NUF*, p. 50).

•

How did Cather move from an aesthetic theory of force to one of suggestion and approach Woolf's view of Kipling? Her college writings introduce another version of the artist which mediates between the aesthetics of force and the aesthetics of suggestion, between Kipling and Woolf, between masculine and feminine—that of the inspired Poet, the divine creator. Introducing this model in her 1891 college essays on Carlyle and Shakespeare, Cather elaborated on it in commentaries and reviews in Lincoln and Pittsburgh papers through the 1890s. Because she knew less about Carlyle and Shakespeare than about her contemporary Kipling, she was freer to project herself into these writers, and so in reading her literary opinions we learn more about Cather than her subjects. Although the aesthetic paradigms she employed were derived from male writers, this was a less aggressively and exclusively masculine vision of the artist than that of the warrior–hero, less closely tied to the decade's cult of virility. Hence it contained more potential for redefinition by a woman writer who ultimately found it easier to imagine herself a literary descendant of Shakespeare than of Rudyard Kipling.

Writing Mariel Gere in 1896, Cather declared that art was now her religion and she intended to follow it to the end.[14] In conflating art and religion—a recurring theme in her fiction—Cather drew on her readings in myth, romance, the British Romantic poets, Carlyle, Ruskin, Emerson, her childhood favorite *Pilgrim's Progress*, and her family's Protestant heritage, for the self-proclaimed atheist was influenced by her inheritance of rural piety as well as Romantic literary theory in developing a religion of art. Defining the artist's holy vocation in her college writings, Cather constructed an elaborate framework of religious and chivalric motifs that permeates her undergraduate essays on Carlyle and Shakespeare as well as her commentaries on other admired writers.[15] Seeking truth and beauty in the "kingdom of art," Cather's questing artist commits himself to a harsh Old Testament deity:

> In the kingdom of art there is no God, but one God, and his service is so exacting that there are few men born of woman who are strong enough to take the vows. There is no paradise offered for a reward to the faithful, no celestial bowers, no houris, no scented wines; only death and the truth. (*KA*, p. 417)

This Romantic artist is a more complex and contradictory being than Cather's manly writer. On the one hand, this artist must deny or severely regulate human relationships, sacrificing the part of life traditionally associated with the domestic and the feminine in pledging himself to the pilgrim's celibate and lonely vocation. In her college writings Cather frequently refers to the "loveliness and lovelessness" of the artist, the "isolation" of creative genius, the "loneliness which besets all mortals who are shut up alone with God, of the gloom which is the shadow of God's hand consecrating his elect" (*KA*, p. 153).

On the other hand, Cather imagined this artist as a passionate, deeply feeling soul, motivated not by intellect, ambition, or power but by love. Hence her divine creator is a divided soul, drawn to fellow human beings by love and sympathy, pulled away by his commitment to art. As she observed in writing of Shakespeare (whom she imagined as a lonely, wayfaring *isolato*):

> Authors are not made of marble or of ice, and human sympathy is a sweet thing. There is much to suffer, much to undergo: the awful loneliness, the longing for human fellowship and for human love. The terrible realization of the soul that no one knows it, no one sees it, no one understands it; that it is barred from the perceptions of other souls; that it is always alone. It is a hard thing to endure. (*KA*, p. 435)

And yet the artist must endure this painful isolation, she thought, for the creed Cather derived from Keats and nineteenth-century Romanticism demanded fealty: "That beauty alone is truth, and truth is only beauty; that art is supreme; that it is the highest, the only expression of whatever divinity there may be in man" (*KA*, p. 402).

Agreeing with Shelley, Coleridge, and Emerson that the Poet's creative act echoes God's original formation of the universe from nothing, Cather located art's divinity in its mimetic power. "The greatest perfection a work of art can ever attain is when it ceases to be a work of art and becomes a living fact," she commented in reviewing a performance of *Camille*. "Art and science may make a creation perfect in symmetry and form, but it is only the genius which forever evades analysis that can breathe into it a living soul and make it great" (*KA*, p. 263).

From one perspective, Cather's fealty to the Romantics' divine Artist enmeshed her in a masculine aesthetic that excluded women even more firmly that did Kipling's. As Gilbert and Gubar have shown, the God-like creator is explicitly and implicitly male, the Father of a fictive universe, the Speaker of the Word, the progenitor of the text.[16] If the pen is a penis and the creative act one of fathering, it would seem that a woman writer could imagine herself a member of the Western literary tradition only by repudiating her gender. But in Cather's depiction of the creative process in her first essays we find a creative power that could potentially be imagined as feminine. Her divine artist did not have to exchange the love that drew him to other human beings for a worship of force; he just had to channel that love into his art. Whereas "force" and "power" are the key concepts in Cather's portrait of the manly artist–warrior, the central words in understanding the creative process experienced by her artist–god are "passion," "love," and "sympathy"—qualities that Cather associated with women writers even in the 1890s, and later attributed both to Sarah Orne Jewett and to herself.

The imagination provided the divine inspiration that could ''breathe'' vitality into characters and make them living souls, Cather thought, but another force first sparked the imagination—the ''exuberant passion'' that once inspired God to create the world (*KA,* p. 45). Questioning the mysterious origins of creativity, in her college writings Cather gave variations of the same answer she would offer in *The Song of the Lark,* the novel celebrating her own birth as an artist; in it Harsanyi tells Dr. Archie that the ''secret'' of Thea's creativity is ''desire'' (p. 477). In her undergraduate essays on Carlyle and Shakespeare, Cather associated the writer's desire to create with his need and ability to enter into other lives—an empathic power she generally described as ''sympathy.'' Carlyle abundantly possessed this emotional and aesthetic gift: as the historian and biographer he could bring others' lives to life in his words. Because his literary power lay in the ''strength of his great heart,'' not his intellect, Cather observed, Carlyle ''posed but poorly as a political economist.'' In his ''boundless'' love and sympathy for humanity lay ''his power as a biographer and as a historian,'' she thought. He could understand ''how the Marseillaise might set men's hearts on fire; the storming of the Bastille, and the revolt of the women [were] pictures after his own heart'' (*KA,* p. 422). So while Carlyle devoted himself to a merciless God of art who demanded ''human sacrifices'' and made his wife a ''secondary consideration,'' he poured his ''love and sympathy'' into writings that became ''pictures after his own heart'' (*KA,* p. 423.).

Cather articulated her view of the creative process as a drama of sympathetic identification most fully in her 1891 essay on ''Shakespeare and Hamlet.'' The ''great secret'' of Shakespeare's power was ''supreme love, rather than supreme intellect,'' she contended; he achieved dramatic greatness because he understood the ''few elementary emotions which are the keystone of life.'' Speculating on the genesis of *Hamlet,* Cather suggested that Shakespeare probably had no definite purpose or abstract goal in mind when he began the play: ''He probably read the legend and felt sorry for the young prince, and as an expression of his sympathy wrote about him. . . . It may be his feeling and individuality were wrought up intensely, and crept out into the play which he happened to be writing.'' Stressing identification, reciprocity, and interconnection rather than subjugation and domination, in imagining how Shakespeare came to write *Hamlet* Cather envisioned a creator markedly different from Kipling. Because Shakespeare was so moved by the legend of Hamlet, Cather proposed, he in turn infused the character with his own ''feeling'' and ''individuality.'' As both reader and writer, Cather's Shakespeare erases the boundaries between self and other. Despite the egotism implied by the Romantic exaltation of the divine artist, Cather's Shakespeare divests himself of ego in the creative process: the act of writing she describes in ''Shakespeare and Hamlet'' is simultaneously a form of self-expression and an ''expression of . . . sympathy'' for an imagined other.

The relationship Cather imagines between the playwright and the audience in ''Shakespeare and Hamlet'' as well as in other commentaries on the theater during the 1890s is similarly marked by love, sympathy, and collaboration rather than force, conflict, and subordination. Because Shakespeare understood Hamlet, the audience could ''only sympathize'' with the character's plight as well, she thought, entering into the same bond with the character that Shakespeare experienced first when he read Hamlet's legend and later when he transformed the story in his play. As a result, the "legacy of . . . love" which the author gave to his character was passed on to the audience. Motivated by emotion rather than intellect, Shakespeare "did not write to make men think," Cather speculated, but "to make men feel" (*KA,* pp. 426–36). In her other writings of the 1890s, Cather frequently mentions the engendering of feeling in the reader or audience as the mark of literary vitality: an artist had created life in his work only if he elicited a passionate response that matched his inspiration. "It is all a thing of feeling, you cannot apprehend it intellectually at all. What do you know of a grief till it is in your heart, of a passion till it is in your veins" (*KA,* p. 120)? As Cather suggests in several reviews and commentaries, these empathic links between writer and character, actor and role, were essential for the reader's or playgoer's sympathy to be elicted by the book, play, or performance. And without sympathy, drama could not transport the audience out of themselves into others' lives and experiences.

Master of drama's transformative and transporting power, Cather's Shakespeare recalls both Keats' "chameleon poet" and Virginia Woolf's androgynous, "man-womanly" artist. Shakespeare's harmonious integration of masculine and feminine gave the "suggestive power" to his art that Kipling's lacked, Woolf thought; the "marriage of opposites" that took place in his mind allowed his plays to root and "grow" in the minds of others.[17] Although Cather did not mention gender in her discussion of Shakespeare's genius, she invested him with qualities women as well as men might possess. Shakespeare's ability to sympathize—to enter into another self (a power also granted him by Keats and Woolf)—corresponds to the selflessness and empathy traditionally associated with women and which Cather explicitly allotted to women writers when, in her 1895 commentary on Barrett Browning, Christina Rossetti, and Sappho she attributed to women poets the "power of loving" (*KA,* p. 348).

This second model of the artist and the creative process was closer to Cather's own experience as woman and writer and contributed more to her reconciliation of identity and vocation than did the more narrowly masculine model represented by Kipling. We can see this connection in her reworking of the language and ideas of her 1891 Shakespeare essay for her "Preface" to Sarah Orne Jewett's *The Country of the Pointed Firs.* In this major statement of the mature writer's aesthetics (which describes herself as well as Jewett), Cather imagines the artist moved to write by a "gift of sympathy" for subject matter and characters, losing the self in the creative process ("[she] dies of love only to be born again" [*CPF,* p. 71]). Cather's description of the relationship between artist and reader in "The Novel Démeublé" as requiring the "feminine" powers of connection, sympathy, and intuition also resembles the empathic version of the creative process the young writer attributed to Shakespeare in her 1891 essay.

Cather discerned another potentially "feminine" process at the heart of patriarchal poetics. Romantic theory gave her a view of creativity, inherited from classical tradition, in which the writer abandons rational control and submits to an overpowering, nonrational force—whether the power of love, the force of inspiration, the unconscious, or the Muse. Hence she envisioned Carlyle writing only what "must be written," chosen by his subject matter rather than choosing it (*KA*, p. 424), Shakespeare as claimed by Hamlet, and several other writers overpowered by the Muse who "moulds a man to her will" (*KA*, p. 414). She found a memorable description of the artist's submission to greater powers in "The Poet," where Emerson envisioned his creator's dominion over nature arising from his self-abandonment to "celestial life." Just as "the traveler who had lost his way throws his reins on his horse's neck and trusts to the instinct of the animal to find his road," Emerson writes, "so must we do with the divine animal who carries us through this world."[18]

Cather could have found this common view of poetic inspiration articulated by the "ancients" Emerson mentions or by Coleridge, Wordsworth, Shelley, Thoreau, or Whitman, among others. But when she described her own self-abandonment to the creative process in writing *O Pioneers!*, she echoed Emerson's metaphor of the traveler who finds the road by trusting to some instinctive, intuitive power: "he finds that he need have little to do with literary devices; he comes to depend more and more on something else—the thing by which our feet find the road home on a dark night, accounting of themselves for roots and stones which we had never noticed by day" (*AB*, p. ix). In both Emerson's and Cather's formulations, the artist takes the role of the dependent child, not the powerful father or conqueror who rules—he or she is lost, wandering, and must rely upon "something else" to find the way home.

The power of the divine writers Cather admired in the 1890s did not arise solely from ecstasy, inspiration, or self-abandonment. The supreme creators possessed gifts that she feared the woman writer lacked: discipline, shaping intelligence, and craft. She associated these qualities with the artist's control and command of form. The "much talked of emotional qualifications are only the beginning of true greatness," she wrote. "It requires a master intellect to apply strong emotions, or it is the feverish, meaningless passion of a child" (*KA*, p. 73). Art was not simply "emotion" or "thought," she wrote in 1896, but a combination of the two which she defined as "expression." The artist's ability to express—to press out private emotions into public forms that contained and communicated them—guaranteed that he could complete what Cather termed the "voyage perilous" from the "brain to the hand," the journey from private to public, inspiration to expression (*KA*, p. 417). Cather was fond of quoting Dumas' dictum that drama required only "one passion and four walls," and the walls—those of form as well as of the theater—were as essential to the production of art, she thought, as the passion (*KA*, p. 83). Women writers might possess the artist's passion, but in Cather's view most of them could not construct the walls to contain it.

In drawing on the Romantic model of the poet as divine creator—seemingly the furthest from a woman's reach—Cather nevertheless envisioned a creative process at least potentially available to a woman writer. Her failure to term Shakespeare, Carlyle, and other semi-divine creators "manly" or "virile"—as she had Kipling and Stevenson—suggests that, to use Virginia Woolf's terms, the pattern of creativity displayed by such "man-womanly" male writers might be separated from the masculine as socially defined and hence ultimately be inherited by a "woman-manly" female writer. Even if the artist's social role was male and some aspects of creativity metaphorically associated with paternity, the sympathy, identification, and receptive submission to inspiration Cather attributed to male writers in the 1890s and finally claimed for herself in fact were considered female attributes by her society.

It is also possible that Cather was helped in her struggle to integrate creativity and femaleness by a sexual model of artistic inspiration which on the surface seems firmly to have excluded the woman writer—the vision of creativity as an erotic union between a male artist and a female Muse. As a lesbian, Cather could participate in this sexual and creative scenario more easily than could a heterosexual woman writer, which is perhaps why her college writing reveals her attraction to the idea of the Muse, connected in Cather's imagination with Cleopatra and Aphrodite (*KA*, p. 414).[19]

•

As critic and editor who commanded respectful attention from her college and Lincoln colleagues, Cather seemingly enjoyed unusual power, freedom, and self-expression for a woman writer in Victorian America. The "meatax" judge of others was reversing the linguistic and gender roles in her childhood memory of the little girl who was expected to say what the judge wanted to hear. But since the beginning writer was apprenticing herself to a profession she associated both socially and metaphorically with masculinity, her reconciliation of gender and vocation was necessarily delayed. When she equated the power and force she admired with writers who were not only male but manly, Cather revealed her acceptance of the Victorian ideology of masculinity.

Other nineteenth-century women writers engaged in metaphoric transvestism that at first seems similar to Cather's identification of creativity with masculinity. But when Sand, Eliot, and the Brontës took male (or male-seeming) names, they did not signify their alienation from their gender but their desire to reject social definitions of the feminine. Trying to escape the woman writer's social and literary subordination, they wanted to ensure that their books would not be dismissed as frivolous or second-rate novels by "lady writers." By contrast, Cather's praise of art's virile force reveals her internalization of social categories of gender that other women writers were challenging.

And yet even in the 1890s Cather's views of the artist, the art work, and the creative process were far more complex than her back-slapping praise of football and Kipling alone suggests. There was room for subversion of social categories and stereotypes even in conceptions of art she inherited from the Western literary tradition. The possibility that questioning and transforming such concepts as "artist" and "masculine" could take place can be seen in the second model of creativity we have considered here, which was finally the more useful to Cather in imagining and constructing an artistic identity. Although the writers Cather praised the most highly were male, in drawing on love and sympathy, as she thought they did, Shakespeare and Carlyle found their power in the domain Victorian Americans associated with women.

Throughout the 1890s Cather became more and more drawn to male writers and artists who did not fit socially approved definitions of masculinity. The largely fictional memoir she wrote in 1900—"When I Knew Stephen Crane"—reveals this progress as it sums up the ambiguities and contradictions we have been seeing in Cather's views of gender and creativity.

On the surface, "Stephen Crane" tells the story of a male-identified woman writer eager to be a literary father's son, signified by Cather's use of a male pseudonym—"Henry Nicklemann"—when the article appeared in *The Library,* a Pittsburgh literary magazine. But she published the piece three weeks later in the Lincoln *Courier* under her own name. Although there were pragmatic reasons for avoiding pseudonyms in a hometown journal, Cather's second by-line reveals a less obvious pattern. At first it seems that the Stephen Crane she wants to celebrate is a brother of Kipling, but as her story progresses he increasingly resembles both the "man-womanly" Shakespeare she described in her 1891 essay and Cather herself.

Written shortly after Crane's death, this curious piece purports to be a reminiscence of Cather's meeting with Crane during an 1895 visit to Lincoln, which he was visiting to report on the Nebraska drought for the Bacheller-Johnson syndicate. It is likely that Cather saw Crane in the offices of the Lincoln *Journal* and possible that they talked about literature, although not certain: as Bernice Slote has pointed out, "When I Knew Stephen Crane" is fictional in several details and cannot be taken as biographical fact.[20] But whether it is wholly or largely fabricated, like Cather's early memories the essay tells us what she wished, or imagined, had happened. The story she tells is a rite of literary succession in which a male writer passes on the secret of creativity to the narrator, thus including him (or her) in the American literary tradition.

The narrator is a college student, part-time newspaper reporter, and devotee of arts and letters awed by a brush with literary greatness in the person of Stephen Crane, famous as the author of *The Red Badge of Courage*. The narrator admits that the emaciated, unkempt, impecunious writer was not "impressive" to look at, but he nonetheless inspired hero worship: "I cut my classes to lie in wait for him, confident that in some unwary moment I could trap him into serious conversation, that if one burned incense long enough and ardently enough, the oracle would not be dumb." In some ways the Stephen Crane imagined in this reminiscence is a Kiplingesque hero—the author of stirring war stories and a rough-hewn, manly writer on the move. Having won widespread praise for *Red Badge* shortly before his death, Crane exemplified the fame to which Cather aspired and which she associated with her uncle William Boak's death in the war Crane took as his subject. "His ancestors had been soldiers, and he had been imagining war stories ever since he was out of knickerbockers," Crane supposedly confided to "Henry Nicklemann," describing a childhood much like Willa Cather's (*WP* 2, p. 776). Cather's identification with Crane, the tough-minded literary roustabout, can be seen in the narrator's self-description: "I was just off the range; I knew a little Greek and something about cattle and a good horse when I saw one, and beyond horses and cattle I considered nothing of vital importance except good stories and the people who wrote them."

When the oracle speaks and the narrator receives the desired revelation, however, Crane seems both less tough-minded and less masculine. The secret of the creativity demands remarkably feminine virtues, patience and receptivity: "The detail of a thing has to filter through my blood, and then it comes out like a native product, but it takes forever," Crane supposedly remarked. "I distinctly remember the illustration," Henry Nicklemann continues, "for it rather took hold of me" (*WP* 2, pp. 772–78).

Whether or not Crane ever spoke to the young Cather, it is questionable that he freely offered literary advice. Years later, after she had established her personal and literary identity and had no need to imagine male precursors who included her in rituals of literary inheritance and identity, Cather painted a very different portrait of Crane. Asked by Alfred Knopf to write the preface to *Wounds in the Rain,* she praised Crane's impressionist handling of detail but described him as an unlikely oracle: "there is every evidence that he was a reticent and unhelpful man, with no warmhearted love of giving out opinions" (*OW,* p. 74). In fact the "opinion" she assigns to Crane in 1900 was her own literary aesthetic, a vision of the writer's personal and emotional investment in her material that Sarah Orne Jewett—Cather's real mentor—put in words a few years later, as we know from a more reliable literary reminiscence, the "Preface" to *The Country of the Pointed Firs.*

> In reading over a package of letters from Sarah Orne Jewett, I find this observation: *"The thing that teases the mind over and over for years, and at last gets itself put down rightly on paper—whether little or great, it belongs to Literature."* (*CPF,* p. 6)

The question that naturally arises—why, in 1900, did Cather invest Stephen Crane with her own literary secret and imagine him as the precursor who passes on the mantle of fiction to her?—leads us back to the novice writer's complex relationship to male literary precursors and contemporaries. By 1900 Cather wanted to imagine a male precursor and mentor who was not too uncomfortably masculine. Like her uncle William Boak—also a hero who died in youth—Crane filled this role. Cather identified him with some of the masculine enterprises she admired, and yet he did not possess—at least in Cather's imagination—the manliness and virility which she saw in Kipling. As an artist, Cather's Crane corresponds to the second pattern of creativity discussed in this chapter. As a man, he is no chest-thumping Kipling but a moody, slender youth with "delicately shaped" hands and "thin, nervous fingers," more closely resembling the sensitive, androgynous, and perhaps homosexual hero of "Paul's Case" (written five years later) than the virile warrior–artist Cather portrayed in her college journalism.[21] Like her uncle William Boak, Stephen Crane could be fashioned to suit the requirements of a woman writer who had begun to question the extreme model of masculinity Kipling represented. Cather later satirized Kipling's "outworn" and "chesty" manner, but Crane still impressed her in 1936: "One can read him today" (*NUF,* p. 91).

The two names with which Cather signed the two separately published texts of "When I Knew Stephen Crane"—the male pseudonym and the female given name—reflect her ambivalent attitude toward gender in the memoir. On the surface, Henry Nicklemann's account of his meeting with a tough-minded roving journalist, author of war stories, and respected literary craftsman reveals the male-identified woman writer's need to envision a heroic literary father who would include her in a patrilineal tradition. But Willa Cather's account of her meeting with a sensitive "man-womanly" writer who eventually seems more a brother, or an alternate self, than a father tells the story of Cather's identification with a male precursor whose words could be heard both by Henry Nicklemann and Willa Cather. "When I Knew Stephen Crane" establishes a bond with a literary ancestor whose example and advice did not necessarily exclude a woman writer.[22]

•

Like Cather's other commentaries on male writers in the 1890s, "When I Knew Stephen Crane" reveals how a woman writer differs from those literary sons presented in Harold Bloom's Freudian reading of literary psychohistory.[23] Far from engaging in an oedipal struggle with forefathers she wanted to supplant or overpower, during the 1890s Cather was trying to forge, not break, chains of inheritance and legitimacy. Yet even while she was investing male artists with privilege and authority, she was looking for evidence of female creative power. Cather's early journalism tells us still another story—that of the young woman's search for a female artistic tradition. Although her family ties with women writers and performers brought as many problems as possibilities, the attention Cather gave women artists in her college years and throughout the 1890s reveals that she was not complacently settled in her belief that creativity was God's patriarchal gift to man. Questioning in some articles the masculine values she was affirming in others, during the same decade when she was praising the "brute force" she saw in art and athletics, celebrating the sympathetic imagination she found in *Hamlet,* and imagining Stephen Crane as her first mentor, Cather was searching for something neither Kipling, Shakespeare, nor Crane could offer: proof that "woman" and "artist" could be compatible identities.

NOTES

1. For a comprehensive presentation of the young writer's aesthetic theory, see Bernice Slote, *KA,* pp. 43–81.

2. Sandra M. Gilbert and Susan Gubar trace the interrelationships among these tools of destruction, creation, and procreation and observe that in Western literature "the text's author is a father. . . . An aesthetic partiarch whose pen is an instrument of generative power like his penis." The penis/pen easily transmutes into a weapon, particularly in satire where the "writer's manly rage transforms 'his' pen into a figurative sword." See *The Madwoman in the Attic: The Woman Writer and the Nineteenth-Century Literary Imagination* (New Haven: Yale University Press, 1978), pp. 5–6, 68.

3. Virginia Woolf, *A Room of One's Own* (New York: Harcourt, Brace, 1929), p. 80.

4. Cather's opposition of primitivism and civilization reveals the contemporary interest in atavism, the inversion of optimistic notions of cultural progress based on evolutionary theory ("They may tame the tiger for centuries and think they have made a kitten of it, but in the hot blood of some descendant it breaks out sooner or later"). See *KA,* pp. 230–32.

5. Jackson Lears discusses Guiney's attraction to the male role as well as to the knightly ideal in *No Place of Grace: Antimodernism and the Transformation of American Culture, 1880–1920* (New York: Pantheon, 1981), pp. 124–28.

6. For an analysis of these interrelated cultural pheonomena and their relationship to an emergent capitalism, see Lears, *No Place of Grace*. Larzer Ziff discusses the connections among Americans' fears of cultural decline, Roosevelt's militarism, and Owen Wister's transformation of the knight into the cowboy in *The American 1890s: Life and Times of a Lost Generation* (New York: Viking, 1966), pp. 206–28. In *The Return to Camelot: Chivalry and the English Gentleman* (New Haven: Yale University Press, 1981, Mark Girouard traces a similar cultural synthesis of codes of chivalry, antimodernist impulses, and masculine ideology in nineteenth-century Britain.

7. Mitchell is quoted in Lears, p. 104.

8. *Selections from Ralph Waldo Emerson* ed. Stephen E. Whicher (Boston: Houghton Mifflin [Riverside ed.], 1957), pp. 240–41.

9. Cather's belief in the mimetic power of the imagination contributed to her distaste for realism. Since the writer's loftiest vocation was to create life—not to mirror it—the aesthetics of realism, with its stress on reportorial accuracy and faithfulness to the probable course of events, in her view placed too many constraints on the writer's imagination; the realist's art was circumscribed by the social environment he should be transcending. The novelist should "make," not analyze or preach, she thought, creating literature, not tracts on "heredity, or divorce, or the vexed problems of society" (*KA*, p. 311).

10. For an analysis of the ways in which romance functioned in the Victorian period to express Utopian longings denied an outlet in late capitalism, see Fredric Jameson, *The Political Unconscious: Narrative as a Socially Symbolic Act* (Ithaca: Cornell University Press, 1981), pp. 103–50, esp. pp. 103–10.

11. John G. Cawelti analyzes how popular and formulaic literature reinforces the social structures they seem to challenge in *Adventure, Mystery, and Romance: Formula Stories as Art and Popular Culture* (Chicago: University of Chicago Press, 1976).

12. See Ziff, pp. 224–28.

13. Woolf, p. 106.

14. Willa Cather to Mariel Gere, August 4, 1896, Gere Collection, Nebraska State Historical Society, Lincoln, Nebr.

15. For a discussion of the undergraduate writer's use of chivalric and religious motifs, see Bernice Slote, *KA*, pp. 43–45.

16. See Gilbert and Gubar, pp. 3–7.

17. "Perhaps a mind that is purely masculine cannot create, any more than a mind that is purely feminine," Virginia Woolf reflects. "In fact one goes back to Shakespeare's mind as the type of the androgynous, of the man-womanly mind" (*Room*, p. 102).

18. Emerson, p. 233 (see n. 8 above).

19. Women writers who reverse the Muse's gender in imagining or experiencing creative inspiration face a dilemma, Joan Feit Diehl argues; because the male Muse then can be identified with a male precursor, the woman writer may find the creative process involving a frightening submission to male authority (" 'Come Slowly—Eden': An Exploration of Women Poets and Their Muse," *Signs: Journal of Women in Culture and Society* 3, no. 3 [Spring 1978]: 572–87). See also Gilbert and Gubar, pp. 607–11, and Albert Gelpi, *The Tenth Muse: The Psyche of the American Poet* (Cambridge: Harvard University Press, 1975), pp. 246–60. As Lillian Faderman and Louise Bernikow observe in their comments on Diehl's article, a lesbian writer could experience the creative process differently (*Signs: Journal of Women in Culture and Society* 4, no. 1 [Autumn 1978]: 188–91 and 191–94). For Cather, whose imagination returned to female presences, the creative process did not

involve this double submission to masculine force. Her precursors were male, but her muses were not.

20. For an analysis of the fictional elements in "When I Knew Stephen Crane," see Bernice Slote, "Stephen Crane and Willa Cather," *The Serif* (December 1969): 3–15.

21. Crane was also an unthreatening literary precursor because Cather viewed his work with respect but not reverence; she was capable not only of discerning literary flaws (noting the "careless sentence-structure" as well as the "wonder" of *Red Badge*, for example [*WP* 2, p. 773]) but also of producing the same "meatax" criticism she used to attack female writers. Reviewing his collection of poems *War Is Kind* in 1899, she speculated that "Either Mr. Crane is insulting the public or insulting himself, or he has developed a case of atavism and is chattering the primeval nonsense of the apes." In contrast to some of his puerile verses, she contended, " 'Jack and Jill' or 'Hickity, Pickity, My Black Hen' are exquisite lyrics" (*WP* 2, p. 701).

22. In "(E)Merging Identities: The Dynamics of Female Friendships in Contemporary Fiction by Women" (*Signs: Journal of Women in Culture and Society* 6, no. 3 [Spring 1981]: 413–35), Elizabeth Abel suggests that the woman writer's "relation to the male tradition reflects women's oedipal issues, the relation to female tradition reflects the preoedipal" (433–34). Although Cather's relationship to the male literary tradition reveals the preoedipal issues of merging and separation as well as the oedipal issues of struggle and conflict, her relationship to the female literary tradition more strongly replays the preoedipal issues characterizing the early mother–daughter bond.

23. Harold Bloom, *The Anxiety of Influence* (New York: Oxford University Press, 1973).

8

Women's Voices, Women's Stories

> Théy were listening to a Mexican part-song; the tenor, then the soprano, then both together; the barytone joins them, rages, is extinguished; the tenor expires in sobs, and the soprano finishes alone. . . . Then at the appointed, at the acute, moment, the soprano voice, like a fountain jet, shot up into the light. . . . How it leaped from among those dusky male voices! How it played in and about and around and over them, like a goldfish darting among creek minnows, like a yellow butterfly soaring above a swarm of dark ones.
>
> *The Song of the Lark*

> I have not much faith in women in fiction. They have a sort of sex consciousness that is abominable. They are so limited to one string and they lie so about that. . . .When a woman writes a story of adventure, a stout sea tale, a manly battle yarn, anything without wine, women and love, then I will begin to hope for something great from them, not before.
>
> *The Kingdom of Art*

When she was visiting London in 1909 to secure manuscripts for *McClure's*, Willa Cather attended a meeting of expatriated Russians who were celebrating the release of political prisoner Vera Figner, freed after spending over twenty years in a Russian prison, much of them in solitary confinement. Since Figner spoke to her compatriots in her native language, Cather ignored verbal content and listened only to the voice itself. She found it "one of marvellous resonance and power, beautifully modulated in spite of the fact that it was mute for so many years."[1] Cather was particularly affected by the captive's transition from muteness to speech. In her account of the meeting this verbal awakening becomes the symbol of Figner's imprisonment and release, perhaps because it reflected Cather's own struggle to establish her voice as a writer, a matter of particular conflict in 1909 when she was simultaneously an apprentice to S. S. McClure and to Henry James. But her response to Figner also reflects an interest in the female voice first evident in her college years.

Even during the 1890s when Kipling's strident tones, Whitman's barbaric yawp, and Homer's epic chant had temporarily drowned out the female storytellers of her childhood, Cather was listening to women's voices speaking, singing, and writing. They did not overwhelm baritones and tenors as did Thea Kronborg's powerful soprano, but the individual voices of actresses, opera singers, poets, and novelists provided an important counterpoint to the masculine chorus. In the female voices she heard on the stage, Cather recognized a dramatic incarnation of femininity and creativity: the opera singer's voice in particular became her recurrent metonymy for the woman artist. The female voices she read on the page were less inspiring. The stories women writers told, and the stylistic and narrative strategies they employed to tell them, made Cather question whether literary women could ever speak with authenticity and authority.

•

"Has any woman ever really had the art instinct, the art necessity?" Willa Cather inquired in commenting on Mary Anderson's retirement from the stage. "Is it not with them a substitute, a transferred enthusiasm, an escape valve for what has sought or is seeking another channel?" (*KA*, p. 158). Among the actresses and opera singers Cather encountered during her college years, only a sanctified few, she thought, possessed the "art necessity," an undeniable drive to create complemented by talent and technique. Others were at least competent professional women who challenged Cather's stereotypes of Lincoln females as silly sorority girls and culture-hungry clubwomen. Such women demonstrated that professional artistry was not exclusively a male domain; in their dramatic and operatic roles, they claimed the equality with men denied them in their social roles.

As emancipated women unconstrained by middle-class ideals of domesticity, piety, and perhaps even chastity, actresses and opera singers held a symbolic importance in Victorian society far in excess of their numbers.[2] Certainly actresses were important symbols of female autonomy and creativity to Willa Cather. As a child in Red Cloud she worshipped these deities from across the footlights, while as Lincoln drama critic she drew closer to the performers' charmed world, reviewing their work, earning their friendship, and sharing their champagne.[3] Opera was a slightly later discovery: Cather saw her first full-scale opera in 1895 when she traveled to Chicago for a visit to the Metropolitan. This trip marked the beginning of her life-long passion for the art form and for the divas who dominated it. Her admiration for women opera singers later culminated in her friendship with Wagnerian soprano Olive Fremstad and in the novel celebrating her own artistic birth and Fremstad's vocation, *The Song of the Lark*.

Seeing and developing friendships with some women performers, Willa Cather encountered female artistry personified before she committed herself to a literary vocation. Actresses and opera singers did not simply represent female art, as did the woman author's name on the title page. They incarnated it. An abstraction—female creativity—became tangible and credible in their presence. Writing to Dorothy Canfield in 1916, Cather explained that she had decided to make her heroine in *The Song of the Lark* an opera singer because her artistry was concrete; people saw and heard a living voice before them, she continued, and could not help being captured by a Jenny Lind.[4]

The columns Cather wrote for Lincoln and Pittsburgh newspapers in the 1890s reveal her fascination with women performers. Eager to learn what made the strange cohabitation between "woman" and "artist" possible, Cather filled her drama and music reviews with questions. Did women artists have innate limitations relegating them to secondary status in the arts as in society? Were women's emotional needs so strong that they would inevitably abandon the artist's lonely vocation for human relationships? Querying herself as well as her readers, at the same time that she was celebrating Kipling's manly force, Cather was not happily settled in assuming that creativity was a patrilineal inheritance that she could receive only in her William Cather costume. Even though she neither fully realized nor admitted it in the 1890s, her artistic inheritance would have to come, at least in part, from the female line.

As a drama critic Willa Cather respected the creativity and the strenuousness of the actor's profession. An actor was "an author who writes a book every night," she thought, "an artist who every evening paints a picture in the gaslight" (*KA*, p. 215). Cather's commentaries on specific performances and performers create no gender-based hierarchies of talent. Indeed, the two actresses who most impressed her in the early 1890s—Sarah Bernhardt and Eleanora Duse—not only equalled their male colleagues; they also possessed the divine attributes Cather worshipped in male writers. Frequently opposing the two women in her journalism, Cather attributed their creative power to different sources. Bernhardt was passionate and fiery, Duse spiritual and lofty; Bernhardt's art was one of "disclosure," Duse's one of "concealment" (*KA*, p. 119). Such women were legitimate inheritors of the male tradition without being masculine. Duse was, Cather thought, a "daughter of Dante" (*KA*, p. 154).

Cather's tributes to the great actresses of her day frequently grant them the force and potency she attributed to male conquerors, warriors, and heroes. Mary Anderson could occasionally "aspire and create and conquer . . . strike fire from flint . . . make the blind see and the deaf hear," while Clara Morris could stun her audience with "thunder-bursts of passion and pain" (*KA*, pp. 155, 154). Such gifted performers knew the artist's "secret" Cather defined in *The Song of the Lark:* passion, that "power of supreme love" she found most fully expressed in Shakespeare. Watching Bernhardt was a particularly wrenching emotional experience for the audience, Cather thought, because she communicated feeling so intensely that she aroused it in the viewer (*KA*, p. 120). Transforming emotion into art, private feeling into common experience, the great actresses experienced the creative process as did Cather's divine artist. Sharing the creator's sympathetic imagination with writers like Carlyle and Shakespeare, in Cather's view Duse and Bernhardt could also vivify a world. At their finest, the great actresses created performances so convincing they seemed to be "more lifelike than life," creations Cather regarded as "nature's work."

Because actresses like Duse and Bernhardt also possessed the male writer's formal control, they could safely make Cather's "voyage perilous," that short but treacherous journey from the "brain to the hand." The fiery Sarah Bernhardt communicated intense feeling to her audience because years of "rigorous training" gave her the "technique that in itself is enough when for a moment her inspiration fails." Like the great male artist, Bernhardt balanced inspiration and control, emotional power and finely honed craft: "She never fails or disappoints you because under all those thousand little things that seem so spontaneous there is a system as fixed and definite as the laws of musical composition'' (*KA*, p. 121). And yet she was not masculine. In 1896 Cather declared confidently that Bernhardt would never play men's parts: ''she knows so well the power of sex in art,'' she thought—the power both of sexuality and of her own sex (*KA*, p. 126).

Yet female performers who possessed creative genius were uncommon, Cather concluded. Commenting on the personal lives of actresses and singers in her 1890s journalism, Cather saw ''woman'' and ''artist'' in almost irreconcilable conflict. Most women were unwilling to sacrifice human relationships for the artist's demanding vocation. Love and work, marriage and career, could not mix, she thought, and so what ''ennobled'' one actress as a woman (her decision to leave the stage for marriage) ''sadly limited'' her as an artist (*KA*, p. 158). Or, as she put it elsewhere, perhaps thinking of the raped and silenced Philomela of Greek legend, ''Married nightingales seldom sing'' (*KA*, p. 176).[5]

Willa Cather was not the only woman writer who saw a conflict between woman's role and an artistic vocation in late nineteenth-century America. The metaphoric association of childbirth with artistic production favored by male artists was not a resource for women writers who generally portrayed motherhood and creativity, reproduction and literary production, as highly incompatible. Rebecca Harding Davis' *Earthen Pitchers,* Elizabeth Stuart Phelps' *The Story of Avis,* Kate Chopin's *The Awakening* and ''Wiser Than a God,'' Louisa May Alcott's *Little Women,* and May Austin's *A Woman of Genius* tell the same story: to create, the woman must sacrifice marriage; to marry, she must sacrifice art. The dedication, autonomy, and self-expression required of the artist were, these women writers agreed, inconsistent with the roles of wife and mother, which demanded selfless, time-consuming, draining devotion to others. Recognizing this opposition, Chopin's pianist–heroine in ''Wiser Than a God'' refuses to marry the man she loves ''Because it doesn't enter into the purpose of my life.'' Chopin demonstrates the other alternative in her ironic portrayal of the sumptuously maternal Adele Ratignolle in *The Awakening;* the nineteenth century's idealized ''mother–woman,'' Adele channels her creativity into designing and sewing ''a diminutive pair of night drawers.'' Adele's recompense for this socially approved task is to become art–object rather than artist, the heroine of a script that is, Chopin hints, male-authored: she is the ''bygone heroine of romance,'' the ''fair lady of our dreams.''[6]

Seeing this conflict in her real-life heroines and later portraying it in *Troll Garden* stories like ''A Wagner Matinee,'' Cather did not scorn female artists who chose love over vocation. They were making the choice to which she was, in part, strongly drawn when they chose the ''human relationships'' she described as the ''tragic necessity'' of life in ''Katherine Mansfield'' (*NUF*, p. 136). The emotional and sexual power of her attachments to women threatened her own independence, so Cather understood the conflict heterosexual women experienced between art and love. Writing on Mary Anderson's decision to leave the stage for the altar and the home, Cather praised the actress' ''clear vision'' in exchanging fame and fortune for marriage. And yet the price was clear: as a married nightingale, she would not sing.

The silencing of a woman's voice meant a double loss to Cather at this time.

She valued the female voice metaphorically as a sign of self-expression and creativity—literally as the natural instrument of theatrical and musical power. In her drama reviews she frequently tried to capture the shadings and tonalities of actresses' voices. In opera, the power and importance of the female voice were even more prominent, and Cather's enduring passion for the art form must be seen as part of her fascination with the female voice.

•

Willa Cather's love for music and musicians is readily apparent to a reader of her fiction. Two novels *(The Song of the Lark* and *Lucy Gayheart)* and many short stories (among them "Nanette: An Aside," "A Singer's Romance, "A Wagner Matinee," " 'A Death in the Desert,' " "The Garden Lodge," "The Diamond Mine," "A Gold Slipper," and "Uncle Valentine") explore the worlds of musicians and opera singers. Elsewhere in her fiction musical references contribute to a rich, intricate web of allusion, theme, and structure.[7] The musicians in her novels are not merely generic artists or convenient stand-ins for writers. Cather was interested in music and opera as specific as well as representative art forms. From her earliest days in Red Cloud, where she listened to Julia Miner's piano playing and bombarded Professor Schindelmeisser with questions, to her last days in New York as "Aunt Willa" to the Menuhin children, she prized close friendships with music-lovers and musicians. As Edith Lewis observes, music for Cather was not an "intellectual interest" but an "emotional experience" that had a "potent influence" on the creative process. Cather's readers had not "sufficiently emphasized, or possibly recognized, how much musical forms influenced her composition, and how her style, her beauty of cadence and rhythm, were the result of a sort of transposed musical feeling," Lewis thought, "and were arrived at almost unconsciously" (*WCL,* pp. 47–48).

Why did music have such a "potent influence" on Cather and her creative process? Certainly music offered her the fusion of passion and form she valued in great art, presenting a symbolic as well as experiential reconciliation of the tension between disorder and order, emotion and control, desire and limitation. Even the most tempestuous passage from Tchaikovsky or Wagner follows what she called the "fixed and definite . . . laws of musical composition." Although the laws of composition are "definite," music's impact on the listener is imprecise and suggestive; because it does not rely on words to communicate, music enters the realm beyond language Cather associated with the "overtone divined by the ear but not heard by it" in "The Novel Démeublé," conveying the "thing not named" to the listener (*NUF,* p. 50). Music, Cather thought, was the "speech of the soul," the discourse of the ineffable and the transcendent. Music was also a "notable emotional language," the discourse of the feelings and passions Cather associated with our primitive origins: "Yet, forget it not," she wrote in 1899, "music first came to us many a century ago, before we had concerned ourselves with science, when we were but creatures of desire and before we had quite parted with our hairy coats, indeed; and then it comes to us a religious chant and a love song" (*WP* 2, p. 654).

As a young woman discovering that her intense friendships with other women were socially defined as "unnatural," Cather may also have found music compelling because it offered her a text without words—or, in the case of opera, a musical text accompanied by a verbal text in a language other than English. Whereas Cather the reader had to translate literary texts into her own "emotional language," Cather the listener could project "unnatural" passions directly into the aesthetic experience. In music nothing is named and so everything can be imagined by the listener who is less constrained by the text than the reader. Of course in opera, which Cather loved for its fusion of music, drama, and narrative, passion is inscribed as heterosexual in the narrative line. But the artifice of a performance of an opera like *Fidelio* visually offers the possibility for transformation and inversion of gender and sexuality: when a woman dons male attire and becomes the object of another woman's love, the lines between heterosexual and homosexual, sexual and nonsexual, male and female—fixed in the social world—become blurred upon the stage. Yet the potentially disruptive passion is contained: the listener's emotional release is made pleasurable by the "four walls" of the theater as well as by the "laws" of musical composition.

In attempting to distinguish opera singers from actresses, Cather observed that "Singing is idealized speech" (*KA*, p. 217). The women who spoke opera's idealized speech, far more than the men, struck her as embodiments of opera's aesthetic, emotional, and imaginative power, and Cather's commentaries on women singers reveal a set of metaphoric equivalences that counterpoint and revise the associations of

sword/penis/pen/male/artist.

She began to develop another pattern in her 1890s journalism that flowers in her later fiction, culminating in *The Song of the Lark:*

vessel/womb/throat/voice/woman/artist.

What allowed her to form this revisionist and potentially woman-centered set of equivalences?

In opera the soprano generally dominates the stage and her male counterparts. She is queen of the performance and ruler of the audience, as Cather suggested when she referred to the Wagnerian soprano Nordica as a "splendid Amazon warrior" (*WP* 2, p. 644). Most operas pay musical homage to the soprano's vocal range. In contrast to the social world, in the opera house the higher female voice signifies authority and power rather than insubstantiality and lightness, as Cather suggests in this chapter's epigraph from *The Song of the Lark* where Thea's vibrant soprano voice triumphs over the "extinguished" baritone and the expiring tenor, easily ascending to prominence as it soars above the "dusky male voices" (*SL*, p. 235).[8] Part of the opera singer's appeal was her ability to please mass audiences as well as discriminating critics; she bridged the gap between high- and middle-brow culture that began widening in the nineteenth-century literary world. Writing Dorothy Canfield about her decision to make her artist–heroine in *The Song of the Lark* a diva, Cather said that a singer was the only artist who could impress Moonstone (Thea's hometown). After all, Cather pointed out, Red Cloud people would go to Kansas City to hear Geraldine Farrar.[9]

To Willa Cather in the 1890s the singing voice also signified a woman's possession of an individual artistry. Reviewing a performance by Melba in 1898, Cather observed that "Una Voce Poco Far" was "sung as just one voice in all this world can sing it. . . . One could travel the earth over without finding another organ of such exquisite mechanism; it is a thing apart and unique" (*WP* 1, p. 416). In Cather's reviews throughout the decade, the opera singer's voice frequently seems to take on a life of its own and represent the woman singer in particular, female creativity and originality in general. Consider this 1895 review of a performance by Melba:

> Her voice can act just as it can do everything else. Those vibrant tones can plead and thrill and suffer, can love and hate and renounce of themselves, alone, without aid of any kind. She swells a B flat that is all the triumphant anguish of resignation, she runs cadenzas that are the very swooning ecstasy of love, sings allegros that are the black bottom of hopeless despair. . . . It is not like other voices, it is an individual living thing which can feel and exult and experience. (*KA*, p. 132)

Gifted with a "thing apart and unique," the female singer embodied an artistic self-sufficiency Cather found attractive. Not only did each prima donna possess a voice "not like other voices," but also, unlike the woman writer, she had no direct male artistic models or precursors. In her 1911 short story, "The Joy of Nelly Deane," Cather demonstrates the female singer's autonomy and natural power. The hometown girl who "never had a singing lesson" triumphs in a performance of *Queen Esther* by overpowering the tenor, her schoolmaster, who vainly tries to "eclipse" his student in "his dolorous solos about the rivers of Babylon." But it is "Nelly's night." Becoming as well as enacting a queen, she easily surpasses the representative of male social authority, the teacher, when her natural musical authority is given prominence on the stage (*CSF,* pp. 55–68). Later Cather dramatically portrays the opera singer's independence, ensured by her possession of her own instrument, in the recognition scene in *The Song of the Lark* where Thea's teacher Harsanyi realizes what she has always known: she is a singer, not a pianist. After she sings to him for the first time and he recognizes her gift Harsanyi can only play the role of catalyst to her awakening voice.

Although Cather recognized that a great voice was the product of intelligence, talent, and hard work, because it was rooted in the body—in the physical and the natural—the singer's voice signified the union of femininity and creativity to her.[10] Reviewing the sinuous, serpentine Clara Butt's performance of "The Enchantress," Cather observed that when the singer was "in her element" the tones of her "deep, sonorous, self-sufficient voice" became a natural, unsocialized force: it was an "ocean of voice . . . like the moan of the sea or the sighing of the forest in the night wind" (*WP* 2, p. 649). Separated from linguistic systems, linking body and spirit, the voice allowed some singers to enter the creative realm of the "thing not named," the communicative sphere apart from and beyond language that the woman writer had to struggle to find.

By the time she wrote *The Song of the Lark,* Cather had found the female voice a powerful symbol to counter her earlier identification of creativity with masculinity. Thea Kronborg's artistic power springs from the female body. "Why had he never guessed it before?'' Harsanyi wonders. ''Everything about her'' indicated a great voice, he reflects; and the signs he retrospectively discerns are all physical, nature's gifts along with Thea's body: ''the big mouth, the wide jaw and chin, the strong white teeth, the deep laugh'' (p. 188). Just as the Indian women made beautifully decorated vessels to hold grain, so Thea comes to understand that a singer makes a ''vessel of [her] throat and nostrils.'' Like the Indian women's vessels and the womb with which they are metaphorically associated, Thea's throat can pour forth as well as contain. Her ability to release the voice to the world demonstrates Cather's transformation of a symbol traditionally used to signify female passivity and immanence: Thea's vessel is not a receptacle for male power, but a container and vehicle for female power. Combining nature with culture, biological with artistic gifts, unconscious, inborn endowment with technique and discipline, the woman singer's voice/vessel/womb is Cather's final rival to the male writer's pen/sword/penis, a transformation hinted at in the 1890s writing and achieved in *Song*.

So even while she was praising manly writers, Cather found examples of women's creativity. The soprano's voice suggested to her that femininity and creativity might be naturally—if not socially—compatible. And because the great prima donnas of the late nineteenth and early twentieth century were beginning a tradition, someday, Cather hoped, social recognition of female artistry would follow. ''American prima donnas of the future will look back upon your memory with pride and gratitude,'' she informed Nordica. ''You seem to me to embody all that is best in American womanhood'' (*WP* 2, p. 646).

•

Important as women performers were to Willa Cather, they were nonetheless remote models for a woman writer who had to speak with her pen. The arts of performance and original composition are not the same. No matter how individual her voice, the woman performer interprets preexisting texts and scripts, often enacting stereotypic female roles and presenting male images of women, as did Clara Butt when she sang the enchantress' role. Moreover, a woman performer, then as now, does not invade a male sphere. Operas require sopranos and plays actresses, but American letters has no preordained place for women writers.

And yet the voices of actresses and opera singers helped Cather to imagine the woman writer's creativity. In later comments on her own writing and that of other women writers, Cather frequently uses the term ''voice'' to describe original, unselfconscious prose. During her apprenticeship period when she was enthralled with Henry James, she admitted that ''I was trying to sing a song that did not lie in my voice.''[11] By contrast, she thought that Sarah Orne Jewett possessed that ''cadence'' or ''quality of voice that is exclusively the writer's own, individual, unique,'' and so her stories seemed to Cather to have ''inherent, individual beauty; the kind of beauty we feel when a beautiful song is sung by a beautiful voice that is exactly suited to the song'' (*CPF*, p. 7). Cather associated Katherine Mansfield's ''very individual talent'' with developing her ''very own'' voice, a ''timbre'' which ''cannot be defined or explained any more than the quality of a beautiful speaking voice can be'' (*NUF*, pp. 134–35).

In using the concept of ''voice'' to describe the writer's art, in addition to suggesting a command of the usual literary resources of style, syntax, tone, diction, and subject matter, Cather implies individuality, originality, and identity. Although Cather did not equate voice with language, she knew that—unlike the opera singer—the woman writer only had the resources of language to exploit in developing a voice. This significant difference between writer and performer suggests the difficulty Cather confronted in seeking her own "timbre" within a linguistic system she shared with others—and a system, as feminist critics have pointed out, that both reflects and inculcates masculine values.[12]

Although Cather did not think that literary women could equal the opera singer's oceanic power, she was nevertheless reading their work throughout the 1890s. Not self-consciously thinking back through her literary mothers during the 1890s, as she did after meeting Sarah Orne Jewett, Cather was still intensely engaged with female predecessors in her newspaper columns—chastising, rebuking, encouraging, mocking, celebrating, advising. Cather's complex, ambivalent relationship with her female literary precursors and contemporaries reveals her searching, however unconsciously at this point, for a female literary tradition to which she could belong.[13]

Although Cather found the female literary tradition problematic, at least it was there. By the end of the nineteenth century Cather was reading fiction by such predecessors as Jane Austen, Charlotte Brontë, George Eliot, and George Sand; by contemporaries like Kate Chopin, Sarah Orne Jewett, and Gertrude Atherton; and by the best-selling women novelists whose works filled her mother's bookshelves: Harriet Beecher Stowe, Mrs. E.D.E.N. Southworth, Augusta Evans, Ouida, and Marie Corelli. As for poets, there were Elizabeth Barrett Browning and Christina Rossetti in the previous generation, Ella Wheeler Wilcox and Louise Imogen Guiney in the contemporary one. And yet as she faced the female lineage that Virginia Woolf revered and sought to recover in *A Room of One's Own,* Cather was frequently dismayed. "Sometimes I wonder why God ever trusts talent in the hands of women," she wrote in 1895, "they usually make such an infernal mess of it. I think He must do it as a sort of ghastly joke" (*WP* 1, p. 275).

In her reviews and columns, Cather plays her own ghastly joke on the literary woman, placing her in an almost irresolvable double bind: either she could write as a woman, in which case she created a limited art, or she could write as a man, in which case she created an inauthentic art. Writing in 1895 in response to Christina Rossetti's death, Cather chose to raise the "very grave question whether women have any place in poetry at all." If they did, woman's place was as circumscribed in poetry, she thought, as it was in society:

> Certainly they have only been successful in poetry of the most highly subjective nature. If a woman writes any poetry at all worth reading it must be emotional in the extreme, self-centered, self-absorbed, centrifugal. Generally it is confined either to reverence or love. . . . A woman has only one gift and out of the wealth of that one thing she must sing and move with song. . . . When he [Zeus] came to woman he had nothing of that kind left to give [the gift of reason that he gave to man], so he gave her the power of loving. . . . Out of the fulness of that power a woman must do her work. . . . A woman can be great only in proportion as God put feeling in her. (*WP* 1, p. 146)

Cather's woman poet cannot win. To write she must draw upon her one gift, the "power of loving," yet in doing so she creates a minor art. She must be constrained by the lyric, which Cather viewed as a lesser genre than those her brothers commanded—epic and drama—and produce inward-turning and solipsistic ("centrifugal") texts. (By contrast, when a male artist like Shakespeare drew on his "power of loving," Cather thought he connected to others through the sympathetic imagination.)

On the other hand, the woman who aspires to the male artist's higher range and replaces the power of loving with learning loses all chance of developing an original voice. "Learning and a wide knowledge of things does [sic] not seem to help women poets much," Cather continued. "It seems rather to cripple their naturalness, burden their fancy and cloud their imagination with pedantic metaphors and vague illusions." Using Rossetti and Barrett Browning to illustrate the woman poet's two choices, Cather cast Rossetti as the supreme but limited lyricist who kept to woman's poetic place and Barrett Browning as the overreacher who sacrificed naturalness for knowledge.

Until this point in Cather's commentary, the double bind seems inescapable: a woman poet could fail either by writing as a woman or by imitating men, the two possibilities she also held out for herself. But Cather granted one exception. Sappho transcends her mutually exclusive categories:

> There is one woman poet whom all the world calls great, though of her work there remains now only a few disconnected fragments and that one wonderful hymn to Aphrodite. Small things upon which to rest so great a fame, but they tell us so much. If of all the lost richness we could have one master restored to us, one of all the philosophers and poets, the choice of the world would be for the lost nine books of Sappho. Those broken fragments have burned themselves into the consciousness of the world like fire. All great poets have wondered at them, all inferior poets have imitated them. Twenty centuries have not cooled the passion in them. Sappho wrote only of one theme, sang it, laughed it, sighed it, wept it, sobbed it. Save for her knowledge of human love she was unlearned, save for her perception of beauty she was blind, save for the fullness of her passions she was empty-handed. She was probably not a student of prosody, yet she invented the most wonderfully emotional meter in literature, the sapphic meter with its three full, resonant lines, and then that short, sharp one that comes in like a gasp when feeling flows too swift for speech. She could not sing of Atrides, nor of Cadmus, nor of the labors of Hercules, for her lyre, like Anacreon's, responded only to a song of love. (*WP* 1, p. 147)

In drawing on the woman poet's "one gift," Cather's Sappho managed to avoid both Barrett Browning's inauthenticity and Rossetti's subordination, becoming a "master" without imitating men; her expression of passions so potent that "twenty centuries have not cooled . . . them" grants her the power Cather found in opera singers, that of an original voice. Nor did Sappho sacrifice range and force for a limited lyric art. Cather conveys her sense of Sappho's range by the verbs she chooses as synonyms for "wrote"—"sang," "laughed," "sighed," "wept," "sobbed"—which evoke a complex, varied presentation of feeling and a multitude of voices and cadences. And perhaps most important, Cather's Sappho is not an inspired but unregulated, untutored female poet: she places passion within the four walls of form, inventing a meter that both reveals and contains emotion.

Cather's Sappho also eludes the limitations constricting other women poets by finding an enduring readership and literary descendants (one of whom would be Willa Cather in "The Star Dial" [1907]). Whereas Rossetti, in Cather's view, was not sufficiently "fervid and impassioned" to gain immortality or "potent" enough to "greatly influence or guide the poets of the future," Sappho's explosive "broken fragments" were still moving readers after twenty centuries. Having created a form that transcended her own lyrics, Sappho reversed the traditional paradigm of male poet/female imitator. One of her followers was Swinburne, to Cather a "great living English poet" and an inheritor of Sappho who "imitated the Sapphic measures perfectly in stubborn, unyielding English syllables" (*KA*, p. 350).

There may have been other reasons that Cather considered Sappho a magnificent exception to the opposition between "woman" and "writer." As a Greek poet, she belonged to the male-dominated classical tradition that Cather venerated; as a poet whose work consisted of fragments and commentaries by male poets and critics, Sappho was easy to revere (Cather could imagine her "lost richness" as even greater than what was left); and as the poet of "love and maidens," as we have seen, Sappho offered Cather a mirror in which she could see herself and her love for women as natural. Cather may also have been attracted by one important difference. Fearing her own "power of loving," in a great poet whom she thought combined passion with craft Cather could view her conflicts glorified and reconciled. Sappho conveyed her abandonment in controlled, sharply etched lines and suggested realms of feeling beyond language through her mastery of form and meter.

•

The poet's vocation has struck some feminist critics as even more contradictory to feminine identity and role than the novelist's. According to this reasoning, the woman poet invades classical realms of learning and literature traditionally off limits to her sex; because the poetic process is associated with divine inspiration, the poet is a secular priest (one of God's "elect," as Cather phrased it) and hence occupies a literary role too elevated for women; and the post-Romantic lyric's focus on the self and the awakened imagination makes poetry an egoistic genre inconsistent with the self-effacement culturally required of women. By contrast, the novel was a more recent and less prestigious genre than poetry, connected, from its bourgeois beginnings in the eighteenth century, with women writers and readers; novel-writing could be a proper, even if a conflict-ridden, occupation for middle-class women forced to support families since—unlike poetry—it could provide income; the stories women writers told were frequently didactic or romantic, thus reinforcing accepted definitions of woman's role and nature; fiction writers drew on the supposedly female talent for accurate social observation and emotional insight; and in creating characters apart from themselves, unlike the lyric poet, female novelists displayed the feminine quality of self-effacement, modestly directing the reader's gaze toward others, not themselves.[14]

But Willa Cather thought fiction the loftier genre. To her, "woman" and "novelist" seemed even more contradictory identities than "woman" and "poet." The fiction writer's ability to vivify a world was to Cather a closer analog to God's original creative act than the poet's introspection. Enraptured since her childhood with storytelling, Cather continued to find the novelist's narrative art supreme throughout the 1890s. Writing in 1895, she declared it a

> solemn and terrible thing to write a novel. Ink and paper are so rigidly exacting. One may lie to one's self, lie to the world, lie to God, but to one's pen one cannot lie. . . . You are then a translator, without a lexicon, without notes, and you are to translate—God. (*KA*, p. 409)

Yet the woman novelist, in Cather's view, did not fare well in this supreme art. Unlike the actress, the opera singer, or the poet, she had to find her voice within the social world not only because she had to use language, but also because she had to tell stories. Rather than "translate God," she would then lie by writing one of the sentimental and romantic tales that Cather considered dishonest: "They [women novelists] are so limited to one string and they lie so about that" (*KA*, p. 409). Since most women novelists lied, Cather could not trust them. "I have not a great deal of faith in women in literature," she proclaimed in 1897. "As a rule, if I see the announcement of a new book by a woman, I—well, I take one by a man instead. . . . I have noticed that the great masters of letters are men, and I prefer to take no chances when I read" (*WP* 1, p. 362).

But this comment itself is not fully honest. Cather may have tried "great masters" first, but she kept reading novels by women. Her columns and reviews of the 1890s mention, among others, Jane Austen, George Eliot, Mrs. Humphrey Ward, George Sand, Madame de Staël, Charlotte Brontë, Augusta Evans, Mary Jane Holmes, Harriet Beecher Stowe, Louisa May Alcott, Sarah Grand, Ouida, Marie Corelli, Amelie Rives, the "Duchess," Frances Hodgson Burnett, Gertrude Atherton, and Kate Chopin. Cather's relationship to this female literary tradition was characterized by complex attachment rather than outright dismissal. The "great masters of letters" may have been men, but she nevertheless took "chances" throughout the decade, reading novels even by women writers she professed to despise.

•

The popular Victorian women novelists Willa Cather read can be divided into two schools: the home-centered American novelists known as the sentimentalists, or, more recently, as the "literary domestics," and the racier, largely British group of writers who wrote "sensational" fiction.[15] The first group flourished largely from 1840 to 1870, the second from the Civil War to the turn of the century. The lines dividing the two groups are not always clear. Mrs. E.D.E.N. Southworth, one of Mary Kelley's literary domestics, is included by Dee Garrison with the sensationalist writers of "immoral fiction" whose books were banned by some libraries in the late nineteenth century.[16] Although Cather did not employ this terminology or seek self-consciously to categorize women writers, her commentaries of the 1890s reveal that she distinguished among writers roughly according to these two groupings.

By now, the story of the rise of women's popular fiction in Victorian America is a familiar one. As the publishing industry became a profitable business in the nineteenth century and the book a salable commodity, women dominated both readership and authorship. Publishers eagerly sought writers whose stories of domestic and romantic life captured the hearts of female readers; writing became an acceptable, even if at times a conflict-ridden, profession for Victorian women who could earn money to support families without literally leaving the home. These prolific female novelists and their best-sellers prompted Hawthorne's famous diatribe against the "scribbling women" whose literary "trash" was depriving him of readership.[17] Hawthorne's economic, if not his literary, analysis was correct. Any one best-seller by a female novelist outsold all the writings of Whitman, Hawthorne, Melville, and Thoreau combined in the 1850s.[18] Scorning these writers' literary talents yet fearing their economic power, male writers confronted their rivals with a mixture of contempt and anxiety. Mark Twain's hilarious Emmeline Grangerford—the inspired, untutored poetess who could "rattle off" mortuary verse without ever stopping to think—pined away and died when she failed to come up with a rhyme for "Whistler," but her real-life sisters were more robust.[19] Popular and prosperous, they brought about a revolutionary change in American literature and society. As Nina Baym notes, as a result of their success "authorship in America was established as a woman's profession."[20] Yet Cather and many other American readers, writers, and critics made an important distinction. Middle-class women novelists made *popular* authorship a woman's profession. Serious authorship remained a male preserve.

Devoted to her religion of art, the young Willa Cather was not impressed by popular women authors, whether British or American. They might be successful professionals, but in her view they were not artists. In several columns and reviews of the 1890s she expresses her distaste for these undesirable literary relatives as firmly as did George Eliot in "Silly Novels by Lady Novelists." How deplorable that "maudlin," "sensational," and "trashy" fiction by women should outsell works by respectable male writers, she thought. "Why such people find publishers and readers is the difficult thing to understand," but in any case she knew that these maddeningly prolific women would "continue to produce literature for the news agents to sell long after people of respectable ability have abandoned literature because their clothes are worn out." And their trashy fiction was "considerably more lucrative than that of George Meredith, or the late Mr. Stevenson. There's the justice of literature for you!" (*KA*, p. 194).

Cather had not changed her opinion of female-authored best-sellers by 1903 when she summed up her view of Victorian popular fiction in "The Hundred Worst Books and They That Wrote Them" (*WP* 2, 961–64). Connecting the "debauching" of public taste with the rise of publishing companies eager to satisfy the "lowest element of the reading public" and libraries dedicated to pleasing the masses, Cather declared that "estimable women who turn[ed] out their two novels a year" and sold books "by the thousands" made her ashamed of her country.

Cather held a specific grievance against best-selling American women writers: if they were not motivated by crass commercialism, they were doubtless inspired by the feminine, and inartistic, need to preach. Anyone who used the pen as a pulpit was not, in her estimation, an artist. "The mind that can follow a 'mission' is not an artistic one,'' Cather asserted in 1894. ''For this reason *Uncle Tom's Cabin* will never have a place in the highest ranks of literature.'' Cather found the use of literature as a platform for advocating social reforms reprehensible in those male writers who wanted to cleanse industrial society's Augean stables, but she found in Stowe's novels an equally annoying female trait: ''The feminine mind has a hankering for hobbies and missions, consequently there have been but two real creators among women authors, George Sand and George Eliot'' (*KA*, p. 406).

Although nineteenth-century women writers in America took themselves and their work seriously, authors who considered themselves moralists or spokeswomen for female culture could not be adequate mentors for a young writer who wanted to turn literature into a religion, not religion into literature. Because Cather's predecessors remained tied—even if ambivalently at times—to the female role that she had rejected outright, the Victorian ideology of femininity affected their view of themselves as professional writers. Striving to reconcile their professional with their social roles and to occupy the masculine public sphere and the feminine private sphere simultaneously, literary women like Augusta Evans, Harriet Beecher Stowe, and Mrs. Southworth never fully surmounted the ambivalence with which they regarded their craft. Although publishers competed for their novels, reviewers praised their work, and readers began to regard authorship as a woman's profession, many of these writers could not escape conflict for engaging in professional labors they feared were unwomanly.[21]

Even though they became competent professional writers and effective competitors in the nineteenth-century literary marketplace, many American women novelists embraced a self-effacing authorship: not the writer's dissolution of self in subject matter that Cather valued, but a modest reluctance to take full credit for their labors or to claim the authority of authorship. Their self-abnegation was frequently signified by title pages where given names frequently do not appear. Scores of literary women chose to publish decorously veiled by pseudonyms like ''Elizabeth Wetherell,'' ''Marion Harland,'' and ''Fanny Fern''; unlike the Brontës or Marian Evans, they chose feminine, often flowery names to ressure readers (and perhaps themselves) that their literary aspirations were compatible with the feminine role. Other women were even more unwilling to claim authorship; writers like Catherine Maria Sedgwick, Maria McIntosh, Augusta Evans, and Maria Cummins, among others, at first published without naming themselves at all.[22]

Although historians and literary critics disagree about the social meaning of nineteenth-century American women's fiction, these novels ultimately support the ideology of romantic love and view the home as the appropriate realm for female fulfillment.[23] Their authors' commitment to what Cather termed the ''one string'' of female narrative meant that these diseased texts were infected with what she termed an ''abominable sex consciousness.'' Reflecting the ''hateful distinction'' Cather found between girls' and boys' books, these novels tell versions of the tale Nine Baym describes as

> the story of a young girl who is deprived of the supports she had rightly or wrongly depended on to sustain her throughout life and is faced with the necessity of winning her own way in the world. . . . The happy marriages with which

> most—though not all—of this fiction conclude are symbols of the successful accomplishment of the required task and resolutions of the basic problems in the story, which is in most primitive terms the story of the formation and assertion of a feminine ego.[24]

Like the realists and naturalists, the literary domestics were telling Willa Cather the wrong story. For a woman reader whose dreams of success were marked by unfeminine ambition—and who was uninterested in the heterosexuality assumed by the marriage plot—the American authors of domestic and sentimental fiction did not constitute an acceptable literary family. Willa Cather did not wish to develop a "feminine ego" along the lines these writers had laid out.

•

The British sensationalist writers were popular in post-Civil War America. Among the best known were Ouida, Mary E. Braddon, Rhoda Broughton, and Marie Corelli. In their stirringly melodramatic plots, these writers told a more subversive female story than did the literary domestics, exploring such unfeminine subjects as murder, bigamy, and adultery; these deviant actions were frequently perpetrated by the passionate and angry heroines. As Elaine Showalter observes, these writers were subverting both the formulas and the ideology of the domestic novel, offering fantasies of "protest and escape" to female readers who covertly shared their heroines' eagerness to kill the angel in the house.[25]

Sensationalist fictions frequently featured sensual and passionate heroines who flamboyantly scorned chastity and piety. Mrs. Oliphant, the author of domestic novels whose values and plots the sensationalists were subverting, reacted with genteel horror to unconventional heroines who married their grooms "in fits of sensual passion," prayed for lovers "to carry them off from husbands and homes they hate," and "at the very least of it" gave and received "burning kisses and frantic embraces," living in a "voluptuous dream."[26] These salacious, corrupting foreign imports were frequently censured by American libraries, and in 1881 the American Library Association—self-appointed guardian of feminine morals—banned works by several authors including Ouida, whose affronts to public decency gave her a contemporary reputation as an "apostle of insidious immorality." But Ouida's notoriety only increased the sales of her books: everybody knew she was "smutty and 'not nice,' " and therefore "everybody read her."[27]

No exception, Willa Cather read Ouida and her shocking cohorts. Why, one might wonder, did she not regard women authors who created defiant and passionate heroines, challenged Victorian pieties, scorned religion, and portrayed the repressiveness of the female role as literary relatives she wanted to acknowledge? But these "maudlin" and "trashy" writers roused her anger during the 1890s, despite the fact that the adolescent girl had named Ouida's hero Tricotrin her favorite character in romance (*KA*, p. 193).

Cather gave two reasons for her dislike when she castigated the sensationalists for writing novels drenched with "contemptible feminine weakness" and "mawkish sentimentality" (*KA*, p. 408). "Feminine weakness" suggests Cather's assessment of the sensationalists' typical plot. As Elaine Showalter has shown, the angry and defiant heroines inevitably conclude their stories "punished, repentant, and drained of all energy." Unable to develop the radical implications of the stories they began to tell, Showalter observes, these writers backed off from their volatile subject matter and sought to recontain protest by reasserting, however unconvincingly to us, the social and sexual ideology they also challenged.[28]

But Cather's second charge—"mawkish sentimentality"—directs us toward her most significant reason for attacking the sensationalists' stories: even as they defied the ideology of domesticity, they reinforced the ideology of romantic love. The liberated, defiant heroine who deserts an uninteresting husband for a dashing lover is still dependent on men for fulfillment, no matter how sexually emancipated she appears. Female-authored romances differed from the male-authored romances Cather loved. The romantic hero's visionary imagination or dazzling strength becomes the romantic heroine's passive and "voluptuous dream," as Mrs. Oliphant called it, of sensual and erotic gratification.

Cather was distressed when she found the sensationalist's alliance of romantic love and feminine weakness in a novel by a woman writer she respected. Her important review of Kate Chopin's *The Awakening* (1899) tells us why she wanted to dissociate herself from writers who treated the sensationalists' subjects of adultery and female sexual desire. Along with Emma Bovary and Anna Karenina, Edna Pontellier belonged, Cather thought, to a common "feminine type": the love-obsessed heroine who "demands more romance out of life than God put into it." Cather found Edna and her sisters "victims of the over-idealization of love" and attacked these "self-limited" heroines in a highly self-revealing diatribe:

> They are the spoil of the poets, the Iphigenias of sentiment. The unfortunate feature of their disease is that it attacks only women of brains, at least of rudimentary brains, but whose development is one-sided. . . . With them, everything begins with fancy, and passions rise in the brain rather than in the blood, the poor, neglected, limited one-sided brain that might do so much better things than badgering itself into frantic endeavors to love. . . . They insist upon making it [the "passion of love"] stand for all the emotional pleasures of life and art; expecting an individual and self-limited passion to yield infinite variety, pleasure, and distraction, to contribute to their lives what the arts and the pleasurable exercise of the intellect gives to less limited and less intense idealists. So this passion, when set up against Shakespeare, Balzac, Wagner, Raphael, fails them. . . . Any relaxation short of absolute annihilation is impossible. (*WP* 2, pp. 698–99)

Cather is displeased by two flaws she finds in Edna's character: her subjection to the poets' exaltation of romantic love and her narcissism. In Cather's view Edna's seemingly "natural" awakening of body and soul (associated by Chopin with the elemental force of the sea) and her defiance of the repressive institutions of marriage, motherhood, and the female role are social, not natural, behaviors. Edna's rebellion, Cather suggests, is a product of the "brain" rather than the "blood," the "fancy" not the body; her imagination is possessed by the "poets" who purvey stories of romantic love to female readers.

When Cather complained that such women were ''Iphigenias of sentiment,'' she implies that they were the heroines of a script ultimately written by men, not women. Iphigenia is lured to Aulis, as the classicist Cather would have known, by her father, who has told her that she will play the leading role in the heroine's plot: she thinks she is to marry Achilles. Like Iphigenia, the romantic heroines Cather attacks are victimized not by the direct exercise of male power but by their own romantic imaginations that make them dependent on men. Because they identify personal fulfillment with a self-indulgent romanticism, Cather observes, they are headed for annihilation, not awakening.

These ''self-limited'' heroines roused Cather's anger for another reason. Possessing emotional and imaginative power, these women were potential artists. Yet because they channeled ''passion'' and ''fancy'' into dreams of erotic fulfillment—like the adolescent Edna Pontellier daydreaming about the picture of a famous tragedian—they did not create. Structuring their fantasies by a script that was already written, women of the ''Bovary type'' were not poets but ''spoil[s] of the poets,'' inferior to male artists whose imaginations were not similarly corrupted by the ''over-idealization of love.'' In describing the romantic heroine as ''self-limited,'' Cather further suggests that her creative potential is undermined by narcissism. Enraptured with her image of the beloved, the romantic heroine resembles the ''horribly subjective'' women novelists and the ''self-absorbed'' women poets who could only write of love and passion. Whereas Cather's male artist used the ''power of love'' to transcend the self and create a world, the love-obsessed heroine pursued self-destruction. Her suicide was the logical outcome of her deadly narcissism.

Like the failed ''women of brains'' she attacked in her review of Chopin's novel, the sensationalists struck Cather as potential but failed artists. Because they possessed the prerequisites of great art—passion and imagination—these authors annoyed Cather far more than did writers like Stowe, whom she viewed as simply didactic. Marie Corelli and Ouida in particular impressed her as writers gifted with a crude imaginative power. Why weren't they better than they were? This question had a special urgency for Cather in the 1890s.

•

In her journalism of the nineties, Cather makes Marie Corelli a parodic version of the sensationalist female writer. Breathless, passionate, melodramatic, her specialties were, as Cather observed, the ''dash and the exclamation point'' (*KA*, p. 194). Corelli demeaned the religion of art, and Cather needed to ridicule her in part because she saw in Corelli's unregulated effusiveness a distortion of her own intensity.

Although scorned by professional reviewers, Corelli was beloved by the public, who made pilgrimages to her home in Stratford-on-Avon; the town became associated for a time in the popular imagination with Corelli as well as Shakespeare. Reveling in her roles of literary luminary and Shakespeare's daughter, the ''Swan of Avon'' enjoyed floating down the river in a gondola, imported from Venice complete with gondolier.[29] Cather could not resist raising her meatax against this ''Stratford Sappho'' and ''Ashtoreth of the pen'':

> Probably the most glaring and inexplicable instance of successful fraud that we have to admire today is that of . . . Marie Corelli. . . . Her residence at Stratford, her championship of "the bard" who sometime inhabited there, her fanciful portrayals of herself in several of her novels, all indicate that here we have a female genius of the good old school—rapt, ethereal, art-dedicated. Among all the estimable women who turn out their two novels a year and eat the bread of toil, there is no second to this inspired and raving sibyl, who could have been fitly described and adjectived only by Ouida in her vanished prime. . . . In all the dull grind of contemporary literature we have nothing else so rare as this Stratford Sappho. (*WP* 2, p. 963–64)

In viewing Corelli as desecrating both male and female representatives of art and creativity—Ashtoreth (the matriarchal Phoenician deity of sexuality and fertility), Shakespeare, the sibyl, Sappho—Cather seems to be granting a popular writer unusual power. That Cather's angry dismissals of sensationalists like Corelli may have owed something to unconscious feelings of identification and connection is suggested by the several commentaries on Ouida in her journalism of the 1890s. Cather had read Ouida as an adolescent, perhaps introduced to her fiction by her mother; at that time, one of Ouida's heroes was her favorite character in fiction, but by the 1890s Cather had abandoned her youthful enthusiasm for Ouida's work.

A British expatriate living in Florence, although practically unknown today, Ouida was a prominent, best-selling novelist who published forty-six books between 1863 and 1908. Praised by the young Max Beerbohm, who dedicated a book to her, Ouida did not pass Henry James' more rigorous inspection. He found hers a "childishly primitive 'art' " and the author a "finally pathetic *grotesque*."[30] But despite James' condescending dismissal, Ouida enjoyed a legendary, if temporary, fame during the 1880s and 1890s when

> everybody, from peers to pantry-maids, read Ouida. Her books were masked under impromptu covers of newspapers; deposited in secret cupboards; fetched home from the library by demurely veiled ladies themselves ashamed to send their maids; read indeed in open defiance of parental authority. But they were read and re-read, and no logical argument could stop their popularity.[31]

To Cather, Ouida was the typical female writer. Like the romantic "women of brains" corrupted by the poets, she was a talented woman who never developed or controlled her emotional and imaginative intensity. Allowing "hectic emotionalism" to dominate her work as it did her heroines' lives, Ouida indulged in "debauches of the imagination," Cather charged, never subjecting her fiction to the artist's shaping power (*WP* 2, p. 696). Given to maudlin excess, Ouida committed "sins of form and sense. Adjectives and sentimentality ran away with her, as they do with most women's pens" (*KA*, p. 408). Lacking "verbal precision and restraint," Ouida possessed no faculties of selection or discrimination: "She gives you absolutely everything that comes into her mind, good and bad" (*WP* 2, p. 696).

Cather was personally preoccupied with the issues of passion and control, and in an 1895 review of *Under Two Flags* she broke into a diatribe against Ouida's work that suggests the anger stemming from identification, not indifference:

> I hate to read them [Ouida's novels]. I hate to see the pitiable waste and shameful weaknesses in them. They fill me with the same sense of disgust that Oscar Wilde's books do. They are one rank morass of misguided genius and wasted power. They are sinful, not for what they do, but for what they do not do. They are the work of a brilliant mind that never matured, of hectic emotions that never settled into simplicity and naturalness. (*KA*, p. 409)

The language Cather uses in condemning Ouida—"hate," "pitiable," "shameful," "disgust," "rank," "sinful"—betrays her emotional investment in her subject. Ouida evidently aroused Cather's anger because she failed to be the female precursor she could have been if she had only combined passion and imagination with craft and control, as had Sappho. Instead she "wasted" her power.

At the same time that she was commenting on Ouida's impassioned fiction, Cather was also discovering that she possessed a passionate nature that had to be regulated, if not concealed, if she were to write. Simultaneously she feared that her style was overly emotional and undisciplined, saturated with the same adjectives that "spoiled" Ouida's fiction. Ouida's example thus raised troubling questions. Would Cather be able to transform emotion into art? Or would she betray herself and her gifts, falling into the same "rank morass of misguided genius" that swallowed Ouida and other women writers? Ouida was a distorting mirror in which she saw one version of her future artistic self as well as, perhaps, of her novel-reading mother. Not wanting to resemble either woman, Cather needed to smash the reflection.

•

While neither Stowe nor Ouida could provide Cather with a female literary tradition in the 1890s, it is not immediately clear why she did not find acceptable forebears and models among the American women authors of local-color fiction who flourished in the 1870s, 1880s, and 1890s. Writers like Alice Brown, Rose Terry Cooke, Mary Murfree, Kate Chopin, Mary Wilkins Freeman, and Cather's future mentor Sarah Orne Jewett were publishing in the *Atlantic* and other respected periodicals while Cather was in college. Praised by arbiters of taste like William Dean Howells and Henry James, this was the first group of American women to create "serious" rather than "popular" fiction. Unlike the domestic and sensationalist writers, these women considered themselves artists as well as professional writers.[32]

During the 1890s, however, Cather was not impressed enough by Jewett or by other local-color writers to commend them in her journalism. Commenting in 1895 on Scottish regional novelist Ian Maclaren, Cather announced her distaste for local-color fiction, citing the objections she had to realism as a whole: "limited powers and limited imagination." Hence it was an appropriate genre for the limited female mind: local color was the "element of women," Cather observed, who "seldom write about anything else." Elaborating on her objections, she could have been giving an uncharitable assessment of Freeman's and Jewett's New England stories.

> One likes to read about sound, active, healthy men of the world sometimes, and not always about a collection of melancholy freaks. There is a wearisome sameness about the romances of old men and old women and boys and spinsters, who should have married and did not. The world is really not responsible for age or celibacy and gets tired of having the romances of these sad old people thrust forever in its face. (*WP* 1, p. 278)

Enthralled by the cult of virility and looking to male heroes for the power and autonomy she wanted in her own life, Cather was not ready fully to appreciate Jewett's work in the early 1890s. Although she read *The Country of the Pointed Firs* shortly after its publication in 1896 and respected the book enough to give it to her literary friend George Seibel as a Christmas present, Cather most likely did not fully appreciate the literary power she later recognized in Jewett's understated depiction of a communal tradition of female storytelling. During the nineties she rated Jewett's ''austere and unsentimental'' *A Country Doctor* even more highly than *Country,* but by the 1920s, when she was arranging her edition of Jewett's fiction for Houghton Mifflin, she ranked *Country* with *Huckleberry Finn* and *The Scarlet Letter*.[33]

When Cather constructed her female literary canon, the list was short. After chastising women for their ''abominable'' sex consciousness in her commentary on Ouida's fiction, she acknowledged four exceptions:

> They are so few, the ones who really did anything worthwhile; there were the great Georges, George Eliot and George Sand, and they were anything but women, and there was Miss Bronte [Charlotte] who kept her sentimentality under control, and there was Jane Austen who certainly had more common sense than any of them and was in some respects the greatest of them all.

In selecting these four novelists, however, Cather still suggests that ''woman'' and ''writer'' are contradictory identities. To be sure, both Brontë and Austen—who, Cather thought, regulated the feminine tendency toward sentimental excess—devoted themselves to recording female experience, placing their heroine's maturation in the context of courtship, love, and marriage. Cather's admiration for these writers might thus be taken to signify her conviction that femininity and literary creativity were compatible. Yet at this point Cather still could not escape the male identification displayed in her veneration of Kipling. Characterizing the ''great Georges'' as ''anything but women'' is ambiguous, since the phrase might mean either that she considered them exceptional females or that she thought them unfeminine, but Cather's concluding advice to the woman writer who wanted to be as ''great'' as the Georges reveals that she connected literary excellence with the masculinity their names signified to her:

> Women are so horribly subjective and they have such scorn for the healthy commonplace. When a woman writes a story of adventure, a stout sea tale, a manly

> battle yarn, anything without wine, women and love, then I will begin to hope for something great from them, not before. (*KA*, p. 409)

In advising future writers Cather ignores the fact that none of the novelists she admired, not even the supposedly unwomanly Georges, devoted themselves to stout sea tales and manly battle yarns. The aspiring female writer who read this column might well despair at the mutually exclusive options she faced: if she wrote as a woman, she would fail; if she succeeded (by adopting male plots), she would be "anything but" a woman. Asking women novelists to adapt their voices and stories to masculine values, Cather envisions no partner for Sappho, the poet who transcended the culture's polarization of feminine and masculine.

The contradictions and confusions in Cather's aesthetics should not obscure the importance of her four exceptions, however; even if their great example had not been followed by Cather's contemporaries, Austen, Brontë, Sand, and Eliot were there: monumental, inspiring, validating. The "great Georges" were the most important to Cather of the four, perhaps because in her view they made their desire to be "anything but women" manifest in their pseudonyms. Cather mentions Austen and Brontë infrequently in her journalism, but she often refers to Sand and Eliot, often linking the two writers. Sand and Eliot were the "two great creators" who escaped the woman writer's bent for "hobbies and missions" as well as the only novelists to whom Cather attributed the "art necessity" (*KA*, p. 158).

George Sand was an important model for many nineteenth-century literary women. Praised by such diverse sensibilities as Walt Whitman, Matthew Arnold, George Henry Lewes, and Dostoevsky, Sand inspired women who wished to transcend the second-class literary status to which their sex seemed to consign them. George Eliot felt like "bowing before her in eternal gratitude," and Elizabeth Barrett Browning found Sand a literary "sister" whose "woman's heart" disproved her man's name.[34] A confusingly androgynous figure who adopted a male pseudonym, at times wore male attire, and scandalized the bourgeoisie with her affairs with famous men, Sand could be interpreted according to the needs of her female disciples. Hence while Barrett Browning stressed the woman's nature behind the male facade, Cather could imagine Sand as "anything but" a woman.

Sand's personal example was more important to Cather than was her fiction, which she viewed with a critic's eye. In 1895 she queried whether anyone would "wade through the nine hundred and nine pages of *Consuelo* and its interminable sequel for all her greatness." But Sand's autobiography impressed her: "Of course the novels are all masterly and the pastoral ones supremely beautiful," she acknowledged, but "sometimes the workman is above his [sic] works"; in *Histore de ma vie* Sand revealed a "wonderful personality, much greater than any she ever created in her books" (*KA*, p. 210). When she placed Sand's photograph over the fireplace in her New York apartment, Cather was honoring the "wonderful personality" more than the fiction.

With the British George, however, the reverse was true. Cather was particularly impressed by "La Grande George['s]" *The Mill on the Floss* and *Adam Bede*. The former was her favorite Eliot novel, perhaps because of the correspondences she saw between the tomboy Maggie Tulliver and her childhood self; she thought readers would associate the book with "all that is best in our lives and as dear to us as the places and memories of our childhood." (*WP* 1, pp. 362–63).[35] But *Adam Bede* was the novel that convinced Cather of Eliot's greatness because it demonstrated that she, like the great male writers, could bring a thickly peopled world to life. This middle-class woman could write of the "common people," Cather noted approvingly, laying aside her "own traditions" and confining herself to "those simple, elementary emotions and needs that exist beneath the blouse of a laborer, as well as under the gown of a scholar." Appealing to the readers' "sympathies" as well as their intelligence, like Sappho Eliot combined the woman writer's emotional power with the male artist's technical skill and literary self-discipline. "Unquestionably" one of the "masters of the craft, pre-eminently great," Eliot was an important model for a young writer who was beginning to write about Nebraska's "common people" in her own fiction (*KA*, p. 376).

Embarking on her career as a writer in the 1890s, Cather was fortunate to read the works of middle-class women who had been writing since the eighteenth century. Yet with the exception of a few anomalous writers like Sappho, Sand, and Eliot, the powerful, natural voices of women opera singers seemed to vanish in the sentimental stories women novelists told, to diminish in the "centrifugal" verse of women poets. Cather's temporary solution to the conflict between femininity and literary creativity—to ask the woman writer to tell "manly battle yarns"—was the literary equivalent of her William Cather act. As long as she was listening to what Virginia Woolf called "that persistent voice . . . which cannot let women alone"—the voice of male-dominated society and literary culture—the voices of most women writers were as muted as her own.[36]

And yet she wanted to hear even the writers she supposedly despised. Cather fumed at Ouida and her adjectives, excoriated Corelli and her chambermaid of a muse, all the while reading their books. Even while she was overtly viewing literary art as a masculine enterprise, she was covertly engaged in the complex relationship with the female literary tradition we have seen in this chapter, one remarkably similar to her ambivalent bond with her mother. Cather's rage at Ouida's wasted talent and sentimental weakness may then be connected with the daughter's repudiation not only of her mother's social roles as wife and mother, but also of Virginia Cather's role as reader of romantic stories, which reconfirmed what the daughter perceived as woman's subordinate status.

The mother–daughter relationship may have affected Cather's relationship to women writers who preceded her in another way. As a daughter looking back to literary mothers, she was engaging in a drama of inheritance and identity similar to the one she enacted with her mother and the women in her family, seeking to define herself both in opposition to and in connection with female predecessors. The intensity—both positive and negative—of Cather's responses to literary women suggests the conflicts of the mother–daughter bond. Her denunciation of Ouida and Corelli dramatizes the daughter's and writer's need to differentiate herself from the literary mother she does not want to resemble (perhaps intensified by the fact that Virginia Cather enjoyed these conventional female narratives). Meanwhile her praise of Eliot and Sand demonstrates the daughter's equally compelling drive to define her literary self through identification with a literary mother she admires, in this case one who is "anything but" a woman.[37]

•

Yet Cather's attitude toward women writers was gradually becoming more sympathetic throughout the 1890s as she proceeded to disentangle creativity from masculinity. By 1901, she could even see some value in a woman writer of her mother's generation—Mrs. E.D.E.N. Southworth. In that year Cather described an unusual pilgrimage she had undertaken: a visit to Southworth's Georgetown home. Her record of this literary encounter, particularly when juxtaposed to "When I Knew Stephen Crane," which appeared a year earlier, demonstrates Cather's increasing desire to see "woman" and "writer" as compatible identities. And yet the account of her visit to Southworth's "humble dwelling" simultaneously reflects the conflicts we have considered in this chapter, revealing the daughter/writer praising a maternal precursor from whom she simultaneously wants to distinguish herself.

A hardy and prolific writer—she published about fifty novels from the 1850s to the 1890s—Southworth was phenomenonally popular in nineteenth-century America. A strong-minded woman who survived a failed marriage to enjoy literary fame and fortune, Southworth created a rich variety of admirable female characters in her novels whom she contrasts to the overbearing, unreliable, and otherwise defective men they eventually reform. Her heroines frequently challenge Victorian expectations of submissive femininity, and in *The Hidden Hand,* perhaps her most popular work, Southworth introduced one of her most independent heroines: Capitola, who dresses as a boy and succeeds in the unfeminine art of dueling before achieving the heroine's traditional reward of marriage.[38]

That Cather associated the previous generation of women writers with her own mother, hinted at in her commentaries on other writers, becomes overt here as she associates Mrs. Southworth and her fiction (much of which was set in rural Virginia) with a Southern girlhood like her mother's. "Most of us had mothers who in their youth considered this woman the inspired priestess of the softer emotions," Cather confided, women who found her style the "most poetic and intoxicating in the world." It was "very much the fashion," Cather recalled, for young Southern ladies to recuperate from the "strain of rural gaieties" by reading Southworth's latest novel "as they reclined in hammocks on their wide verandas."

Cather approached Mrs. Southworth's home "in a spirit of jest," she admitted, with "the cries of queenly servant girls, who were spirited away in cabs and married to disinherited lords, ringing in my ears"—a comment which suggests she may not have read much of Southworth's fiction, where the heroines are not invariably as passive as Cather describes them. But after a few minutes, Cather continued, she found herself in a "more respectful frame of mind." The rest of the article recounts her conversion.

Accompanied by a well-read friend who knew something of Southworth's life, Cather learned of the writer's "herculean labors" and her commitment to her profession, which led her to make an important concession for a Jamesian: "I doubt whether Mr. Henry James himself is more sincere, or whether his literary conscience is more exacting than was hers, according to her light. She took herself and her work with entire seriousness, and strange as it may seem, some of her novels were rewritten many times, were hoped and dreamed and prayed over." Cather was particularly touched by what she learned of Mrs. Southworth's relationship with her readers. Each year, Cather discovered, the writer received "appreciative letters from thousands of admirers"—from young women who aspired to write, from readers who found "spiritual and intellectual food" in Southworth's novels. The writer's importance to so many readers made Cather ask

> If this is not fame, what is it, please? How many of us ever think of writing to Henry James when we approve of him, or beg him to be merciful and recall his heroines to a life when they perish, or care very much whether they perish or not?

But while Cather could admire Southworth's dedication to her work and her female readers' devotion to Southworth, she could not, despite her newly awakened respect, change her evaluation of the novelist's literary achievement. In defending Southworth's work against the "jests" of the unsympathetic, Cather could only say that "these tales are quite too pitifully wanting to be subjects for mirth" and reflect that "unabashed romance" thrives in the "untutored mind." So this pilgrimage is not finally the successful meeting with a predecessor imagined in "When I Knew Stephen Crane." Nor is it the daughter's independent discovery of a female tradition described in the Panther Canyon sequence in *The Song of the Lark*. Not only is the mother's work flawed, but also the daughter's eyes are opened by an unnamed male companion—a "man of such profound literary knowledge that he could afford to be charitable"—whose superior understanding legitimates the female reader's newfound appreciation of a writer she comes to pity as well as to respect (*WP* 2, pp. 830–32).

The story of a failed quest, Cather's description of her visit to Southworth's home nevertheless illuminates the search for a female literary tradition evident both in her dismissals of Stowe, Ouida, and Corelli and in her praise of Sand and Eliot. Even while she was worshipping the manly writer, she was looking for signs that women could be literary masters. Because these signs were still faint in the 1890s, Cather began to write stories when she did not have "much faith in women in fiction" and preferred to read male-authored narratives—an unpromising point of departure for a woman writer.

NOTES

1. "Introduction" to David Soskice, "The Secrets of the Schluesselburg: Chapters from the Secret History of Russia's Most Terrible Political Prison," *McClure's* (December 1909): 144–45. Cather's description itself is filled with metaphors of sound and stillness, speech and silence: she found it remarkable that the "shrill night-noises of a crowded London quarter" did not disturb the speaker, particularly "when one remembers how much she still suffers from sudden or unexpected sounds." Cather also remarked on Figner's

learning languages other than Russian, another means of moving from silence to speech and developing a voice: "She studied English and Italian grammars in prison, memorizing a large vocabulary in both languages. . . . Arriving in London at the age of fifty-eight, and never having spoken the tongue, she was able in three weeks to converse in English with considerable fluency, and is rapidly perfecting herself in the language."

2. For an analysis of the significance of the actress in Victorian society, see Christopher Kent, "Image and Reality: The Actress and Society" in *A Widening Sphere: Changing Roles of Victorian Women,* ed. Martha Vicinus (Bloomington: Indiana University Press, 1977), pp. 94–116.

3. See "Willa Cather Mourns Old Opera House" for a reminiscence of the actresses who entranced her in childhood and adolescence (*WP* 2, pp. 955–58).

4. Ellen Moers established the importance of women writers' bonds with literary predecessors in *Literary Women* (New York: Doubleday, 1976); see esp. pp. 42–66. Willa Cather to Dorothy Canfield, March 15 [1916], Dorothy Canfield Papers, Bailey Library, University of Vermont, Burlington, Vt.

5. Cheryl Walker discusses the significance of the Philomela/Procne story as a paradigm for the woman poet in *The Nightingale's Burden: Women Poets and American Culture before 1900* (Bloomington: Indiana University Press, 1982).

6. Kate Chopin, *The Awakening and Other Stories,* ed. Lewis Leary (New York: Holt, Rinehart and Winston, 1970), pp. 9, 208.

7. Richard Giannone discusses Cather's thematic use of music and musical references in *Music in Willa Cather's Fiction* (Lincoln: University of Nebraska Press, 1968).

8. Cather's female literary predecessors on the Continent—Madame de Staël in *Corinne* and George Sand in *Consuelo*—likewise exalted the transporting, transforming power of the female voice. As Ellen Moers observes, "Intimacy without rivalry" explains why women writers could make the diva the "heroic stand-in for the woman of the letters" (p. 190).

9. Cather to Canfield, March 15 [1916].

10. Elaine Showalter reviews recent feminist scholarship on the metaphoric associations between women's body and women's writing in "Feminist Criticism in the Wilderness," *Critical Inquiry* 8, no. 2 (Winter 1981): 179–205, esp. 187–190.

11. Eva Mahoney, "How Willa Cather Found Herself," *Omaha World-Herald,* November 27, 1921, p. 7.

12. See *Women and Language in Literature and Society,* ed. Sally McConnell-Ginet, Ruth Borker, and Nelly Furman (New York: Praeger, 1980).

13. As Elaine Showalter observes, the woman writer is the product of two cultural heritages, two literary traditions: the male (or "dominant") and the female (or "muted") ("Feminist Criticism," p. 204). Even though Cather wanted to belong to the first, she could not help acknowledging her implication in the second, bearing out Showalter's contention that the woman writer is always engaged in a "double-voiced discourse" with two literary traditions, the male and the female.

14. See Sandra M. Gilbert and Susan Gubar, *The Madwoman in the Attic: The Woman Writer and the Nineteenth-Century Literary Imagination* (New Haven: Yale University Press, 1979), pp. 539–46, and "Introduction: Gender, Creativity, and the Woman Poet," *Shakespeare's Sisters: Feminist Essays on Women Poets* (Bloomington: Indiana University Press, 1979), ed. Gilbert and Gubar, pp. xv–xxvi; Walker, pp. 1–3, 138–142. "It is almost inevitable," Gilbert and Gubar argue, "that a talented woman would feel more comfortable—that is, less guilty—writing novels than poems" (p. 549). Mary Kelley, however, basing her interpretation on extensive research in nineteenth-century women novelist's diaries, journals, letters, and other private papers, as well as their fiction, finds that women

fiction writers experienced considerable conflict as professional women being simultaneously wedded to literary careers and the ideology of domesticity. See *Private Woman, Public Stage: Literary Domesticity in Nineteenth-Century America* (New York: Oxford University Press, 1984).

15. See Kelley, p. 345. Elaine Showalter discusses the sensationalist writers in *A Literature of Their Own: British Women Novelists from Brontë to Lessing* (Princeton: Princeton University Press, 1977), pp. 153–81.

16. Dee Garrison, "Immoral Fiction in the Late Victorian Library," *American Quarterly* 28 (Spring 1976): 71–89.

17. Hawthorne to William D. Ticknor, January 1855, quoted in Caroline Ticknor, *Hawthorne and His Publisher* (Boston: Houghton Mifflin, 1913), pp. 141–42.

18. James Hart, *The Popular Book: A History of America's Literary Taste* (New York: Oxford University Press, 1950), pp. 91–92.

19. Mark Twain, *Adventures of Huckleberry Finn,* 2nd ed. (New York: Norton [Norton Critical Edition], 1977), pp. 85–86.

20. Nina Baym, *Woman's Fiction: A Guide to Novels by and about Women in America, 1820–1870* (Ithaca: Cornell University Press, 1978), p. 11.

21. Mary Kelley cites the example of the sixteen-year-old Caroline Howard Gilman who discovers that she has become a published author; a relative had sent a poem she had written to a Boston paper, which printed it without her knowledge, and Gilman "cried half the night with a kind of shame," she remembered: it was as if she had been "detected in man's apparel" (p. 180). In her studies of nineteenth-century American women writers, however, Susan Coultrap-McQuin finds less conflict than does Kelley. Analyzing the correspondence between women writers and their publishers, she argues that the ambivalence toward professional roles suggested by the fiction did not characterize women writers' management of their careers, which they did quite successfully by "[taking] advantage of opportunities in the expanding nineteenth-century publishing industry" ("Why Their Success? Some Observations on Publishing by Popular Nineteenth-Century Women Writers," *Legacy* 1, no. 2 [Fall 1984]: 8). As Janice Radway has suggested, the discrepancy between Kelley's and Coultrap-McQuin's accounts may be connected with their data: Kelley concentrates on the fiction and the personal papers of the writers, Coultrap-McQuin on the professional correspondence (Commentary on the session "Women Writers and Their Male Publishers" at the Modern Language Association, December 1985).

22. See Kelley's discussion of the writers' motivations for choosing anonymous or pseudonymous publication, pp. 125–27, and Baym's chronological listing of novels published between 1822 and 1869, which includes title page information, pp. 302–8. Unlike the literary domestics, Louisa May Alcott chose the androgynous "A. M. Barnard" as her pseudonym, reflecting the unfeminine rage revealed in works like *Behind the Mask.*

23. Some view this literature as conservative, others as feminist, others as revealing a complex mixture of self-effacement and self-assertion. Henry Nash Smith articulates the conservative viewpoint in "The Scribbling Women and the Cosmic Success Story," *Critical Inquiry* 1, no. 1 (September 1974): 47–70, and in *Democracy and the Novel: Popular Resistance to Classic American Writers* (New York: Oxford University Press, 1978); Helen Waite Papishvily views the domestic writers as subversive feminists in *All the Happy Endings: A Study of the Domestic Novel in America, the Women Who Wrote It, the Women Who Read It, in the Nineteenth Century* (New York: Harper, 1956); Baym argues that the fiction celebrates female power within the constraints of Victorian ideology while Kelley stresses the ambiguities and tensions in the novels as well as in the writer's lives.

24. Baym, pp. 11–12.

25. Showalter, *A Literature of Their Own,* p. 159.

26. Quoted in Showalter, *A Literature of Their Own,* p. 100.

27. Garrison, p. 87.

28. Showalter, *A Literature of Their Own,* p. 181.

29. This description is derived from William Stuart Scott, *Marie Corelli: The Story of a Friendship* (London: Hutchinson, 1955).

30. Henry James is quoted in Eileen Bigland, *Ouida: The Passionate Victorian* (London: Jarrolds, 1950; rpt., 1951), p. 106. See also Monaca Stirling, *The Fine and the Wicked: The Life and Times of Ouida* (New York: Coward-McCann, 1958), and Yvonne Ffrench, *Ouida: A Study in Ostentation* (London: Cobden-Sanderson, 1938).

31. Ffrench, pp. 49–50.

32. For a discussion of these writers, see Josephine Donovan, *New England Local Color Literature: A Women's Tradition* (New York: Frederick Ungar, 1983). Donovan argues that the local-color writers constituted a "counter-tradition to the sentimental/domestic convention" (p. 2).

33. George Seibel, "Miss Willa Cather from Nebraska," *New Colophon* 2, Part 7 (September 1949): 202. Cather's appreciation of *A Country Doctor* was evidently not strong enough for her to mention Jewett in her 1890s journalism.

34. See Moers, pp. 52–53.

35. The emotional and psychological resonance of the ending, in which Maggie and Tom drown locked in each others' arms, may have contributed to the novel's grip on Cather's imagination, since she found drowning such a compelling representation of loss of self and autonomy.

36. Woolf, p. 78.

37. Cather's relationship to her female precursors accords with Elizabeth Abel's suggestion that the woman writer's relationship to the female tradition reflects the preoedipal issues of merging, identification, and separation. See Abel's "(E)Merging Identities: The Dynamics of Female Friendships in Contemporary Fiction by Women," *Signs: Journal of Women in Culture and Society* 6, no. 3 (Spring 1981): 413–35.

38. For a brief biographical sketch of Southworth and an analysis of the somewhat subversive female story she told, see Baym, pp. 111–26.

II

APPRENTICESHIP

9

Disclosure and Concealment: The First Stories

The great art of other women is disclosure. Hers [Duse's] is concealment.

The Kingdom of Art

"Her secret? It is every artist's secret,"—he waved his hand,—"passion. That is all. It is an open secret, and perfectly safe."

The Song of the Lark

Before Willa Cather left Red Cloud for Pittsburgh in the summer of 1896 to assume editorship of the *Home Monthly,* she had published nine short stories, most of them in college publications. Later she overlooked these first efforts when she claimed that she did not find her "own material" until *O Pioneers!,* the novel in which she took the "road home" to her native prairies (*AB* [1922], pp. viii–ix). Yet four of her first stories focus on immigrants struggling to survive on the Nebraska plains and introduce the themes of loss, transplantation, and inhabitation to which she often returned in her novels: "Peter" (1892), her first published story (later reworked as the death of Mr. Shimerda in *My Ántonia*); "Lou, the Prophet" (1892); "The Clemency of the Court" (1893); and "On the Divide" (1896), Cather's first publication in a national journal *(The Overland Monthly).* Seemingly unconnected to Cather's later work, three other stories feature exotic settings derived from reading rather than observation: "A Tale of the White Pyramid" (1892)—ancient Egypt; "A Son of the Celestial" (1893)—China and San Francisco; and "A Night at Greenway Court" (1896)—eighteenth-century Virginia. In "The Elopement of Allen Poole" (1893) Cather drew on her memories of Back Creek, a source of fictional material to which she did not return until *Sapphira and the Slave Girl* (1940), and " 'The Fear That Walks by Noonday' " (1894) was a ghost story she co-authored with Dorothy Canfield.

Willa Cather later disowned this apprenticeship work. Writing literary critic Edward Wagenknecht in 1938, she used the metaphor of Robert Frost's apple-picker to describe the mature artist's view of her early fiction. Some apples were worth preserving, she wrote, but others—the immature or imperfect fruit—should be left to decay in obscurity.[1] Anxious to control her literary reputation during her lifetime, Cather never republished these first stories; hoping to extend control after her death, in her will she also forbade republication. So eager was she to dissociate herself from her unwelcome literary progeny that she even disclaimed sole authorship, telling Wagenknecht that some stories were partially written by college mentors, a claim that has been effectively disputed by Mildred Bennett.[2]

It is understandable that Cather would want to detach herself from this early fiction. Like many writers, the accomplished author was embarrassed by the beginner's stylistic and technical misadventures. In addition, even though she was working with her "own material" from the start, the undergraduate writer's perspective on Nebraska life was far more negative than that of the the older Cather, who found much to praise in her Nebraska past once she had left it behind. This transformation is reflected in *My Ántonia,* where the father's suicide, the focus and ending of "Peter," is outweighed by Cather's celebration of Ántonia's life-giving power. Once Cather had found the elegiac voice of her novels, she did not like to be reminded of her youthful complaints.[3]

It is also possible that Cather sought to suppress her early work because the mature writer could see how much it revealed. Telling us more about the young writer than about her ostensible subjects, the first stories are important biographical and literary documents: in their aesthetic deficiencies are hidden the sexual and creative conflicts Cather would resolve before becoming the author of *O Pioneers!* and *My Ántonia.*[4]

Commenting later on her 1890s short stories, Cather gave two seemingly contradictory reasons for their inadequacies. On the one hand, she said, they were imitative and artificial. "It is not always easy for the inexperienced writer to distinguish between his own material and that which he would like to make his own," she explained in her 1922 preface to *Alexander's Bridge;* as the literary acolyte eager to "follow the masters," she had strained and falsified her voice by singing a song incompatible with her range and cadence (*AB* [1922], p. viii). But at other times Cather attributed the inadequacies of this early fiction to its source in deeply felt emotions. Because it was "hard to write about the things that are near to your heart," she told an interviewer in 1913, she had been forced to find "self-protection" by trying to "distort and disguise" the "personal feeling" that inspired her work (*KA,* p. 449). In another interview Cather similarly condemned her first stories as "bald, clumsy and emotional" (*CSF,* p. xxvi).

These contradictory assessments of her college work were not meant to mislead. Simultaneously imitative and authentic, stylized and clumsy, self-protective and self-revealing, the early fiction reveals the difficulties confronting the beginning writer. Like most literary novices, Cather had not mastered the basic tools of her craft. Her excessive indebtedness to literary models is most evident in her exotic, artificial settings and self-conscious, obtrusive allusions—like the Kiplingesque poem which introduces "A Son of the Celestial" ("Ah lie me dead in the sunrise land,/ Where the sky is blue and the hills are gray,/ Where the camels doze in the desert sun,/ And the sea gulls scream o'er the big blue bay" [*CSF,* p. 523]). In addition, the plots are frequently melodramatic, the characters flat, the dialogue awkward, and the style overblown. There is little trace of the complex art of suggestion based in apparent simplicity and understatement that Cather recommended in "The Novel Démeublé" and achieved in her novels. The college student was immersed in what she later termed her "purple flurry" period when she was indulging in a "florid, exaggerated, foamy-at-the-mouth, adjective-spree" style. "I knew even then it was a crime to write like I did, but I had to get the adjectives and the youthful fervor worked off" (*KA*, p. 450).

Cather confronted another, and potentially a more crippling, dilemma than the apprentice writer's struggle to master the trade: the woman writer's anxiety of authorship. Recall the fierce contradictions she experienced as she began to write. At the same time that she was describing fiction as a manly art and literary talent as a "ghastly joke" God played on women, Cather aspired to the "solemn and terrible" act of writing fiction in which most women were exposed as frauds, liars, or sentimental fools (*KA*, p. 409). Cather knew that she had to draw on the "power of loving" she considered the source of female literary power, and yet she had to avoid sentimentality and self-limitation, traps for women who could not control or transcend emotion in their work (*WP* 1, p. 146). Self-conscious about feminine limitations and eager to emulate the "great Georges" she thought "anything but women" (*KA*, p. 409), Cather understandably did not possess that "cadence" and "quality of voice that is exclusively the writer's own" which she later attributed to Sarah Orne Jewett and finally achieved herself (*CPF*, p. 7).

In turning from criticism to fiction, from writing which evaluates others to that which draws on the self, Cather also confronted the danger that she might reveal too much. The art of Eleanora Duse was one of "concealment," Cather thought, and yet most women's art was one of "disclosure" (*KA*, p. 119). And what did women artists disclose? When she referred to the source of female art as the "power of loving," Cather revealed her view that the woman writer is inspired by the passionate and the erotic. And we might add, using terminology to which she did not have access, that such a writer also draws on the unconscious.[5] This vision of literary production invests the creative process with two opposed dangers. A writer who suppresses or represses too much psychic or emotional material may produce fiction that is contorted, false, or artificial. Yet the failure to repress can be terrifying, particularly for a writer like Cather who found her raw material in emotional more than in intellectual responses to the world. The novice writer realized that she faced the possibility of profound and perhaps uncontrollable self-revelation in the creative process: "One may lie to one's self, lie to the world, lie to God, even, but to one's pen one cannot lie. . . . [w]hen one comes to write, . . . [e]very artificial aid fails you. All that you have been taught leaves you, all that you have stolen lies discovered" (*KA*, p. 409).

In a 1913 interview Cather struck a similar note when she observed that fiction had to draw on deep emotion—the source, she later thought, of her best work: "Nothing was really worth while that did not cut pretty deep," she contended, and "the main thing was always to be honest" (*KA*, p. 409). All writers who draw deeply on the self and value emotional and literary honesty face the risks of exposure: Hawthorne, for example, continually worried whether too much of what he called the "inmost Me" was revealed even through the veil of art.[6] But a lesbian writer finds additional inhibitions to self-expression since the sources of art include socially forbidden—and hence unnameable—desires and emotions. Because Cather had "stolen" at least some of her material from her "unnatural" love for women as well as from the writers she was imitating, in turning to fiction she thus confronted a dilemma that was social, historical, and psychological as well as literary: How could she both express and repress, disclose and conceal?

Some forms of literary camouflage that homosexual writers use to surmount this difficulty—allusion, encoding, metaphor—can enrich fiction, and much of Cather's mature work is so enriched by the tension between disclosure and concealment.[7] But this is not the case with her early work. No fiction is a direct, unmediated expression of the self, but when Cather referred to the ways in which she concealed the "personal feeling" underlying her first stories, she described a continuum ranging from a form of representation (disguise) to one of misrepresentation (distortion). Whereas disguise can be a subtle form of concealment, the young writer's inability to resolve the conflicts embedded in the creative process frequently results in exaggerated forms of literary disguise which approach distortion. Cather later contended that the writer "ought to get into his copy as he really is, in his everyday clothes" (*KA,* p. 447). But just as the "everyday clothes" Cather wore during her first years in college were a form of dress no one could ignore, so her first stories strain so hard to mask the conflicts at their source that they call attention to the patterns they seek to conceal.

Returning to Cather's contradictory descriptions of her early fiction as both imitative and self-revelatory, we can now see how these statements are compatible. In much of her early fiction, including her apprenticeship phase of the late 1890s and early 1900s when "no one seemed so wonderful as Henry James" (*CSF,* p. xxvi), imitation for Cather was simultaneously a means of disguise and defense. Cloaking her fictions in allusions and pastiches, she concealed the unnameable and the unnatural; preoccupied with literary models, however, she allowed unconscious material to enter her house of fiction through the back door. Hence while her early work is frequently imitative, "clumsy," and wooden, it is also startlingly self-revealing in ways Cather did not recognize—at least at the time. Below the conventional surface are several interlocking subtexts that reveal the sexual and creative conflicts the young writer experienced when she attempted to transform life into art.

•

"A Tale of the White Pyramid" epitomizes Cather's conflict in drawing on emotional and erotic material that she simultaneously wanted to keep hidden. This bizarre story, apparently derived only from her reading, at first glance reveals no connection to her later work. Set in the Egypt of the pharaohs—the Egypt Cather found in Shakespeare, Daudet, and Gautier—the tale is a stage set complete with pyramids, mummy cases, stifling heat, black mud, and the "tawny serpent" of the Nile. The central action is the burial of the pharaoh by his successor, King Kufu, who directs the funeral ceremony. The pharaoh's body is placed in a pyramid which is to be sealed, Cather tells us, by a "great stone" placed over the top of the immense tomb. As the stone is being hoisted up the pyramid, it suddenly begins to slip. But the king's architect and tomb-builder—a "stranger" of the "Shepherd people of the north"—leaps on the stone and balances it. Catastrophe is averted: instead of plunging down the pyramid into the crowd, the stone settles firmly over the tomb's mouth. The king praises his "master builder" and asks him to build his own pyramid, one hopes with a less perilous design.[8] But this news does not please the crowd, whose "faces were dark" at the prospect of the stranger's continued employment. The story ends with an enigmatic remark by the scribe who is narrating/writing the story: "Of the great pyramid and of the mystery thereof, and of the strange builder, and of the sin of the king, I may not speak, for my lips are sealed" (*CSF*, pp. 529–33).

The story's flaws are so glaring that it has not attracted attention even among critics interested in Cather's apprenticeship fiction. Its derivativeness is obvious: What did a girl from Nebraska know about pyramids and pharaohs? The narrative technique is also curious, seemingly amateurish. Cather employs a first-person narrator, as she did in "A Night at Greenway Court," and as she would again, with far greater success, in *My Ántonia*. The son of a high priest, the narrator witnesses the event and later decides to write down what he saw; hence, as in *My Ántonia,* we are supposedly reading a text written by the narrator. But in "White Pyramid" Cather offers no overt reason why the narrator felt compelled to record the story. Because the emotional and psychological links between writer and text seem absent, the first-person narration appears to be a misguided technical experiment.

If we consider the "inexplicable presence" of what is unnamed, however, and scrutinize the narrator's role, the story becomes more interesting and its place in Cather's work less anomalous (*NUF,* p. 50). His final comment suggests an equation between the story he tells and the pyramid. Because there is a "mystery" of which he cannot speak or write, the preceding story becomes imbued with the same secrecy he ascribes to the pyramid. The comment also equates the narrator and the pyramid: with his lips sealed, containing a mystery he will not or cannot reveal, the cryptic narrator is also the crypt sealed with a stone. Whatever his reasons for telling us about the White Pyramid, we know that he will not tell any more tales. These multiple equations raise further questions. Why cannot the narrator speak further? What is the "mystery" that demands silence? The most obvious possibility is implied in the text where Cather hints at a homosexual relationship between the king and his builder. The only possible referent in the story for "the sin of the king" is his unpopular relationship with the stranger to whom no other man can compare "for beauty of face or form." Cather's Nebraska readers might have assumed that the crowd disapproved of the king's friendship because the builder was a foreigner and outsider, but the loaded word "sin" implies a stronger deviation. And since the setting of the story is Egypt, which Cather associated with sensual indulgence and unbridled hedonism—this was the "oriental" world of "unconventionalized" passions she found in such texts as Daudet's *One of Cleopatra's Nights* and Flaubert's *Salammbô*—it seems likely that she had in mind the sin whose name neither she nor her narrator could speak (*KA,* p. 138).[9] Thus both Cather and the narrator tell a story of concealment and repression in which the creative is intertwined with the sexual.

If we look again at the story's central action, the closing of the tomb, we can discern a related subtext that reflects the author's unconscious conflicts rather than her conscious strategies of concealment and suggests further resemblances among author, architect, and narrator. The obvious act of concealment is the sealing of the tomb, mirrored, as the narrator's last comment suggests, by his inability or unwillingness to divulge a ''mystery'' associated with sexuality. In fact, both artists in the story—the architect and the scribe—enclose their artifacts, sealing tomb and text. And because Cather chose to make the narrator the sole speaker/writer in the story (in contrast to *My Ántonia,* where she uses the preface to create an author separate from Jim Burden), his closing and enclosure of the story are also hers.

The sealing of the tomb spatially embodies the text's movement from openness to enclosure, disclosure to concealment: the partially opened tomb, shaped like a receptacle with a continuous flow between inside and outside, is transformed by the stone into a completely enclosed space with clear boundaries separating inner from outer. This central action has several implications for the creative process. After the stone descends, the inside is transformed into the hidden, the secret, and the mysterious, suggesting the writer's desire to keep the inner realm protected and concealed, a desire explicitly possessed by the scribe and implicitly shared by Cather. Another result of the transformation of the open into the closed pyramid applies more to the woman writer than to the male scribe, however, for it hints at the encasement and repression of female sexuality as well as of creativity: the aperture is closed, the lips are sealed. As these various enclosings and disguisings suggest, in guarding the private the writer may also be sealing the tomb of the unconscious, closing her lips, and blocking the source of art. If extended too far, concealment becomes repression. ''White Pyramid'' thus hints that the writer might conspire in undermining the sources of her own creativity since it is the ''master builder,'' the architect who builds the edifice, who also closes it—just as the narrator/scribe closes the story with a vow of silence.[10]

''A Tale of the White Pyramid'' dramatizes the central literary and psychological dilemma Cather experienced in beginning to write fiction. If the writer cannot draw on the inner, the secret, and the passionate but must enclose this realm just as the pyramid encases the king's body and the narrator hides a ''mystery,'' she may find herself silenced and her creativity as concealed—or, perhaps, as dead—as the body in the tomb.

•

This interpretation may seem too ponderous to impose on such a slight text, as potentially unstable as the wavering stone itself. But like most of Cather's aesthetically crude yet psychologically revealing first stories, ''White Pyramid'' anticipates important patterns more explicitly developed in the later fiction. Cather's reworkings and transformations of the imagery and action of ''White Pyramid'' suggest that this story was her first attempt to portray her own creative process and that, long before she began to write explicitly of artists in *The Troll Garden* (1905), she was concerned with the psychodynamics of creativity. Like Henry James and Edith Wharton, Cather frequently used the architectural metaphors this story introduces to explore the art of fiction and fiction-making. She envisioned the novel as an unfurnished house in ''The Novel Démeublé,'' and in ''Behind the Singer Tower'' (1911) and *Alexander's Bridge* (1912) she reveals artistic anxieties and ambitions in stories ostensibly concerned with the building of skyscrapers and bridges.

Cather even more directly used the imagery of "White Pyramid" to speak of passion, expression, and repression in an 1895 commentary on Eleanora Duse's creative power. After declaring that Duse had to suffer "in secret and in silence" because her art was one of "concealment" rather than of "disclosure," Cather duplicated the central action in "White Pyramid": "She takes her great anguish and lays it in a tomb and rolls a stone before the door, walls it up and hides it away in the earth. And it is of this that she is dying, this stifled pain that is killing her" (*KA*, p. 119). In "Nanette: An Aside" (1897), Cather employed the same imagery of entombment, repression, and death in describing the opera singer Tradutorri's creative power: While some singers "vent their suffering," this singer "holds back her suffering within herself. . . . She takes this great anguish of hers and lays it in a tomb and rolls a stone before the door and walls it up. . . . And now she is dying of it, they say" (*CSF*, p. 408).

"White Pyramid" also anticipates a central metaphoric pattern in Cather's fiction that links spatial configurations with silence and self-expression and mirrors Cather's artistic emergence. We have already briefly explored the symbol of female creativity Cather used in *The Song of the Lark:* the vase/vessel which anticipates Thea's unleashing of her voice, the stream contained within the throat's vessel. In discovering her creative power, Thea transforms the passion which is the source of her art into an "open secret," connecting inner and outer, interior and exterior worlds (*SL*, p. 477). Similarly the vase/vessel is an enclosed yet open space that links inside and outside, like the pyramid before it is sealed. But the stories Cather wrote during the 1890s and early 1900s—before she removed the social and psychological forces blocking creativity and self-expression—are filled with the imagery of enclosure, signifying silence, repression, and death. There are the coffins in "Peter" and "The Clemency of the Court"; the oppressive cell in "Clemency"; the suffocating, encircling mother's bedroom in "The Burglar's Christmas"; the dying Katharine Gaylord's bedroom in " 'A Death in the Desert' " (1903); the overheated hotel room in "Paul's Case" (1905).

In somewhat later stories, most of them written between 1896 and 1913, the year Cather found her "own material" in *O Pioneers!,* the intermediate metaphor of the box is featured. Although seemingly as closed as the sealed tomb or coffin, the box can be opened and the distinction between "inside" and "outside" erased. Mediating between the enclosed and the open space, the box is a female symbol suggesting both Pandora's box of sexuality and the container of unexpressed but potential creativity. There are the black boxes the Chinese carver in "A Son of the Celestial" receives from his native land, which open to reveal the raw materials of his art—ivory, silk, and ebony; the steel box in "The Count of Crow's Nest" (1896), filled with private family documents associated with sexual secrets; the jewel box in "The Burglar's Christmas" which discloses a prophetic symbol of creativity, the drinking cup which becomes the vessel in *Song;* the buried box in "El Dorado: A Kansas Recessional" (1901), containing the hidden record of a sexual and familial attachment; the buried treasure in "The Treasure of Far Island" (1902), momentoes of childhood creativity and play; the trunk hidden in the attic in "The Namesake" (1907), containing the uncle's treasured possessions which inspire the nephew's artistic birth; the box the dying woman gives the narrator in "On the Gull's Road" (1908), filled with symbols of a lost love.

In *The Song of the Lark* the drawer where Thea hides the secret of creativity until she is ready to release it is another version of the box: "From that day on, she felt there was a secret between her and Wunsch [her piano teacher]. Together they had lifted a lid, pulled out a drawer, and looked at something. They hid it away and never spoke of what they had seen; but neither of them forgot it" (p. 79). When the box or drawer is opened or the buried treasure unveiled in this fiction, speech, creative revelation or expression, or sexual flowering results, reflecting Cather's increasing ability to roll the stone away from the tomb of repression and to draw upon the artist's "open secret."

Beginning with *O Pioneers!,* the boxes and drawers are transformed into enclosed yet continually open spaces which, like Thea's voice/vessel, are without a lid: the furrows where Alexandra's heart is hiding in the cornfields; the sheltered draw-bottom where Jim Burden feels completely happy; Thea's sunny cave in the cliffs; the city hidden in the mesa in *The Professor's House*. These natural or constructed spaces offer shelter for life and creation without enclosure or suffocation and reflect the balance between disclosure and concealment Cather ultimately found in the creative process, a harmonious integration of the secret and the open, the private and the public.

The fearful, unsettling aspects of creative expression never totally dissipated for Cather, however, and at times the imagery of enclosure employed in the first stories returns. In a strange episode in *Death Comes for the Archbishop*—the "Stone Lips" chapter where the Archbishop enters a cave associated with Indian rituals and sacrifices—Cather once again conveyed the fear of self-exposure and self-annihilation she associated with tapping the primitive, erotic, and unconscious sources of art. Since this anomalous sequence, which contains disturbing psychic and sexual material that is not assimilated into the larger and more reassuring thematic patterns of the novel, develops the patterns of imagery and meaning associated with the creative process Cather first introduced in "White Pyramid," it will be useful to consider it before returning to the early stories.

•

Bishop Latour and his Indian guide Jacinto are caught in a blinding snowstorm. Death threatens until Jacinto leads the priest to a cave hidden in the mountains, a secret, ceremonial space where the two find shelter until the storm is over. Even though it offers protection, the cave strikes the Bishop as fearful and threatening, a response connected with Cather's investment of this female space with terrifying sexual and creative powers. Entering the cave between the "two great stone lips," the priest simultaneously enters the mouth and the womb of the natural world, sliding through an "orifice" into the "throat of the cave." Both womb and throat, the cave mingles procreative and creative power like the pyramid and the cave/vessel/womb/throat in *The Song of the Lark*. Once he enters this concealed realm, the Archbishop finds another spatial secret: a hole that seems to lead to an even more hidden cavern which Jacinto walls up with stones collected "within the mouth of this orifice," a gesture of repression and immurement recalling both "White Pyramid" and Cather's description of Duse's self-destructive power.

But Jacinto's attempts at concealment do not succeed. The power of the larger cave is too strong for the priest. Even though the unpleasant "fetid odour" disappears after the second aperature is enclosed, the Bishop still feels an "extreme distaste for the place" which increases when he hears, through a third aperature, a "terrible" sound, one of the "oldest voices of the earth": the "source of a great underground river." In this space of seeming safety and shelter, then, the Archbishop confronts something fearful that he has not incorporated into his waking life, a sound associated with the "voice" of nature, Indian rituals, serpent-worship, the unfamiliar, and the matriarchal power of female sexuality and creativity which his patriarchal religion domesticates in the figure of the Virgin. Because he has repressed this primal force, the cave—which speaks through the voice of the river—has the power to silence him, and the ossified lips, it becomes increasingly clear, are his.

In *O Pioneers!* Cather uses the concealed river to symbolize the unconscious when she describes Alexandra Bergson's submerged passion flowing under, and nourishing, her fields. Since Alexandra integrates unconscious and conscious forces, channeling her erotic energies into cultivating the soil, like Thoreau the farmer/artist in *Walden* she makes the earth speak crops instead of grass.[11] By contrast, the Archbishop is too fearful of the cave's contents to unite inner and outer by bringing the submerged to the surface. He will resist the lure of telling, as well as of hearing, disturbing narratives: he "never spoke of Jacinto's cave to anyone. . . . No tales of wonder, he told himself, would ever tempt him into a cavern hereafter."

This conclusion recalls both "White Pyramid" and "A Night at Greenway Court," which also end with a silencing of the narrative impulse. The narrator of the latter story discovers Lord Fairfax's sexual secret—his affair with the mysterious lady of rank whose slashed portrait conceals her identity even while it betrays a mystery. Asked by the lady to keep silent, the narrator agrees not to speak and so must tell the king who asks him to "go on with [his] story" that he has "no story to tell" (*CSF*, p. 491). The implication for the writer who imagined these silenced narrators is serious: if Cather could not find a way to tell tales from the underground, the art of "concealment" would become one of silence, and thus no art at all.

•

Another of Cather's first stories—"The Elopement of Allen Poole"—suggests that disclosure could be even more perilous than concealment, however, and that telling stories might lead to a more disastrous end than not telling them. Whereas the narrators of "White Pyramid" and "Greenway Court" silenced themselves, suffering a metaphoric death, here the protagonist is literally killed by others when his secret is revealed. Allen has two secrets that are overtly disclosed to the reader: he is a moonshiner with a hidden still and a lover who plans to elope. The story ends with the discovery of Allen's crime by the revenue officers, his shooting, and his death in Nell's maternal arms. In contrast to the enclosures in "White Pyramid," the story's open spaces (the setting is a fertile Virginia valley with a view of the distant Blue Ridge mountains) both signify and cause Allen's discovery and death. When he comes out of hiding, the revenue officers easily discover him before he can make his sexual secret open and marry Nell.

Cather concludes with an allusion to *Hamlet* that at once reveals the story's overt concern with sexuality and its covert concern with creativity: ''He drew a long sigh, and the rest was silence.'' Although Allen seems more a lover than an artist, Cather makes a point of telling us that he possesses a form of aesthetic expression: the ''naturally musical'' Allen is ''noted for his whistling.'' Allen's repertoire of folk tunes and arias is further associated with the forbidden activity of moonshining. His tone is as ''sweet and ''mellow'' as his apple brandy, and he is shot by the revenue officers just after he breaks into a ''passion'' of whistling on his way to meet Nell. Like the writer risking self-exposure, Allen finds that ''all that [he has] stolen lies discovered,'' and the death that follows—equated by Cather with silence—seems punishment for both his expression of desire and his distilling.

The submerged preoccupation with creativity, discovery, and punishment in ''The Elopement of Allen Poole'' surfaces in a later story where Cather also employs the imagery of *Death Comes for the Archbishop*—the cavern and the underground river. In ''The Forty Lovers of the Queen,'' an interpolated story in ''Coming, Aphrodite!'' (1921) the metaphoric death and silence in the ''Stone Lips'' chapter become literal for the explorer who engages with sexual and creative forces hidden in the cavern.

Don Hedger tells Eden Bower the story of an Aztec princess who becomes enamored of another stranger from the north, here a chieftain her father captured on a raiding party to the kingdom's northern border.[12] The Captive is associated both with primitive, feral sexuality and with artistic power. Both artist and art-object, like Melville's Queequeg he has tatooed his flesh, thus turning his body into a text, a sign of the ''great unsystematized forces'' Cather associated with creativity: ''his arms and breast were covered with the figures of wild animals, bitten into the skin and coloured.'' Desiring to be similarly transfigured, the Princess asks the Captive to ''practise his art'' upon her, but the relationship between artist and art-object quickly moves from the creative to the sexual, and the Captive ''[falls] upon the Princess to violate her honour.'' Discovered by the King, the Captive suffers a double silencing. Punished both for his erotic and his creative power, he is castrated and his tongue is torn out. The mute Captive then becomes the Princess' servant and procurer, conducting her lovers to the royal bedchamber by a ''secret way'' which winds ''underneath the chambers of the fortress.'' These subsequent lovers also pay for sexual pleasure with death, plunging into a cavern ''that was the bed of an underground river.'' The tale ends with the King's discovery of the Princess' secret: jealous of the Princess' love for a man she decides to let live, the Captive conducts the King to her bedchamber; he then executes the lover, the Princess, and the Captive. The story thus recalls the punishment Cather also associated with sexual and creative expression in ''Allen Poole,'' brutally signified in ''Coming, Aphrodite!'' by the Captive's castration, muting, and death (*YBM*, pp. 42–46).

•

When we consider the implications of early stories like ''The Tale of the White Pyramid'' and ''The Elopement of Allen Poole'' as they are developed in ''Stone Lips'' and ''Forty Lovers,'' it seems that Cather was facing two perilous alternatives as she began to write fiction. Any engagement with the underground cavern of the erotic and the unconscious has its risks, these stories suggest: either sealing it up or entering it results in silencing. If the writer chooses "concealment"—if the cavern/tomb is closed or its power evaded, as in "White Pyramid" and "Stone Lips"—the voice is stilled and the story is ended. But if the writer chooses "disclosure"—if the cavern is entered, as it is metaphorically in "Allen Poole" and literally in "Forty Lovers"—discovery of the secret and the sexual leads to punishment by external powers, silence, and death. Cather's first stories, in particular "White Pyramid," thus suggest that she was trying to contain the powerful but potentially fearful source of her art even as she was beginning to draw upon it. This creative process was a risk-laden venture, what she termed elsewhere the "voyage perilous" (*KA*, p. 417).

The power and the danger Cather associated with the act of writing are concisely revealed in the paradoxical meanings she attaches to the knife in her first stories. The artist's tool for imaginative self-expression, the knife is used by the Chinese opium-smoker in "A Son of the Celestial" to fashion ivory into "strange images" (*CSF*, p. 524) and by the farmer in "On the Divide" to carve images into his windowsills that reveal his inner demons and desires—death's heads, serpents, luxuriant and sensuous foliage. For these artist/carvers, as for the surgeons whom Cather associated with novelists in their power to "enter" other people's skins, the knife extends the shaping power of the hand, which to Cather embodied and signified the writer's creative force. But, as Cather knew from her childhood encounter with the boy who threatened to cut off her hand and from her adolescent experiments in vivisection and dissection, the knife can mutilate and kill as well as create. We see this negative association in the serpent-headed knife the vicious Frenchman carries in "Greenway Court," its effects mysteriously echoed in the slashed painting. In "Peter," by contrast, stabbing becomes erotically charged and masochistically pleasurable, although equally deadly; in his native Bohemia, Peter saw Bernhardt stab a man on the stage, and he thinks that "he would like to die too, if he could . . . have her stab him so" (*CSF*, p. 542).

If the tool of art is also a weapon of death, this pattern implies, then writing—unlike surgery—could harm the creator as well as the subject. Cather developed the paradoxical meanings of the knife and linked them more explicitly with the dangers of unleashing creative force in "Forty Lovers," where the artist who carves mysterious designs into his own body—both decorating and mutilating himself—finds the knife used against him when his designs become sexual. Other writers, of course, have linked the creative process with violence and power since the transformation of life into art can be seen as a sacrifice of the vitality and mutability of disorder for the deathlike purity and fixity of order.[13] As a young writer, however, Cather was more fearful of violence to the artist than to the subject; the creative force might rebound to destroy the creator.

Cather's fear of the creative process, connected with the opposed yet linked perils of concealment and disclosure, was in part connected with the lesbianism she could not name. If the artist's "secret" included passions the culture deemed unnatural, silence might be the only way to avoid uttering the unspeakable. But if Cather did, in some way, "name" the unnameable or reveal what she knew she must hide, might not punishment follow self-expression? When Cather later said that she "knew even then that it was a crime to write like I did," she was referring to the "crime" of her youthful stylistic excesses (*KA*, p. 450). But since the first stories feature so many crimes, criminals, and outsiders (some of whom are versions of the artist), the phrase also suggests that it may have seemed a crime to write at all, as does the fact that seven of the nine stories end with the silencing or death of the protagonist.

Yet the cultural identification of lesbianism with deviance was not the only cause of the conflicts Cather faced as she began to write. Writing is a psychological as well as social process, and the young author was gripped by a fear of self-exposure and self-annihilation that cannot be attributed solely to her awareness that her sexual identity was a punishable offense. This fear had many sources. Like many other writers, Cather found writing "perilous" since she was drawing on emotion, a less easily regulated creative source than intellect; she was also drawing on the unconscious, another source of imaginative power not susceptible to rational ordering. In addition, she feared the loss of formal and stylistic control that might result from unleashing the woman writer's "one gift" of passion (*KA*, p. 348). If she could not emulate Sappho, whose mastery of form gave her the "four walls" that contained but did not conceal her passion, she might find herself in the disreputable company of the overly emotive adjective-users Ouida and Marie Corelli (*KA*, p. 249).

Her metaphoric use of the knife, which suggests her association of the pen, the penis, and the hand, conveys two additional and seemingly opposed, conflicts. As a woman writer accommodating herself to male-defined aesthetics, she was agreeing to mutilation. But as a woman writer aspiring to the creative power she associated with the masculine, she feared mutilation by others, retaliation for such phallic overreaching.[14]

Finally, and most important for her literary development, Cather experienced the creative process as a replay of the psychological and emotional dynamics of the mother–daughter relationship. Several of the stories reveal a covert plot: the child's search for a mother, later to be the explicit and consciously chosen subject of novels like *My Ántonia* and *Sapphira and the Slave Girl*. Where do we find this hidden story?

•

When Cather concluded "The Elopement of Allen Poole" with the phrase "the rest was silence," she was the woman writer ostentatiously showing her familiarity with—and desire to belong to—the male literary tradition. But since the quotation is from *Hamlet,* Shakespeare's oedipal drama, the allusion has a maternal subtext that we can illuminate by considering Cather's 1891 essay, "Shakespeare and Hamlet," where she twice used the same quotation. This essay makes explicit the associations among creativity, sexuality, secrecy, and the mother–daughter bond which are implicit in "The Elopement of Allen Poole" and other early stories.

Cather found *Hamlet* a "cryptogram" to which she thought she had the key. The "great secret of Shakespeare's power" as a dramatist was his "supreme love," she contended, a love that inspired him to create. Hamlet was also driven by love, but he could not create because, in Cather's view, his discovery of his mother's secret—her sexual and emotional treachery—led to his obsessive quest for revenge, a perversion of the creative impulse. Had Hamlet been spared the "son's first sorrow"—the mother's betrayal—"he might have been a great artist of some kind." But as it was, he could find no way to transform his passionate inner world into an acceptable public form. Consumed by a distorted love, "crushed" by the loss of his mother, Hamlet resembled the charcoal that

> feels within its black breast every throb, every aspiration of the kindred gem [the diamond, Cather's metaphor for the successful artist]. It feels its very being throb and break with light, light quiver through its every atom, but it is all latent, men do not see it. (*KA*, pp. 426–36)

Her response to *Hamlet* was thus filtered through her preoccupation with the mother's power to reduce a child—a potential artist—to silence, the subtext of the Nebraska stories.

Shortly after completing this self-revealing essay, Cather began to write stories in which the dynamics of the mother–child bond, "latent" in *Hamlet,* were central if concealed. On the surface "Peter," "Lou the Prophet," "The Clemency of the Court," and "On the Divide" constitute the young writer's attempt at social realism. Concerned with protagonists weakened, maddened, or destroyed by social, economic, or natural forces, these stories sharply contrast with Cather's college literary aesthetics, which exalted the worship of force and the power of romance. Turning from theory to practice, writing in the midst of the economic and agricultural depression of the 1890s—when droughts and crop failures were plaguing Nebraska farmers—Cather resembled Hamlin Garland more closely than Rudyard Kipling.[15] On the level of plot, external agencies similar to those in Garland's fiction cause the protagonists' defeat, but if we consider Cather's imagery as well as the events, the oppressive forces are emotional as well as economic. The subjects of this fiction—loss, isolation, exile, dislocation, death—suggest not only the immigrants' despairing response to the terrible Nebraska drought, but also a bleak world where the self has lost a sustaining connection with a maternal presence, variously represented as a lover, a mother, a homeland, and a fertile, responsive environment.

"The Clemency of the Court" most fully reveals this maternal subtext. The story concerns Serge, an orphaned Russian boy, apprenticed to a vicious farmer and his wife. When the farmer kills Serge's dog, his only companion and link to his dead mother, he murders his employer with the same hatchet. Sentenced to life imprisonment, Serge dies in his prison cell, tortured by a guard who strings him up by the arms; he is strangled when he can no longer bear the pain.

On one level the story is stringent social criticism: Cather attacks an inhumane legal system that crushes the individual, a purpose somewhat ponderously signaled in the ironic title. But if we consider the psychological and emotional subtext, another story emerges. Like William in "The Burglar's Christmas," Serge is an abandoned child seeking his mother. Having lost both his homeland and his natural mother, he transfers his affection to his dog, to whom he gives her name.

After this second loss, he imagines the State as the good mother who will compensate for his deprivation: a "woman with kind eyes, dressed in white with a yellow light about her head, and a little child in her arms, like the picture of the virgin in the Church." Separating this maternal State from her masculine embodiment, the court that sentences him, Serge trusts that "His mother, the State" will protect him from the prison guard's brutality. But the imagery Cather uses to describe his imprisonment, suffering, and death reveals that it is his mother who kills him. The "dark cell" where he is confined and tortured is associated with a deadly maternal embrace (one prisoner talks "to his mother all night long, begging her not to hug him so hard, for she hurt him"). In this womb/tomb Serge begins a regressive, self-destructive journey; after he is strung up, his arms are "paralyzed from the shoulder down" and the guard must "feed him like a baby." He dies in the fetal position:

> The rope grew tighter and tighter. The State must come soon now. He thought he felt the dog's cold nose against his throat. He tried to call its name, but the sound only came in an inarticulate gurgle. He drew his knees up to his chin and died. And so it was that this great mother, the State, took this wilful, restless child of hers and put him to sleep in her bosom. (*CSF*, pp. 515–22)

Like "The Burglar's Christmas," the story demonstrates the dangers, as well as the allure, of returning to the mother; the only way the adult can recapture the infant's preoedipal oneness is to die, the final erasure of the boundaries between self and other. Although in *The Professor's House* (1925) Cather associated the womb/tomb—the attic room—with creative expression for part of the protagonist's life, in "Clemency" Serge progressively loses the power of language as he reverses the stages of human development in returning to his "great mother": first his arms are paralyzed and the hand Cather associated with creative power crippled; then he is unable to name the mother whom he longs to join; then he produces only the infant's "inarticulate gurgle"; then, like the narrators/writers in "White Pyramid" and "Greenway Court," he is silenced.

The psychological and emotional double bind Cather's first stories reveal—that self-annihilation can result either from losing or from gaining the mother/lover—is conveyed by their depiction of space and landscape. On the one hand there are the vast, barren, endlessly stretching plains of the Divide where the self is exposed, vulnerable, and abandoned to die "in a country as flat and as gray and as naked as the sea" (*CSF*, p. 496). On the other hand there are the enclosed, suffocating environments that also kill, like Serge's coffinlike cell or Lou's cave. Cather imagines no space where the self is neither isolated nor incorporated. Hence there is no interior that anticipates the structures Cather later used to represent creativity, the vases, vessels, and cups that provide Gaston Bachelard's "intimate immensity": a harmonious flow between inner and outer and a balance between containment and release.[16]

While the psychological and emotional patterns we have been investigating in these first stories recall the preoedipal issues of merging and separation, fusion and loss, "Clemency" also suggests the child's oedipal drive to possess and be possessed by the mother, a desire that becomes overt in "The Elopement of Allen

Willow Shade, Back Creek, Virginia
Nebraska State Historical Society

William and Caroline Cather
Nebraska State Historical Society

Sidney Cather Gore
Nebraska State Historical Society

Rachel Seibert Boak

From Willa Cather: A Pictorial Memoir, *by permission*

Charles Fectigue Cather

Nebraska State Historical Society

Virginia Boak Cather

Nebraska State Historical Society

Willa Cather in Virginia

Nebraska State Historical Society

Jessica Cather

Willa Cather Pioneer Memorial Collection, Nebraska State Historical Society

Willa Cather as Hiawatha

Willa Cather Pioneer Memorial Collection, Nebraska State Historical Society

Bird's eye view of Red Cloud, 1881

Nebraska State Historical Society

Downtown Red Cloud in the early 1900s

Nebraska State Historical Society

Willa Cather's home in Red Cloud

Nebraska State Historical Society

Willa Cather at age 17, with two Red Cloud friends

Willa Cather Pioneer Memorial Collection, Nebraska State Historical Society

Willa Cather Pioneer Memorial Collection, Nebraska State Historical Society

Willa as William Cather, Jr.

Nebraska State Historical Society

Nebraska State Historical Society

Willa Cather Pioneer Memorial Collection, Nebraska State Historical Society

Willa Cather (with top hat) and cast of *The Beauty and the Beast* (1888)

Willa Cather Pioneer Memorial Collection, Nebraska State Historical Society

Louise Pound

Nebraska State Historical Society

Louise Pound and Willa Cather as undergraduates

Willa Cather Pioneer Memorial Collection, Nebraska State Historical Society

Willa Cather during her second year at the University of Nebraska

Nebraska State Historical Society

Willa Cather in college theatricals

Nebraska State Historical Society

Willa Cather dressed for the opera in 1895

Willa Cather Pioneer Memorial Collection, Nebraska State Historical Society

Willa Cather in her Graduation Ball gown, 1895

Willa Cather Pioneer Memorial Collection, Nebraska State Historical Society

Willa Cather at her desk at the *Journal*

Bailey/Howe Library, University of Vermont

University of Nebraska campus, 1894

Nebraska State Historical Society

Willa Cather and the *Hesperian* staff

Willa Cather Pioneer Memorial Collection, Nebraska State Historical Society

Union Station in Pittsburgh, with Central High School on the hill to the right

Carnegie Library of Pittsburgh

Home Monthly cover during Willa Cather's tenure as managing editor

Carnegie Library of Pittsburgh

Willa Cather during her Pittsburgh teaching days

Willa Cather Pioneer Memorial Collection, Nebraska State Historical Society

Henry James in 1905

Smith College Archives, Smith College

S. S. McClure at his desk

Lilly Library, Indiana University

Isabelle McClung

From Willa Cather: A Pictorial Memoir, *by permission*

Mrs. Fields and Sarah Orne Jewett in the library at 148 Charles Street

Boston Atheneum

Mrs. Fields on the porch of her Manchester summer home

By permission of the Houghton Library, Harvard University

Sarah Orne Jewett at her writing desk in South Berwick

By permission of the Houghton Library, Harvard University

Edith Lewis in the Smith College Yearbook, 1902

Smith College Archives, Smith College

Dorothy Canfield as an undergraduate

Bailey/Howe Library, University of Vermont

Elizabeth Sergeant

From Willa Cather: A Pictorial Memoir, *by permission*

Zoë Akins

By permission of the Huntington Library, San Marino, California

Cliff Palace, Mesa Verde

Courtesy Laura Gilpin Collection, Amon Carter Museum, Fort Worth, Texas

Willa Cather in Wyoming, summer 1905

Willa Cather Pioneer Memorial Collection, Nebraska State Historical Society

Isabelle McClung (left) and Willa Cather in Wyoming, summer 1905

Willa Cather Pioneer Memorial Collection, Nebraska State Historical Society

Illustration for *Alexander's Bridge*
Spahr Library, Dickinson College

C Sherwood.

For Carrie Sherwood

[This was the first time I walked off on my own feet—everything before was half real and half an imitation of writers whom I admired. In this one I hit the home pasture and found that I was Yance Sorgeson and not Henry James.]

Willa Cather

1928

Willa Cather's inscription in Carrie Miner Sherwood's copy of *O Pioneers!*

Willa Cather Pioneer Memorial Collection, Nebraska State Historical Society

Willa Cather in 1915

By permission of the Houghton Library, Harvard University

Poole.'' The ''crime'' of sexuality, portrayed as homosexual in ''White Pyramid,'' is here transcribed into the child–mother bond. Cather published ''Allen Poole'' anonymously in the *Hesperian* in April 1893. She thus wrote it in the midst of her love affair with Louise Pound, and the story's mingling of sexual desire with disappointment, pain, and death may owe something to the conflicts aroused by that relationship.

From the setting alone, the reader might have expected a more satisfying outcome. In sharp contrast to the bleak, rejecting landscapes in the Nebraska stories, Cather recreates the nurturant pastoral world from which she imagined being uprooted at the age of nine. Allen journeys toward his lover through a fertile paradise of desire and anticipation:

> The red harvest moon was just rising; on the one side of the road the tall, green corn stood whispering and rustling in the moonrise, sighing fretfully now and then when the hot south breeze swept over it. On the other side lay the long fields of wheat where the poppies dropped among the stubble and the sheaves gave out the odor of indescribable richness and ripeness which newly cut grain always has.

When Cather reworked this story of love and death in *O Pioneers!*, she created a similar Keatsian landscape of fruition on the edge of decay and foreshadowed the lovers' murder by juxtaposing Emil with the scythe, the graveyard, and the ''brown hole'' in the earth where his friend Amedee is to lie (p. 257). She signals the lover's fate less subtly in ''Allen Poole'': her hero pauses by the graveyard and wishes that he could donate some of his own ''warm blood'' to the corpses ''shut up down there in their narrow boxes.''

Allen must leave Virginia's pastoral landscape for his own ''narrow box,'' since like the other children/lovers in Cather's early stories he is guilty of something (William is a thief, Serge a murderer, Allen a moonshiner). As in the other stories, the protagonist's ''crime'' is sexual as well as social: the unlawful desire for the mother. After Allen is shot, his environment becomes hostile and wounding as he drags himself toward the meeting place, tearing the flesh on his hands—a milder form of the dismemberment that hands often suffer in Cather's fiction. The goal of the quest is Nell, both the reward and, indirectly, the cause of his suffering. With the moonlight ''about her in a halo'' this ''little Madonna of the hills'' is a fictional representation of a maternal and ''divine'' femininity similar to the portrait Cather admired in the art gallery. But Allen has already been punished for achieving this long-awaited union, and he enjoys the maternal embrace only when he cannot be conscious of it; Nell rocks her dying lover as ''a mother does . . . a little baby that is in pain'' (*CSF*, pp. 573–78).

The punishments Allen receives for his desire—injury to his hands and death—suggest an additional reason why Cather associated the knife with the pen and mutilation with artistic expression. If a central source of her art was the daughter's desire for the mother, then like the oedipal son she deserved castration for her sexual and literary crime at the hands of the father, who is represented in the stories we have considered by the male authority figures (the jailers, the revenue officers, the King) who mutilate and silence the guilty artist/sons.

And yet the daughter/artist's desire for a maternal presence was a source of great artistic power to Cather, once her fears of exposure and reprisal dissipated. To understand how the silenced protagonists in the early short stories could be replaced by characters like Alexandra Bergson and Thea Kronborg—daughters whose expressive power does not remain "latent"—we need to consider the connections between the mother–child bond and the creative process. How does the mother enhance—or impede—the child's creativity?

•

The work of British object-relations theorist D. W. Winnicott can help us to see the points of intersection among the loss of the mother, the desire to regain her, the child's imaginative development, and adult creativity. Although establishing a separate sense of self requires the preoedipal child to lose the pleasurable fusion with a maternal presence, according to Winnicott the child resourcefully finds compensation for this deprivation through play and fantasy. The "transitional objects" which Winnicott has studied—blankets, toys, and other items that the child invests with emotional and psychological significance—are replacements for the mother, intermediary attachments that help the child through the difficult process of separation and individuation. As Winnicott observes, it is not the object but the infant who is transitional, moving from a "state of being merged with the mother to a state of being in relation to the mother as something outside and separate." By creating replacements for the mother, the child enters a psychic territory midway between fusion and separation that Winnicott calls "potential space"; there, in imaginative play, the child connects to and separates from the mother at the same time.

In a sense, transitional objects are the first symbols human beings create—objects that both replace and signify something that is absent. The mother may aid the child/artist's symbol-making ability if she accepts and validates the real or imaginary object he or she has invested with meaning, thus "allowing the infant the illusion that what the infant creates really exists." Such a mother is thus the future creator's first audience who suspends disbelief.[17]

In Winnicott's version of human development, losing the original oneness with the mother—in Dorothy Dinnerstein's more pessimistic view, our "original and basic human grief"—is an essential component both of childhood play and of adult creativity.[18] If the self is not experienced as separate from the mother, Winnicott suggests, then there is no urge to regain or replace her, whether through the infant's attachment to a favorite blanket, a child's creative play, or the artist's aesthetic endeavor. Certainly Willa Cather's experience of creativity accords with Winnicott's theory. As the uprooted child, she responded to loss by finding replacements for her mother and her homeland among Nebraska's immigrant women; as the adult novelist, she often created heroines to regain maternal figures she had lost. Several heroines were based on real-life models: Cather transformed the Bohemian girl Annie Sadilek into Ántonia, the playful and enticing Mrs. Silas Garber into Marian Forrester, and her own mother into Victoria Templeton and Sapphira Colbert. Removed by time and distance from the female figures to which her imagination so often returned, in the "potential space" of fiction-writing Cather imaginatively returned to maternal presences without fearing fusion or absorption;

her novels resemble the child's transitional objects, symbolic intermediaries that allow the self simultaneously to connect with and to separate from an object of desire.

If we drew only on Winnicott's work, we might assume—at least from a psychoanalytic perspective—that Cather's creative emergence was unproblematic, the transition from the child who loved play-acting to the artist who created *My Ántonia* an easy one. And yet this was not the case. Since Cather experienced the mother–daughter relationship as a threat, as well as a support, to the self, not surprisingly the beginning writer found that the creative process—informed by the strains of the mother–child bond—endangered identity and autonomy. In Cather's first stories we see characters who desire to regain the mother but find silence or death, not imaginative self-expression. Her college essay "Shakespeare and Hamlet" further reveals her awareness that the child/artist's development could be delayed or destroyed if preoccupation with the mother's emotional and erotic power were too strong.

Supplementing Winnicott's somewhat optimistic portrait of the mother's unqualified support for the child's imaginative development, psychoanalytic theories of adult creativity may help us understand why the mother–child relationship has the potential to hinder the artist's work. While the child moves from the state of preoedipal fusion with the mother to separation and differentiation, the adult artist momentarily reverses this journey in the creative process by returning to an undifferentiated state, annihilating the ego and re-joining the self and the world. Hence he or she regains the oneness with the mother that we all yearn to recapture but gain only in heightened moments: sexual passion, intoxication, spiritual exaltation, creative inspiration. As psychoanalyst Anthony Storr observes, the "persistent desire of human beings to 'lose themselves' in a larger whole" can be traced back to our common origins as infants who begin life unable to distinguish ourselves from our mothers. "To be 'carried away'; to lose one's identity; to be relieved of the burden of striving, even to die, and thus be reunited with the stuff of the universe from which one sprang are universal, if regressive, longings," Storr contends, which "have often been the motive force for art."[19]

If such "regressive" longings are the "motive force for art," as they were for Willa Cather, then they are not, finally, regressive because they lead to active, fruitful endeavor in adult life, not the passive return to infancy we saw in "The Burglar's Christmas." In re-attaching the self to the world, the artist experiences what Storr views as healthful regression, a temporary dissolution of ego boundaries similar to the self-transcendence Cather described in *My Ántonia* as being "dissolved into something complete and great" and becoming "part of something entire" (*MA*, p. 18). As psychiatrist and critic Ernst Kris describes it, the artist undergoes "regression in the service of the ego," or as Cather eventually experienced the process, she returned home in order to advance.[20]

Cather's "road home"—her journey into the past in *O Pioneers!*—would lead her forward into literary adulthood (*AB* [1922], p. ix). But why did it take her so long to find this road?

Dissolution of the self's boundaries can be a terrifying as well as an exhilarating experience, as we saw in our consideration of Cather's ambivalent response to Nebraska's open spaces. Returning, however imaginatively, to preoedipal oneness can threaten annihilation as well as offer a blissful escape from the self's boundaries. And the danger of self-dissolution might seem particularly strong to a woman writer. Since the daughter's separation from the mother is never as complete as the son's, her return to a state of fusion potentially can be more complete and more perilous. Certainly Willa Cather feared "erasure of personality." The woman who refused to fall asleep on the prairies because she might never awaken at first found the association between regression and creativity disturbing, perhaps because, as a lesbian writer, Cather never dissociated the dynamics of separation and connection, differentiation and identification, from eroticism. Returning to a maternal presence that retained primal sexual power doubly endangered her separateness.

We see the threat of annihilation for the young writer most dramatically in "The Burglar's Christmas," which is, in part, a story about the desire for the mother, regression, and creativity. When William enters the maternal bedroom he discovers symbols that Cather eventually identified with artistic power as well as with maternal sexuality and nurturance: hidden in the mother's jewel case is the silver cup from which he drank milk as a child. In Cather's major fiction, written when she found her maternal origins strengthening rather than frightening, the box and the cup/vessel/vase reflect her ability both to draw on and to tell the artist's "open secret" of passion. But in "The Burglar's Christmas" the protagonist cannot transform his regressive and erotic desires into creative ones. Re-absorbed by his mother, William returns to an undifferentiated state from which he never emerges. His story, like that of most protagonists in Cather's early fiction, ends in silence. "The Burglar's Christmas" does hint, however, that creativity could potentially be a gift, or an inheritance, from the mother. The cup that William finds in the jewel box is his; like the trunk Lyon Hartwell discovers in "The Namesake," it has his name on it. Yet in "The Burglar's Christmas" the emphasis is on theft, not inheritance: overwhelmed with shame and guilt when his mother discovers him, William is the thief who has no right to this legacy.

The artist's imaginative return to psychic origins characterizes the phase of the creative process in which the ego's conscious control is relaxed so that fantasies can emerge. This is the phase that Ernst Kris and others call "inspiration"; at this time the artist seems to lack volition and autonomy. This self-abandonment is associated by some writers with the release of unconscious energies, by others with submission to the erotic power of the muse. Creative inspiration is characterized, Kris writes, by the "feeling of being driven, the experience of rapture, and the conviction that an outside agent acts through the creator." Or, as critic Anthony Storr puts it, the artist then has the "feeling of being controlled rather than controlling."[21] Of course the artist also shapes, forms, and orders in a subsequent phase of the creative process. Art is not automatic writing or dream-vision, and as Kris also observes, creativity requires a period when "the experience of purposeful organization and the intent to solve a problem predominate."[22]

The young Willa Cather found the artist's self-abandonment (what she later called "letting go with the heart") difficult to attain, or rather, to allow to happen (*WC:AM,* p. 215). It was hard for her to "let go" until her sense of self, both personal and artistic, was more fully established; as "White Pyramid" suggests, she could not allow unconscious material to emerge until she did not excessively fear what was hidden within the self. And since she associated the women writers she despised with passionate, rhapsodic effusions, during the 1890s and early 1900s she had political as well as psychological reasons for distrusting inspiration and valuing control: she identified the first creative method with that devalued figure, the woman writer.

In composing *O Pioneers!*, however, Cather became "another writer," as she explained in the "Preface" to the 1922 edition of *Alexander's Bridge,* not only because she returned to childhood experiences but also because she found she had "less and less power of choice about the moulding" of her material (*AB* [1922], p. viii). In letters and essays, Cather frequently described her major novels as having written themselves, claiming, for example, that *My Ántonia* "came along, quite of itself and with no direction from me" (*OW,* p. 96). Becoming the artist who finds herself "controlled rather than controlling," Cather eventually could enjoy the periods of dreamy receptivity she attributes to the creative protagonists in her first novels: Alexandra responding to the Divide "with love and yearning," Thea soaking up sun and sounds in Panther Canyon, Jim Burden drowsily meditating in his grandmother's garden (*OP,* p. 65). Surer of a separate identity and more confident of her artistic powers than she had been in the 1890s, the author of *O Pioneers!, The Song of the Lark,* and *My Ántonia* could "let go" in writing her novels without fearing self-annihilation. The connections between the creative process and the mother–child relationship finally were more enriching than disturbing, as we will see in more detail when we explore Cather's apparently miraculous transformation from the imitative author of *Alexander's Bridge* to the spontaneous composer of *O Pioneers!,* the subject of Chapter 18.

To be sure, the conflicts the young writer encountered in the act of writing never completely left the mature artist, as the "Stone Lips" chapter in *Death Comes for the Archbishop* demonstrates. At times writing even seemed an unlawful pastime to Cather: in a 1923 letter to Dorothy Canfield, Cather equated writing with stealing, a comparison recalling the crimes committed in her first short stories. When a story was forming itself in her mind she had to be alone, she told Canfield; at this crucial time she felt like a thief hiding from the police with the precious, cursed stuff she had stolen and needed to protect.[23] The guilt evident in this statement can be traced to her experience of the creative process. As she "dissolved" into her art, Cather was regaining the mother–daughter fusion that had been taken away by separation; in creating characters, she was "stealing" back the exclusive mother–daughter attachment she had lost. Moreover, in attempting to speak of female desire for the mother (or for an erotically powerful woman), she was committing both the child's earliest crime and her culture's newly defined crime—articulating "unnatural" love. After *O Pioneers!,* however, she more often viewed creativity as a gift than as a theft—as the daughter's legitimate inheritance and possession.

On the whole, after her creative emergence in *O Pioneers!,* Cather found writing a happy, carefree activity in which she resolved many polarities and conflicts—balancing separation and connection, autonomy and nurturance. She often compared writing to play, sport, and recreation rather than to work; while she was engaged in a novel, she knew she possessed a sanctuary where life's sorrows and disappointments did not intrude (*WC:AM,* p. 117). In 1931, while her mother lay dying in a California sanitarium, Cather's only escape was *Shadows on the Rock:* the novel had been her rock of refuge for three years, she wrote to Dorothy Canfield, implying that writing sheltered her spirit just as the rock in her narrative protects Quebec's citizens from a harsh environment. She felt so grateful to her book, Cather told Canfield; it was like a coat warming her during an Atlantic crossing, or a tapestry she could take to hotels, sanitariums, and other strange places—in working on it, she forgot her bleak surroundings.[24]

As Cather's comments to Canfield suggest, her writing remained strongly interwoven with the strands of the mother–daughter bond. Ultimately, however, Cather could be both daughter and mother in the act of writing, enjoying nurturance and attachment as she preserved power and autonomy. We see these psychological and emotional polarities unified in *Shadows on the Rock.* Just as Cather was losing her own mother, her novel became a kind of maternal presence, offering refuge, warmth, and care; simultaneously Cather was mother to her novel (and herself), creating an independent child.

Although many novelists compare writing fiction to giving birth (Cather among them), she may have found that writing allowed her to be both mother and daughter because she identified her books so strongly with their protagonists. In writing she was both creating and enjoying a relationship with another human being: during the composition of *The Song of the Lark* Cather felt that she was living with Thea Kronborg, whereas in writing *One of Ours* she enjoyed a glorious companionship with Claude Wheeler.[25] So intimately did she identify her books with her characters that in letters Cather generally refers to a novel as "him" or "her," depending on the gender of the protagonist. In creating these beings or texts who could live apart from her, Cather was mother to her novels; in developing profound attachments to her characters, however, she was also the daughter relying upon a maternal presence. She had so "depended on Thea," Cather told Elizabeth Sergeant, that when the book was finished and the "close inner tie was severed, she felt the pang and emptiness of one deserted"—as if the book were both the child being born and the mother leaving the child (*WC:AM,* p. 137).

Cather's adopting the double role of mother and daughter in writing her major novels perhaps can best be seen in the imagery of eating and nurturing she often used to describe the creative process. If, as Anthony Storr contends, creative desire is one example of the human impulse to recapture the lost connection with the mother, then writing can be thought of as a symbolic reunion with the mother's life-sustaining body—an equation the adult writer made explicit in declaring that the writer must know her subject as intimately as "the babe knows its own mother's breast" (*WWC,* p. 195). In essays and interviews in the 1920s and 1930s she frequently portrayed the imitative or unsuccessful writer as "starving" and the writer who had found her subject as "fed" by her creativity; she once told an interviewer that creative inspiration drove her to write "exactly as food makes a hungry person want to eat."[26] Cather's most explicit identification of writing with eating occurred in a 1925 interview for the *Century.* Asked about her love for cooking, she stressed the importance of good cuisine to her creative well-being and went on to say "My mind and stomach are one! I think and work with whatever it is that digests."[27] And yet the novelist Cather was not only the nurtured child: she was also the mother who nourished her characters. Writing Dorothy Canfield when she was completing *One of Ours,* she reversed the metaphor of feeding she commonly used to describe writing, telling her friend that despite her exhaustion she knew she had enough to feed Claude until he was done.[28]

Central to the adult novelist's delight in writing was her discovery that her narrative voice was retrospective. Like Proust, Cather wrote both about memory and from memories; hence distance from the past in both time and place was central to her art. And the fact that her imagination was retrospective is connected to the conflicts revealed by her first stories. It was safer for her to "[let] go with the heart" when time and distance intervened between herself and the female figures to whom she was drawn.[29] And since the creative process itself replayed the dynamics of the mother–daughter bond, the distance she finally gained on her material allowed writing to provide an exhilarating rather than a frightening loss of self and to grant her the roles of sustaining mother as well as cared-for daughter.

But writing fiction—the adult novelist's blissful recreation—intensified rather than resolved conflicts for the beginning author. In addition to suffering the woman writer's uncertainty about creative authority, Cather feared the crime of erotic and aesthetic disclosure, the artist's risky journey into the unconscious, and the dangers of regression. And yet the same forces that made writing perilous made it a compelling form of self-affirmation and self-expression. So although she wrote of protagonists and narrators who ceased to tell their stories, she still continued to write.

Unlike her silenced characters, Cather kept speaking because writing itself was a paradoxical activity that mediated between the unacceptable poles of concealment and disclosure. As she noted in referring to her early work, there was a possible alternative to concealing or disclosing "personal feeling": disguise, a form of symbolic representation and expression central to all fiction which allows the writer to keep and to tell a secret at the same time—just as the symbol, or the child's transitional object, both displaces and replaces, both hides and signifies, an invisible presence. Cather's first stories disclose her aesthetic, psychological, and sexual conflicts, but they also disguise them. And so—unlike her protagonists, who do not find symbolic forms to express their desire—Cather retained the power of speech.

•

All writers disguise the self when they dress up in fiction, but Cather's writing, according to recent readings which take into account her lesbianism, wears two layers of clothing. Throughout her literary career, Cather was both the writer transforming the self in art and the lesbian writer at times forced to conceal "unnatural" love by projecting herself into male disguises. This gender transformation is mirrored in the closing of the tomb in "A Tale of the White Pyramid"; a potentially female symbol, the partially open receptacle, is changed by the addition of the stone into the masculine pyramid. Cather's inability to write directly of love between women—at least until late in her career when she let the mother–daughter relationship emerge from the background to the foreground of her fiction—raises several interpretive difficulties for her readers, who find themselves confronted with what may seem to be patriarchal views of women as well as male narrators or centers of consciousness whose gender is suspect.[30]

"The Elopement of Allen Poole" introduces a problematic pattern that some readers have interpreted as signifying Cather's male identification: the woman writer's uncritical use of limiting, male-defined views of women. The "little Madonna" who rocks Allen in her arms anticipates the frequent appearance in Cather's fiction of female stereotypes and archetypes that traditionally reflect men's ambivalence toward women: virgin/temptress, nurturing mother/devouring mother, angel/demon. Cather's use of such stereotypes in her early fiction does suggest that she had not freed herself from male images of femininity, not a surprising failure for a woman whose youthful imagination was structured so strongly by the works of male writers who in part defined the terms in which her love for women could be expressed. In her novels, however, Cather often appropriates and redefines the male-authored stereotypes she employed uncritically in the first stories, as she does when she gives Ántonia the power of storytelling as well as of childbearing, or when she examines the unconscious motivations that lead Neil Herbert to apply the virgin/whore dichotomy to Marian Forrester, a complex woman who eludes such categorization. Moreover, in *O Pioneers!, The Song of the Lark,* and other novels, Cather creates female characters who are subjects imbued with consciousness and power, not merely objects of male desire and yearning.

Cather's use of male characters to explore the sexual and emotional currents in romantic love, introduced in "Allen Poole" and continued throughout her literary career, raises a complex interpretive problem her critics have only begun to address: When is a male character not a male character but a mask for a lesbian consciousness? It has been argued that we should consider all Cather's male characters involved in heterosexual love plots masks for women and the author a celebrant of patriarchal values.[31] But before we adopt such a sweeping translation or deduce the author's male identification from the fiction, a shaky methodological leap, we need to consider Cather's motivations, both conscious and unconscious, for employing male protagonists in stories like "Allen Poole" and "The Burglar's Christmas."

Certainly Cather's use of male characters in her early fiction reflects the young writer's infatuation with "manly" values. Because she associated the masculine with cultural power and privilege, she chose to focus on male characters because she had internalized a male aesthetic. But Cather's choice of male protagonists in her first stories should also be attributed to the social and psychological pressures that required her to conceal her emotional and erotic orientation. Cather was drawing on her own response to women in "Allen Poole," but like the narrator in "White Pyramid," her lips were sealed since certain intimacies could not be mentioned in the 1890s. And there was another, doubtless unconscious reason why Cather chose to tell stories like "The Burglar's Christmas" from a male point of view; fearing absorption and obliteration by a mother or lover, she adopted male personas to place the barrier of gender between herself and the female presences to whom her imagination returned, a literary and psychological means of defending against the annihilation that engulfs her protagonists.

As these several possibilities suggest, we should not assume either Cather's subservience to patriarchal values or her male identification from her use of male personas. During the 1890s she did identify more strongly with men than she did with women, but by 1913—when she reworked "Allen Poole" for the lovers' subplot in *O Pioneers!*, her first novel featuring an artist–heroine—she did not. Cather may have impressed some critics as male-identified because her solution to the tension between disclosure and concealment—using male characters to explore lesbian desire—obscures her resolution of the opposition between female identity and artistic vocation. Her continued use of male narrators and characters to write of woman's maternal/erotic power in the major fiction signifies Cather's continued inability to name the "thing not named," not her male identification.

Cather's projection of self into male characters, her portrayal of romantic love as heterosexual, and her occasional creation of unconvincing male masks whose gender seems indeterminate or female raise another interpretive difficulty for the reader of her fiction. Those critics who view Cather's portrayal of heterosexual passion as an encoded transcription of lesbian love assume that the overt text conceals a covert text which is the "real" story Cather would have written had she been able. But assuming that all Cather's heterosexual plots are cover stories for homosexual ones and that her male lovers are invariably masks for women simplifies both text and writer. To argue that most of her male characters engaged in love affairs are not male at all is to question the writer's ability to transcend self, gender, and sexuality to adopt other selves; it is also to assume that whenever Cather wrote of heterosexual love she was encoding a lesbian attachment, a reductive view of her fiction as well as of her imagination.

In reading Cather's fiction we need to develop interpretive strategies for determining when a male character is a "mask" rather than an opposite-sex character whom Cather created by drawing, in part, on herself. On what grounds do we conclude that a male character is not "really" male but female or that a heterosexual love plot conceals a homosexual one? On the basis of the stories we have examined, I suggest that only if textual clues contradict or question the male character's assigned gender can we proclaim him a "mask" or "cover." In "The Burglar's Christmas" the mother's overidentification with the "son" suggests the mother–daughter relationship, while in "Allen Poole" and "Clemency" the association of the child's return to the mother with fusion and annihilation recalls the intense, mutually engulfing identifications between mother and daughter, defended against by the use of an opposite-sex character. By contrast, a character like Godfrey St. Peter in *The Professor's House*, however "feminine" his sensibility, should be considered male. Not to do so would be to overlook Cather's redefinition of male as well as female roles and identities.

In stories where textual clues direct us to consider the male characters as masks for a female perceiver, the heterosexual "cover" story functions simultaneously as disguise and defense and serves a social as well as a psychological function. Of course this does not mean that Cather was not invested in the surface story:

like the costumes she chose to wear in life, her literary disguises revealed as well as concealed the self. Hence Cather's early stories call for the same interpretive strategy required by most of her fiction. Since at times she was writing two stories at once, a heterosexual and a homosexual one, just as she projected herself into both male and female characters, we are frequently faced with indeterminate meaning rather than with a clearly encoded subtext that constitutes the "real" message of the text. The heterosexual "cover" story, although socially necessary as a way of naming indirectly the "thing not named," is not invariably the false one, the hidden lesbian story the real. Instead, meaning and authorial intention oscillate back and forth between the two. The spatial metaphor that best captures this pattern is the continual interplay between figure and ground rather than a hierarchical opposition between surface and hidden, overt and covert meaning.

Cather's need to imply the presence of the "not named" may also have contributed to her fiction the allusive, suggestive qualities we associate with modernism. The aesthetic of indirection she espoused in "The Novel Démeublé" evokes at once the lesbian writer forced to conceal and the twentieth-century writer aware both of the inadequacies and the possibilities of language. Even though she was not a self-conscious modernist and in fact espoused an antimodernist rhetoric, because Cather could not tell the truth directly she was, at times, forced to tell it slant, and the resulting creative tension between expression and suppression produced novels that are subtle, richly symbolic, and ambiguous, enriched by the repressed, the hidden, and the covert. Ultimately her need both to disclose and to conceal lesbianism—one of the conflicts that threatened to silence the beginning writer—contributed to the pleasure she found in the creative process.

Unlike her novels, however, Cather's first stories are weakened rather than enriched by the "inexplicable presence" of the unnamed. The young writer did not possess enough control over the psychological and emotional conflicts revealed in the unassimilated subtexts. As a result, unlike her novels the short stories are not separate enough from their unconscious sources to be fictions that both draw upon the self and exist apart from it. Although there are occasionally powerful moments that anticipate the later fiction—her understated description of Peter's suicide, for example—the technical flaws reflect not only the inexperienced writer's failure to command the tools of her trade, but also the young woman's failure to control and shape her recurrent psychological preoccupations.[32] Indeed, these two failures were connected. Cather could not master either her craft or the psychic material from which her fiction sprang until she could momentarily abandon control and make the artist's closed secret open. When she could, the reworked story of the child's search for the mother became *My Ántonia*. Hence Cather's twenty-year journey from her first short stories to her novels reveals psychological as well as technical growth.

•

Even though her fascination with women actresses and opera singers foreshadowed her eventual reconciliation of "woman" and "artist," Cather was writing her first stories at the same time that she was praising manly writers and encouraging women to write like men. Her fascination with masculine values is evident both in her indebtedness to male-authored texts and in her preference for male protagonists. Of course a writer's ability to create characters of the opposite gender testifies to the emphatic power of the imagination; we should not expect a woman writer necessarily to focus on female characters. Nevertheless, since most apprentice women writers create at least some female characters who embody their own conflicts or successes in combining femininity with creativity (as did Cather's contemporaries Virginia Woolf and Edith Wharton), it is surprising that none of Cather's first stories features a female character with an independent consciousness. In them we see no literary sisters to Alexandra Bergson, Thea Kronborg, or Lucy Gayheart. To some extent we can attribute Cather's use of male protagonists to her need to place the barrier of gender between herself and the "great mothers" to whom characters like Serge and Allen are drawn. But these male characters also reflect Cather's repudiation of her gender during her Lincoln years, the college student's response to Victorian restrictions on female identity and role.

During the ten years she spent in Pittsburgh—from 1896 to 1906—even though she wrote little imaginative literature, Cather began to resolve the beginning writer's conflict between gender and vocation. Her success in the male world of journalism strengthened her separation from her mother and the feminine role she embodied, and this process, supported by her love for Isabelle McClung, in turn made it easier to attribute positive values to female experience, to risk self-exposure in her fiction, and to create more complex and sympathetic women characters. At the beginning of her Pittsburgh decade Cather announced her dedication to art while assuming that creativity was a masculine privilege; at the end of the decade she was writing her first stories of gifted and artistic women. She signed these with the name signifying the next stage in her evolution as woman and writer—Willa Sibert Cather. Discarding the purely masculine "William Cather" and placing the sign of her maternal uncle's legacy ("Sibert") between her given and her family names, Cather combined male and female inheritances in fashioning the artist's name that marked the apprenticeship stories of the 1900s and the title pages of her first four novels.

NOTES

1. Willa Cather to Edward Wagenknecht, December 31, 1938, Pierpont Morgan Library, New York.

2. See *CSF*, pp. xxvii, xxxv. According to Bennett, "Peter" was "touched up" by Cather's English teacher Herbert Bates, who submitted it to *The Mahogany Tree* without her knowledge (p. xxvii). But there is no evidence linking Bates with "On the Divide," which appeared after Cather's graduation, and even less reason for assuming a college mentor's involvement in some of the later fiction for which Cather also disclaimed sole authorship ("The Professor's Commencement," "The Treasure of Far Island," "Jack-a-boy," and "Eric Hermannson's Soul").

3. For a fuller discussion of this motivation, see Curtis Bradford, "Willa Cather's Uncollected Short Stories," *American Literature* 26 (January 1955): 547–50.

4. Mildred Bennett provides a comprehensive literary history of Cather's short fiction in the "Introduction" to *CSF*, pp. xxiii–xli. David Stouck has a brief but insightful dis-

cussion of the ways in which Cather's first stories reveal characteristic patterns in the fiction in *Willa Cather's Imagination* (Lincoln: University of Nebraska Press, 1975). See in particular his analysis of "The Clemency of the Court" and "The Elopement of Allen Poole" where he notes that the "orphan is more than a romantic convention in Willa Cather's fiction; it suggests a psychological state central to her art" (pp. 37–39, esp. p. 38).

5. By the 1920s Cather was aware of Freudian and Jungian theories of the unconscious (her friend Elizabeth Sergeant was in Jungian analysis). Although she was skeptical of psychoanalytic theory, she wondered whether Sergeant—simultaneously undergoing analysis and writing a novel—would find her creative power affected, a concern revealing Cather's suspicion that artists drew on unconscious energies (*WC:AM*, p. 238).

6. Nathaniel Hawthorne, *The Scarlet Letter, The Centenary Edition of the Works of Nathaniel Hawthorne,* Vol. I, ed. William Charvat and Roy Harvey Pearce (Athens: Ohio State University Press, 1962), p. 4.

7. As Jeffrey Meyers observes in referring to male homosexual writers, "The clandestine predilections of homosexual novelists are both an obstacle and a stimulus to art, and lead to a creative tension between repression and expression. The novels become a raid on inarticulate feelings, and force the authors to find a language of reticence and evasion, obliqueness and indirection, to convey their theme" (*Homosexuality and Literature 1890–1930* [Montreal: McGill-Queen's University Press, 1977], p. 1).

8. William Curtin traces the parallels between Ibsen's *The Master Builder* and *Alexander's Bridge* in "Ibsen's *The Master Builder* and Cather's *Alexander's Bridge,*" paper presented at the Northeast Modern Language Association meeting in Hartford, Conn., March 1979. He observes that Cather "probably read the five-volume edition (1891–92) of Ibsen's works by William Archer" (p. 7); in any case, since *The Master Builder* was published separately in 1892, she had read at least that play before writing "White Pyramid."

9. In an article on Flaubert's response to Egypt, John Finlay points out that for Flaubert, as for Rimbaud and other nineteenth-century Western travelers, Egypt "eliminated the sexual neurosis of the nineteenth century"; a place where sex could be released from bourgeois restrictions, Egypt was a country of the mind, associated with the primitive urges repressed in capitalist, industrialized Western cultures ("Flaubert in Egypt," *Hudson Review* 36, no. 3 [Autumn 1983]: 496–509, esp. 501.)

10. The silence with which this story ends does not resemble, I think, that which French feminist theorists associate with the preoedipal phase: a kind of authentic and powerful "speech" which precedes the female speaker's subjection to patriarchal language. Instead, it is the silence of one who commands language but, fearing discovery, decides not to speak. For a discussion of the links between the preoedipal mother–daughter bond and female speech free of patriarchal language, see Lucy Irigary, "When Our Lips Speak Together," *Signs: Journal of Women in Culture and Society* 6, no. 1 (Autumn 1980): 69–79, and "And the One Doesn't Stir Without the Other," *Signs* 7, no. 1 (Autumn 1981): 60–70.

11. See in particular "The Bean-Field," where Thoreau is dedicated to "making the yellow soil express its summer thought in bean leaves and blossoms rather than in wormwood and piper and millet grass, making the earth say beans instead of grass." Henry David Thoreau, *Walden,* ed. Sherman Paul (Boston: Houghton Mifflin [Riverside ed.], 1959), p. 108.

12. Cather originally heard this "story of an Aztec Cleopatra" from Julio, the young Mexican who entranced her during her 1912 visit to the Southwest (*WC:AM,* p. 81). See Chapter 18, pp. 404–418.

13. An obvious example of a writer who associates violence, death, and the purity of art is Ernest Hemingway, whose bullfighter/artists enact a ritual of domination. For a discussion of traditional connections between the pen and the sword, the power to create and

the power to kill, see Sandra M. Gilbert and Susan Gubar, *The Madwoman in the Attic: The Woman Writer and the Nineteenth Century Literary Imagination* (New Haven: Yale University Press, 1979), p. 14; see also their discussion of the male writer who "kills" his female subject into art (pp. 14–17).

14. Like Cather's memory of the boy who threatened her with a knife, this pattern suggests both the reality of castration she faced as a woman and the threat of castration she encountered as a woman who temporarily identified with the male imagination. Cather's investment in the knife as a symbol of power and mutilation suggests that the Lacanian position as articulated by Helene Cixous may need to be modified for the male-identified woman. As Cixous points out, the "woman does not have the advantage of the castration complex" to mark her entrance into the symbolic; but Cather's early memory and her fiction suggest the fear of castration supposedly unique to men. See Helen Cixous, "Castration or Decapitation?" *Signs: Journal of Women in Culture and Society* 7, no. 1 (Autumn 1981): 41–55, esp. 46.

15. Cather was well aware of the devastating impact the drought of the early 1890s had on her immigrant neighbors. Writing her *McClure's* colleague Witter Bynner in 1905, she said that during her college years the country was burned up by the drought, and that her neighbors began to go insane and commit suicide in the most heart-rending way (Willa Cather to Witter Bynner, June 7, 1905, Houghton Library, Harvard University, Cambridge, Mass.). She made the same observation in "On the Divide": "Insanity and suicide are very common things on the Divide. . . . It causes no great sensation there when a Dane is found swinging to [*sic*] his own windmill tower, and most of the Poles after they have become too careless and discouraged to shave themselves keep their razors to cut their throats with" (*CSF*, p. 495).

16. Gaston Bachelard, *The Poetics of Space,* translated by Maria Jolas (Boston: Beacon Press, 1969), p. 183.

17. D. W. Winnicott, *Playing and Reality* (London: Tavistock Publications, 1971), pp. 1–25, 112–18.

18. Dorothy Dinnerstein, *The Mermaid and the Minotaur: Sexual Arrangements and the Human Malaise* (New York: Harper & Row, 1976), p. 60.

19. Anthony Storr, *The Dynamics of Creation* (New York: Atheneum, 1972), p. 181.

20. Ernst Kris, *Psychoanalytic Explorations in Art* (New York: International Universities Press, 1959), p. 253.

21. Storr, p. 65. As Storr observes, "There is an element of passivity, or dependence, even of humility in the creative process" (p. 200).

22. Kris, p. 59.

23. Willa Cather to Dorothy Canfield, April 8, 1923, Dorothy Canfield Fisher Papers, Bailey Library, University of Vermont, Burlington, Vt.

24. Cather to Canfield, May 1, 1931, and n.d. As Jane Flax argues, reconciling the polarities of autonomy and nurturance can be particularly difficult for women because of the ambivalences of the mother–daughter bond ("The Conflict between Nurturance and Autonomy in Mother/Daughter Relationships and Within Feminism," *Feminist Studies* 4 [February 1978]: 171–89).

25. Cather to Canfield, n.d.

26. For references to starving and feeding, see the "Preface" to *Alexander's Bridge* (1922 ed.) and "Katherine Mansfield" in *NUF*. Flora Merrill, "A Short Story Course Can Only Delay, It Cannot Kill an Artist, Says Willa Cather," *New York World,* April 19, 1925, section 3, p. 6.

27. Walter Tittle, "Glimpses of Interesting Americans," *The Century Magazine* (July 1925), p. 312.

28. Cather to Canfield, n.d.

29. As Judith Kegan Gardiner observes in discussing female friendships in contemporary fiction by women, a "living process of interaction between women, with its exhilarating fusions and frightening threats to autonomy, often yields to a safer relationship with an absent other who can be recreated in imagination and memory." See "The (US)es of (I)dentity: A Response to Abel on '(E)merging Identities,' " *Signs: Journal of Women in Culture and Society* 6, no. 3 (Spring 1981): 436–42, esp. 441.

30. Some recent commentators on Cather's fiction have concluded that her use of male narrators signifies male identification, not taking into account the lesbian writer's need for disguise. See Deborah Lambert, "The Defeat of a Hero: Autonomy and Sexuality in *My Ántonia,*" *American Literature* 53, no. 4 (January 1982): 676–90, esp. 676; Carolyn G. Heilbrun, *Reinventing Womanhood* (New York: Norton, 1979), pp. 79–81; Phyllis Rose, "The Point of View Was Masculine," *New York Times Book Review,* September 11, 1983, p. 15.

31. Lambert, pp. 677, 680.

32. David Stouck makes a similar observation (p. 38).

10

Presbyteria, Bohemia, and the "Garden Fair"

> The business of an artist's life is not Bohemianism for or against, but ceaseless and unremitting labor. *The Kingdom of Art*

> A garden fair/ . . . Its guest invited me to be.
> Dedicatory poem to *The Song of the Lark*

After graduating from the University of Nebraska in 1895, Willa Cather remained in Lincoln for several months. She spent her time writing for the *Courier* and the *Journal,* attending the theater, and seeing friends. When her funds grew short she moved back home to Red Cloud and commuted back and forth to Lincoln several times during the spring of 1896.[1] Although this was a productive year—in addition to her journalism, Cather published "On the Divide" in the January 1896 issue of the *Overland Monthly*—she felt restless and depressed, particularly after her return to Red Cloud. Confessing her boredom and isolation to Mariel Gere, she labeled her hometown the "Province" and datelined one letter "Siberia." After the freedom and sophistication of her Lincoln years, Red Cloud mores seemed unbearably primitive; at a New Year's dance she attended with her brother Douglass, Cather told Mariel Gere in mock horror, the natives actually passed around their refreshments in a potato basket.[2]

Having begun the process of separation from home and family, Cather found it difficult to return. She did not feel as close to her family as she had before leaving home, she told Gere; she no longer shared their way of looking at things. Especially trying was the contrast between her family's high expectations and her as yet unfulfilled promise. She had the unpleasant feeling that they were waiting for her to accomplish something extraordinary, Cather wrote, but they believed in her more than she did. She didn't feel that she had the resources to do anything—and certainly not in Red Cloud. She had not yet seen enough of the world, and besides, Cather continued accusingly, she was a terribly superficial person.[3]

Escape was not far off. With the help of Lincoln friends, in the spring Cather secured her first full-time job: the editorship of the *Home Monthly,* a new women's magazine based in Pittsburgh. She left Nebraska in the summer of 1896 for the city she later described as pulsing with "the incandescence of human energy"

(*CSF*, p. 140). Writing Mariel Gere shortly after her arrival, Cather portrayed her eastward train journey as a gleeful flight from restriction to freedom.[4] In many ways her Pittsburgh years justified her eager hopes. During her ten years there as magazine editor, newspaperwoman, and high school teacher she poured the same energy into her writing that the city was channeling into its industrial and commercial growth. She produced hundreds of columns, articles, and reviews for the *Home Monthly,* the Pittsburgh *Leader,* and the *Library,* a short-lived literary journal, as well as sending material back to Lincoln papers; she published several stories and poems, some in national publications; and her first two books appeared, *April Twilights* in 1903 and *The Troll Garden* in 1905.[5] During these years of apprenticeship and experimentation, Cather grew in technical skill and literary self-confidence as she continued to separate the personal and artistic qualities she admired from masculinity. At the beginning of her Pittsburgh decade she wrote "Tommy the Unsentimental," a short story whose admirable heroine is "aggressively masculine and professional" (*CSF,* p. 478); at the end, she wrote *The Troll Garden* stories, some of whose male and female protagonists elude the traditional gender roles and identities Cather accepted during her Lincoln years.

Although Cather had praised the artist's devotion to a celibate quest in her college journalism, her literary and professional development did not require loneliness and isolation. On the contrary, her emergence as a writer and her growing reconciliation of gender and vocation coincided with her love affair with Isabelle McClung, whom she met in 1899. McClung played many roles for Cather—lover, mother, friend, patron, muse, reader; as Cather later told her friend and colleague, the playwright Zoë Akins, Isabelle was the one person for whom all her books had been written.[6]

•

By the late 1890s, Pittsburgh was a city of 300,000, a major industrial and steel-producing center. Here, as in Red Cloud, the dominant creed was business, but Pittsburgh's capitalists—Carnegie, Mellon, Westinghouse, Frick—far outranked their Nebraska counterparts. Because of the fortunes these men amassed, the city had a thriving cultural as well as business life. The Carnegie Library, Music Hall, and Art Gallery opened in 1895; the Pittsburgh Symphony was founded in 1896. Major dramatic and operatic touring companies visited the city's several theaters, and Cather soon found her way into Pittsburgh's artistic and musical circles.

Although she knew it depended on business-created wealth, Cather viewed the artistic world as the locus for alternative values—passion, creativity, spontaneity—unrecognized, repressed, or devalued in the larger society. Writing on Pittsburgh life for the *Journal* (where she had the freedom to criticize her new home), Cather observed that "All Pittsburgh is divided into two parts. Presbyteria and Bohemia, and the former is much the larger and the more influential kingdom of the two." By "Presbyteria" Cather meant the fusion of religious and commercial values that ruled the city, the mutually reinforcing Protestant ethic and spirit of capitalism that denigrated emotion and art. Presbyteria viewed "aesthetic enjoyment" as sinful, she thought, because art aroused feeling: "There is nothing on the earth that a Pittsburgh Presbyterian fears and hates as he does the 'human emotion,' '' she contended, which was his ''synonym for wrong.'' Suspicious of ''any public gathering except a funeral,'' the Presbyterian businessman had no objections to ''undemonstrative'' sins like greed or selfishness—these could increase profits. But passion, like art, was wasteful as well as immoral (*WP* 2, pp. 505–7).

''Bohemia,'' by contrast, was Cather's alternative, marginal world of art and artists who worshipped truth and beauty with aesthetic and emotional fervor rather than mercantile piety. ''Bohemia'' is a densely coded word in Cather's vocabulary during this period. In addition to art, culture, and the imagination, it signifies a free realm of feeling and sensation opposed to conventional values and bourgeois morality, a place where, as in French fiction, the great passions were not ''wholly conventionalized'' (*KA*, p. 138). In her letters to Mariel and Ellen Gere, Cather uses ''Bohemia'' and ''Bohemian'' in this double sense, referring both to the kingdom of art and to the land of unrestrained emotion.[7]

At first she gloried in her difference from James Axtell, publisher of the *Home Monthly,* and his respectable Presbyterian family. Axtell invited Cather to stay with him when she arrived, doubtless unaware that he was allowing a serpent to enter his Calvanist sanctuary and welcoming as editor of his respectable ladies' magazine a reader of French fiction, a self-proclaimed atheist, and a former male impersonator. Cather seems to have enjoyed her brief stay in the enemy camp because she could feel superior to her stodgy hosts and satirize them in letters home. She imagined herself as Coleridge's Lamia and mocked the repressive propriety symbolized by their parlor, complete with haircloth furniture and portrait of grim, humorless, Presbyterian Grandpa.[8]

While we might assume that Cather would have claimed residence solely in Bohemia, her letters to Gere reveal a growing desire to acquire dual citizenship in conventional and unconventional realms. The correspondence shows Cather thoroughly enjoying her naughtiness—being in a Puritanical atmosphere reinforced her sense of herself as a romantic rebel—but it also reveals her new delight in donning the disguise of propriety. Writing Mariel Gere in July, she told of a recent triumph: visited by three prim Presbyterian ladies, she had demurely discussed gardening and church music, and they never suspected that she was not what she appeared.[9]

As Cather continued to be attracted by the pleasures of acceptance, she came to associate Bohemia with isolation, exile, and loneliness. Social life was more pleasant than she had ever thought it could be, she wrote Gere. Now that she didn't have short hair or dubious friends to mark her as queer, it was like beginning a new life away from the old mistakes, and she found she agreed with Charles Lamb: ''Gad, how we like to be liked!'' She was leaving Bohemia behind, she assured Gere several times; she had never intended to dwell forever in that disreputable realm. Letters home refer to a round of parties, picnics, and boat trips, to romantic attentions from young men (which she may have overemphasized to impress Mariel Gere), and to memberships in various women's clubs.

Cather's introduction to the women's social scene was a great triumph, which she relayed to Gere with amusement and pride. On one occasion a friend brought her to a tea given by the federated women's clubs. The subject for discussion was Carlyle, and Cather rose to deliver a seemingly impromptu oration that she really derived from her college essay. Unaware of this self-plagiarism, the ladies were impressed and besieged her with invitations, some of which Cather accepted. Even though she thought the ladies stupid and sophomoric to be so deceived, she spoke proudly of their acceptance in letters to Gere; it was unusual to be called on so early, Cather informed her friend.[10]

Entering a social context where she was unknown, Cather was free to create a different self from the one she had dramatized in Lincoln, and the mistakes she wanted to leave behind seem connected with her pose of defiance. Although her desire to enter Pittsburgh social circles did not signify her adoption of the ladies' club mentality she attacked during her college years, Cather increasingly wanted to triumph in public as well as in personally chosen terms. She could best surprise her friends and pain her enemies, she told Gere, by living a more conventional existence. The Pittsburgh visit of Dorothy Canfield's mother, who had been one of Cather's Lincoln critics, gave her a chance to do this. It was very satisfying, she confided to Gere, to introduce Mrs. Canfield to leaders of the local women's clubs: Cather knew that her new social status came as a surprise and she gloried in her revenge.[11]

In later years Cather continued to modify the overt signs of rebellion she had cherished as an adolescent. Although she eventually lived in Greenwich Village, Bohemia's symbolic center in New York, she shared a quiet, domestic life with Edith Lewis, carefully protecting her privacy and shunning the Village's avant-garde. Living less flamboyantly than she had in college became a form of self-protection when she was ready to channel her energy into writing rather than self-definition or dramatization. As she observed in an 1896 commentary on the Bohemian life, "to openly defy the accepted conventionalities of any generation requires an even greater expenditure of time [than strictly observing the conventions] and that way lies anarchy. For the business of an artist's life is not Bohemianism for or against, but ceaseless and unremitting labor" (*KA*, p. 413).

Cather's retreat from Bohemia was not merely a conscious strategy or a convenient pose. As a professional woman, a lesbian, and an artist she was a rebel against social norms; yet in other ways she was deeply conservative, coming to value even her family inheritance of religion, though she redefined her Baptist heritage when she and her parents joined the Episcopal Church in 1922. Her growing discomfort with Bohemia and interest in social acceptance during her Pittsburgh years reveal her attempt to combine seemingly opposed personal values and her desire—once the exuberance of adolescent rebellion had faded—to keep "unnatural" emotions and relationships protected and private. Cather's response to the invitations from the women's clubs she had once scorned also reflects her growing interest in the constellation of social values symbolized by the idea and role of the lady—not the Victorian angel in the house, but the aristocratic figure she defined in her essay "Miss Jewett" as the lady "in the old high sense" (*NUF*, p. 85). Cultured, wealthy, and privileged, this figure resembled her own mother but possessed the social and economic power Virginia Cather lacked.[12]

•

Cather's regard for social acceptance grew during her first year in Pittsburgh, but she continued to ridicule the Victorian conventions defining womanhood, including those her magazine was expected to endorse, and she scoffed at the *Home Monthly*'s domestic content. It was the worst trash in the world, she told Mariel and Ellen Gere, just great rot: home and fireside stuff, all babies and mince pies.[13] More than a few ironies surround this first appointment: the college writer who despised the *Ladies' Home Journal* was now in charge of its competitor; the former male impersonator was writing and editing columns on women's fashions; and the ambitious journalist who once scorned domesticity was hoping to succeed in the male world outside the home by producing a magazine that focused on the female worlds of kitchen, nursery, and parlor. In the first issue the editor announced that the magazine's purpose was to explore "every phase of home life" and to "furnish entertainment for the idle hour," reassuring female readers that they would encounter no "sensational" or "unwholesome" fiction within its pages. Proclaiming a moral as well as a domestic mission, the *Home Monthly* pledged to be "pure and clean in tone" and to "educate" and to "elevate" its readers while entertaining them. The articles and columns Cather selected for the first issue included domestic notes ("Table Points"), advice to mothers ("Care of Children's First Teeth"), and moral guidance ("Christian Endeavor"), while the short stories ("A Modern Elaine" and "The Lovely Malincourt") were conventional romances suitably "pure and clean" for the female audience.[14]

Cather was determined to succeed even at such uncongenial work. Of course it was hard for her to write gentle home and fireside material, she told Mrs. Gere, but she simply would do it—and do it well. The magazine's policy was rather namby-pamby, she admitted, but that was the publisher's choice, not hers, and she had to carry out his wishes as best she could.[15] Writing Mariel, she was more blunt: if it was trash they wanted and trash they paid for, then trash they would have—the very best she could produce.[16] In her letters to Mariel Gere and Mrs. Gere, Cather speaks frequently of her desire to prove herself to folks back home as well as to her employer and herself. In one letter she describes her work in terms of struggle, conquest, and victory that recall her identification with heroes of epic and romance. She wanted, she told Mariel, to triumph over her enemies and give satisfaction to her friends.[17]

Producing the first issue of the *Home Monthly* was a rigorous test: Axtell had left town expecting his new editor to handle the August issue alone. In addition to editing, supervising the printing and composing, and helping with layout, Cather had to write almost half the copy because the magazine had no file of manuscripts on hand. The responsibility was awesome, she wrote Mrs. Gere, and she was dreaming about the magazine every night.[18]

The new editor encountered a conflict between her commitments to journalism and to fiction-writing that would endure—with the exception of the years she spent as a high school teacher—until 1911 when she took a leave from *McClure's* to write *Alexander's Bridge*. Cather's creative energies were sapped by newspaper and magazine work, and her letters to friends frequently mention feelings of exhaustion and depression. After a day at the office she was often too drained to work on her own writing. Her years at the Pittsburgh *Leader*, where she felt oppressed by the daily grind of newspaper work, were particularly hard, and she often grew discouraged with her slow progress as a fiction writer.

Another source of the conflict between the journalist's and the novelist's vocation was more insidious. In discussing her friend's apprenticeship years, Edith Lewis observes that in her early stories Cather could not develop her gifts as an "original writer" because she was forced to conform to the mechanical requirements of formulaic popular fiction "which had little relation to her own thinking and feeling" (*WCL,* p. 42). Certainly the *Home Monthly* editor experienced many constraints on self-expression. Producing "pure and clean" entertainment for the publisher and promoting wholesome domestic advice for the reader, Cather was implicated in the distortion of writing for commercial and ideological ends that she herself criticized in her college journalism (*KA,* pp. 281–82). In one sense an emancipated professional woman, in another Cather was speaking the inauthentic female language she had rejected in her childhood rebellion against Southern platitudes. Far from finding the opera singer's idealized speech, the young editor was conversing politely with ladies in Pittsburgh parlors. Dedicated to art on the one hand and to professional success on the other, she experienced internally the division between "Bohemia" and "Presbyteria" she saw in the larger society.

During her brief tenure at the *Home Monthly* Cather displayed her facility in speaking Presbyteria's language and produced a range of articles in several popular modes—the "wives of famous men" article ("Two Women the World Is Watching"), the domestic hints column ("Fashions"), the feature ("Nursing as a Profession for Women"), the topical piece ("The Origins of Thanksgiving"). "Two Women," a sketch of Mrs. William McKinley and Mrs. William Jennings Bryan, reveals how Cather adapted her journalistic voice to literary and social conventions. The Victorian reader of the "wives of famous men" article expected to see domestic values reconfirmed, and in portraying Mrs. McKinley's married life Cather could scarcely have uttered more appropriate sentiments: "For the last twenty-five years Mrs. McKinley's life and interests have been those of her husband, as, in one way or another, actively or passively, every woman's must be" (*WP* 1, pp. 309–10).

In some ways Cather had more opportunity for self-expression as a reviewer than as a feature-writer because she did not have to conform to the formulas regulating popular genres. Yet her Pittsburgh reviews also show her speaking to Presbyteria. Adopting different voices for different audiences, she considerably tones down her acerbic wit and fin-de-siècle enthusiasms for the *Home Monthly* book column. In the Lincoln papers she praised the portrayal of unconventional passions she found in Gautier and Daudet, but in the *Monthly* she adopts a confiding, maiden-aunt tone in promoting books that had, she commented reassuringly, "the delightful atmosphere of home in them." Trying to make French fiction acceptable to her pious readers by stressing its delicacy, Cather informed female Presbyteria that Daudet's *Kings in Exile* was a "story of maternal love" that handled the "sacred theme" more "reverently" than any other novel she knew (*WP* 1, p. 341). In a similar vein, she promised that Dinah Maria Mulock's *John Halifax, Gentleman* was a book for "fireside and family" that a tired housewife would find restful and a young girl think "the sweetest romance" (*WP* 1, pp. 335, 341).

The "entertainment for the idle hour" Cather wrote for the *Monthly* likewise reveals her adjusting to the demands of popular genres and female readers. In the first few issues we see her experiments with the children's story ("The Princess Baladina: Her Adventure"), the sentimental melodrama ("The Burglar's Christmas"), and the hymn to childhood purity ("Thine Eyes So Blue and Tender"). Most of these stories and poems are indistinguishable from the work of other contributors. In "Thine Eyes," for example, the mocker of Presbyterian piety adopts the requisite reverent tone in exalting a bright-cheeked little boy who "brings God nearer,/ mightier than all the creeds." But at times Cather either subverted literary conventions or found the opportunity for conscious and unconscious self-expression by choosing a formula corresponding to her own desires or conflicts. If we look more closely at the *Home Monthly* fiction, we can see that her imagination was not quite as stifled by the conventions of magazine fiction as Lewis believed.

•

Editing the *Home Monthly* may not have been as thrilling a venture as the quests undergone by the protagonists of the romances Cather loved, but it was still, despite the magazine's domestic focus, a male narrative. As a young woman from the provinces recently arrived in a city and eager to succeed in the professional world, Cather was living a plot she could not tell to her female readers. Or at least not too ostentatiously or directly: one story Cather wrote for the first issue shows the young editor's intention to satirize the feminine virtues she was required to endorse elsewhere and to place a female protagonist in the hero's plot.

"Tommy, the Unsentimental" (1896) mocks the Victorian ideal of womanhood to which the *Home Monthly* was supposedly dedicated. The heroine is the boyish Tommy, not the clinging, dependent Miss Jessica, and Tommy's favorite pastimes, as Cather must have known, were a particular challenge to Presbyteria: playing whist and mixing cocktails were not the feminine skills Axtell wanted to recommend to his teetotaling readers. In subverting Victorian conventions of gender and genre, Cather goes even further by substituting a male plot for a female one, paying tribute to the narrative the author—not her readers—was living. Opening with Tommy's fondness for Jay Ellington Harper, the story seems to introduce a conventional female narrative, preparing the reader for the romantic formula to follow: complications and misunderstandings followed by a happy ending with marriage either achieved or intended. But Cather quickly transforms "Tommy" into a masculine genre—the adventure story—with a female hero as Tommy mounts her bicycle and sets off to save Harper from financial ruin. Cather then completes her subversion of female narrative conventions by turning the love story into a humorous subplot. She allows Miss Jessica and Harper to recognize their love—with Tommy's prompting—but since the lovers are so silly and inept, and Tommy so competent and likable, the traditional ending is ironic. We are glad that Tommy leaves the story unmarried.

In publishing this iconoclastic story in her first issue Cather risked alienating Axtell as well as her readers. Why did she do so? Despite her assurances to Mariel Gere that she could control her antipathy to the magazine's domestic content, she seems to have wanted to declare herself under the cover of fiction, to tell the truth but tell it slant. Her alter ego Tommy's triumph over Miss Jessica then represents the author's self-declared superiority to the *Home Monthly*'s ideal reader and Victorian society's ideal woman. Tommy's rescue of the inept Jay Ellington Harper from financial disaster further suggests Cather's desire to write of the professional woman's ability to outdistance less competent men.

The final hint that the story was the editor's veiled but rebellious self-disclosure is the signature she attached to it. Cather employed a range of pseudonyms in the *Home Monthly* as she did in other Pittsburgh publications. There were several reasons why she masqueraded as Mildred Beardslee, Helen Delay, Emily Vantall, John Esten, Charles Douglass, John Charles Asten, Mary Temple Bayard, George Overing, Clara Wood Shipman, Gilberta S. Whittle, Mary K. Hawley, W. Bert Foster, and Henry Nicklemann, among others. She needed to conceal her substantial contributions to the *Home Monthly* and later to the *Library,* a Pittsburgh literary magazine; she was producing material she did not particularly value; and she took a continuing delight in disguise for its own sake. Her two other literary contributions to the first issue were also pseudonymous: "The Princess Baladina: Her Adventure," a children's story written by "Charles Douglass," and "My Little Boy," a sentimental poem by "John Esten." But "Tommy" was signed "Willa Cather." She may have been able to risk disclosure because she assumed that the romantic subplot would function as the cover story for Tommy's rebellious tale. Missing the satire in the author's portrayal of Miss Jessica and her union with the useless Jay Ellington, the reader might assume that "Tommy" was only a variation on the sentimental formula rather than a subversion of it.

While "Tommy" reveals Cather's conscious overturning of Victorian gender stereotypes by manipulating the conventions of popular fiction, in other *Home Monthly* stories she found room for unconscious self-revelation while following the rules for women's and children's fiction. Having just left Nebraska, the young editor was living in a strange city and feeling lonely and homesick, as her letters to Mariel Gere reveal. The uprooted daughter was editing a magazine dedicated to the home, perhaps one reason why the domestic formulas she had to employ were not too grating. The conventional and sentimental subjects of the popular fiction she was expected to write—centering on home, family, and the mother–child bond—corresponded to the emotional conflicts preoccupying her at the time. Although determined to succeed on her own, Cather still yearned for old attachments.

The two children's stories Cather wrote for the magazine reveal that she could invest herself emotionally as well as professionally in the "home-and-fireside" material she professed to despise. Accustomed to telling stories to her younger siblings, Cather drew on the older sister's narrative skills in "The Princess Baladina: Her Adventure" (1896) and "The Strategy of the Were-Wolf Dog" (1896). Cather used the "running away from home" motif in both stories and as well as in "Wee Winkie's Wanderings" (1896), written for another magazine published by Axtell, *The National Stockman and Farmer*. In all three stories Cather undermines the protagonists' urge for escape, independence, and adventure: the princess who runs away to discover a prince discovers a "dearth of Princes" in the real world and returns home (*CSF,* pp. 567–72); the reindeer in "Were-Wolf Dog" escape Santa's warm but confining house only to drown in a Polar sea; and Wee Winkie, scratched and bruised, goes back to mother. Of course the return of a wandering child is a formula in children's fiction, but Cather's continued use of the pattern suggests that the stories reflected the ambivalence about leaving home that she expressed in letters to Mariel Gere.

"The Burglar's Christmas" (1896), which Cather also wrote for the December issue, even more overtly reveals how, in adhering to popular conventions, the young writer allowed unconscious material to surface. At once formulaic and dreamlike, in "The Burglar's Christmas" the young author uses the prodigal son plot to write her own fairy tale of wish-fulfillment, the desire to return home to a nurturing and seductive mother. Evoking the terrors and the allure of regression, the story also suggests the young editor's fear of professional failure. William is a former magazine editor who knows that those who expected great things from him—as Cather's family and friends did from her—had "read him wrongly," for he "knew now that he never had the essentials of success, only the superficial agility that is often mistaken for it" (*CSF*, pp. 557–66). I am a terribly superficial person, Cather told Mariel Gere in May of 1896, while in August she confessed that she doubted she would ever do anything very good.[19]

Writing to formula, then, was one way of resolving the tension between disclosure and concealment Cather experienced in her college fiction. Concentrating on the rules governing plot, characterization, and style required by popular genres, she could disguise the private sources of her narratives for herself as well as for her readers.

•

Even while circumscribed by the magazine's conservative editorial policy, Cather wrote three stories for her adult readers that mark a new direction in her writing—in "The Prodigies," "A Resurrection," and "Nanette: An Aside," she turns from male to female protagonists who become more and more complex. The heroines of "The Prodigies" (1897) and "A Resurrection" (1897) admittedly have stereotypic traits. But the bad mother/good mother opposition in these stories reflects Cather's emotional preoccupations as well as the literary conventions she was using, so the stories are not merely formulaic exercises. "A Resurrection" even challenges Presbyteria's ideology of femininity. On the surface the story extols a familiar Victorian figure, the angel in the house, represented by the maternal and self-sacrificing Margie. Yet Cather chose (and in the magazine's Easter issue, at that) to flout convention by granting her heroine quite a pagan resurrection: reunited with her lover after years of misunderstanding, Margie allows the "flood gates" of suppressed passion and emotion to open, symbolized here by the river's swirling currents, swollen by spring rains (*CSF*, pp. 425–40).

Commenting in "A Resurrection" on gifted, creative women like Margie who find themselves stifled in parochial, limiting environments, Cather observes that some women "have the wrong parts assigned them" (*CSF*, p. 426). As the writer of fictions, however, she could assign parts to female characters that women did not often play in the social world, and she was beginning to discover that transforming power. In "Nanette: An Aside" Cather makes her heroine an opera singer, a forerunner of Thea Kronborg. Exploring Tradutorri's conflict between human relationships and the austere demands of art, Cather portrays the woman artist's sacrifices as well as her triumphs. Although the story is neither thematically nor stylistically subtle, it reflects Cather's growing fascination with women artists and her gradual separation of creativity from masculinity. The story also shows her understanding of the close bonds uniting women that are not recognized in any conventional way: in this case, the affection and dependence connecting Tradutorri to her maid Nanette, who is both servant and daughter to her imperious but loving mistress (*CSF,* pp. 405–10).

One *Home Monthly* story points in another literary direction, however, anticipating not Cather's artist–heroines but her Jamesian apprenticeship of the early 1900s. The literature Willa Cather considered ''art'' had its conventions too, and Henry James' influence on Cather is first evident in ''The Count of Crow's Nest,'' the story of the conflict between the honorable but impecunious Count De Koch, the last of a noble European family, and his avaricious daughter, a third-rate opera singer. The subject—a quarrel over the disposition of family letters—suggests *The Aspern Papers,* while Cather's experiment with point of view reveals the presence of James' narrative technique: the story is told by an observer who sympathizes with the Count's stubborn integrity. Cather had not made James' method her own at this point, as she did in *My Ántonia* and *My Mortal Enemy,* but her control of point of view is nevertheless more adept than in ''A Tale of the White Pyramid.'' Imitation of the master could be rewarded, Cather discovered: *Cosmopolitan* wanted to publish the story, but she needed to fill space in the *Home Monthly* and rejected the offer.

While Cather was home for a visit in the summer of 1897, the *Home Monthly* changed ownership and she resigned.[20] She soon secured a newspaper job, however, editing wire copy on the telegraph desk of the Pittsburgh *Leader* (she had been writing drama reviews for the paper since the fall of 1896). Returning to the city in the fall, she continued to write her column ''Old Books and New'' for the *Home Monthly,* sent ''The Passing Show''—a chatty commentary on the Pittsburgh arts scene—back to the *Journal* (later the *Courier*), and reviewed plays, concerts, and opera for the *Leader.* Eventually she transferred her book column from the *Home Monthly* to that newspaper.

Cather continued at the *Leader* until early 1900 when she left to freelance for the *Library,* a new and short-lived literary journal. She spent the winter of 1900–1901 in Washington, where she visited her cousin Howard Gore, translated government documents, and wrote articles on the Washington scene for Lincoln and Pittsburgh papers. Then in March 1901 Cather returned to Pittsburgh and a job teaching Latin at Central High School. The next fall she moved into the English department, where she stayed for two years before moving on to Allegheny High School; she remained at Allegheny until she left Pittsburgh in 1906 to begin her job at *McClure's.* Although Cather was not profoundly committed to teaching, her work could be more easily accommodated with writing than could journalism; at least she had her evenings and summers free. By most accounts she was a successful teacher, although one who did better with the exceptional than with the average student.

When Cather returned to Pittsburgh and teaching she moved in with Isabelle McClung and her family, ending a rootless existence and beginning both a lifelong love affair and a five-year period of emotional well-being and literary productivity. Having published only two minor stories in 1898 and 1899—"The Way of the World" and "The Westbound Train"—in her new home within five years she wrote ten stories and published *April Twilights* (1903) and *The Troll Garden* (1905). Since Cather had often claimed that human attachments and artistic creation were incompatible, particularly for women, a question naturally arises: Why was this friendship a catalyst for her work?

•

Always distressed by change—particularly dislocations which replayed her childhood uprooting from Virginia—even with her busy social life Cather was deeply lonely during her first years in Pittsburgh, living in four boardinghouses in five years and existing on "miserable food" (*WCL,* p. 43). Although her letters to Mariel and Mrs. Gere alternate between excitement at the beginning of her new life and sadness at the ending of her old, the thread that runs throughout is a sense of loss. Why don't your heartless daughters write to me, Cather asked of Mrs. Gere shortly after arriving in Pittsburgh. If only they knew how lonely she was, she knew they would. She could stand the loneliness during the day, she confided, but at night she yearned for her friends.[21]

Surprisingly these feelings increased instead of dissipating as Cather made new friends in Pittsburgh. In April 1897, she wrote Mariel that Dorothy Canfield's visit had left her lonelier than before.[22] Later that year, after returning to take up her job at the *Leader,* she told Mariel that she could not be away from Nebraska any longer; she was a fool to be an exile, and her only solace was reading Housman's melancholy poems. What good were money and professional success, she wondered, if she were unhappy?[23] Other letters reveal her fear of losing old attachments. Writing Frances Gere in 1898, she confessed her terror that she might drift out of her old friends' lives and become just another stranger. In the same year, Cather told another friend that living away from home was an unpleasant duty.[24] She visited Nebraska for long periods in the summers of 1897, 1898, and 1899. While it might seem unusual for a recently emancipated daughter to return so frequently to her childhood home, as long as Cather felt homeless in Pittsburgh she needed to reestablish her Nebraska roots and to rekindle friendships and family attachments.

Cather's feelings of exile and isolation surprisingly persisted as she developed close friendships in her new home. She might have been lonely in Pittsburgh, but she was not alone. Dorothy Canfield, who frequently stopped in Pittsburgh en route to her family home in Vermont, alleviated loneliness temporarily. It was so comforting, Cather confided to Mariel Gere, to have Dorothy waiting for her when she got home from work.[25] One Christmas Cather brought Canfield to meet new acquaintances, George and Helen Seibel. George Seibel, a freelance journalist (later a librarian), had brought Cather an article for the *Home Monthly* and thus began an important friendship. Soon she was visiting the Seibels once or twice a week, sharing meals, and reading French and German literature with them. Although the Seibels were new friends, the role they filled in Cather's life was a familiar one: they represented the warm, cultivated, European family to which she wanted to belong (the Seibels were of German descent). Offering the domestic comforts she lacked in her boardinghouse, the Seibels provided food for the body and the mind. In their home Cather could indulge first in "noodle soup, plebeian but nourishing potato salad," and "cookies of crisp and crackling texture" before reading the romantic French literature Cather loved and Seibel termed "devil's food"—Daudet, de Musset, Gautier, Hugo, Flaubert. These times, Cather told Mrs. Seibel in an 1897 letter, were the pleasantest she spent in Pittsburgh.[26]

As Cather became a familiar member of the city's artistic, intellectual, and musical circles, she established a busy social life and made other friends: Lizzie Hudson Collier, the leading actress at the local repertory theater, who took a motherly interest in the young woman; the composer Ethelbert Nevin; Edwin Anderson, head librarian at the Carnegie; Mrs. John Slack, a patron of the arts who gave musical parties Cather frequently attended. Her closest emotional ties were with three other young women: May Willard, a librarian; Ethel Litchfield, an amateur pianist married to a local doctor; and Isabelle McClung. After hearing of May Willard's death in 1941, Cather reminisced in a letter to a mutual friend about the happy days when the four young women had just met; they were so compatible that they melted together, she remembered, and people who came later could never be part of that first design the quartet had made.[27] But the three friends were not indistinguishable. As the center of Cather's design, Isabelle McClung satisfied her friend's yearnings for closeness and intimacy. After their first meeting, the references to loneliness and isolation in the letters to Gere diminish and disappear, as do the letters themselves.

Isabelle's father Samuel McClung was an eminent resident of Presbyteria. A wealthy and conservative judge, like many other Americans of his class and social position, McClung disliked Bohemians, radicals, and other deviants who threatened the moral, social, or political order. His wife was a less forbidding person whom Edith Lewis portrays as "quite opposed to the Judge in temperament, naturally fond of pleasure and gaiety" (*WCL,* p. 52). Members of Pittsburgh's economic and social elite, the McClungs lived on Murray Hill Avenue in the posh Squirrel Hill area. Dorothy Canfield describes their residence as a "great rich house, conducted in the lavish style of half a century ago" where the McClungs indulged in "stately entertaining."[28]

A less flagrant rebel than Willa Cather, Isabelle McClung in some ways defied her parents and their way of life. Scorning the decorous, restricted life of an upper-class lady (although refusing to abandon the privileges her social status provided), Isabelle escaped to the enticing and forbidden world of Bohemia, cultivating as friends and acquaintances "writers, painters, musicians, actors, foreigners," and other exotics (*WCL,* p. 52). Elizabeth Moorehead, Pittsburgh resident and a friend of both women in the early 1900s, describes McClung as an "imaginative and outgoing" person "impatient with the narrow round of social life." Constricted by her parents' social world, she began to "make excursions into other less conventional realms," including the dressing rooms of actresses she admired. Hence Isabelle McClung and Willa Cather, despite their social differences, were traveling in the same avant-garde circles. The two young women met sometime during 1899 in the dressing room of Lizzie Hudson Collier. Drawn together by their common rebellion against Presbyteria's values, the two young women quickly became close friends. By October 1899, Cather was a frequent guest at the McClung mansion.[29]

In 1901 Isabelle ended Cather's transient boardinghouse life when she asked her friend to live with her at Murray Hill Avenue. There Cather enjoyed an ordered, protected life in comfortable—even luxurious—surroundings until she left to assume her job with *McClure's* in 1906. Reversing her earliest remembered trauma, the uprooting from Virginia, this move had immense psychological, emotional, and creative resonance. As Elizabeth Sergeant notes, despite her rejection of the bourgeoisie, Cather always had more affinity for the family than for Bohemia (*WC:AM*, p. 25). Although she romantically liked to envision the artist as a lonely, dedicated soul, Cather needed to be rooted in a warm, familiar domestic space in order to write, and throughout her life, in addition to making individual friends, she often adopted families: in addition to the McClungs, there were the Miners, the Canfields, the Westermanns, the Seibels, and later the Knopfs and the Menuhins.

The McClungs' reactions to their daughter's invitation and to their permanent house guest are different in the several versions of the story. According to E. K. Brown, they "wondered at the propriety" of welcoming Cather into their household, but their daughter reportedly threatened to leave home and they conceded.[30] A McClung grandson remembers that Cather's presence "caused a strain between the judge and his children" which "never vanished," although according to other accounts Cather eventually developed amicable relationships with Isabelle's parents, if not with her siblings.[31] But Willa Cather's sister Elsie and her niece Helen Cather Southwick portray the household arrangement as harmonious and satisfactory to everyone.

It is difficult to sort out these conflicting stories and to determine why the parents might have initially opposed Cather's entrance into their household. They need not have interpreted their daughter's relationship with Cather as lesbian to have been disturbed by it; Isabelle was at a marriageable age, and they must have realized that her new intimacy signified a lack of interest in men. The McClungs may also simply not have wanted their lives and family patterns disrupted by a stranger, particularly one who did not always defer to others' wishes. Whether or not the parents were initially opposed to her inclusion, Cather eventually became an adopted member of the family.[32] But even while living *en famille*, she and Isabelle found time for privacy and intimacy, as Elizabeth Moorehead's portrait of their shared life suggests:

> The two young women would forsake the family group soon after dinner, and evening after evening would go upstairs to the bedroom they shared to read together in quiet. This room was at the back of the house and its wide low window gave on a downward slope across gardens and shaded streets toward the Monongahela river and green hills rising beyond. There were no close neighbours to destroy their sense of privacy. Here the friends . . . devoured the novels of Tolstoi, Turgenev, Balzac, and Flaubert.[33]

A gracious hostess who possessed the social skills that Cather lacked, Isabelle McClung impressed observers with her beauty as well as her poise. Edith Lewis describes her as "handsome" (*WCL,* p. 51), and Dorothy Canfield remembers her as "very beautiful . . . in a sumptuous sort of way." To Elizabeth Moorehead she was "the most beautiful girl I had ever seen." Cather, on the other hand, struck Moorehead as possessing "strength of character" rather than conventional beauty: "You saw at once that here was a person who couldn't easily be diverted from her chosen course."[34]

Lacking Cather's correspondence with Isabelle McClung, destroyed when Jan Hambourg returned the letters after his wife's death, we cannot reconstruct their relationship in any detail. But it is still possible to suggest why this bond was central to Cather emotionally and imaginatively. Beautiful, gracious, well-born, Isabelle possessed the "innate aristocratic refinement" Cather worshipped in the "divine femininities" she saw in the Lincoln art gallery (*WP* 1, pp. 124–25). Isabelle was a member of an elevated social class which Cather, despite her democratic sympathies with Nebraska immigrants, revered. Like Virginia Cather, she was a lady; unlike Virginia Cather, she possessed the wealth and privilege Cather associated with that lofty rank.[35] Combining the power of money and status with an interest in Tolstoi and Flaubert, Wagner and Ibsen, Isabelle satisfied her friend's needs to live simultaneously in conventional and unconventional worlds. And unlike Dorothy Canfield, Elizabeth Sergeant, and Zoë Akins, Cather's other important women friends, she was not a writer and hence not a competitor. Isabelle was interested in fostering someone else's literary talent, not in developing her own.

Able to express with Isabelle aspects of the self she had suppressed in her struggle to separate from her mother and the feminine role, during her residence at Murray Hill Avenue Cather became more interested in women's dress. Isabelle loved beautiful clothes and enjoyed buying them for her friend, sparking Cather's attraction to the "lovely fabrics, furs, and beautiful hats" she would later have custom-made at Bergdorf Goodman's, where she also had shoes dyed to match her evening wear ("frequently green or any shade of red").[36] Just as she admired the "old, high" sense of the lady, Cather liked royal and theatrical clothing, signs of female power and autonomy; she eventually preferred satins and velvets, generally brightly-colored, like the "Pasha-like turreted turban" she wore in New York, "so heavy with gold lace, velvet, and aigrettes," Sergeant recalls, that it "might have come right off the opera stage" (*WC:AM,* p. 121). Her niece Helen Cather Southwick remembers the ritual of unpacking Aunt Willa's trunk when she visited Red Cloud in the 1920s. Out would come a variety of female dress: "In addition to the cotton skirts and middy blouses that she wore in the morning, she brought beautiful clothes to wear in the afternoon and evening—heavy silk crepe skirts and chiffon and crepe de Chine blouses and dresses. In the winter there were fine wool crepes and velvets."[37]

During the Pittsburgh years Isabelle evidently did not impose her own fashion preferences on Cather, who also continued to enjoy the professional woman's tailored dress. According to Elizabeth Moorehead, when they attended the symphony each "dressed to please herself," Isabelle wearing "lace and chiffon and feather" with Cather at her side in "shirtwaist and skirt." During her years of schoolteaching Cather dressed plainly, wearing tailored suits and white blouses, white shirtwaists with starched collars and ties. *(See photograph section.)* Occasionally, though, she would abandon the tailored look and disconcert her students by appearing in a "frilly blouse and draped skirt" or, even more daring, in "blue stockings with white polka dots visible above high shoes."[38]

Discussing Olive Fremstad's use of costumes in "Three American Singers," in 1913 Cather suggested why the female dress she found confining and unacceptable in adolescence could be redefined and reimagined along with the female identity it symbolized—the process Isabelle aided. Just as Fremstad defined her own voice, developing an upper range so that she could sing soprano roles, so the singer found in designing her operatic costumes that Cather termed a "power of transformation." Able to give her costumes "exactly the lines she desire[d]," Fremstad made dress a means of self-expression like her voice.[39] Cather's growing interest in female dress during her Pittsburgh years, like her later commentary on Fremstad, demonstrates her discovery of some flexibility within the supposedly monolithic and unvarying concept "woman." Finding new possibilities for fashioning a self, Cather gradually discovered the variety in women's costumes that William Cather could not recognize.

As Cather's new interest in female dress suggests, her relationship with Isabelle offered her a satisfactory reenactment of the mother–daughter relationship—the central reason why the friendship was such a creative catalyst. If, as Adrienne Rich observes, the loss of the mother to the daughter is the "essential female tragedy," with Isabelle Cather found compensation for that primal loss.[40] She called attention to the maternal role Isabelle played in an 1899 letter to Dorothy Canfield. Isabelle was being so good to her she was feeling positively kiddish, Cather confided. Soon Isabelle would have her playing with dolls. In the same letter Cather associated Isabelle with classical power and perfection: Here I am chez the goddess, she told Canfield exultingly. Isabelle had met her at the train station, she explained in a tone of amazed delight, looking as if the frieze of the Parthenon ought to have been with her.[41] Unlike her own mother or Louise Pound, however, the classically beautiful Isabelle was an attainable goddess who, instead of requiring worship from a distance, descended to earth and devoted herself to mothering her friend. And since she was four years younger than Cather and deferential to her talent, Isabelle was not a dominating mother. At times she even needed protection herself: Isabelle was so frightened of reformers, Cather confided to Elizabeth Sergeant.[42]

In general, Cather viewed romantic relationships with the same deep ambivalence she brought to the mother–daughter bond—she needed intimacy and attachment yet feared incorporation and engulfment. But with Isabelle, Cather balanced her conflicting needs for autonomy and connection. Unlike the devouring mother in "The Burglar's Christmas," Isabelle acknowledged her friend's separate existence and wanted to enhance her creativity, offering her a third-floor study that remained Cather's private realm even after she left for New York. During Isabelle's Sunday afternoon teas even though Cather was the "chief attraction," she would sometimes remain in her study, Moorehead remembers, "leagues away from us all in that rich world of the imagination which Isabelle always understood and protected.'' The return to childhood Cather experienced in Isabelle's presence was thus the artist's creative regression which Ernst Kris describes. Safe in the sheltering space Isabelle provided—which resembled her Red Cloud attic bedroom—like the child who plays, Cather could find that ''intermediate area'' portrayed by D. W. Winnicott, halfway between inner and outer reality, where dreaming, symbol-making, and creativity can take place.[43] Isabelle may have allowed Cather to feel ''kiddish,'' but—in contrast to the silenced protagonists in the first short stories—in her presence Cather was playing with words.

•

Not only did Cather become more prolific during the these years but she also took more command of her craft, selling several stories to national magazines and beginning to sign her stories ''Willa Sibert Cather,'' signifying her more confident adoption of the artist's role. Even after Cather moved to New York and was living with Edith Lewis at 5 Bank Street, Pittsburgh remained Cather's emotional home and Isabelle the mother of her fiction. Cather would return to Murray Hill Avenue to write, often staying for several weeks at a time. When Cather left her job at *McClure's* to work on *Alexander's Bridge* and ''The Bohemian Girl,'' Isabelle presided over the difficult transition, taking her friend to a country retreat in Cherry Valley, New York, where she could be undisturbed. Later Cather wrote substantial portions of *O Pioneers!* and *The Song of the Lark* in Pittsburgh.

In telling Zoë Akins that Isabelle was the one person for whom all her books were written, Cather later acknowledged the connection between her writing and her love for Isabelle. Cather dedicated two books to her: *The Troll Garden,* the book of short stories she wrote under her protection, and *The Song of the Lark,* her semi-autobiographical novel of a woman artist who also discovers her creative powers in a female environment. The dedications reveal the emotional reciprocity Cather eventually experienced in the creative process. Isabelle inspired her to write, and Cather's books were, in a sense, gifts which she gave back to her. Writing a novel was thus an act of communication as well as self-expression, a letter she was sending to Isabelle, her muse and ideal reader.

The dedication Cather included with *The Song of the Lark* metaphorically conveys Isabelle's central role in her creative life:

TO
ISABELLE MCCLUNG

On uplands,
At morning,
The world was young, the winds were free;
A garden fair,
In that blue desert air,
Its guest invited me to be.

The juxtaposition of the wind's freedom and the garden's protection suggests the paradoxical union of expansiveness and enclosure Cather gained in this important friendship. The writer portrays her friend as an oasis in a desert, a garden where the "guest" can root and grow like the novel's heroine (Thea becomes a "tree bursting into bloom" when singing her triumphant Sieglinde [*SL,* p. 478]). Regaining the psychological and emotional landscape she imagined she left in the Virginia of her childhood—a protected, fertile mother country—Cather found with Isabelle the balance between self-abandonment and self-expression she would later experience in the creative process.

Cather's decision to leave Pittsburgh and Isabelle for New York and a job at *McClure's* may have been difficult, but she assumed that the relationship would endure. And her bond with Isabelle did remain central for several years. They took vacations together, and Cather would travel to Pittsburgh for long visits. Isabelle also came to New York; Sergeant remembers seeing her at Bank Street gatherings where she "cast a glow which was reflected in Willa's shining face and quick stumbling rushes of talk and color" (*WC:AM,* p. 129). Cather seems to have assumed that her "garden fair" would always remain in Pittsburgh, ready to welcome her back whenever she wished to return. But when Isabelle's father died in 1915 and the family mansion was sold, Cather lost this secure and protected space. Writing her Aunt Franc, she mourned the loss of the safety and care Murray Hill Avenue symbolized in her life; no other place could ever be home, she thought.[44] And shortly she lost Isabelle, who married violinist Jan Hambourg in April 1916, soon after her father's death. This event seems to have been an unexpected shock for Cather. Isabelle was then thirty-eight, and the two women had been intimate friends for seventeen years; Cather had not anticipated the marriage and was not emotionally prepared for the ensuing loss.

After 1916 Cather retained Isabelle as reader, friend, and muse but not as the lover whose love and company were always available. At first she was devastated. When she told Elizabeth Sergeant of Isabelle's wedding, Cather's face was "bleak," her eyes "vacant" (*WC:AM,* p. 140). Writing Dorothy Canfield at the same time, Cather described the change in her life as irrevocable, the loss as overwhelming. Things couldn't go on as they were, she knew, and she felt that a good many doors had been closing—a reference to the death of Mrs. Fields in the same year and the loss of another refuge, 148 Charles Street. This image of exclusion—the door shutting, leaving the child outside the house—recalls the sense of exile she first experienced in Pittsburgh. But now Cather experienced this deprivation as a direct blow to her creativity. Her next book was there, she told Canfield (she had begun thinking about *My Ántonia*), but she couldn't work, she was indifferent toward it.[45]

She did return to her writing, and to a novel exploring the power of a strong, nurturant woman who could inspire creativity in someone who loves but cannot marry her. If Isabelle's maternal presence had aided Cather's work in the early years, her absence did in later ones; if Cather's first stories were letters, then several later novels, including *My Ántonia* and *A Lost Lady,* were Winnicott's transitional objects, symbolic means of replacing—and connecting to—the lady (or ladies) she had lost.

Cather resolved her feelings toward Jan Hambourg in a more aggressive way. Her nasty portrayals of predatory Jewish men in stories written after she knew of

Isabelle's marriage plans ("The Diamond Mine" and "Scandal") were in part motivated by the unfortunate anti-Semitism Cather shared with many American writers of her generation (Hambourg was Jewish) and in part by her hostility to her victorious rival. Later Cather seemed to announce her affection for Hambourg when she dedicated *A Lost Lady* and *The Professor's House* to him. But a closer look suggests that at least the first dedication was an ambiguous gesture. In *A Lost Lady* Cather describes a beautiful woman's sexual and financial dependence on two equally objectionable men (Frank Ellinger and Ivy Peters). In *The Professor's House,* however, she offers a more flattering portrait of a Jewish male and in later editions she dropped the dedication to *A Lost Lady.* In life, if not in fiction, Cather achieved an amicable relationship with Hambourg, possibly because she recognized that—if she were to retain a close friendship with Isabelle—she simply had to accept her husband. He, in turn, acknowledged Cather as his wife's dearest friend, and the three spent a good deal of time together over the years.[46]

In 1921 Cather tried to recapture the creative sanctuary she had enjoyed in Pittsburgh during a lengthy stay with the couple in their Toronto home. After the Hambourgs moved to France and bought a house at Ville d'Avray, they wrote "repeatedly begging her to come and stay with them," even hoping that she might "make Ville d'Avray her permanent home." But, Edith Lewis reports, this was not possible, as Cather discovered during a 1923 visit:

> They had arranged a little study in their house for her to work in. . . . But although the little study was charming, and all the surroundings were attractive, and the Hambourgs themselves devoted and solicitous, she found herself unable to work at Ville d'Avray. She felt, indeed, that she would never be able to do any work there. (*WCL,* p. 131)

It is understandable that Cather would "never be able to do any work" at the Hambourgs' house. Indeed, that summer she suffered a debilitating neuritis in her right arm and shoulder, a physical sign of her creative paralysis. Instead of being the cared-for child, there she was the excluded one, the orphan—or, perhaps, the daughter who had to relinquish her mother to the father. For some reason Cather experienced the permanence of her loss most fully during the disastrous 1923 visit; it may have taken her that long to acknowledge that the marriage was going to last, its endurance symbolized by the home the Hambourgs had created together at Ville d'Avray. Given the importance Cather attributed to space and architecture, seeing the couple in their permanent residence could have sharpened her sense of exclusion. (In fact the marriage lasted, apparently happily, until Isabelle's death in 1938.) The emotional impact the visit had on Cather can be glimpsed in the grim portrait Leon Bakst painted of her that summer, in the profound spiritual and emotional crisis reflected in *The Professor's House,* and in her often-quoted remark that the "world broke in two in 1922 or thereabouts" (*NUF,* n.p.).

Yet while the break with Isabelle was cruel, it was not complete. Some connections between Cather's worlds remained. The love between the two women persisted despite marriage and separation, being maintained through visits and letters. Although we do not have Cather's letters to Isabelle, a few letters from Isabelle escaped destruction and were discovered in the Grand Manan cottage several years after Cather's death. Filled with references to books, music, travel, nature, and friends, they attest to an enduring love. Isabelle takes a solicitous attitude toward her friend's health, urging her to see a new doctor ("I am sure she would be good for you") and wishing her a productive summer on Grand Manan ("Oh, I hope and hope for a happy summer of good work for you my dear, darling Molly" [Isabelle's private name for Cather]). References to the activities she was sharing with her husband are intermixed with memories of her life with Cather, as when Isabelle writes "The smell of the box[wood] was strong—That always takes me straight back to Willowshade and the little square bit of ground behind your great grandparents' house. That was where I got its real smell," or tells Cather she longs to "again have that wonderful afternoon" when the two visited Louisa May Alcott's house. The salutations and closings also convey the writer's deep affection: "My darling Willa," "Darling Willa," "So lovingly to you, Isabelle," "A loving heart to you, I.," "So lovingly, Isabelle."[47]

The passage of time brought more flexibility in the two women's roles. As age, illness, and the threat of death entered both lives, claiming Isabelle first, Cather gradually took the part of mother; writing Dorothy Canfield in the 1930s, she confided that one of her greatest pleasures was taking care of Isabelle. In 1935, caring for Isabelle became more poignant when she came to New York, gravely ill with a kidney disease. Cather did no writing that year, spending her time visiting Isabelle in the hospital every day and traveling with her when her health improved. The final loss was Isabelle's death in October 1938. Writing Zoë Akins in November, Cather said that she was comatose; the power to feel seemed totally gone, and she didn't think that she ever wanted to feel again. In December she told her old friend Carrie Miner that, if only Isabelle could have lived a little longer, things would have been so much easier for her.[48] But once again Isabelle's absence caused her to write. At first she had to send scores of letters to mutual friends, mourning and celebrating her lost love; and then, after several months, Cather returned to the novel on which she had been working—*Sapphira and the Slave Girl* (1940), the story of a mother–daughter reunion in which art repairs the losses experienced in life.

In moving from silence to speech after Isabelle's loss, Cather was repeating a creative pattern also evident in her response to the death of her own mother in 1931. Virginia Cather's long illness had been even more painful than her death; paralyzed with a stroke in December 1928, she lay helpless and speechless for two and a half years in a California sanitorium. This illness had a profound effect on her daughter, Edith Lewis thought:

> She realized with complete imagination what it meant for a proud woman like her mother to lie month after month quite helpless, unable to speak articulately, although her mind was perfectly clear. In Willa Cather's long stays in Pasadena . . . she had to watch her continually growing weaker, more ailing, yet unable to die. It was one of those experiences that make a lasting change in the climate of one's mind. *(WCL,* pp. 156–57)

Cather spoke of her pain to Dorothy Canfield, telling her that to see a strong woman paralyzed was the most terrible thing in the world; viewing her mother's pointless suffering, she felt as if she had lost her bearings and could not write.[49] But, eventually, Cather did write—and when she did, she painted a sympathetic portrait of her silenced mother in "Old Mrs. Harris," giving her back the voice she had lost.

After Isabelle's death, Cather once again found compensation for the loss of a maternal figure through her art. In *Sapphira and the Slave Girl* we can see Isabelle's enduring creative gift: her absence as well as her presence inspired Cather to write.

Notes

1. Willa Cather to Mariel Gere, May 2, 1896, Gere Collection, Nebraska State Historical Society, Lincoln, Nebr.

2. Cather to Gere, January 2, 1896.

3. Cather to Gere, May 2, 1896.

4. Cather to Gere, n.d. Some confusion surrounds Cather's obtaining this first appointment. E. K. Brown claims that the publisher offered her the job at the home of Charles Gere, while Bernice Slote argues that Lincoln friend George Gerwig was the go-between (*KA,* p. 29). For an analysis of the competing stories as well as a well-researched discussion of Cather's tenure at the *Home Monthly,* see Peter Benson, "Willa Cather at *Home Monthly,*" *Biography* (Summer 1981): 227–48.

5. For fuller biographical information on Cather's Pittsburgh years, see Kathleen D. Byrne and Richard C. Snyder, *Chrysalis: Willa Cather in Pittsburgh, 1896–1906* (Pittsburgh: Historical Society of Western Pennsylvania, 1980); Mildred Bennett, "Willa Cather in Pittsburgh," *Prairie Schooner* 33 (Spring 1959): 64–76; John P. Hinz, "Willa Cather in Pittsburgh," *The New Colophon* 3 (1950): 198–207; George Seibel, "Miss Willa Cather from Nebraska," *The New Colophon* 2, Part 7 (September 1949): 195–207; Mildred Bennett, "Introduction" to *CSF*; William Curtin, notes and commentaries in *WP*.

6. Willa Cather to Zoë Akins, May 20, [1939], Akins Collection, Huntington Library, San Marino, Calif.

7. See in particular Willa Cather to "Neddius," Gere Collection, n.d.

8. Cather to Gere, n.d. Axtell's name is given as "Charles" by previous biographers, but Byrne and Snyder, who researched Pittsburgh city directories, discovered the mistake (p. 97). See also Benson, pp. 230–31.

9. Cather to Gere, July 13, 1896.

10. Cather to Gere, April 25, 1897; August 4, 1896; August 10, 1896.

11. Cather to Gere, January 10, 1897. Cather had also appeared "quite consistently" in Lincoln society news in 1895–96 when she was often the guest of the "Patriarchs," Lincoln's "most exclusive social club" (*KA,* p. 28).

12. For a discussion of Cather's investment in the ideal of the lady and, more generally, in aristocratic culture, see Patricia Lee Yongue, "Willa Cather's Aristocrats," *Southern Humanities Review* 14 (Winter 1980): 43–56 and (Spring 1980): 111–25.

13. Cather to Gere, April 25, 1897; Willa Cather to "Neddius," n.d.

14. See E. K. Brown, *Willa Cather: A Critical Biography,* completed by Leon Edel (New York: Knopf, 1953), p. 78.

15. Cather to Mrs. Gere, July 13, 1896.

16. Cather to Gere, April 25, 1897.

17. Cather to Gere, August 4, 1896.

18. Cather to Mrs. Gere, July 13, 1896.

19. Cather to Gere, May 2, 1896; August 4, 1896. For a view of Cather's *Home*

Monthly short stories that stresses their formulaic qualities, see Marilyn Arnold, *Willa Cather's Short Fiction* (Athens: Ohio University Press, 1984), pp. 9–14; for one that attributes unconscious self-revelation to them, see David Stouck, *Willa Cather's Imagination* (Lincoln: University of Nebraska Press, 1975), pp. 38–39, 79–80. Peter Benson also finds "Tommy, the Unsentimental" to be a self-expressive story, although in his view "The Burglar's Christmas" is completely "programmed by the requirements of its sentimental genre" (p. 238).

20. The accepted view is that Cather resigned voluntarily, although Byrne and Snyder raise the possibility that she may have been discharged (p. 7).

21. Cather to Mrs. Gere, July 13, 1896.

22. Cather to Gere, April 25, 1897.

23. Cather to Gere, n.d.

24. Cather to Frances Gere, June 23, 1898; Cather to Mrs. Seibel, August 20, 1898.

25. Cather to Gere, April 25, 1897; Cather to Gere, n.d.

26. Cather to Mrs. Seibel, June 21, 1897. See also George Seibel, "Miss Willa Cather."

27. Cather to "Dear Friend," May 6, 1941, University of Virginia Library, Charlottesville, Va.

28. Brown, p. 96.

29. Elizabeth Moorehead, *These Two Were Here: Louise Homer and Willa Cather* (Pittsburgh: University of Pittsburgh Press, 1950), p. 48.

30. Brown, p. 97.

31. See Byrne and Snyder, pp. 40–43, esp. p. 42.

32. According to one friend of the McClung family (who thus may not be a reliable source), the household tension was the result of "Cather's unreasonableness, her bad disposition, and her irritability" (Byrne and Snyder, p. 41). Helen Cather Southwick effectively questions the reliability of Moorehead's memoir (the source for E. K. Brown's portrait of the McClung/Cather ménage) in "Willa Cather's Early Career: Origins of a Legend," *Western Pennsylvania Historical Magazine* 65, no. 2 (April 1982): 85–94. Pointing out a number of factual errors in Moorehead's story, Southwick particularly takes issue with Moorehead's statement that the two young women shared a bedroom, countering with Elsie Cather's testimony that Willa had her own bedroom as well as a study (p. 91). If Cather had wanted to conceal a lesbian attachment to family members, however, it is unlikely that she would have told them that she and Isabelle shared a bedroom. Southwick convinces me that Moorehead's account may contain some factual errors, yet Southwick's purpose in writing the article is so clearly to refute any suspicion that Cather could have been lesbian that her use of evidence is also suspect.

33. Moorehead, pp. 49–50. Southwick views this description as romantic and inaccurate (p. 92).

34. Brown, p. 96; Moorehead, p. 47.

35. Yongue connects Cather's attraction to Isabelle with her admiration for "people of great wealth and aesthetic sensibility" (p. 47).

36. Byrne and Snyder, pp. 43, 34.

37. Southwick, "Willa Cather's Early Career," p. 12.

38. Byrne and Snyder, pp. 49, 55.

39. "Three American Singers," *McClure's* 42, no. 2 (December 1913): 33–48, esp. 47.

40. Adrienne Rich, *Of Woman Born: Motherhood as Experience and Institution* (New York: Norton, 1976), p. 237.

41. Willa Cather to Dorothy Canfield, October 10, 1899, Dorothy Canfield Fisher Papers, Bailey Library, University of Vermont, Burlington, Vt.

42. Willa Cather to Elizabeth Sergeant, June 4, 1911, Barrett-Cather Collection, University of Virginia Library, Charlottesville, Va.

43. Moorehead, p. 50; D. W. Winnicott, *Playing and Reality* (London: Tavistock Publications, 1971), p. 15. See also Ernst Kris, *Psychoanalytic Explorations in Art* (New York: International Universities Press, 1959), p. 253, and Ruth Perry, "Introduction," *Mothering the Mind: Twelve Studies of Writers and Their Silent Partners,* ed. Ruth Perry and Martine Watson Brownley (New York: Holmes and Meier, 1984), pp. 3–24. As Perry observes, an artist's creative "mother" can take many shapes and is not necessarily female; the collected essays explore the influence of "lovers, sisters, patrons, mothers, aunts, friends, husbands" on writers' lives and work (p. 4). Mothering figures, as she points out, can play many roles: "intercepting the world, conferring unconditional approval, regulating the environment, supplying missing psychic elements, and mirroring certain aspects of the self of the artist" (pp. 5–6).

44. James Woodress, *Willa Cather: Her Life and Art* (New York: Pegasus, 1970; rpt. Lincoln: University of Nebraska Press [Bison ed.], 1975), p. 173.

45. Cather to Canfield, March 15, [1916].

46. For discussions of Cather's relationship with Jan Hambourg, see Brown, p. 112; Woodress, p. 178; and Phyllis C. Robinson, *Willa: The Life of Willa Cather* (New York: Doubleday, 1983), pp. 205–8. "The Old Beauty," written after Isabelle's death—the final loss—suggests that unconscious resentments may have reemerged; the story reintroduces the predatory Jewish male, here a gross sexual plunderer.

47. The letters are included in the maddeningly undocumented book by Marion Marsh Brown and Ruth Crone, the investigators who found the letters, *Only One Point of the Compass: Willa Cather in the Northeast* (Danbury, Conn.: Archer Editions Press, 1980), pp. 66–69, 79–84.

48. Cather to Canfield, n.d.; Cather to Akins, November 13, 1938; Willa Cather to Carrie Miner Sherwood, December 7, 1930, Willa Cather Pioneer Memorial, Red Cloud, Nebr. Cather describes the extensive letter-writing she engaged in after Isabelle's death both in the 1938 letter to Akins and in her letter to Canfield of March 5, 1939.

49. Cather to Canfield, n.d. [summer 1929].

11

Hard Apprenticeship: European Travels and *April Twilights*

It was Balzac himself who used to wander in the Pere-Lachaise in the days of his hard apprenticeship, reading the names on the tombs of the great. ''Single names,'' he wrote his sister, ''Racine, Moliere, etc.; names that make one dream.''

The World and the Parish

Service of gods is hard.

''Winter at Delphi''

''Life began for me,'' Willa Cather told Elizabeth Sergeant, ''when I ceased to admire and began to remember'' (*WC:AM,* p. 107). Considering this reference to her birth as a novelist in *O Pioneers!* can help us to read her writings of the early 1900s, which include travel letters published in the *Nebraska State Journal, April Twilights,* and several short stories. Like many retrospective statements about her literary career, this comment exaggerates the opposition between her youthful and her mature work. It is more accurate to think of ''admiration'' and ''memory'' as points on a continuum rather than as polar opposites. Even after her literary ''birth'' Cather continued to admire, and to refer to, literary predecessors (as she did in titling *O Pioneers!* after a Whitman poem). By 1913 Cather had attained a comfortable relationship with the dominant literary tradition; secure enough in her own creative power and artistic identity, she did not have to borrow prestige and acceptance along with her titles. During the early 1900s, however, she regarded the literary past with less self-assurance: male writers constituted an exclusive club she wanted to join, a club that kept denying her membership.

Cather's contrast between ''admiration'' and ''memory'' has psychological as well as literary meaning. The admiring self is the child who invests another with the adult's authority, while the remembering self is the adult who incorporates the child. What is admired is external, seemingly not possessed by the self; what is remembered is internal, a possession integrated with the structures of consciousness. Willa Cather's growth from apprentice to accomplished writer did not merely require the refining of technique, then, for the forces requiring a woman writer to take the child's admiring role and impute authority to male predecessors rather than to herself were social as well as aesthetic. Her transition from admiration to memory thus demanded that she grant herself the adult's literary authority, and this was a difficult process. Like Balzac's, Cather's apprenticeship was hard; unlike his, hers was long: almost twenty years separate her first published short story from her first novel, *Alexander's Bridge* (1912).

As we have seen, there were other reasons why Cather's apprenticeship was so lengthy; as a woman writer, she confronted the potentially silencing conflict between gender and vocation and, as a lesbian writer, that between disclosure and concealment. Moreover, when she was working as a journalist she simply did not have the time or energy to devote herself to the writing of fiction; she published only two stories after leaving the *Home Monthly* for the Pittsburgh *Leader*, one in 1898, the other in 1899. But during the early 1900s—freed from the newspaperwoman's year-round work and cared for by Isabelle McClung—Cather entered an unprecedented period of productivity. Between 1900 and 1905 she published her travel letters, eighteen poems, and thirteen stories as well as *April Twilights* (1903) and *The Troll Garden* (1905).

Yet in most of these writings Cather did not attain that "quality of voice," as she phrased it in "Miss Jewett," which is "exclusively the writer's own, individual, unique" (*NUF*, p. 79). The young writer characteristically takes the admirer's stance, demonstrating the daughter/writer's allegiance to the literary fathers who simultaneously inspired and inhibited her. She did not yet fully see the literary possibilities in her own personal past spent among Virginia farmers, European immigrants, and rural housewives and mothers. And yet, as we will see, there are faint signs of Cather's eventual transition from admiration to memory, from apprentice to accomplished artist, even in this early work.

More evidence of Cather's literary emergence during this period can be found in her life than in her writing. Although she did not fully transform the family dynamics of literary influence into a female drama until she met Sarah Orne Jewett, she prepared for this moment by continuing to disentangle creativity from masculinity in the early 1900s. To be sure, Cather continued to equate "male" and "artist" during this period, but she more often imagined the artist as a brother than as a father. Although the brother could exert a powerful literary influence, as Cather suggested in commenting on Christina and Dante Gabriel Rossetti's literary relationship, nevertheless the brother–sister bond contained the potential for the reciprocity that she later found in her friendship with Jewett. The fathers were the artistic predecessors whose books dominated her imagination, but the brothers were contemporaries whose accomplishments, while greater than her own, were not as intimidating. They represented similarity rather than difference. Gazing at Balzac's tomb, Cather saw a hero "second only to Napoleon himself" (*WP* 2, p. 928). Reading A. E. Housman's poetry or listening to Ethelbert Nevin's music, she saw versions of herself.

•

Cather's most unsettling encounter with literary fathers occurred during her first pilgrimage to Europe. Accompanied by Isabelle McClung, she left Pittsburgh in June of 1902 to spend the summer in England and France. Dorothy Canfield, who was already in Paris, joined the two friends abroad and traveled with them for part of the time. Cather and Isabelle began their tour in Liverpool, moving on to Chester, Shropshire, and London; then they crossed the Channel to Dieppe, Rouen, Paris, Barbizon, Avignon, Marseilles, and Hyères, ending their trip in the "heart of Daudet's country" of Provence (*WP* 2, p. 947). Although all three young women were vacationing, Cather possessed a different traveler's identity from her two friends. Like Hawthorne and James before her, she was the American writer embarking on a journey to discover her cultural origins, to confront history and the past.[1]

Although Cather's journey into the European past was a profound experience, it was not the unqualifiedly joyous "home-coming" Edith Lewis portrays in her memoir. Lewis' account—unreliable because it is partial—has influenced later biographers to view Cather's first encounter with Europe solely as positive:

> This first European journey was, of course, a great imaginative experience. For an artist, who lives intensely in ideas, for whom all the important adventures happen in the mind and spirit, there is nothing quite like that first encounter with European culture on its own soil, in its age-old stronghold—it is a home-coming more deeply moving and transfiguring than any home-coming to friends and family, to physical surroundings, can ever be. (*WCL*, p. 55)[2]

But Cather's travel letters (published in the summer of 1902 in the *Nebraska State Journal*), her fictional reworking of the European trip in *One of Ours* (1922), and her correspondence with fellow-traveler Dorothy Canfield tell a less triumphant story: one of exclusion in which the daughter feels she is not welcome in the ancestral home. After spending several weeks as the admiring traveler, Cather did not end her trip with the confident sense of imaginative and aesthetic possession she later portrays in *The Song of the Lark*. In the American Southwest, Thea Kronborg discovers "things which seemed destined for her," but Cather returned from Europe feeling dispossessed rather than replenished (*SL*, p. 301).

Hoping to record the ebb and flow of daily life for her Nebraska readers, Cather sent back to the *Journal* her impressions of English canal boats and river families, London shop girls, an Italian feast day procession, French farmlands, and a Provençal fishing village.[3] Interspersed in her travelogue are descriptions of inns and hotels and accounts of meals, including a mouth-watering evocation of a ten-course dinner in Avignon, featuring "lamb chops with a wonderful sauce of spinach, big yellow melons and figs and grapes, cream of carrot soup and patties of rice, broiled larks on toast and marvellous little cakes made of honey and spice and flour" (*WP* 2, p. 936). But the descriptions most useful for understanding why Cather's journey into the European past made her feel like a disinherited daughter are those in which she contemplates the "long past" beneath the "mask of the immediate present" (*WP* 2, p. 907).

Drawn to sites of what seemed individual rather than collective or institutional power, Cather made numerous pilgrimages to the birthplaces and graves of the artists she admired. Reading the inscriptions on monuments and gravestones, she contemplated what Emerson calls "the sepulchres of the fathers"—represented in her travel notes by Burne-Jones, Rossetti, Flaubert, de Maupassant, Balzac, Dumas, Heine, and de Musset.[4] As its Latin derivation suggests, the monument is expected to inspire and ensure remembrance, but for Cather the memories of literary fathers, incarnated and signified in their monuments, inspired admiration rather than memory: they directed her toward patriarchal society's definition of literary greatness, not toward her own material or voice (*WP* 2, p. 925). Although the monuments seemed to honor individual men—and thus to be signs of the artistic tradition the woman writer wanted to join and the greatness she hoped to achieve—they were in fact public signs of male literary and social power, and so represented a system of meaning and significance in which Cather feared she could never fully participate.

Most of Cather's encounters with the fathers' sepulchres occurred in France, the country and civilization which, even more than England, she honored as the repository of culture. In France, everything—not just literature, music, painting, but also wine, food, gardens—had aesthetic significance. As she later told Elizabeth Sergeant, the French "have values, aims, a point of view, and have acquired wisdom from the enduring verities. One did not find anything of the sort in the Middle West" (*WC:AM,* p. 145).[5] She devoted one dispatch to a description of the cemeteries of Paris and the graves of its artists. In Montmartre she found the tombs of Dumas, Heine, and the Goncourt brothers; in Père-Lachaise, those of Balzac and de Musset. Cather was particularly prepared to be impressed by Père-Lachaise. In 1896 she had written admiringly of this cemetery where Balzac used to read the inscriptions on the "tombs of the great" and honor those marked by a single name. It was fitting that he should rest there, she then thought, having "won his longed-for chrism, a grave marked by a single name" (*KA,* p. 168).

The single-name grave, which often appears in Cather's European dispatches, has several meanings: it suggests the artist's fame, which Cather equates to that of the warrior or statesman; the triumph of his name and his work (another form of monument) over mortality and mutability; and his ability to engender literary progeny through the literary monuments he left for his sons. Cather encountered all these interlocking meanings in her visit to Rouen, Flaubert's birthplace, where she found the "beautiful monument erected to Flaubert" matched, "just across the road," by the statue of "his friend and pupil" de Maupassant. She thought this joint commemoration of mentor and protégé, literary father and son, "very fitting" (*WP* 2, p. 923). The patrilineal succession displayed in the neighboring statues of Flaubert and de Maupassant was not solely a mark of high culture. Among the Provençal peasants, Cather discovered, even the folk arts of singing and poetry were passed on from the "grandfather [who] sings by the fire on winter nights the songs he made in his youth" to the "grandson [who] sings to some Arlesienne the song he made yesterday" (*WP* 2, p. 948).

The traveler's meditations on inscriptions, monuments, and tombstones reveal the associations she was making among artistic achievement, cultural remembrance, personal immortality, and naming. The monument or tomb is society's way of naming, by isolating the "single name" of the artist, the dominant literary and cultural heritage, and Cather found this concern with tradition particularly evident in France where there seemed to be a "monument in every square and a dozen in every park." But the French had a better way of "commemorating their great men than building monuments to them," Cather thought—naming their streets after them. This form of recognition placed individual names on the map and in the "mouths of the living." Cather liked this way of integrating great artists into the ongoing life and collective memory of the community: the names of France's great men could "still mean something in the big, stirring life of the common [people]" who might live close by a "rue Balzac" or a "rue Racine" or a "rue Moliere." "Nearly every street in Paris bears the name of a victory," Cather reflected, "either of arms or intellect," revealing her association of the artist and the conqueror (*WP* 2, p. 929).

We can see the significance that Cather attached to the artist's single name in the title of the last book of *The Song of the Lark*—"Kronborg." By then Cather could imagine a woman artist achieving this solitary stature. But during her 1902 trip to Europe, she did not derive literary self-confidence from the names she saw carved on the tombs and monuments. Filled with awe and admiration as she looked at inscriptions that corresponded to her own inscription with the European literary tradition, Cather felt excluded more than included in culture, oppressed by the burden of the past rather than finding her own past lengthened and extended.

In her travel letters Cather does not overtly mention this sense of exclusion, although she reveals it covertly: she characteristically portrays herself as separated from the signs of the great men but included in the ongoing life of ordinary folk. During a long, happy, communal dinner in a Barbizon hotel, for example, she came to feel part of French culture: "Oh, we were a motly, clever, self-appreciative lot of people at Barbizon that night, and it was a good world we lived in" (*WP* 2, p. 932). Cather may not have experienced her exclusion from France's high culture consciously herself, or—more likely—may have been unable to acknowledge it publicly. Several years later, however, she spoke of the suffering she endured during this trip in two letters to Dorothy Canfield written shortly after she had completed *One of Ours* (1922).

The personal source of Claude Wheeler's humbling encounter with France, Cather explained in the first letter, was her trip to France with Canfield in the summer of 1902. In *One of Ours* she had tried to convey, she told Canfield, the feeling a sensitive roughneck has when he is plunged into the midst of a rich culture. It was not only his vanity that suffered, but he felt as if he had been cheated out of everything, deprived of the whole treasure of the ages just because he didn't know some language. This was a terrible experience, Cather wrote—the confrontation of an uncultivated, even primitive self with an older, tradition-filled civilization. She had tried to express this contrast in the relationship between Claude and David (Claude's sophisticated friend who is at home in French culture). That friendship, Cather confessed, was an emotional picture of herself and Canfield in France, twenty years previously. Reworking this painful experience in *One of Ours* was her revenge, she continued, since Canfield had admitted how moved she had been by Claude's and David's relationship. A month later Cather told Canfield what she had not been able to say during their 1902 trip: Canfield had never been a "roughneck" like herself because Canfield's mother and father—unlike Cather's parents—were intellectuals. Canfield had never understood why she had suffered so in the past, Cather wrote, but maybe now—with Claude as an intermediary—she could.[6]

The wounds of a rural childhood heal slowly. By 1902 Cather was doubtless as well read as Canfield and had been imbued with a love for French literature since college days. But since her parents had not given her a comparable cultural and intellectual inheritance Cather felt excluded from the European heritage which she assumed Canfield, the child of educated, upper-class parents, enjoyed as a birthright. Canfield had received a maternal gift that Cather particularly envied. Her friend's ability to speak fluent French, to Cather the sign of inclusion in the culture she loved and admired, had first been encouraged by her mother Flavia Canfield, an artist who had taken her young daughter on several trips to Paris. Meanwhile, as Cather then thought, she had been given nothing by the women in her family, who did not paint, sculpt, or take her to foreign countries.[7]

Although Cather was not unversed in French in 1902—she could read and write the language, even if she could not converse as easily as Canfield—when she recalled the "emotional picture" of her trip in *One of Ours* she made Claude Wheeler almost completely ignorant of French, reflecting her own feelings of exclusion during the 1902 trip. In a painful scene he looks into the "admiring eyes" of a French child who has asked him a question he does not understand (*OO*, p. 279). Unable to answer, failing the child who admires him, Claude in turn takes the child's stance in relation to the French-speaking David Gerhardt, to whom Cather attributes Canfield's fluency.

During her European trip Cather experienced another drama of exclusion in which Dorothy Canfield was favored with inclusion: the pilgrimage Cather, Isabelle McClung, and Dorothy Canfield made to A. E. Housman's London home. Too distressing to record in her letters home, this disastrous visit may also have contributed to Cather's conviction that Canfield spoke the language of culture more fluently than she.

Cather had reviewed *A Shropshire Lad* in 1897. She admired Housman's craft and control, but was particularly drawn to his themes of exile, loneliness, and loss that corresponded to her own dislocation and homesickness during her first years in Pittsburgh. She thought of him as a literary soul-mate, and often quoted from his poetry in letters home to Mariel Gere, finding her own yearning for Nebraska expressed in Housman's nostalgic, elegiac evocation of the Shropshire countryside. He was a poet she had discovered herself, not one inherited from the established literary tradition, and her comments on Housman in the late 1890s reflect the excitement of discovery and the sureness of mutual understanding. After seeking "far and wide" for information about him, Cather confessed failure: "I have decided to let him remain a literary mystery," she wrote in 1900, "and the theme of many imaginings." Unconstrained by biographical details, Cather was free to grant Housman a life like hers: "But at least I know that he has eaten the bitter bread of exile, and trod the hostile streets of great cities and hungered for the little village where he was a boy" (*WP* 2, pp. 707–9).

Once she arrived in England, however, Cather was not content to let this literary mystery alone. She became a detective. Visiting Housman's Shropshire, she viewed all the landmarks mentioned in Housman's poems and saw a scene that exactly matched the requirements of her imagination: "a company of lads with their pigskin ball" came "racing out over the green" as if stepping out of a

Housman poem (*WP* 2, p. 897). But, as she later told her *McClure*'s colleague Viola Roseboro', there was no trace of the man himself—not even a legend. So, she recalled, she battered on the doors of his publishers until they gave her his London address.

Cather's subsequent pilgrimage to see Housman in his London exile was not as pleasurable as her visit to Shropshire. Once she met the man in person, she could no longer envision him as her other self, her poet/brother. Joined by Dorothy Canfield in London, Cather and Isabelle McClung found Housman in his Highgate lodgings, which Cather found shabby and depressing, not the idyllic setting she had imagined. A strained and embarrassing visit followed. Shy and reserved, Houseman could not converse easily with the uninvited pilgrims or respond as a spiritual sibling when Cather tried to tell him how important his poetry was to her. Poetry was not really his business, he told them; he took more interest in Latin texts and classical philology. Dorothy Canfield, who was working on a dissertation in seventeenth-century French literature, could converse on these academic subjects, however, and the two of them chatted about philological matters for the rest of the visit. When she left, Cather confessed to Roseboro', she was crying.[8]

Given Cather's high expectations for the visit, based on her imagined similarity with Housman, the contrast between her exclusion and Canfield's inclusion, however unsought, must have intensified her sense of estrangement. Housman was not yet a single-named artist: like her, he was embarking on a literary career. But this contemporary did not offer her the literary kinship she had imagined Stephen Crane giving her in ''When I Knew Stephen Crane.''

•

Although Cather did not receive a sense of imaginative possession either in contemplating the sepulchres of the fathers or in visiting a brother/poet, her travel letters show how fully her imagination was possessed by male writers who had gazed at the European landscape before her. Cather's account of the trip is filled with admiring references to writers who had already described the scenes she was viewing. Seeing London caused her to think Kipling a ''greater man than I had ever thought him before'' since she now saw how he had captured the city in his work; contemplating the tunnels in the chalk cliff at Dieppe, she thought of Robert Louis Stevenson; in Marseilles, she knew she was in ''Monte Cristo's country'' (*WP* 2, pp. 910, 939).

In many ways Cather's encounter with Europe was enriched by her previous literary encounters. Had she not read *The Count of Monte Cristo*, seeing the Chateau d'If would not have been as ''moving to contemplate,'' or at least not moving in the same way (*WP* 2, p. 939). And yet the literary associations informing her response to European cities and landscapes were constricting as well as enhancing, as one seemingly anomalous episode suggests: Cather's visit to Le Lavandou, a tiny fishing village in southern France, located in a rough, deserted coast marked by pine-covered cliffs and narrow beaches. She and Isabelle stayed in the one good hotel, where they ate their meals on a straw-thatched, vine-covered veranda facing a beautiful Mediterranean bay. ''Out of every wandering in which people and places come and go in long successions, there is always one place remembered above the rest because there the external or internal conditions were such that they most nearly produced happiness,'' Cather reflected. For her, that one place was Lavandou: ''Nothing else in England or France has given me anything like this sense of immeasurable possession.'' Why did Lavandou offer Cather a sense of possession rather than of exclusion?

She and Isabelle decided to visit Lavandou, Cather informed her readers, ''chiefly because we could not find anyone who had ever been here, and because in Paris people seemed never to have heard of the place. It does not exist on the ordinary map of France.'' In this unmarked place—off the maps labeled with the single names of great men—Cather found a realm without tombs or monuments, a place where admiration was not required. She found this cultural absence liberating: ''No books have ever been written about Lavandou, no music or pictures ever came from here, but I know well enough that I shall yearn for it long after I have forgotten London and Paris'' (*WP* 2, pp. 942–44). Boasting no literary or historical associations, Lavandou could be hers as Père-Lachaise or Marseilles could not because it had not already been possessed, marked, and appropriated by the dominant culture.

In a sense, of course, Lavandou had already been ''marked'' for Cather by books and writers she had read. Even though Cather had not encountered the specific locale in any text, she had read Wordsworth's poetry, *Treasure Island,* Emerson's essays, *Huckleberry Finn,* and other romantic narratives in which self-discovery, illumination, or transfiguration occurs in locations off the ''ordinary map'' of culture and society. Since she liked to see her own life in terms of such fictions, she may have been predisposed to feel a strong attachment to a place where other travelers had not been.

In explaining why Lavandou would be engraved upon her imagination, Cather mentioned both ''external'' and ''internal'' reasons. She was drawn to Lavandou not only because it lacked monuments, the external signs of social and literary power, but also because its culture and topography corresponded, in ways she could not fully describe, to her own internal and unconscious desires. This correspondence between outer and inner worlds contributed to the sense of possession she enjoyed there. Although Cather never articulated the ''internal'' power of Lavandou, her letters suggest that the region allowed her to experience a restorative, invigorating return to familial and ultimately to maternal origins.

Shortly before visiting Lavandou the traveler began to feel as if she were in a familiar landscape. Paying a visit to Millet's country near Barbizon, Cather surprisingly found herself in the realm of memory rather than admiration. The wheat fields, which reminded her of the ''country about Campbell and Bladen,'' were ''more familiar than anything I have seen on this side the Atlantic.'' Away from cities, graveyards, and monuments to great men, in the midst of a rural landscape which gave her an ''almost unbearable homesickness for Nebraska,'' Cather found the familial as well as the familiar:

> To complete the resemblance [between the landscapes of Provence and Nebraska] there stood a reaper of a well known American make, very like the one on which

> I have acted as a super-cargo many a time. There was a comfortable little place where a child might sit happily enough between its father's feet, and perhaps, if I had waited long enough, I might have seen a little French girl sitting in that happy, sheltered place, the delights of which I have known so well. (*WP* 2, p. 931)

Recalling Cather's happy years in Virginia, this is a vision of a father–daughter relationship where the daughter is included rather than excluded. In the travel letters that follow, this paternal memory leads not to public reminders of patriarchy, however, but to Cather's discovery of Lavandou, a landscape imbued with female presence; it is as if in going from Barbizon to Lavandou Cather were traveling even further into her psychic and emotional past, back to maternal powers.

Cather's descriptions of the village are dominated by her references to gardening, cooking, and eating, activities which suggest both women's life-sustaining work and the mother's emotional role in the family. Recalling her contrast of Virginia's fertile paradise and Nebraska's barren desert, she portrayed Lavandou as a "semi-desert country" where, nevertheless, the villagers managed to eat well: "We have good fish," Cather reported, "plenty of fresh figs and peaches, and the fine little French lobster called langouste." Although the region's deprivation struck her ("Potatoes, figs, olives and grapes are almost the only things that will grow at all in this dry, sandy soil"), she was even more impressed by the villagers' ability to "make a savory dish of almost anything that grows," even deriving a salad from the local sea-grass which they "dressed with the oil they get from their olives." Even though Cather did not directly acknowledge that these were signs of women's creativity, the descriptions of Lavandou anticipate her later association of the farm woman and the artist.

Close by the village Cather discovered a landscape for dreaming, not admiring: a "little plateau on the flat top of a cliff extending out into the sea, brown with pine needles and shaded by one tall, straight pine tree that grows on the very tip of the little promontory." In this intermediate landscape—where the borders of sea and land, feminine and masculine intermingle—Cather spent days enjoying the pleasures of idleness and reverie, "wrapped in a steamer rug" when the sea breeze was strong, gazing into the "great water" and becoming "part of the foam that drifts, of the wind that blows." Finding this diffusion of self pleasurable rather than fearful, Cather describes a joyous sense of inclusion in the natural world that counters the exclusion and separateness she felt when viewing the monuments of French culture. In this landscape—possibly because the masculine tree made the sea's maternal power seem less overwhelming—Cather did not experience the terror of drowsing and dreaming that overwhelmed her on the Nebraska prairies. Here the dreamer's regression to a psychological moment when the distinctions between self and other disappear was blissful rather than disturbing (*WP* 2, pp. 944–45).

While Cather thought her response to Lavandou an original, unmediated one, it was prepared for both by her reading in literary romanticism and, more profoundly, by the mother–daughter bond. The state of dreaming and reverie she enjoyed there—inspired by the maternal landscape—more closely foreshadows her mature fiction than do the accounts of her visits to the great artists' monuments and tombs. As the dreamer who became "part of the foam that drifts," Cather was momentarily the adult who releases memory and imagination in the creative process, connecting to the child within—as she would in writing *O Pioneers!* By contrast, the dreamer who admired the tomb of Balzac was the child who grants deference to the father. As this contrast suggests, Cather's shift from admiration to memory would be connected with a redefinition of her role as child: moving from the daughter who reveres the father to the daughter who connects to the mother, a psychological, literary, and emotional transformation that would allow her to be simultaneously child and adult, daughter and mother.

Since her days at Lavandou allowed Cather to possess part of herself, unlike the other European sites she visited this was a possession she did not lose. The "emotional picture" of Lavandou and its neighboring cliffs reappears in the Panther Canyon sequence in *The Song of the Lark* where Thea Kronborg enjoys the same luxurious idleness in an even more obviously maternal setting. The specific features of Lavandou—the pine-covered promontory and the ocean—reappear in *My Mortal Enemy* and "Before Breakfast." In these fictions this is the landscape of revelation where the dreaming, contemplating protagonists escape the structures and expectations of ordinary social life. Similarly, Lavandou offered the traveler a realm—precisely because she thought it was off the cultural map—where she could escape the fathers' psychological and literary power.

•

Although the visit to Lavandou foreshadows Cather's reexperience of the mother–daughter bond in the creative process as well as her discovery of a female artistic inheritance, there are no signs in the account of her European trip of any interest in literary woman. Eager to discover a fatherland, not a mother country, Cather did not visit the birthplaces or graves even of the female authors she admired. She toured Housman's Shropshire, Kipling's London, Balzac's Paris, Flaubert's Rouen, and Daudet's Provence, but not Brontë's Yorkshire, Eliot's London, or Sand's Paris. The only monument to a woman artist she mentions—to Rosa Bonheur—she encountered in passing and thought it the only tasteless one she had seen.

Cather does give, however, one brief account of a bond between women that anticipates her experience of "immense possession" of a literary past when she became friends with Annie Fields and Sarah Orne Jewett. In recounting the history of England's Chester Cathedral, a stop early on her trip, Cather tells a story of retrieval and recovery in which women rescue and preserve female predecessors from obscurity.

Like Cather's other symbolic encounters with the past, this story involves ruins and tombs. During her stay in Chester Cather heard the legend of St. Werburgh. A seventh-century abbess, Werburgh was the "daughter of a heathen king of the Mercians" (a king whose name we never learn). Her remains were buried in Herefordshire, but

> during the second Danish invasion, when the Danes were ravaging all the churches in their path, the daughter of Alfred the Great, who had devoted herself to a religious life and had founded a church at Chester, hearing that the Danes were

> approaching Hereford, dug up St. Werburgh's remains and mounting her palfrey rode over with her train to Chester. Here she reinterred St. Werburgh's remains in her own church. (*WP* 2, p. 895)

In recounting this drama of recovery, excavation, and preservation, Cather momentarily places herself in a female lineage. In telling the story to her readers, she re-enacts the daughter's act of fealty to her precursor, passing on the stories both of St. Werburgh and of her female rescuer. But this is a fleeting moment in Cather's account of her trip, just as it is in medieval history. The diminishment of the female tradition, as well as its preservation, is suggested in the story itself. The heroic rescuer is identified only as the nameless ''daughter of King Alfred,'' reversing the relationship between St. Werburgh and her unnamed father, just as women's marginality and vulnerability are demonstrated by the necessity of moving the saint's remains.

That the fathers reasserted their power over Cather's imagination is demonstrated not only by the visits to Parisian cemeteries (which follow her account of St. Werburgh's rescue) but also by the concluding pages of the final letter, ''In the Country of Daudet,'' in which the travelers are back on the literary map after their sojourn at Lavandou. Cather recalls a day spent wandering among Roman ruins, examining ''broken columns and fragments,'' and reflecting on the ''suicidal vastness'' of Rome. Among the fissures and the cracks of the fallen empire, we might have hoped Cather would find space to imagine a new social, political, or literary structure in which women might be included, but she ends the last letter reaffirming the values of empire and patriarchy.

Seeing the crumbling monuments of the ''empire and all it meant'' caused her to remember a moment in England ''several months ago'' when she had seen Lord Kitchener and his victorious troops return from Africa and the Boer wars. The ''spirit'' of the day Kitchener marched through London could be seen, Cather thought, in the admiring faces of the children:

> Before those thousands of horses there were rows and rows of children, children who . . . were all petting and stroking the animals with a pride, an earnestness, a wistfulness touching to see. . . . Each [was] whispering a vow to the horses of the cavalry. One felt in a flash of conviction from what blood the world's masters were to come. (*WP* 2, pp. 951–52)

Cather thus concludes her travel letters by turning from Rome to England, from fallen empire to existing empire. In doing so she again pays tribute to the power of the fathers, evident here in the image of patrilineal and military succession that by implication leaves the female viewer excluded from the scene she witnesses.

While there are hints in Cather's account of her European trip of her transition to memory—a few signs of the strength she would eventually draw from a psychological and literary return to origins—overall her travel letters record the woman writer's service to the ''world's masters.'' Social, literary, and cultural authority—like military power—belong to the fathers, and all a daughter can do, it seems, is to join the throng of admiring children.

•

Contemplating the sepulchres of the fathers may have given the young writer a sense of urgency as well as exclusion: her first book, *April Twilights*, was published in April 1903. The collection includes several poems she had published between 1900 and 1902 as well as new ones inspired by her European trip. Cather had been deeply affected by the death of composer Ethelbert Nevin in 1901, which warned that artists could die young. This fear, combined with her exposure to literary mortality and immortality in the cemeteries of France, may have inspired her to place her work in a less ephemeral form than the newspaper or the magazine. But the poems do not display artistic self-confidence: on the contrary, the major subjects of *April Twilights* are death, mutability, and failure in love and art.

Although Cather thought fiction a more elevated genre than poetry, she had been writing and publishing poems since 1892 when "Shakespeare: A Freshman Theme" was accepted by the *Hesperian*. In the 1890s her poetry appeared in a variety of forms and contexts. There are epigraphs to short stories and articles, translations (from Horace, Anacreon, de Musset, and Heine), and sentimental verses written for the *Home Monthly*. Then in 1900 her output suddenly increased. Her relationship with Isabelle was a catalyst; moreover, she was assured of publication in the *Library* (seven poems appeared there that year, two ostensibly by "Clara Wood Shipman"). In 1901 Cather placed two poems in national magazines, in 1902 four more, and then *April Twilights* appeared the next year with twenty-seven new poems.

"I do not take myself seriously as a poet," Cather told an interviewer in 1925.[9] The established writer repudiated her "book of callow verse" even more vociferously than she did her first short stories, eventually trying to suppress *April Twilights* by buying up all copies she could retrieve (*WC:AM*, pp. 62, 182). In 1923 Cather published *April Twilights and Other Poems* with several poems expunged and some new ones added, and she included a revised version of the 1923 edition along with *Alexander's Bridge* in Volume Three of the Library Edition, not a place of honor.[10]

The poems do not contradict her later assessment, but they are useful documents for charting her literary evolution. "Service to gods is hard," laments the speaker in "Winter at Delphi." *April Twilights* shows Cather serving many gods, the literary fathers whose ghosts permeate the collection—Virgil, Shakespeare, Keats, Rossetti, Yeats, Swinburne, Tennyson, FitzGerald, Housman. Writing Viola Roseboro' a few years later, Cather acknowledged a special debt to Housman. Her poems openly revealed her knowledge of *A Shropshire Lad*, she admitted; she had been Housman's mental bond slave ever since his book appeared.[11] Later she recognized her indebtedness to at least two other late nineteenth-century poets. "I can still remember the velvet cadence of her voice as she read 'Winter at Delphi' and 'Aftermath,' " her friend George Seibel recalled, "sugared sonnets which neither of us then realized were begotten by Swinburne and Rossetti."[12]

What is most striking about *April Twilights* is its derivative nature. Better read, more self-conscious about the literary tradition as a high school teacher in 1903 than she had been as a "roughneck" college student in 1893, in this volume

Cather deserts the embittered farmers, crop failures, and cornfields we see in her first gritty stories of Nebraska life. Instead she evokes an abstract, golden Arcadia or a misty, chivalric medieval world, replacing Peter, Lou, and Eric with shepherds and maids, minstrels and gods, knights and warriors. Although in a few poems Cather momentarily drops the minstrel's pose and draws on her personal and Western roots (''Night Express,'' ''Prairie Dawn''), the overwhelming majority of the poems take classical, medieval, or late-Victorian poets as models, as the titles suggest: ''Arcadian Winter,'' ''Winter at Delphi,'' ''Provençal Legend,'' ''The Tavern,'' ''L'Envoi.''

Cather's imitations are eclectic: she mixes Greek and Latin references with fin-de-siècle ennui, Provençal springs with Celtic twilights, as her pensive speakers contemplate love, time, and the fading of roses. Most of the poems are ostentatiously allusive as well as derivative, showing the young writer's desire to convince the reader of her familiarity with classical models. Straining to dissociate herself from the minor mode of woman's poetry, Cather makes no overt connections to female predecessors or contemporaries. Paradoxically, however, she commits the same literary sin she had attributed to Elizabeth Barrett Browning, the aspiring poet who ''read Greek for ten years to perfect her style'' and then produced only ''tedious imitations'' of ''gentlemen some centuries dead'' (*WP* 1, p. 146).

Speaking of Housman in 1900, Cather observed that his power depended on his ability to ''say the oldest thing in the world'' in a ''new voice'' (*WP* 2, p. 708). In *April Twilights,* however, instead of rephrasing old themes in a new voice, Cather adopts the cadences of her male mentors and masters, as in this Housman-like lyric: ''Lads and their sweethearts lying/ In the cleft o' the windy hill;/ Hearts that hushed of their sighing,/ Lips that are tender and still'' (''In Media Vita''). Far from displaying the sympathetic imagination she attributed to Shakespeare—the artist's self-transcending power to enter into another identity—in these lyrics Cather is invaded by other selves, becoming Keats, FitzGerald, or Housman rather than a character of her own imagining.

The extent to which her deference to the male literary tradition undermined Cather's literary authority and originality is most evident in her allusions to Shakespeare. Even though Shakespeare presented her with a less aggressively masculine model of poetic identity than did Kipling, his achievements made Cather fear that she could never speak powerfully herself. In her first published poem, ''Shakespeare: A Freshman Theme,'' the college student had contrasted Shakespeare's richness with her own impotence by using metaphors of paternity and genealogy. A ''sun born bard'' who inherited Apollo's power, Shakespeare the son was also an imposing literary father whose achievements doomed his descendants to be the ''dwarfed children of earth's sterile age'' (*AT,* p. 61). In *April Twilights,* Cather continues to cast the poet as Shakespeare's deformed child, an image which has special resonance for a woman writer since female creativity is so often imagined as monstrous.[13] The speaker in ''Paradox'' expresses her fear that she will not attain her youthful desire for the ''power/ Of melody'' through allusions to *The Tempest*—Miranda's isle is the realm of youth, Ariel the poet–prince, and Caliban the failed poet. The fact that Shakespeare had preceded her accounts for the speaker's fear that she cannot create, and the poem ends with her recognition that Caliban, the misshapen child, is her other self.

We see many father–son relationships in Cather's Shakespeare poems: Apollo and Shakespeare, Prospero and Caliban, Shakespeare and his "dwarfed" literary descendants. As this pattern suggests, Cather could only imagine literary inheritance as patrilineal. Indeed, father–son relationships provide the paradigm of creative inheritance that informs several poems in *April Twilights:* Apollo granting or denying his troubador and shepherd sons the gifts of song. Throughout the collection male myths and legends (primarily those of Apollo, Orpheus, and the Grail) provide the narrative structures Cather uses to convey her speakers' desires for creative expression. Given this omnipresent association of creativity with masculinity and the father–son bond, it seems that for the woman writer to inherit from the literary tradition she must deny her gender—an implication borne out by "The Namesake" in which Cather imagines herself her uncle's double and heir.

There are no mothers who might empower daughters to speak in these poems, no Demeter to rescue Persephone from the underworld and renew the land. The only women mentioned are objects of male desire or characters in male-authored texts—Helen ("Aftermath") and Cleopatra ("Aftermath," "On Cydnus"), Miranda ("Paradox") and Brunhilda ("White Birch in Wyoming"). Even the land—a feminine presence here as in Cather's later fiction—possesses no creative power. Although the pastoral traditionally celebrates a timeless, golden age where unity between the self and nature is possible—a mythic version of preoedipal union with a nurturing mother—Cather's singers inhabit a sterile, wintry Arcadia, a wasteland where nothing can grow.[14] Like the young Cather who was "thrown out" of Virginia's pastoral landscape, the speakers in these poems are children cast out from Eden, separated from a natural world which itself possesses no powers of growth or fertility (*KA*, p. 448).

The image of the dead, snow-covered land mirrors the speaker's fear that her own creative powers might be suppressed or dead. These poems create a world in which spring—even if it returns to nature—may not revive the poet's imagination. As the speaker asks in "Winter at Delphi," "Birds will come back with the spring,—but Apollo, the god, Apollo?" The speaker cannot be sure that the god of poetry is allied with nature's female powers of regeneration, for Apollo does not serve other gods. And so she must wait, "Frozen, dumb," while lesser voices are "Babbling May."

Cather's portrayal of nature in *April Twilights* represents the dilemma of a daughter/poet who cannot connect to any creative powers that can be imagined as female. Paternal power is present but inhibiting; maternal power is absent or inhibited. Although the volume's title refers to the season of pastoral harmony and creative emergence, the winter mood and imagery in most poems suggests that "twilight"—with its connotations of diminishment and death—is a better guide to the volume than "April."[15]

Since Cather did possess one maternal predecessor "whom all the world calls great," as she observed in 1895, it may seem odd that she drew no significant support from Sappho in a first volume of poetry that is so dependent on classical models and myths (*KA*, p. 349). During the early twentieth century, as Susan Gubar observes, Sappho's "status as a female precursor empowered a number of female modernists." Poets as diverse as Renée Vivien, H. D., and Amy Lowell formed imaginative attachments to Sappho, whose poetic achievement was particularly validating to lesbian poets who searched "in vain for a native lesbian poetic tradition."[16] As early as 1895, Sappho had impressed Cather as a powerful literary progenitor whose incandescent love poems had inspired many descendants; in 1907 Cather would become one herself when she published "The Star Dial," a poem in which she speaks both for and as Sappho.[17] But in *April Twilights* she neither alludes to the poet whose "broken fragments [had] burned themselves into the consciousness of the world like fire" nor envisions a matrilineal inheritance being passed down the centuries (*KA*, p. 349).

Cather was not able to find Sappho an empowering predecessor in *April Twilights* for several reasons. Having just returned from her European tour preoccupied with the power of the fathers, she directed her gaze toward male, not female predecessors. Moreover, unlike Vivien, H. D., and Lowell, Cather wanted to conceal the desire she thought "unnatural" and may well have feared that overt references to Sappho might categorize herself and her poetry as lesbian—and thus condemn her to marginality and obscurity. Desiring acceptance by the dominant literary culture, Cather was not self-consciously seeking to place herself in a lesbian poetic tradition.

Her inability to acknowledge Sappho reflects the uncomfortable dilemma she faced in the volume's love poems. Although drawing, in part, on her own experience of love, in transforming such feelings in poetry, Cather faced severe inhibitions to self-expression. Unlike Sappho, she could not portray both the speaker and the object of desire as female. Consequently Cather either translates emotion into heterosexual terms by drawing on earlier literary models or retreats into cloudy diffuseness. In a few love poems the speaker is female, the loved one male; in several others the speaker's gender is indeterminate (and so would be assumed male by the reader), while the loved one is female; but in most the gender of both speaker and love object are indeterminate, veiled by abstractions and personifications like "Pleasure" and "Passion" ("In Rose Time").

Cather's choice of genre in her first book was not a happy one, as she later realized. The lyric, with its emphasis on self-exposure through the single voice, was not a congenial form. She could not divide herself, as she could in her fiction, among several characters—a means for simultaneously achieving disclosure and concealment. Although the emotions inspiring the poems were doubtless deeply felt, the result is stylized, detached, and unconvincing love poetry in which Cather attempts to convey the pain of loss in formulaic pastoral images like "the snow lies drifted white/ In the bower of our delight" ("Arcadian Winter").

As in Cather's account of her European trip, however, we can discern in *April Twilights* a few signs of female creative power and lineage that are either undeveloped or undermined by the surrounding context. She alludes to the Amazons in "White Birch" and to the moon in "Dedicatory," although the moon-goddess' influence wanes as the "sun-born" Apollo appears in the poems that follow. Cather makes a fleeting moment of connection to her own female heritage in "Grandmither, Think Not I Forget,'' but once again this is not a sustaining tie—not only does the speaker desire death because of unrequited love, but she also expresses herself in an unconvincing Scotch dialect (''Grandmither, gie me your peaceful lips, white as the kirkyard snow,/ For mine be red wi' burnin' thirst, an' he must never know'').

More promising signs of Cather's desire to integrate femaleness and creative power can be seen in the metaphors that simultaneously disclose and conceal sexual and aesthetic desire. She makes numerous references to the moon, jewels, and flowers as well as to interior spaces (bowers, rooms, chambers, hearths, and houses) that recall the protected spaces Cather associated with maternal care and creativity (like the attic room or the ''garden fair''). The cup or the grail, the symbol of maternal and erotic power in ''The Burglar's Christmas,'' appears several times, associated with jewels and flowers in ''Thou Art the Pearl'' and with Cleopatra's ''milk-white'' breast in ''On Cydnus.'' Simultaneously an enclosed and an open space, the cup represents Cather's attempt to replace the male metaphors of creativity (penis/pen/knife) with a sign of female artistic power—a transformation she would not fully accomplish until *The Song of the Lark,* where one of the feminine landscapes Thea loves is called ''Pedro's Cup.''

•

But these undeveloped references to female creativity and inheritance do not offset the collection's emphasis on male authority. Unlike the male poets in Harold Bloom's psychodrama of literary influence, the author of *April Twilights* did not want to defeat her literary fathers and brothers but to be adopted by them. Cather's service to the gods in *April Twilights* is understandable—this was her first book—and may even have been useful to her literary evolution. When the adolescent girl emulated men by becoming ''William Cather,'' she was adopting an external sign of social power that allowed her to imagine a professional future; similarly, the woman writer who spoke in Housman's voice was adopting a sign of artistic power that allowed her to write. In a way, then, by patterning her verse after that of male predecessors, Cather could assume the poet's authority and voice—even if the voice was not hers. In *April Twilights* she imagines her speakers ''Frozen, dumb,'' and yet she was speaking, writing, and publishing.

Nevertheless, imitating literary fathers undermined Cather's authority as a woman writer. She was borrowing prestige and privilege from the male literary tradition, just as she was borrowing voice, form, and genre. In one sense she was speaking, but in another she was as silent as the winter landscapes that dominate her poems. From this point of view, the most disturbing poem in the collection is not a Housman imitation like ''In Media Vita,'' but ''White Birch in Wyoming.'' This is a portrait of the Western landscape, but Cather cannot even see her own country except through a male predecessor's eyes (''Stark as a Burne-Jones vision of despair''). This was not bringing the muse into her own country, as she would in *My Ántonia.* This was offering her own country up to someone else's muse. Separated by gender from the literary masters she admired (a barrier she did not want to acknowledge), in *April Twilights* Cather is a ''dwarfed [child]'' amidst giants. Only after she abandoned such strict ''service to gods'' did she rewrite the pastoral in *O Pioneers!,* where she grants her artist/heroine the power to bring the natural world and her own creative force to life, thereby redeeming the sorrowful, self-pitying speakers in these poems.

In her subsequent revisions of *April Twilights,* Cather tried to make the collection conform to the course her fiction was taking. She deleted some of the more derivative poems and added a few new ones that drew on familial, personal, or Western roots. In the 1923 edition—entitled *April Twilights and Other Poems*—she included ''The Swedish Mother,'' ''Spanish Johnny,'' ''Prairie Spring,'' ''Macon Prairie,'' and ''Going Home,'' and in a 1933 reissue she added ''Poor Marty.'' She retained all of these in the Library Edition, but was evidently still not pleased enough with the volume as a whole to place *April Twilights* in Volume I.

•

Cather's poem of identification with her fallen uncle, ''The Namesake,'' is an anomalous presence in the misty world of *April Twilights,* where it is one of the few poems that point toward her major fiction. The poet is still worshipping at sepulchres—here the uncle's grave—but in this simple elegy she frees herself from the conventions of the pastoral, chivalric romance, and late Victorian poetry:

Two by two and three by three
Missouri lies by Tennesee;
Row on row, an hundred deep,
Maryland and Georgia sleep;
Wistfully the poplars sigh
Where Virginia's thousands lie.

Somewhere there among the stones,
All alike, that mark their bones,
Lies a lad beneath the pine
Who once bore a name like mine,—
Flung his splendid life away
Long before I saw the day.

On the surface, this poem seems to create a father–son lineage that would be inhibiting to a woman writer, like the Apollo–Shakespeare references that Cather scatters throughout the collection. But something else is going on here: the male relative in ''The Namesake'' so resembles the speaker that he becomes more the brother than the father. When the adolescent girl identified with her heroic ancestor William Boak and became William Cather, Jr., her uncle was then her symbolic progenitor and father. But by 1903, when she was thirty, he could assume another familial role. Cather was then older than her uncle had been at his death and could regard him as a brother, just as she could imagine Stephen Crane as a compatriot who shared a similar childhood in her 1901 essay.

''The Namesake'' demonstrates the creative force Cather derived from imagining the ancestor as a brother rather than a father. In contrast to her imitative pastoral poems, where creativity seems impossible and failure inevitable, the speaker here is not only invigorated by the identification but even imagines surpassing her predecessor, who has left his work unfinished (''And I'll be winner at the game/ Enough for two who bore the name''). The relative's incompleteness, which inspires rather than intimidates, is symbolized here by the unmarked grave (the uninscribed stones are ''All alike,'' contrasting to the single-named tombs Cather found in France). With his grave unnamed, he leaves room for his namesake's accomplishments—which may rescue him from obscurity if she achieves enough fame for ''two who bore the name.''

As we have seen, Cather found the brother an important male figure who represented several subversive alternatives to the Victorian ideology of gender: the sister's ''masculine'' qualities, the possibility of androgyny, and a male identity less imbued with patriarchal power than that of the father. During the early 1900s Cather became increasingly fascinated by the figure of the brother and the brother/artist, whom she frequently portrays as possessing traits socially defined as feminine. She acknowledges the familial source of the brother/artist association in the ''Dedicatory'' poem in *April Twilights* and evokes the time when she and her two brothers were joint authors of ''Starry wonder-tales.'' Several other poems and short stories Cather wrote between 1900 and 1903 either draw on memories of her own brothers, depict the artist as a brother, or use the brother–sister relationship to question the culture's polarization of ''masculine'' and ''feminine.'' ''Eric Hermannson's Soul'' is based on an experience Cather shared with her brother Roscoe (*WWC*, p. 33); ''The Treasure of Far Island'' gives her brother Douglass' name to the hero; ''Jack-a-Boy'' was inspired by her youngest brother; and ''The Professor's Commencement'' features a brother who possesses a ''woman's heart'' and a sister who is the ''more alert and masculine of the two'' (*CSF*, pp. 291, 285).

The literary pattern reflects Cather's development of two attachments with brother/artists: her imagined bond with A. E. Housman, whom she considered one of her ''brothers in exile'' until she met him (*WP* 2, p. 709); and her real-life friendship with composer Ethelbert Nevin, whose family owned the Pittsburgh *Leader*. Brother to her sister, Nevin helped Cather to relinquish her identification of artistic power with an aggressive masculinity.

Born in 1862, Nevin grew up at Vineacre, his father's estate near Pittsburgh. Displaying musical talent as a child, he studied in Boston, Dresden, and Berlin, and by the 1890s was a well-known composer of lyric art songs and short, delicate piano pieces. After spending several years abroad—Berlin, Paris, Florence, and Venice—he returned in ill health to his family home in Pittsburgh in 1898, where he stayed for two years before moving to New Haven. He died in 1901 at the age of thirty-eight. Known as the musical poet of ''beautiful little verses'' rather than a ''master of great things,'' Nevin made his mark with songs like ''The Rosary,'' which brought him immediate fame, and piano music like *Water Scenes* and *May in Tuscany*. Viewed by some as sentimental, the strength of his work lies in its tenderness, delicacy, and lyricism.[18] Reportedly a charming friend, a delightful host, and an affectionate husband and father, Nevin was always ''the simplest and least pretentious of men,'' wrote a friend in 1917. ''The most real thing in life to him was his affections and emotions in their relation to those he loved and to his music.''[19]

Certainly he was kind to Willa Cather. She met Nevin in 1898, after he had returned, ill, to his home. She spent many "musical Sundays" at Vineacre and grew to worship the man she considered the greatest American composer (*WP* 2, p. 636). He appears as an idealized romantic hero in the articles she sent back to Lincoln papers—a prince, a troubadour, a delegate from Parnassus. Nevin and his wife entertained Cather at their home, and once Nevin went shopping with her, she reported exultantly to Mariel Gere, and bought her an enormous bunch of violets.[20] Later he sent her a volume of Shakespeare sonnets and dedicated a composition to her—"La Lune Blanche"—as Cather made sure to report to her Lincoln readers (*WP* 2, p. 640).

In contrast to the manly Kipling, whom Cather was also praising in the late 1890s, Nevin disliked physical exercise and preferred the company of women to that of men. He appears as an androgynous figure in Cather's writings. Slight and boyish, with the "slender, sloping shoulders and shapely hips of a girl," he seemed twenty to her even though he was in his thirties (*WP* 2, p. 533). In childhood he had been Cather's opposite, a "girlish little boy" who was "always much concerned about his mother's dresses and fond of masquerading in dresses himself" (*WP* 2, p. 628). The adolescent boy despaired at his encroaching masculinity, she discovered, because his body's maturation meant the loss of his beautiful soprano voice. The younger brother in a family of boys who accepted conventional male identities and roles as "bankers, merchants, editors," Nevin never achieved a male identity in their estimation because he became a musician. His brothers—all "big fellows" who "made their mark"—called him "The Boy," Cather noted (*WP* 2, p. 535). And yet far from criticizing this unmanly man, Cather praised the "boy–girl" Nevin's musical genius and his commitment to art. To her, he was the "monarch" of song (*WP* 2, p. 627).

Cather's friendship with Nevin helped her to see that a male artist might also experience a conflict between gender and vocation in a country where business was a man's profession and music a woman's diversion, an insight that later contributed to her sensitive portrayal of the displaced aesthete in "Paul's Case." In one article Cather even attributes Nevin's artistic power to a source culturally associated with women—"sentiment," by which she meant "feeling" rather than "sentimentality." She connected Nevin's genius with his "exquisite sensitiveness," his ability to be affected by "penetrating and wounding experience" and to retain the "painful susceptibility of youth" in maturity. These receptive, "feminine" attributes contrast sharply to the imperialistic force she once admired in Kipling and may have helped her to see that the most potent source of her imaginative power—love for people and places—had a legitimate role in the creative process.

Thus Nevin's appearance to Cather as a "girl–boy," her opposite-sex twin brother, helped her to disentangle creativity from masculinity and encouraged her to appreciate the supposedly feminine values of sensitivity and emotional vulnerability. She drew on her memories of Nevin for two artists in her fiction—the composers Adriance Hilgarde in " 'A Death in the Desert' " and Valentine Ramsey in "Uncle Valentine." In the latter story, the brother–sister bond that Cather experienced with Nevin becomes overt. As Charlotte Waterford and Valentine

Ramsey, close friends and spiritual siblings, gaze at a river that suddenly becomes luminous in the moonlight, they murmur "the Rhinegold" in one voice, evidence of their intertwined sensibilities (*UV*, p. 25). Later Valentine plays Siegmund's love song from *Die Walküre* for Charlotte. This song of yearning for his twin sister and lover Sieglinde also appears in "A Garden Lodge" and *The Song of the Lark* to represent a male–female relationship based on resemblance rather than opposition—what once Cather called a "blood-identity" in a letter to Dorothy Canfield.[21]

The stories Cather wrote during the early 1900s also show how her attachment to her brother/artists Nevin and Housman and her interpretation of the brother–sister bond helped her to question social definitions of masculinity—and thus of femininity. Whereas her travel letters and poems assert the rule of the fathers, short stories like "Jack-a-Boy," "The Treasure of Far Island," and "The Professor's Commencement" feature male characters who shun patriarchal power. Cather had once thought Kipling too much of a man to linger in the company of Romantic poets dripping with curls and lace collars, but one cannot imagine the sensitive male protagonists in these stories at ease in the barracks room either.

Inspired by her affection for John ("Jack") Cather, "Jack-a-Boy" is a stilted, sentimental story of interest mainly for the insight it provides into Cather's changing views of gender, sexual identity, and artistic vocation. The hero is an impossibly perfect six-year-old boy, a precocious artist who reveres the "holiness of beauty" and thrills to Greek mythology. He is also a boy who likes the "gentle ways of girls" yet dresses in a soldier costume and leaps into battle with a neighborhood bully (*CSF*, pp. 311–22). The child's death provides an appropriately tearful ending for a story published in the *Saturday Evening Post*, but it evades the problem Cather confronted in her own life: What will happen to the androgynous child/artist when he reaches adolescence? Despite the story's flaws, it shows Cather attempting something far more adventurous than she did in *April Twilights:* creating a character who combines stereotypically male and female traits, thus defying gender categories.

In "The Treasure of Far Island," which appeared in *New England Magazine*, Cather allows the imaginative "girlish" boy to retain the child's imagination and love of beauty in adult life by granting him the artist's vocation. The story makes it clear that the sensitive, aesthetic male was becoming Cather's alter ego, brother to her sister. Given her brother Douglass' name, Douglass Burnham is also granted her own childhood pursuits: as a boy he creates a packing-box town and plays on an enchanted island. Ostensibly a love story, the real concern of "Far Island" is the androgyny of childhood, the Edenic time when brothers and sisters share activities and identities before the Fall into sex roles (*CSF*, pp. 265–82).

"The Professor's Commencement" features a "sensitive" male protagonist with "delicate" hands that are "exceedingly small, white as a girl's, and well kept as a pianist's." As the last comparison suggests, the professor is also committed to aesthetic and intellectual values considered unmanly in a "vast manufacturing region given over to sordid and materialistic ideals." His favorite student—the only one who shares his love of art—also possesses "the gentle eyes and manner of a girl" and, like Jack-a-Boy (and, later, Tom Outland), is condemned to an early death. Although at his own graduation from high school the professor could not complete his recitation of "Horatius at the Bridge," a failure he repeats at his retirement dinner, Cather makes it clear that he is a warrior of another sort in fighting ignorance and materialism with devotion to art and literature. She surrounds him with martial imagery that links him with Bunyan's spiritual warriors rather than Rome's military ones. In fact, the story calls into question the identification of the artist and the conqueror she was simultaneously affirming: "The Professor's Commencement" appeared in the summer of 1902, when Cather was celebrating Rome's and England's imperial greatness in her travel letters.

•

The Troll Garden stories, which Cather was also writing during this period, follow the direction of "The Treasure of Far Island" and "The Professor's Commencement" rather than *April Twilights* or the travel letters. In them the artist-figures are sensitive, vulnerable brothers, not oppressive, godlike fathers. And yet these supreme artists are still male, still inhibiting to female creative power. In *The Troll Garden* Cather is unable to identify artistry either with women or with her Nebraska origins. Consequently the literary anxieties evident in *April Twilights,* although they become less intense, do not dissipate in Cather's first book of fiction.

NOTES

1. For a discussion of Hawthorne's and James' imaginative "possession" of Europe's ruins and monuments, see Elissa Greenwald, "The Ruins of Empire: Reading the Monuments in Hawthorne and James," *The CEA Critic* 46, nos. 3 and 4 (Spring–Summer 1984): 48–59.

2. Drawing on Lewis, E. K. Brown describes the trip as a "time of exciting intellectual and aesthetic discovery," a "landmark in the formation of the novelist" (*Willa Cather: A Critical Biography,* completed by Leon Edel [New York: Alfred A. Knopf, 1953], p. 99). See also James Woodress, *Willa Cather: Her Life and Art* (New York: Pegasus, 1970; rpt. Lincoln: University of Nebraska Press [Bison ed.], 1975), pp. 99–104, and Phyllis C. Robinson, *Willa: The Life of Willa Cather* (New York: Doubleday, 1983), pp. 108–16. Also relying on Lewis, Robinson likewise portrays the trip as a deeply moving "homecoming" (p. 108).

3. Cather's travel letters first appeared in the *Nebraska State Journal* in the summer of 1902. Collected and edited by George N. Kates, they were published in book form as *Willa Cather in Europe: Her Own Story of the First Journey* (New York: Alfred A. Knopf, 1956). William Curtin provides a more accurate text in "The European Scene" (*WP* 2, pp. 889–952).

4. "Nature," from *Selections from Ralph Waldo Emerson,* ed. Stephen E. Whicher (Boston: Houghton Mifflin [Riverside ed.], 1957), p. 21.

5. Writing in the *Nebraska State Journal,* the college student had observed: "Most things come from France, chefs and salads, gowns and bonnets, dolls and music boxes, plays and players, scientists and inventors, sculptors and painters, novelists and poets. It is a very little country, this France, and yet if it were to take a landslide in the channel some

day there would not be much creative power of any sort left in the world'' (*WP* 1, p. 223). See also Michael Gervaud, ''Willa Cather and France: Elective Affinities,'' in *The Art of Willa Cather,* ed. Bernice Slote and Virginia Faulkner (Lincoln: University of Nebraska Press, 1974), pp. 65–81.

6. Willa Cather to Dorothy Canfield, [April 7], 1922 and [May 8], 1922, Dorothy Canfield Fisher Papers, Bailey Library, University of Vermont, Burlington, Vt.

7. See Ida H. Washington, *Dorothy Canfield Fisher: A Biography* (Shelburne, Vt.: The New England Press, 1982), p. 17. During Cather's Lincoln years, she had admired Mrs. Canfield's apparent sophistication; as E. K. Brown observes, ''Mrs. Canfield's cult of art, nourished by her association with French artists and her broad reading of French literature, was important to Willa Cather'' (Brown, p. 64). In the scathingly satirical ''Flavia and Her Artists,'' written a year or two after the European trip, Cather gives Mrs. Canfield's name to a shallow, superficial woman who also makes a ''cult of art.'' The story may have in part been Cather's first literary attempt at revenge against Dorothy, and perhaps also an exorcism of Flavia Canfield's influence over her younger self. Sometimes in 1905 (shortly after ''Flavia and Her Artists'' appeared in *The Troll Garden*) Dorothy Canfield broke off her friendship with Willa Cather. In the same year Cather wrote to Flavia Canfield apologizing for her part in a melancholy affair that had caused this sad loss, but she leaves the details vague (Willa Cather to Mrs. Canfield, n.d. [summer 1905], Dorothy Canfield Fisher Papers, Bailey Library, University of Vermont, Burlington, Vt.). The friendship was not firmly reestablished until the early 1920s.

8. Willa Cather to Viola Roseboro', n.d., and Cather to Roseboro', June 14, n.y., Houghton Library, Harvard University, Cambridge, Mass. These letters are copies given by Roseboro' to Witter Bynner, another *McClure's* colleague, who donated them to Harvard along with his own Cather letters. Throughout her life Cather was plagued by mythical versions of her visit to Housman, one of them perpetuated by Ford Madox Ford who claimed that she had presented the poet with a gold laurel wreath. She wrote three letters to Carl Weber, a college professor intrigued by the story, at first giving him information and then, once he became too much of a nuisance, brushing him off. (Willa Cather to Carl Weber, January 10, 1944, December 12, 1944, and January 31, 1945, Colby College Library, Waterville, Me.) See Brown, pp. 105–9 and Woodress, pp. 101–2 for accounts of the Housman visit and its aftermath.

9. Quoted by Alice Hunt Bartlett in ''The Dynamics of American Poetry—XI,'' *Poetry Review* 16 (1925): 408.

10. For detailed background on Cather's first book, see Bernice Slote's introductory essay to *AT* (1903), pp. ix–xlv.

11. Cather to Roseboro', June 14, [n.y.].

12. George Seibel, ''Miss Willa Cather from Nebraska,'' *New Colophon* 2, Part 7 (September 1949): 196–97.

13. See Sandra M. Gilbert and Susan Gubar, *The Madwoman in the Attic: The Woman Writer and the Nineteenth-Century Literary Imagination* (New Haven: Yale University Press, 1979), pp. 16–20. Their discussion of metaphors of literary paternity, in which creativity is linked with male procreative and progenitive power, provides a useful context for considering *April Twilights* (pp. 3–11).

14. David Stouck distinguishes between a ''pastoral of innocence,'' in which the self is united with a maternal world, and a ''pastoral of experience'' in which the self is cast out from Eden, drawn imaginatively toward a myth of union yet recognizing the inevitability of separation. He views *April Twilights* as the latter. See *Willa Cather's Imagination* (Lincoln: University of Nebraska Press, 1975), pp. 35–42.

15. See Stouck, p. 39.

16. Susan Gubar, "Sapphistries," *Signs: Journal of Women in Culture and Society* 10, no. 1 (Autumn 1984): 43–62, esp. 44, 46.

17. As Gubar observes, "Precisely because so many of her original Greek texts were destroyed, the modern woman poet could write 'for' or 'as' Sappho and thereby invent a classical inheritance of her own" ("Sapphistries," pp. 46–47).

18. My portrait of Nevin is based on John Tasker Howard, *Our American Music: Three Hundred Years of It* (New York: Thomas Crowell, 1930), pp. 367–78, esp. p. 378.

19. Francis Rogers, quoted in Howard, p. 369.

20. Willa Cather to Mariel Gere, January 10, 1898, Nebraska State Historical Society, Lincoln, Nebraska.

21. Cather to Canfield, [April 7], 1922. Cather was referring specifically to the identification she felt with her nephew G. P. Cather, who also played the imaginative role of brother to her sister. Charlotte Waterford and Valentine Ramsey share a similar spiritual and emotional kinship, and as Bernice Slote observes, etymologically their names stem from one root meaning "little strong one" ("Introduction," [*UV*, p. xxvii]).

12

The Troll Garden

We must not look at Goblin men,
We must not buy their fruits;
Who knows upon what soil they fed
Their hungry thirsty roots?
"Goblin Market"

A fairy palace, with a fairy garden; . . . inside the trolls dwell, . . . working at their magic forges, making and making always things rare and strange.
Charles Kingsley

While she was working on the poems to appear in *April Twilights,* Cather continued to write and publish short stories. Her new job and new relationship inspired more productivity in her chosen genre as well: she published six stories in 1900 (five in *The Library*), two in 1901, and two in 1902.

Two of these stories—"Jack-a-Boy" (1901) and "The Professor's Commencement" (1902)—suggest that Cather may have been finding her life of admiration a bit burdensome, even deadening. Although both stories overtly celebrate the cultural tradition represented by the classics, covertly both associate Greek and Latin literature with suffocation, decay, and death. The professor in "Jack-a-Boy" is a dusty antiquarian who lives and works in a tomblike room surrounded by maps of "dead and forgotten" cities, "dusty plaster casts of Greek philosophers," models of the Parthenon, and an Egyptian mummy, and we learn that his great unfinished tome on Greek prosody—the work of a lifetime—is unlikely ever to come to life (*CSF,* p. 313). Similarly, the professor in "The Professor's Commencement" feels like the "ruin of some extinct civilization." Worn out from years of teaching literature to disinterested youths, he compares himself to the "mutilated torso" of an antique statue. When he fails for the second time in his recitation of "Horatius at the Bridge"—his memory falters after the lines "Then out spake bold Horatius/ The Captain at the gate"—Cather suggests that his loss of energy may owe something to the professor's somewhat servile approach to the classics: he regards them as texts to be memorized. Had he read past literature more creatively, then he might have been able to continue speaking instead of being silenced when he forgets the words "bold Horatius" was supposed to say (*CSF*, pp. 283–92).

Cather may have been feeling impatient with her own as well as her two professors' subservience to the classics, since three of the stories from this period—"Eric Hermannson's Soul" (1900), "The Sentimentality of William Tavener" (1900), and "The Treasure of Far Island" (1902)—show her gradual turn from memorization to memory. Although all three draw upon her personal or family past and employ Western settings, "The Sentimentality of William Tavener"—a simple, unvarnished narrative—points most directly toward her major fiction.

Cather wrote "The Sentimentality of William Tavener" several years before meeting Sarah Orne Jewett, but since she had read *The Country of the Pointed Firs,* it is possible that Jewett's local-color fiction had suggested a way for her to make use of her rural background. The story appeared in *The Library* along with two strained and artificial stories featuring Western settings, "The Dance at Chevalier's" and "The Affair at Grover Station"; in these Cather tried for the melodramatic and the exotic, featuring jealousy, betrayal, murder, rare poisons, villainous foreigners, and ghosts. But in "William Tavener" she drew unostentatiously on her Nebraska past and possibly on a family story; certainly she made use of her family history of emigration.[1] An understated domestic drama in a rural setting, the tale reveals a Jewett-like concern with the interrelationships among memory, affection, and storytelling.

A Midwestern farm couple have ceased to talk to each other with any intimacy because roles and family loyalties have divided them. The husband is concerned with his farm work, the wife with her children; over the years their conversations have been reduced to negotiations "almost wholly confined to questions of economy and expense." As Hester haggles in behalf of her sons, Cather delicately and understatedly describes the emergence of buried affection between husband and wife. In trying to persuade her penurious husband to let the boys attend a traveling circus, Hester discovers that—unknown to each other—she and William had attended a circus together back in Virginia, their childhood home. This revelation opens the floodgates of memory, and the two begin to share other stories and recollections:

> They talked on and on; of old neighbors, of old familiar faces in the valley where they had grown up, of long forgotten incidents of their youths—weddings, picnics, sleighing parties and baptisms. For years they had talked of nothing else but butter and eggs and the price of things, and now they had as much to say to each other as people who meet after a long separation.

Their conversation leads to a reawakened love based in shared experience, understanding, and memory. As a result of this renewed connection, Cather implies, family allegiances will shift. Hester will no longer always place her sons before her husband: she feels a "throb of allegiance" to William, and her sons sense that they have lost a "powerful ally."

This simple, unaffected early story is one of Cather's best; although emotionally connected to her material, she does not project unresolved conflicts or desires into the text as she did in "A Tale of the White Pyramid" and "The Burglar's Christmas." In fact, "William Tavener" reverses the preoedipal and oedipal fantasies that underlie many of her early stories; here the mother turns from the children to the father with a strengthened attachment. Focusing on the quiet drama in the lives of ordinary people, the story anticipates Cather's Nebraska novels. Just as the release of memory gives rise to the stories William and Hester share, so Cather's personal and family history shapes "The Sentimentality of William Tavener," as she acknowledged by giving Tavener the family name of William and calling her characters' Virginia homeland "Back Creek."

The story further suggests the creative possibility absent in *April Twilights:* if Cather turned to her own past and memories, she could discover there a heritage of female creativity and self-expression. Cather momentarily pays tribute to the female voices in her past—the storytellers who enthralled her in Virginia and Nebraska—when she grants the power of verbal expression to the farm woman. The taciturn William is a man of silence, while Hester, who can "talk in prayer meeting as fluently as a man," possesses a "gift of speech" (*CSF,* pp. 353–57).

Yet Cather did not follow in the direction of "The Sentimentality of William Tavener" in *The Troll Garden.* She turned to the Midwest as a setting in some of the collection's stories, but the portrait she paints of her homeland is a negative one; she was not yet able to imagine herself nourished by her Nebraska roots or to identify the West with artistic power. Moreover, in *The Troll Garden* she does not give her talented women the "gift of speech" she grants Hester. Because they, like her, were aspiring to be artists, Cather could not envision them achieving the self-expression she had not yet attained herself, and so consigns Katharine Gaylord (" 'A Death in the Desert' "), Caroline Noble ("The Garden Lodge"), and Aunt Georgianna ("A Wagner Matinee") to various forms of silence.

•

The publishing history of Cather's second book was markedly different from that of her first. Whereas she had to subsidize the publication of *April Twilights, The Troll Garden* was solicited and published by a major figure in American letters: S. S. McClure, later Cather's employer and mentor at *McClure's Magazine.* Gaining his recognition was a particularly sweet triumph for Cather. Although she had begun to place her work in national magazines—"Eric Hermannson's Soul" in *Cosmopolitan,* "Jack-a-Boy" in the *Saturday Evening Post,* " 'A Death in the Desert' " in *Scribner's,* and several others in *New England Magazine*—she had received only rejections from *McClure's.* But 1903 brought a dramatic reversal in her literary fortunes.

H. H. McClure, head of the McClure Syndicate and S. S. McClure's cousin, was in Lincoln in the spring of 1903 looking for new writers. Will Owen Jones, who had been Cather's colleague at the *Journal,* recommended her to the scout, who passed her name along to S. S. McClure. Always on the lookout for fresh talent, McClure wrote Cather requesting to see some of her efforts, not knowing that his editors had already sent some back. McClure liked what he read, decided that Cather was a valuable prospect, and immediately summoned her to his New York office.

Writing Will Owen Jones shortly after this momentous interview, Cather said that her sense of herself had been transformed in a few hours: at ten o'clock she walked into McClure's offices, not worrying much about streetcar accidents, while at one o'clock she left taking more care. She had become a valuable property. McClure had been enthusiastic about her future, she told Jones exultantly. He wanted to publish her stories as a book, using what he could in his magazine, helping her place others elsewhere; he would ask his friends Howells to publish some of her stories in *Harper's*. He had told her that he wanted to publish as many of her future stories as he could. During their interview he even called in his manuscript readers and demanded to know why her stories had been rejected, a moment she particularly relished. Then he invited her to stay with him for a few days; during this visit, she met Mrs. Robert Louis Stevenson, who had already read her stories. She felt lighthearted, and she thanked Jones for his role in her good fortune.[2]

Thus Willa Cather, obscure Pittsburgh schoolteacher and part-time writer, was discovered by the famous S. S. McClure. She was exhilarated by this long-awaited recognition. Beginning to believe that she might achieve her dreams of literary greatness, she did not consider the obligations that might accompany his sponsorship. McClure fulfilled most of his promises: "A Wagner Matinee" (1904) appeared in *Everybody's Magazine*, "The Sculptor's Funeral" (1905) and "Paul's Case" (1905) in *McClure's*, and *The Troll Garden* was published in 1905 under the imprint of McClure, Phillips, and Co. In addition to the above three stories, the collection included " 'A Death in the Desert' " (published in *Scribner's* in 1903) and three previously unpublished stories—"Flavia and Her Artists," "The Garden Lodge," and "The Marriage of Phaedra." Cather dedicated the book to Isabelle McClung, in whose "garden fair" her imagination had grown.

Although Cather did not write these seven stories knowing that they would appear together, *The Troll Garden* was not a random collection. Critics have variously analyzed the intricate metaphoric and thematic links among the stories, agreeing that Cather structures the volume through juxtaposition and contrast.[3] Concerned with art, artists, and the relationship between art and life, the stories explore the contradictions and contrasts between aesthetic and commercial values, pure and corrupted art, East and West, civilization and primitivism. Some of the tales reveal a heavy Jamesian influence (notably "Flavia and Her Artists" and "The Marriage of Phaedra"), but these aesthetic concerns were Cather's as well as James'.

Although the incompatibility between Presbyteria and Bohemia is the evident subject of several stories, we can discern a subtext woven through the collection that reveals the conflict most troubling to Cather at the time: between gender and vocation. The supreme artists in the stories are male, as in Cather's earlier work, but significantly they are all either absent from the story's narrative time (Adriance Hilgarde of " 'A Death in the Desert' " and Raymond D'Esquerré of "A Garden Lodge") or dead (Harvey Merrick in "The Sculptor's Funeral" and Hugh Treffinger in "The Marriage of Phaedra"). In either case, we only see the artists as they are created in memory or imagination by their admirers. This telling and recurrent absence has two opposed meanings: Cather wanted to convey the power of artistic influence, but she also wanted to dispense with the male artists who were absorbing her imagination. We can see this double purpose in the fact that the stories also portray gifted women whose creative power is suppressed or dissipated by social and psychological forces—Caroline Noble in "The Garden Lodge," Aunt Georgianna in "A Wagner Matinee," and Katharine Gaylord in " 'A Death in the Desert.' "

The epigraphs provide our best introduction to the collection's themes. Cather chose to begin *The Troll Garden* with the quotations from Christina Rossetti and Charles Kingsley that preface this chapter. The stanza from "Goblin Market" (without the poet's name) faced the title page, while the excerpt from Charles Kingsley's *The Roman and the Teuton* was on the title page itself, with the author credited but not the work. Hence Cather placed her first book of fiction in both the female and the male literary traditions, although her debt to Kingsley for the title—made evident by her placing the quotation from his work on the title page—suggests that he was the preferred predecessor. Or at least he was the predecessor she most wanted to claim publicly in 1905.

The Roman and the Teuton is a series of lectures describing the fall of Rome to the Germanic tribes. Cather took her title from the parable of the Trolls and the forest children, leaving out a few phrases as the ellipses indicate. The Trolls are the Romans, the artisans of a civilization; outside the troll garden, gazing at the fairy palace are the wondering, awestruck "forest children," the Germanic tribes who covet the treasures within. Taken in isolation, the epigraph suggests that we should equate the Trolls with artists and their palace and garden with the realms of art and the imagination. The forest children are then the crude outsiders, the barbarians who destroy rather than create things "rare and strange."

But when placed in the context of Kingsley's allegory, the connection between the excerpt and *The Troll Garden* becomes more complete. Kingsley's Trolls are the fabricators of a corrupt civilization who entice the innocent forest children with "fine clothes, jewels, and maddening wine" and tempt them to commit "sins which have no names," a phrase which Cather most likely would have associated with Oscar Wilde's love that dared not speak its name. Thus the Trolls are manipulators of desire, and the wondering forest children are potential artists, corrupted when their desire and imagination are directed from the aesthetic to the sensual, the erotic, and the decadent.

The paradoxical meanings the quotation accrues when it is considered simultaneously in isolation and in context (the Trolls are both artists and seducers, the forest children both crude barbarians and childlike dreamers) introduce the psychological complexity of the stories that follow. Cather is more concerned with complexities and contradictions hidden within the self than with a simple contrast of opposed viewpoints: in *The Troll Garden* a single self can be divided between troll and forest child, artist and tempter, creator and self-destroyer. Boundaries and identities are constantly shifting in these stories; the "troll garden" can be both the magical garden of art and the garden after the fall, the desert of deprivation and loss that appears in several stories as the barren Nebraska plains.

When Cather republished four of her *Troll Garden* stories in *Youth and the Bright Medusa* (1920), she eliminated the epigraph from Kingsley and kept the quotation from "Goblin Market," perhaps because she then realized that her stories were more indebted to "Goblin Market" than to *The Roman and the Teuton*. And the writer who had by then received a female literary inheritance may also have wanted to strengthen her acknowledgment of Christina Rossetti. In any case, Cather's reference to "Goblin Market" illuminates the collection's submerged psychological and aesthetic concerns, which revolve around erotic and romantic temptation, the creative process, and the woman writer's experience of literary influence—all emotional, sexual, and literary acts requiring a relationship between self and other.

To understand what Christina Rossetti and "Goblin Market" meant to Cather, we need to return to her 1895 commentary on women poets where she discussed Rossetti, Barrett Browning, and Sappho. "Goblin Market"—to her a work "pregnant with deep meaning"—was the one poem she described at length. Here is her summary:

> Lizzie and Laura are filling their pitchers at the brook in the evening. Lizzie stops her ears to the goblin song, but "sweet tooth Laura" listens. Lizzie flees home at the sound of their approach, but Laura croaches in the reeds and gazes longingly at the fruit as the swart misshapen goblin men come trooping by. . . . The goblins pile their wares about her and she buys the fruit with one of her golden curls. Then she "sucked their fruit fair or red. . . ." At nightfall she hurries to the brook and listens; Lizzie hears the cry of the goblin men and sees their coming shadow, but Laura cannot hear them, cannot see them. . . . Day by day she goes hungering and thirsting for the goblin fruits, dreaming of them, longing for them, her life is "Thirsty, cankered, goblin-ridden." But once tasted the goblin fruit gives only hunger, not satisfaction; only desire, not fulfillment.

Then Cather describes Lizzie's rescue of her starving, obsessed sister. Confronting the goblin men, who pelt her with fruit, she returns home "with the juices dripping on her face." Laura tastes them, but "they turn to wormwood in her mouth"; in the morning she is her old self again, the goblin men and their fruit forgotten. "Never has the purchase of pleasure, its loss in its own taking, the loathsomeness of our own folly in those we love, been put more quaintly and directly," Cather concluded (*WP* 1, pp. 143–45).

Like Rossetti's poem—in which the boundaries between goblins and sisters, sister and sister, become blurred and obliterated—the *Troll Garden* stories are concerned with disturbances and distortions in the relationship between self and other, represented by patterns of incorporation, possession, identity confusion and fusion, symbiosis, parasitism, and vampirism. In general, the characters who have difficulties in maintaining boundaries, as in Rossetti's poem, are women. Just as the voracious transgressor Laura in "Goblin Market" becomes obsessed with hunger for the mysterious fruit she has consumed, so the female characters in *Troll Garden* are unable to connect to something or someone outside the self without being absorbed. The most enticing form of absorption for Cather's women is a sexual or romantic relationship.

Many readers of "Goblin Market" see Rossetti's story of temptation and fall as sexual, and in her commentary on the poem Cather agrees that Laura's "purchase of pleasure" signifies her succumbing to forbidden desire. "Goblin Market," however, is more concerned with the psychological and imaginative *aftermath* of sexual transgression than with the act; after Laura "sucked and sucked and sucked the more/ Fruits which that unknown orchard bore,'' she enters a state of endless and insatiable desire. Like Jeannie, her ghostly predecessor who had also succumbed to the goblin men's cries, she starves, sickens, and pines away, with ''sunk eyes and faded mouth,'' ''dream[ing] of melons.'' Possessed by an ''absent dream,'' Laura starves because her imagination is corrupted by a desire she can neither fulfill nor dispel. Once she has ''sucked until her lips [are] sore,'' Laura is devoured by the goblin men and their fruit.

In her *Troll Garden* stories, Cather also explores the ways in which women's imaginations can be consumed by insatiable desires. Laura's fate is reenacted in '' 'A Death in the Desert,' '' ''A Wagner Matinee,'' and ''The Garden Lodge'' by other ''thirsty, cankered, goblin-ridden'' women who succumb to the erotic temptations of the garden, feed upon men, and find themselves starving in a desert wasteland where, like Laura, they decline, dwindle, and grow grey. In these stories the enticing goblin fruit is not simple sexual transgression but romantic love—that obsessive dream of fulfillment through another that absorbs women's imaginations so that they cannot act, create, or produce their own fruit.

Aunt Georgianna in ''A Wagner Matinee,'' who gives up a promising musical career for an ''inexplicable infatuation'' with a ''handsome country boy,'' finds that the goblin fruit of romantic love offers insufficient nourishment. Isolated in a grim, sterile Western landscape, she is physically and psychologically twisted and vitiated by deprivation, her plight externalized in the ''crook-backed ash seedlings'' and the ''naked house on the prairie, black and grim as a wooden fortress.'' Both the endless duties of the farm wife and the selflessness demanded by a mother's nurturing role erode selfhood and creativity. When she first reaches Boston her two main worries concern food for others: she can't remember whether she has left instructions about ''feeding half-skimmed milk'' to a weakling calf or if she has told her daughter that a can of mackerel would ''spoil if it were not used directly.'' Georgianna momentarily forgets the demands of feeding others and finds spiritual food for herself when her starved imagination gradually returns to life in the ''hanging gardens'' of art—the ''sunlit landscape'' of the concert hall where her nephew takes her to a performance of Wagner. But the last image in the story (the ''gaunt, moulting turkeys picking up refuse about the kitchen door'') prophesies her return to a landscape of desolation and starvation—the ''silence of the plains'' (*CSF*, pp. 235–42).

An even stronger parallel to ''Goblin Market'' occurs in '' 'A Death in the Desert,' '' where the singer Katharine Gaylord is dying of consumption in a ''bleak, lifeless'' Western landscape of ''spindling trees and sickly vines'' which mirrors her illness. Her consumption is a symbolic disease—the sign, as elsewhere in Cather's work, of her subservience to male power; in this case, consumption is the fitting result of her infatuation with composer Adriance Hilgarde, the absent male artist who consumes her imagination just as the goblin men and their fruit do Laura's. Cather's description of Katharine's romantic obsession is filled with images of starving and feeding (''I drank my doom greedily enough'') that embody the paradoxes of ''Goblin Market'': the goblin fruit gives no nourishment as the ''emaciated'' Katharine is physically, emotionally, and imaginatively consumed by what she has consumed (*CSF*, pp. 199–218).

Women who suppress their own creativity and channel imaginative energy into worshipping men find insubstantial fare, "The Garden Lodge" suggests. The women who become cult followers of Wagnerian tenor Raymond D'Esquerré, "flock[ing] to the Metropolitan" when he performs, participate in a ritual that is more inspired by erotic than aesthetic needs. But they obtain as little nurturance in the concert hall as Laura does from the goblin fruit. That the "mystic bread" the tenor gives them at his "eucharist of sentiment" resembles Rossetti's goblin fruit is suggested by Cather's comparison of D'Esquerré's sexual and aesthetic spell to Klingsor's garden. The evil magician in Wagner's *Parsifal,* Klingsor conjures up a garden of earthly delights—fruits, flowers, seductive women—from the barren sand, but when the temptations are sampled the garden vanishes, leaving only a desert of "hunger, not satisfaction; only desire, not fulfillment" (*CSF,* pp. 187–98).

•

Cather chose "Goblin Market" as an epigraph for another reason: she connected the poem with the lure of art as well as with erotic temptation. As Sandra M. Gilbert and Susan Gubar have argued, the greedy Laura is a stand-in for the woman writer, "metaphorically eating *words* and enjoying the taste of *power,* just as Eve before her did." In their view Laura's punishment, her starving and dwindling from balked desire, is retribution for the woman's overreaching desire for creative self-expression.[4]

But Cather read "Goblin Market" somewhat differently. She did interpret the goblin men's "Melons and raspberries,/ Bloom-down-cheeked peaches,/ Swart-headed mulberries" as the fruits of the creative imagination, but in her *Troll Garden* stories aesthetic fruits grow from male—not female—writers' imaginations, just as the fruit belongs to the goblin men in Rossetti's poem. In Cather's stories the woman artist who absorbs men's words is the reader, not the writer; the consumer, not the producer. Eating the "sugar-baited words" of male predecessors, the "sweet-tooth'd" woman artist is confined by the imaginations of her artistic fathers, brothers, lovers. The danger for such women, Cather shows in these stories, is loss of identity as well as nourishment. Possessed by male gods—the absent creators who haunt the female imagination in "The Garden Lodge" and " 'A Death in the Desert' "—the women characters in these stories do not discover the power of original expression suggested by their regal names ("Noble," "Gaylord"). In making aesthetic desire a female/male drama Cather was drawing directly on her own experience as the "bond slave" of A. E. Housman and, more pertinent to this collection, as the servant of the "mighty master" of fiction, Henry James.[5]

In "The Garden Lodge," musically gifted Caroline Noble chooses an orderly, controlled marriage rather than an artistic career, seeking to regulate both sexuality and creativity. When the enchanter D'Esquerré offers her his "eucharist of sentiment" as he rehearses *Die Walküre* in her garden lodge, he awakens not her desire to create but her urge to adore his creativity, an impulse Cather connects with subordination and silence, not self-expression. D'Esquerré is the soloist, Caroline the accompanist, and when he sings Siegmund's love aria from *Die Walküre,* he holds her as "he always held Sieglinde." Whereas in *The Song of the*

Lark, written after Cather was more convinced of her own creative power, the female singer "burst[s] into bloom" as she sings Sieglinde, here the male singer blossoms as Siegmund. Meanwhile Caroline is a silent Sieglinde, an object of yearning rather than the speaker of desire.

The danger that worshipping a male artist poses for the gifted woman's identity and creativity is satirically explored in "Flavia and Her Artists," where the hostess seeks to live off her artist–guests by a process of "psychic transmission" and so "absorbs rather than produces" (*CSF,* pp. 149–72). The same theme receives a more serious treatment in a " 'A Death in the Desert,' " the story that most resembles Rossetti's poem. The title itself suggests the difficulty in developing an original voice when others have already spoken. Cather borrowed her title from a poem by Robert Browning, but she announced her debt by using the quotation marks—a hint that she did not feel she had really made her predecessor's work her own. (By contrast, when she borrowed the title of *O Pioneers!* from Whitman, the more confident writer dispensed with the marks of indebtedness.) The reference to Browning's poem also recalls "Three Women Poets" and Cather's view of Elizabeth Barrett Browning as subordinate to the male literary tradition her husband represented.

Katharine Gaylord had once had a promising future as an opera singer, but she chooses to worship composer Adriance Hilgarde rather than to cultivate her own art. Her death by consumption in a desert landscape reveals her failure to develop her own creative powers, while her "haunting" music room signifies her failure to construct and preserve a separate identity. Architecture in Cather's fiction is a reliable mirror of the self, and this room tells us that Katharine has been absorbed by Adriance:

> Then it all became clear to him [Adriance's brother Everett]: this was veritably his brother's room. If it were not an exact copy of one of the many studios that Adriance had fitted up in various parts of the world, wearying of them and leaving almost before the renovator's varnish had dried, it was at least in the same tone. In every detail Adriance's taste was so manifest that the room seemed to exhale his personality.

While Katharine inhales and exhales Adriance's "taste" and "personality"—and so dies because, finally, she cannot breathe—he finds self-expression in his work, composing a sonata which contains the "voice of Adriance, his proper speech."[6]

In " 'A Death in the Desert,' " Cather contrasts the fate of the silenced woman and the triumph of the creative man through the symbolic landscapes featured in the collection—the garden and the desert. Starving and suffocating in a "wilderness of sand," Katharine is excluded from the garden with which Adriance is associated, a fertile, sheltered realm "full of the sound of splashing, running water." When she abandons her art for his, she leaves behind the "Garden Theatre" where she once performed, a space presided over by a statue of the "chaste Diana." The reference to the goddess reminds us that only the woman who resists the goblin men's blandishments retains autonomy and creative power.

"Who knows upon what soil they fed/ Their hungry thirsty roots?" asks Rossetti in Cather's epigraph. Several stories in *The Troll Garden* suggest that Cather was asking that question of the relationship between art and life, artist and subject, and thinking of male artists as deriving inspiration and creative force from women's "soil." " 'A Death in the Desert' " juxtaposes the fates of Katharine and Adriance, worshipper and god; as her life wanes, his strength and creativity increase as if, vampirelike, he were living on her gift of devotion and energy. She starves, withers, and ages prematurely, like Rossetti's Laura, while he retains the "face of a boy of twenty." Just as D'Esquerré blooms in the "heat" of his female admirers, so Adriance feeds on her and the other women who inspire and nurture him ("other lives . . . must aspire and suffer to make one such life as his"), turning life into art just as the painter Hugh Treffinger in "The Marriage of Phaedra" derives his greatest work from his tortured bond with his wife.

While women's identities are diffused and their energies drained in service to male creators, the godlike artists in *The Troll Garden* stories profit from women's devotion, finding in them muses as well as audiences. The titles of the two male-authored works mentioned in " 'A Death in the Desert' " and "The Marriage of Phaedra"—Adriance Hilgarde's cantata "Proserpine" and Hugh Treffinger's painting "The Marriage of Phaedra"—suggest how women have traditionally provided male creators with inspiration and subject matter. "Proserpine," which made the youthful Adriance world-famous, implies a further male appropriation of female power since the soil or source of the composer's inspiration is one of the few women-centered Greek myths. Like "Goblin Market," the Demeter–Persephone myth is a tale of combat between masculine and feminine powers in which one woman rescues another from male control. In the context of " 'A Death in the Desert,' " however, the reference is ironic. Adriance uses the myth just as he absorbs Katharine's devotion and imagination, and it is he who is associated with the powers of fertility and creativity controlled by Demeter in the myth.[7]

•

Although the male creators who dominate their disciples' imaginations never appear in the stories, their compelling presences reflect Cather's struggle to define herself as a woman writer in the context of the male-dominated literary tradition. We see this struggle reflected most explicitly in "The Marriage of Phaedra," where she asks a pressing personal question: How does an artist develop an individual voice while continuing to draw upon a preexisting tradition? The problem of literary influence is raised both by the story's subject—artistic originality—and by its form: "Marriage" is the most Jamesian piece in *The Troll Garden*. Possibly because this particular example of literary influence was so close to home, Cather cast all the major characters as male.

As in many of Henry James' stories, the setting is a London drawing room and artist's studio, the dialogue is stilted and lengthy, and the central symbol is a painting, which as in *The Sacred Fount* is a "cryptic index" both to the creative process and to a sexual and emotional relationship which is never adequately decoded. The story's central action concerns the disciple MacMaster's attempt to understand the creative secret of the master Treffinger, who died leaving only his work as a "common legacy to the younger men of the school he had originated."

Although Treffinger is dead when the story begins, his unseen presence pervades the story as MacMaster decides to become his critic and biographer.

Central to Treffinger's creative power, Cather suggests, was his ability to be both an inheritor and originator. He was influenced by his "incentive and guide" Ghillini as well as by medieval writers like Boccaccio whose works "transcribed themselves upon the blank soul of the London street boy." Treffinger absorbs these influences rather than being absorbed by them and so produces original rather than imitative work. But his followers, who paint in the "Treffinger manner," are not so fortunate. After "Treffinger's personality [dies] out in them" they have no private source of inspiration and are silenced when "the hand that had wound them up [is] still."

Treffinger possesses James, the Cockney valet who guards his shrine and legend, and the visiting MacMaster even more fully than these imitators, however; the apprentices return to the "fold" of traditional British art, finding other masters, but James and MacMaster become increasingly obsessed with an unfinished painting, Treffinger's masterwork—"The Marriage of Phaedra." A living monument to Treffinger, James has been imprinted by his employer's stronger personality:

> Many of his very phrases, mannerisms and opinions were impressions that he had taken on like wet plaster in his daily contact with Treffinger. Inwardly he was lined with cast off epithelia, as outwardly he was clad in the painter's discarded coats.

MacMaster similarly becomes possessed by the dead painter in attempting to possess and pass on his legend. After deciding to write his critical biography, like James he becomes the medium through which Treffinger speaks. Indeed the parallel roles James and MacMaster fill—servant and disciple—suggest the ways in which artistic apprenticeship may involve loss of identity and power. Cather's pun in naming MacMaster—for he, like James, serves Treffinger—calls attention to the apprentice writer's concern with literary originality in the story and the collection as a whole. It cannot be accidental that in writing a Jamesian story Cather selected the name of her literary master for the servant James. She may have done so because she wished no longer to be serving James—all the while finding in the master–servant, artist–disciple relationship an analog to her own literary servitude, which she displays as well as comments on.

Cather's concern with her relationship to literary predecessors—those "goblin men" whose words she had read—is evident in her titles. " 'A Death in the Desert' " is taken from a Browning poem whose subject is also discipleship and legacy, and the titles of "The Marriage of Phaedra" and "A Wagner Matinee" refer to other artistic works and artists. Allusion can be an act of strength or of servitude, depending on whether the writer is seeking to borrow stature from a predecessor. Given the context of *The Troll Garden,* Cather's epigraph from Kingsley also suggests a deferential nod to a male precursor, but her quotation from Rossetti is a more ambiguous referential act, linking *The Troll Garden* to a woman writer whom Cather considered both subordinate to the male literary tradition and independent of it. To see more clearly how Cather expresses impatience with her own apprenticeship to male writers in these stories, we need to look more closely at Cather's estimation of Christina Rossetti in her 1895 commentary on women poets.

Shunning epic and drama for the "simplest lyrics," Christina Rossetti possessed the "true artistic consciousness of her limits," Cather observed. In her case familial relationships directly paralleled literary subordination; as her brother Dante Gabriel's little sister, she was, Cather acknowledged, the "lesser light" in a family "richly endowed with genius," and her "simple music" was almost "drowned by the loftier themes and cadences" of her brother, whose "greatness threaten[ed] to absorb hers altogether." Christina Rossetti was menaced, then, by the male creative power that absorbs Laura in "Goblin Market" and the potential women artists in Cather's stories.

And yet her brother's greatness did not, finally, absorb Christina Rossetti's creativity, Cather thought. Because Rossetti did not imitate male writers but drew on her one gift—the "power of loving"—she produced "natural" and original, if limited, poetry. Her example of female literary autonomy intrigued Cather, who reflected that she must have possessed "considerable individuality of thought and inspiration to withstand the constant influences of so winning a personality and so enchanting a style as Gabriel Rossetti's." Because she possessed a "spiritual ecstasy" that freed her from her brother's influence, in Cather's eyes Rossetti wrote with the "mystic, enraptured faith of a Cassandra."

Her allusion to Cassandra captures Cather's contradictory and ambivalent assessment of Rossetti and, by extension, of a woman writer who drew on her "one gift." The virgin who withstood male sexual power when she refused Apollo, just as Rossetti resisted her brother's influence, Cassandra on one level represents female autonomy. Granted the power of prophesy by the god of poetry, she is also a type of divinely inspired artist. But when she spurned Apollo's advances, the god—who could not take away his gift—negated it with a curse: no one would believe Cassandra's truthful prophecies. Deprived of both authority and audience by Apollo, Cassandra resembles Rossetti as the woman poet disempowered by a masculine aesthetic, one whose love poetry is considered "limited" because it does not address male concerns. Of course Rossetti enjoyed a modest audience; however, in her 1895 review Cather imagines the poet as "one of those in which the laurel had not taken root kindly," and so as inferior to male poets more highly favored by Apollo (*WP* 1, p. 143).

In choosing her epigraphs for *The Troll Garden,* Cather was thus making contradictory statements about her relationship to the male and female literary traditions. On the one hand, the references to Kingsley and Rossetti reflect a gender-based hierarchy of literary influence and show Cather allying herself more strongly with the dominant literary tradition, taking her title from the male writer whose subject was epic: the fall of empire. Cather attributes the reference from *The Roman and the Teuton* to "Charles Kingsley" but does not name Christina Rossetti, as if she were unwilling to honor a female predecessor directly. On the other hand, Cather's reference to Rossetti, when placed in the context of her review, shows her linking herself to a woman writer who managed to maintain her own voice and inspiration while listening to her brother's "enchanting" words, an example of female literary autonomy rather than subordination.

As the reference to Rossetti reveals, Cather knew that she was writing within a female as well as a male literary tradition. The allusion to "Goblin Market" suggests an alternative pattern of literary influence in which a woman writer might be an originator as well as an inheritor, like the supreme artist Treffinger. A female predecessor might be able to help Cather find her own voice in a drama like that played out in "Goblin Market," where one sister rescues the other from male influence. None of the stories in *The Troll Garden* enacts this particular female–female plot—the only reference to a similar story is the brief allusion to the Demeter–Persephone myth, which is the source of male rather than female creativity in " 'A Death in the Desert.' " But Cather would enact this female drama when—three years after *The Troll Garden* was published—she met Sarah Orne Jewett, the literary predecessor who at once provided her with a female literary inheritance and helped to break Henry James' potent spell.

•

As most readers of *The Troll Garden* observe, the stories seem to fall into two groupings: those with Eastern settings, which reflect Cather's apprenticeship to Henry James and anticipate *Alexander's Bridge* (1911), and those with Western or Nebraska settings in which Cather looks back to early stories like "Peter" and "Lou the Prophet" and ahead to *O Pioneers!* and *My Ántonia*. In tracing Cather's literary emergence, however, a more useful division is that between stories in which Cather seems to be speaking in another's voice—whether that of Henry James, Edith Wharton, or of genteel magazine fiction of the day—and those in which she seems to be finding her own cadences.

From this perspective, "Flavia and Her Artists," " 'A Death in the Desert,' " "A Garden Lodge," and "The Marriage of Phaedra" anticipate the Jamesian stories Cather published during her sojourn at *McClure's*—"The Willing Muse," "The Profile," and "Eleanor's House"—as well as *Alexander's Bridge,* while "The Sculptor's Funeral," "A Wagner Matinee," and "Paul's Case" anticipate the major fiction. Cather recognized this division herself. When she drew on *The Troll Garden* in selecting stories for *Youth and the Bright Medusa* (1920), she eliminated "Flavia and Her Artists," "The Garden Lodge," and "The Marriage of Phaedra"; refining the collection even further for the Library Edition, she cut " 'A Death in the Desert' " as well.

As Cather's continuing pleasure in "The Sculptor's Funeral," "A Wagner Matinee," and "Paul's Case" suggests, the important distinction in the collection is not between Western and Eastern settings, since by these standards "Paul's Case" (set in Pittsburgh) would not have been selected. All three stories show a significant development in Cather's fiction: inspired by personal experiences or observations rather than by the work of other writers, they are fictions that nevertheless exist apart from their private sources.

In "A Wagner Matinee," for example, Cather integrates a family story into her fiction: Georgianna and her husband measure off their quarter section of land by tying a cotton handkerchief to a wagon wheel and counting off the revolutions, as had Cather's Aunt Franc and her husband George Cather when they first arrived in Nebraska in the 1880s.[8] In the grim portrayal of Georgianna's Nebraska farm, Cather also drew on her own memories of the difficult transition from Virginia to Nebraska, using her memories of her grandparents' farmhouse in describing Georgianna's bleak environment. Her portrayal of the gifted woman starving for music amid the ''silence of the plains'' also owes something to her own experience of aesthetic deprivation in Nebraska, a feeling intensified retrospectively when she discovered what her prairie education lacked. As she told Will Owen Jones, the distaste she had felt for Nebraska was because she had been only half-nourished there.[9] ''The Sculptor's Funeral'' also had real-life sources: Cather's witnessing an artist's funeral in Pittsburgh and, on another occasion, seeing the return of a Nebraska boy—in his coffin—to Red Cloud.

But there were dangers in turning from admiration to memory: the risk of punishment and retribution Cather had associated with self-exposure in her first college stories. After ''A Wagner Matinee'' appeared in *Everybody's Magazine* in 1904, a year before *The Troll Garden,* Cather faced a Nebraska uprising. The resulting controversy over the story was the ''nearest she had come to personal disgrace,'' she told Viola Roseboro' later.[10] Like ''The Sculptor's Funeral'' and Hamlin Garland's *Main-Travelled Roads,* ''A Wagner Matinee'' portrays the Midwest as a harsh, oppressive, and repressive environment. Family members were particularly insulted by the supposed portrait of Aunt Franc in Georgianna and informed Cather that it wasn't ''nice'' to say such things in print. Friends and neighbors found the grim depiction of Nebraska unfair, and Cather was even attacked in her hometown paper by her old friend and colleague Will Owen Jones. ''If the writers of fiction who use western Nebraska as material would look up now and then and not keep their eyes and noses in the cattle yards,'' he complained in the *Journal,* ''they might be more agreeable company'' (*WWC,* p. 254).

In her reply to Jones, Cather tried to explain that she did not intend her stories to be either exposés of Nebraska crudeness or real-estate advertisements. ''A Wagner Matinee'' was a work of the imagination, she explained, derived from intangible impressions and moods. One day she had received a letter from a Nebraska farm woman in the morning and had attended a Wagner matinee in the afternoon, and the story was all worked out before she left the concert hall. It was true that she had used the farmhouse she had once lived in as a setting, but she didn't set out to paint an accurate portrait of any family member. And besides, she continued defensively, life had been hard during the pioneer days, and she had been misguided enough to think ''A Wagner Matinee'' a respectful tribute to the courage of uncomplaining pioneer women who had undergone such hardships.[11]

At first, Nebraska disapproval did not intimidate her. Cather reprinted ''A Wagner Matinee'' in *The Troll Garden* without modification, but when she included the story in *Youth and the Bright Medusa,* she revised it considerably, softening her portrayal of Georgianna's grotesqueness and Nebraska's grimness. By 1920 Cather's view of her Nebraska past was celebratory rather than hostile (in *My Ántonia* [1918] she had envisioned the farm wife as a triumphant embodiment of creative and procreative power), so she made Georgianna a less distant relative of Ántonia by taking out the more brutal descriptive passages. When she revised ''A Wagner Matinee'' for the Library Edition, she made the story even more palatable for hometown readers, replacing the aunt's poignant final protest—''She burst into tears and sobbed pleadingly, 'I don't want to go, Clark, I don't want to go' ''—with the anticlimatic and accepting ''I don't want to go, Clark. I suppose we must.''[12] While Cather's revisions of ''A Wagner Matinee'' in part reflected her own growing nostalgia for her Nebraska past, they also suggest her bending to one of the pressures she faced in drawing on personal materials: some of her Nebraska readers were sure to confuse art with life and take offense, a difficulty that intensified for Cather as she more frequently began to base her characters on people she had known in Red Cloud.

•

In ''Paul's Case,'' however, Cather found a momentary and happy resolution to the tension between disclosure and concealment, life and art. Placed last (and seventh) in *The Troll Garden,* the story eludes categorization. Although set in the East, it does not romanticize Eastern culture and denigrate Western; a psychological study of a romantic temperament—a subject dear to Henry James—it is not Jamesian. In this beautifully crafted story, the high-water mark of her early fiction, Cather draws upon her observations of Presbyteria and Bohemia, her memories of a gifted and unhappy student in her Latin class, and her own adolescent rebellion. The result is neither unmediated self-disclosure not protective imitation, but a controlled work of art that does seem, as she said of Sarah Orne Jewett's Maine stories, to resemble ''life itself'' (*CPF,* p. 6). Long the favorite of critics, this was Cather's first choice as well, and in later years the only short story she allowed to be anthologized.

''Paul's Case'' demonstrates several literary advances. Cather manages to deal with the regressive components of the creative process—the unassimilated material that weakened earlier stories—by distancing herself from the character whose regressiveness becomes self-destructive; her continuing separation from male identification paradoxically allows her to project herself sympathetically into a male character who is not an unconvincing mask; and she integrates psychological and social worlds so subtly that the main character's rebellion and disintegration becomes a social commentary that is not intrusively polemical. Although the story stands beautifully on its own, it brilliantly synthesizes the themes, images, and metaphors of *The Troll Garden,* referring back to the other stories and to the epigraphs from Kingsley and Rossetti without paying undue deference to literary predecessors. Cather would not match this achievement until ''The Bohemian Girl'' (1912) and *O Pioneers!* (1913).

Cather looks back at an earlier self in ''Paul's Case''—William Cather, Jr.—as she portrays another imaginative, sensitive youth at odds with a repressive society. Paul's enemy is the emotionally and aesthetically bankrupt middle-class world devoted to the gods of money and respectability. Cordelia Street, his dreary home where the ugliness of petty-bourgeois life is symbolized by the ''horrible yellow wallpaper,'' the bathroom's grimy zinc tub and dripping spiggots, and his father's hairy legs and carpet slippers, is Cather's Presbyteria. Paul's alternative world of fairy-tale allure—his troll garden—is the concert hall and theater where he feasts on delicious sensations, a version of Cather's Bohemia. Repulsed by the sordid vulgarity he sees around him, Paul plays the dandy's role to signify his citizenship in Bohemia, sporting an opal pin in his frayed coat and wearing a red carnation as his badge of defiance and kinship with the fin-de-siècle aesthetes and decadents.

In many ways Paul is a male version of Willa Cather. In addition to the adolescent rebellion, she grants him her love for music and theater; her distaste for the mundane and the conventional; and her repudiation of the expected gender role, expressed in the story by Paul's distaste for the Horatio Alger values. Another resemblance may be Paul's probable homosexuality, hinted by his similarity to Oscar Wilde and later by his ambiguous night on the town with a ''wild'' freshman from Yale.[13] And yet Paul is a separate fictional creation whom Cather regards critically as well as sympathetically, as when she suggests that there is something unhealthy in his nature. His eyes have a ''hysterical brilliancy,'' the pupils ''abnormally large, as though he were addicted to belladonna.'' Paul is addicted, we discover, although not to belladonna. He is wedded to the artificial world of the theater and concert hall where he indulges in ''orgies of living'' and ''debauches'' of sensation, like the voracious Laura in ''Goblin Market.''

Paul has to be a glutton in Bohemia because he is famished for spiritual, aesthetic, and emotional food in Presbyteria. The portraits of John Calvin and George Washington over his bed embody the grim, repressive patriarchal values of the national religion of financial success, the Protestant ethic allied with a patriotic capitalism. These beliefs—Paul's paternal inheritance—offer him no sustenance, while the one legacy from his dead mother (the framed motto ''Feed my Lambs'' embroidered in red worsted) offers him only the inadequate nurturance of a feminized Christianity. The nourishment he absorbs during his other life as usher at Carnegie Hall and hanger-on at the local repertory company is evident in the transformation he undergoes in these magical realms: he becomes more ''vivacious and animated,'' and the ''colour came to his cheeks and lips.'' Like Katherine Gaylord, however, Paul is the parasite-vampire greedily drawing life from an external source, and his physical appearance reveals that such food is not wholesome. Like the ''thirsty, cankered, goblin-ridden'' Laura in Rossetti's poem or the dying Katherine, Paul is emaciated, with ''high, cramped shoulders and a narrow chest,'' and despite his youth his ''white blue-veined face'' is ''drawn and wrinkled like an old man's.''

The story is subtitled ''A Study in Temperament,'' and as Cather progressively makes clear, Paul's is not the artistic temperament. Delicately and consistently, she separates her point of view from his. She demonstrates the distorting power of Paul's excitable, greedy imagination, as when she shows us the discrepancy between his vision of an opera singer as a ''veritable queen of Romance'' and the mundane reality: the singer is a tired woman well past her first youth, and singing is just another job for her. In addition, Paul is no connoisseur of music; he can derive the escape he desires from anything from an ''orchestra to a barrel organ.'' Symphonies and operas mean nothing to him as art forms. What he wants is to ''lose himself,'' to ''float on the wave of [music], to be carried out, blue league after blue league, away from everything.''

Although Carter associated such loss of self with the creative process, Paul's self-dissolution is not the artist's desire to create, but the child's regressive yearning to regain the preoedipal union with the mother. He uses music to enter a solipsistic realm where he passively absorbs the sensations he craves. Cather creates parallels to the regressive William of "The Burglar's Christmas," who also sought the infant's gratifications, when she has Paul follow the opera singer to her hotel. As he shivers outside in the cold (as did William in "The Burglar's Christmas"), the singer enters the elegant, welcoming building. Paul accompanies her in his imagination:

> He seemed to feel himself go after her up the steps, into the warm, lighted building, into an exotic, a tropical world of shiny glistening surfaces and basking ease. He reflected upon the mysterious dishes that were brought into the dining room, the green bottles in buckets of ice, as he had seen them in the supper party pictures of the Sunday World supplement.

Like William, Paul gains this idyllic space when he becomes a thief: he steals money from his employers to finance his trip to New York, desperate to enter the world of "basking ease" into which the opera singer has vanished. His week-long orgy of gratification further demonstrates Cather's intention to portray his temperament as regressive rather than artistic. Entering his own expensive hotel, Paul gains the house of his dreams, a protected, expensive world where the air is "deliciously soft and fragrant," magically safe from the winter storm outside. Cather connects the hotel's "enchanted palace" both with the fairy palace of the Trolls and with the tempting, seductive garden of "Goblin Market." Commenting on the artificiality and precariousness of Paul's hothouse world through her use of space and metaphor, Cather connects his womblike hotel room—seemingly a refuge from the howling January blizzard—with a flower stand where "flower gardens [were] blooming under glass cases, against the sides of which the snow flakes stuck and melted; violets, roses, carnations, lilies of the valley—somehow vastly more lovely and alluring that they blossomed thus unnaturally in the snow."

The pleasures Paul seeks during his brief stay in the luxurious hotel also reveal the regressiveness of his journey. Despite his love for music he attends the opera only once, since he can attain his real goal—passive gratification—without music's aid. He spends most of his time sleeping, eating, and drinking. His most enjoyable waking moments are those passed in the sitting room, where he drowsily contemplates his delicate flowers, elegant new clothes, and gracious surroundings, or in the hotel dining room where he tastes unfamiliar delicacies and savors champagne.

When the theft is discovered and his money runs out, Paul envisions returning to Pittsburgh—its Sabbath schools, yellowing wallpaper, and cooking smells—in images of drowning, as he senses the "tepid waters of Cordelia Street" about to "close over him finally and forever." He finds this loss of self as terrifying as his floating "blue league after blue league" on the billows of gratification had been delectable. So he chooses to lose himself in death, a decision Cather links metaphorically and psychologically with all the forms of self-dissolution he has pursued. After the artificial garden disappears, Paul's quest for annihilation leads him to a snow-covered wasteland beside the train tracks. He throws himself in front of the train—as had Anna Karenina, whose romanticism Cather also criticized—and "dropped back into the immense design of things" (*CSF*, pp. 243–61).

The differences between "The Burglar's Christmas" and "Paul's Case," both stories where the protagonists seek a seductive, self-destructive nurturance, reveal the growth in Cather's art in the intervening decade. The first is a dreamlike fantasy, the second a rich and complex work of fiction. Cather's narrative technique reveals her psychological and aesthetic distance from her character, an impressive achievement since, as she acknowledged, Paul represented part of herself. Whereas "The Burglar's Christmas" suggests that the author shares William's blissful, self-annihilating desire to return to his mother, the omniscient narrator in "Paul's Case" has sufficient detachment to expose the limitations of his fantasies, while also possessing enough sympathy with them to suggest their appeal.

Although "Paul's Case" does not comment directly on the woman artist's plight, as do several of the other stories, it does show Cather's condemnation of conventional gender arrangements. Paul's alienation is connected with the polarized sex roles that Willa Cather had also confronted in adolescence; he is William Cather, Jr.'s opposite and double, as much in conflict with the male role as she had been with the female. Cather's sympathetic portrayal of this "girl–boy" who resembles Ethelbert Nevin reveals how far she was moving from male identification. Even though Cather was still subordinating herself to the male literary tradition, she had rejected the cult of virility once exemplified by Kipling and football. Androgynous or feminine men could be oppressive to women, too, as in " 'A Death in the Desert' " where Adriance Hilgarde consumes Katharine Gaylord's life and imagination. And yet Cather's ability to create Paul, who is destroyed by the male world she had once wanted to enter, shows her awareness that men as well as women could be victims of masculine—and heterosexual—roles and values.

•

If Cather had expected her life to be transformed when *The Troll Garden* appeared in print, she was disappointed. Most reviewers were encouraging, but not overly impressed. The *Atlantic Monthly*—the voice of a waning New England literary establishment—could not make up its mind, finding the stories "fresh, unhackneyed," yet concluding that no "new ground has been broken." The *Critic* saw "real promise" in *The Troll Garden,* and yet if Cather were looking for direction, the *Critic* would have her writing more Jamesian stories; "Flavia and Her Artists," the reviewer concluded, was "the best of the group." The *Dial,* on the other hand, thought "The Sculptor's Funeral" quite as vivid as "Flavia," and overall praised Cather for her "truly entertaining" stories of "abnormal" people, ambiguous praise for a writer who did not intend to portray abnormality. *The Bookman* was not entertained by a "collection of freak stories that are either lurid, hysterical or unwholesome" and criticized Cather for rummaging in the "ash-heap of the human mind."[14]

But even the praise Cather gleaned in these reviews did not compensate for Nebraska's outrage. *The Troll Garden* had been a "bitter disappointment," she later told an interviewer. Few people bought it, and her Nebraska friends could find no words bad enough for it: "They want me to write propaganda for the commercial club." Cather had put more of herself into *The Troll Garden* than into *April Twilights,* and hence was more vulnerable to attack:

> An author is seldom sensitive except about his first volume. Any criticism of that hurts. Not criticism of its style—that only spurs one to improve it. But the root and branch kind of attack is hard to forget.[15]

Having suffered this attempted uprooting by her Nebraska critics, for the next few years Cather was wary of returning to her native soil for literary inspiration.

Although Cather went back to Allegheny High School in the fall of 1905 as the author of two published books, she had a more muted sense of her literary glory than she had enjoyed in the spring of 1903, when McClure had promised her an exciting future. But the "Chief" soon entered her life again with another offer: he wanted her to work for him. In the spring of 1906 McClure was confronted with a mass exodus. Almost the whole staff, including his prized writers Lincoln Steffens and Ida Tarbell, walked out to form a rival publication, and McClure needed to rebuild his magazine. Losing no time, he went to Pittsburgh in May, had dinner with Cather at the McClungs' mansion, and convinced her to return to journalism. She left Pittsburgh in the summer of 1906, headed for New York and the next stage of her apprenticeship.[16]

NOTES

1. James Woodress speculates that Cather "probably" heard the story from her parents or grandparents (*Willa Cather: Her Life and Work* [New York: Pegasus, 1970; rpt. Lincoln: University of Nebraska Press (Bison ed.), 1975], p. 28). Whether or not she heard this particular story, Cather was drawing on the larger configurations of her family story, the emigration from Virginia to Nebraska, and the resulting gap between past and present that could only be bridged by storytelling.

2. Willa Cather to Will Owen Jones, May 7, 1903, Barrett-Cather Collection, University of Virginia Library, Charlottesville, Va.

3. For analyses of the volume's thematic patterns, see E. K. Brown, *Willa Cather: A Critical Biography,* completed by Leon Edel (New York: Alfred A. Knopf, 1953), pp. 113–15; David Stouck, *Willa Cather's Imagination* (Lincoln: University of Nebraska Press, 1975), pp. 175–81; James Woodress, "Introduction" to *The Troll Garden,* ed. James Woodress (Lincoln: University of Nebraska Press, 1983), pp. xvi–xxiii).

4. See Sandra M. Gilbert and Susan Gubar, *The Madwoman in the Attic: The Woman Writer and the Nineteenth-Century Literary Imagination* (New Haven: Yale University Press, 1979), pp. 564–71.

5. In " 'Come Slowly—Eden': An Exploration of Women Poets and Their Muse," Joanne Feit Diehl examines the difficulty Rossetti and Barrett Browning had in attaining literary independence from literary influence; she argues that the precursor, the muse, and the father—all male--became identified as a "composite father," their "main adversary" (*Signs: Journal of Woman in Culture and Society* 3, no. 3 [Spring 1978]: 572–87, esp. 574). Although she confessed herself the "bond slave" of Housman, because Cather's

muses were women she was less captured by this aesthetic/erotic drama in which women only play supporting roles.

6. Cather returns to the figure of the dying female singer in "Double Birthday," again giving her a symbolic disease: cancer of the throat (*UV*, pp. 41–63).

7. For a discussion of the ways in which women are erased as creative subjects in the male imagination, see Susan Gubar, " 'The Blank Page' and Female Creativity," *Critical Inquiry* 8, no. 2 (Winter 1981): 243–63.

8. Mildred Bennett recounts the family story in *WWC*, p. 12.

9. Cather to Jones, March 6, 1904.

10. Quoted in Woodress, p. 117.

11. Cather to Jones, March 6, 1904. "The Sculptor's Funeral" was also severely criticized by Nebraska readers, who thought Cather's negative portrayal of their homeland distorted and unfair (her setting the story in Kansas evidently did not fool anybody). She defends herself in a letter to Kate McPhelim Cleary, February 13, [1905], Willa Cather Pioneer Memorial and Educational Foundation, Red Cloud, Neb.

12. A typical omission is "Her skin was yellow as a Mongolian's from constant exposure to a pitiless wind, and to the alkaline water, which transforms the most transparent cuticle into a sort of flexible leather." For a complete record of the variants, see Woodress, pp. 145–73.

13. See Larry Rubin, "The Homosexual Motif in Willa Cather's 'Paul's Case,' " *Studies in Short Fiction* 12 (Spring 1975): 127–31.

14. *Atlantic Monthly*, January 1906, p. 46; *Critic*, November 1905, p. 476; *Dial*, June 1, 1905, p. 394; *The Bookman*, July 1905, pp. 456–57.

15. Interview with Eleanor Hinman, *Lincoln Sunday Star*, November 26, 1921, quoted in *WWC*, p. 197.

16. Helen Cather Southwick offers a different timetable, contending that Cather took a leave of absence and left school in the middle of the second semester to work for McClure on a special assignment. When his staff walked out, she was right there and he hired her immediately ("Willa Cather's Early Career: Origins of a Legend," *Western Pennsylvania Historical Magazine* 65, no. 2 [April 1982]: 85–98, esp. 94–95).

13

Paternal Voices

I have been greatly pleased with your letters and your work, as of course you know. S. S. McClure to Willa Cather, June 18, 1909

I find it the hardest thing in the world to read almost *any* new novel. Any is hard enough, but the hardest from the innocent hands of young females, young American females perhaps above all. Henry James to Witter Bynner

Willa Cather's years at *McClure's* as staff writer and managing editor from 1906 to 1911 were, on the surface, years of heady accomplishment. Within a short time she became a powerful figure at one of the nation's best-known and most respected periodicals and infiltrated the New York, Boston, and London literary worlds. Success at *McClure's* meant success according to the rules American society laid out for men, and indeed Elizabeth Sergeant's first impression was of a powerful, expansive masculine presence who filled the "whole space between the door and window to brimming, as a man might do" (*WC:AM*, p. 33).

Cather's pride in her professional accomplishments increased during these years, yet she also suffered from growing self-doubts. Eager both to prove herself as a journalist and to advance as a writer of fiction, Cather found her two professional goals hostile rather than complimentary. With her time and energy consumed by her responsibilities at the magazine, she wrote little; most of the stories she published reveal even heavier traces of Henry James' influence than those in *The Troll Garden.* Cather knew that she was not progressing as a writer. Following the direction of "The Marriage of Phaedra" rather than that of "Paul's Case," she was, as she later observed, trying to "sing a song that did not lie in my voice."[1]

Despite the powerful, assured presence she conveyed to Elizabeth Sergeant, during her *McClure's* years Cather was the deferential daughter to two fathers—S. S. McClure and Henry James. Anxious to please and to emulate these imposing figures, Cather was not singing in her own voice because she was listening so intently to theirs; whether she was following McClure's editorial directives or imitating James' style, she was striving to say and write what she thought others wanted to hear. Both men were mentors from whom she learned much and who contributed to her personal and professional growth. Yet they were also inhibiting figures from whom she would eventually separate as she learned to speak in her own cadences.

•

Attempting to convey S. S. McClure's mesmerizing impact after their first 1903 meeting, Cather told Will Owen Jones that if he were a religious leader people would go to the stake for him. McClure was turning the clock back, Cather told Jones, making her feel as important as she had in her Nebraska heydey when she was editing the *Hesperian*. He even seemed interested in planning the smallest details of her personal life. Since his vision of her was so flattering, she was willing to be convinced, she admitted; this meeting had changed her life.[2]

On that glorious day in 1903 when McClure took charge of her literary life, Cather was contemplating only what she would gain from this relationship. But a mentor who wants to plan the smallest details of his protégée's career can be an oppressive father who may not grant his daughter a separate identity; similarly, a daughter who imagines going to the stake for a fatherly leader is displaying a disturbingly self-destructive allegiance. In S. S. McClure Cather found not only a professional advocate but also a controlling parent who dominated her life for the next few years. When McClure removed Cather from her Pittsburgh obscurity and made her first staff writer, then managing editor of his magazine, he seemed to be offering her the male world of power and authority. Yet even as managing editor Cather was enacting a version of the traditional female role, trying to fulfill a father's expectations. The complex psychological and emotional drama she and McClure were engaged in prolonged her apprenticeship.

In striving to please McClure, in some ways Cather gained a greater sense of her own powers.[3] In 1903 she had felt rejuvenated by McClure's sorcery, and Elizabeth Sergeant's account of her first meeting with Cather in 1910 suggests that this renewal endured. Instead of the middle-aged editor she was expecting, Sergeant saw a "youngish, buoyant" individual who displayed a "boyish, enthusiastic manner" when she heartily shook hands and led the younger woman into her office. As Cather assumed her editorial role, the boyish woman exhibited an "almost masculine personality." Conjuring up "crude oil, red earth, elemental strength and resoluteness," Cather took command and seemed to consume Sergeant's manuscript: her eyes "swam over my paragraphs with hummingbird speed," Sergeant recalled, "her square-tipped fingers flipping the pages of 'my' manuscript—now hers" (*WC:AM*, pp. 33–34).

Sergeant's description of the powerful Cather complements Edith Lewis' observation that McClure's approval gave Cather a "sense of sureness, a happiness of expectation she had not felt before" (*WCL*, p. 69). Certainly Cather's position at *McClure's* brought her great advantages: she was able to publish several stories in the magazine, at times, some staffers thought, without consulting either the fiction editor (then Viola Roseboro') or McClure himself; she traveled to England to recruit new contributors and keep contacts alive with others, meeting such luminaries in the publishing and literary worlds as Ford Madox Ford, H. G. Wells, William Archer, and Lady Gregory; she bought fiction from established American writers like Jack London and Theodore Dreiser and encouraged unknowns like Zoë Akins, who eventually became one of her closest friends. Cather took pride in these editorial triumphs. When she had "pulled off" a coup—like publishing

Saint-Gauden's *Familiar Letters* or Ellen Terry's memoirs—she "felt the job was worth all it cost her," she told Sergeant (*WC:AM*, p. 47).

And yet if these were years of increased power, they were also, as Lewis reminds us, times of "sudden terrifying illnesses" when the dynamic potentate who overwhelmed Sergeant would be weak and bedridden for days (*WC:AM*, p. 40; *WCL*, p. 76). A bout with mastoiditis left Cather in the hospital for over a month early in 1911, and in her letters to McClure, she often apologizes for illnesses that kept her from working.[4] These recurrent illnesses were signs of psychological and emotional distress which Sergeant did not glimpse. Although she struck others as energetic, confident, even domineering, Cather progressively felt depleted, exhausted, and unsure of her literary talents. Although at her first meeting with McClure in 1903 she thought he was reinvigorating her youthful dreams, after two years of working at the magazine she felt that she was prematurely aging, her visions of literary accomplishment unrealized. The covert tensions in her personal and professional relationship with "the Chief" had something to do with this reversal.

Journalism's self-made man and the impresario of muckraking, McClure was a dynamic, charismatic editor who had captured and captivated many co-workers before Willa Cather, including Lincoln Steffens and Ida Tarbell. "Everyone about him caught fire," remembers Ellery Sedgwick. "He would inflame the intelligence of his staff into a molten excitement."[5] Edith Lewis, who also worked at the magazine, likewise describes McClure as a volcanic source of energy who spewed forth ideas like "showers of sparks" and ignited others' imaginations (*WCL*, p. 60). Volatile and erratic, McClure could be a hard man to work for. He created chaos as well as enthusiasm as he stormed around the office, brimming with new ideas which he wanted enacted immediately. At times writers who were particularly desperate to finish stories would squirrel themselves away in rented hotel rooms, safe from McClure's disruptive presence.[6] McClure's defects were "so egregious," Ellen Moers observes, that most of his writers abandoned him after a short time.[7] McClure soon rebounded from such blows, however, as when he hired a new staff after the 1906 walkout.

McClure's relationship with Willa Cather was not marred by the friction that frayed his other professional ties, observers thought. "Only Miss Cather, secure in her respected genius, remained apparently unmindful of office thunderstorms," Sedgwick recalls.[8] Sergeant describes Cather and her mentor as "partners in an alliance" whose "seething inner forces supplemented each other" (*WC:AM*, p. 39). Edith Lewis, who had the opportunity to observe their professional relationship more closely, likewise characterizes it as harmonious (*WCL*, p. 70).

Certainly the association was profitable for McClure. In contrast to writers who kept evading and deserting him, Cather was admiring, steadfast, and loyal; the characteristic closing she employs in her letters to McClure—"Faithfully yours"—captures her feelings for her Chief. Because he "believed absolutely" in Cather's integrity and commitment, Lewis recalls, McClure found her a refreshing "oasis" in the desert of others' "vanities and jealousies and ambitions" (*WCL*, p. 71). Unlike Ida Tarbell, whom McClure had also discovered and encouraged, Cather was the daughter who would not leave home.

McClure genuinely admired Cather's editorial abilities. Writing his associate during her 1909 trip to London, where she was recruiting new writers, McClure regretted that he could not be in New York to welcome her back from her "splendid and successful" journey and assured her that he was "greatly pleased" with her work. Writing the next day, McClure was equally laudatory, telling Cather that he was "awfully proud" of her "splendid work" in England. McClure considered them a team equally dedicated to rebuilding the magazine; when Cather returned from Europe, he told her, they had to be "ready for *our* great work" [italics mine]. Her interests, he assumed, were identical with his, and he expressed his confidence in her abilities and her loyalty when he made her managing editor.[9]

McClure's respect and praise were important to Cather, particularly during her first uncertain months with the magazine. Her uprooting from Pittsburgh and separation from Isabelle McClung must have been painful, and she had scarcely settled in New York—taking a studio apartment at 60 South Washington Square, where Edith Lewis also lived—when she left for several months in Boston. McClure had given her an important job: supervising Georgine Milmine's controversial series, later published as *The Life of Mary Baker Eddy and the History of the Christian Science Church.* The series began publication in January 1907 and appeared at intervals until June 1908. There had been a mistake in the first article, and McClure wanted there to be no more errors if he was going to take on Mary Baker Eddy and Christian Science. The piece needed revising and fact checking. Cather made Boston's Parker House her headquarters (later she moved to an apartment on Chestnut Street) and spent most of 1907 and part of 1908 researching, interviewing, rewriting, and traveling around New England. Faced with such a major assignment, Cather relied upon McClure's trust. Writing him from her new post, she confessed that she had taken his letter along to give her courage; she had read it over many times, she confided, and found it helpful that he valued her and took such an interest in her work.[10] Her work met his expectations, and he promoted her to the editorial desk after her return to New York.

All the letters Cather wrote to McClure—both during and after her tenure at the magazine—support the impression that she gave to Lewis, Sergeant, and other observers: that she considered theirs a happy professional marriage. They reveal her delight in their shared purpose, her eagerness to please him, her gratitude for his confidence.

But McClure's interests were not identical with hers. The publisher's "great work" was rebuilding his magazine, Cather's her dream of writing fiction; these were incompatible goals. Initially, Cather may not have seen this contradiction since she liked to imagine McClure as a crusader for her religion of art. "He reveres genius," she told Sergeant, in an "intimate, admiring tone." "He's lost more money on Joseph Conrad than any editor alive!" (*WC:AM*, p. 40).[11] McClure may have been willing to gamble on Conrad, but he needed Cather first as his staff writer and then as his managing editor; the energies that he had revived he wanted expended in those roles. Pleased with Cather's work on the Eddy series, he kept urging her to do more pieces like it, even though, as she confided to Sergeant, she found investigative reporting somewhat "*infra dig*" (*WC:AM*, p.

55). After he hired Cather, Lewis recalls, McClure's enthusiasm focused "more and more on factual material" (*WCL,* p. 63). Writing Sarah Orne Jewett, whom she met in 1908 at Mrs. Fields' Beacon Hill home, Cather confided that McClure did not want fiction from her but good, clear-cut journalism—articles on popular science and the like. What he wanted, she suspected, was to transform her into as good an imitation of Ida Tarbell as he could.[12]

Cather's observation that McClure associated her with Ida Tarbell was perceptive. Her impression is borne out by a letter McClure wrote her in the fall of 1912, when she was staying with Isabelle McClung in Cherry Valley, New York, and revising *Alexander's Bridge*. McClure fully expected that she would return from this fiction-writing venture to share in his great work once more: Cather should "dismiss the magazine from [her] mind entirely," McClure wrote, and when she came back she should not get "tied up in office machinery" so that she could "work out a series of articles that will give us distinction."[13] McClure was reenacting his relationship with Tarbell, hoping for a happier ending to this second version of the Pygmalion story. He had plucked both Galateas from relative obscurity and placed them in important positions. Yet he wanted them to accept the script he had authored, remaining trustworthy and subordinate. Tarbell had rebelled, but Cather was going to be faithful; as Sergeant recalls, Cather regarded Tarbell's walkout as disloyal to McClure (*WC:AM,* p. 39). Yet as long as she was trying to be the Chief's new Ida Tarbell, Cather faced the dilemma she had portrayed in " 'A Death in the Desert' " where the female protagonist loses a separate identity as she is absorbed and defined by the man she wants to serve.

The editorial tasks that Cather performed for McClure encouraged self-effacement. As the managing editor she was seemingly occupying a powerful role, controlling the writers whose work she was soliciting, rejecting, accepting, and editing. But in another sense, like the female characters in her *Troll Garden* stories, Cather was midwife to others' creativity, allowing others' voices to speak through her. "Every artist makes himself born," Cather commented in *The Song of the Lark,* and as we have seen in 1900 she signified her desire to be "born" as an artist when she became "Willa Sibert Cather" (*SL,* p. 175). The name on the title page signified possession as well as self-expression to Cather; one of the wonderful things about a book, she told an interviewer, was that it was "all yours."[14] But after Cather worked for months on the Mary Baker Eddy series, massively rewriting a poorly structured manuscript, it appeared in the magazine under Georgine Milmine's name. When Cather appeared pseudonymously in the *Home Monthly* and other Pittsburgh publications, the choice to conceal and the means of concealment were hers. By contrast, the editor's anonymity signifies the way in which her work at *McClure's* seemed to be absorbing her identity along with her energy: as Cather confessed to Jewett, her editorial work was leaving her bereft and dispossessed of herself.

Cather's 1908 letter to Jewett reveals a discouraged and exhausted woman. Whereas Sergeant saw Cather embodying "sheer energy," Cather herself felt like a broken circuit from which all the energy was slowly ebbing. She told Jewett that she was a trapeze artist trying to catch the right bar, a comparison suggesting the intense concentration she had to muster on the job and the deadly consequences awaiting failure. With her energy absorbed by work she didn't want to be doing, after a day in the office she simply didn't have the resources to write fiction, she told Jewett, and she knew that she had not progressed artistically since *The Troll Garden*. She was not thirty-four years old and should be a better fiction writer than she was, she confided, but when she tried to write a story she felt like a newborn baby every time—an image which in this context connotes helplessness and ignorance rather than youthful confidence or imaginative power.

Although Cather had felt strengthened and rejuvenated after her 1903 meeting with McClure, after a few years at the magazine she feared that she was aging while her literary abilities were declining. Cather did not hold McClure responsible for her loss of creative energy. Still her loyalty to him was one of the ties binding her to her draining job, and if we read between the lines, McClure emerges as an almost Svengali-like figure who wants to keep his editor/daughter dependent on him. He accomplished this goal by simultaneously flattering Cather as a journalist and undermining her confidence as a fiction writer. At first McClure had been impressed by Cather's literary potential, but once he decided that he wanted her to be his editor, he began dismissing her literary gifts as insubstantial.

McClure had told her, Cather revealed to Jewett, that she would never be much good at writing stories: she was a good magazine executive and should be satisfied with that. Cather's insecurities were naturally roused by McClure's doubts, just as her self-esteem had been strengthened by his approval: he was her mirror. Unable to see how McClure's self-interest might be served by his criticisms, she confessed to Jewett that she often thought he was right. She knew that she was a good editor, but that only required application and discipline; being a better writer of fiction required something more, but she didn't know what that was or where to find it. And she hadn't learned anything about writing since *The Troll Garden,* she confessed. How could she possibly be destined to do something at which she was so inept?

Although Cather did not consciously feel that McClure was profiting at her expense, her letter to Jewett reveals that on some level she experienced theirs as a parasitic rather than a mutually enabling and supportive bond. McClure flourished on journalism, she told Jewett, but she felt as if she couldn't breathe; writing and editing articles offered her about as much food as arithmetic would.[15] This imagery of starving and suffocation recalls " 'A Death in the Desert,' " where a man's creative energy also waxes while a woman's wanes. Lewis and Sergeant saw nothing sinister or debilitating in the relationship, but it is interesting that both mention Cather's role in rejuvenating McClure. Her friend's eagerness restored McClure's "youthful excitement and pride in being an editor," Lewis thought, while Sergeant saw an "inspirational quality about the dynamic unspoiled assistant that kept the older editor afloat on his sea of discovery" (*WCL* p. 71; *WC:AM,* p. 39).

If McClure did feel his youth returning in Cather's invigorating presence, he returned the favor, and perhaps assuaged some guilt, by encouraging her to turn the clock back as well. When she was filling out her entry for the 1909 edition of *Who's Who,* McClure suggested that she cheat: "Willa, let me give you some advice. Knock a year or two off your age—you'll be glad later on." So in 1909

Cather listed her birth date as 1875. In 1919 she subtracted another year to make it 1876, which remained the official birth date until Leon Edel uncovered the fraud while completing E. K. Brown's biography. Edel attributes Cather's motivation to feminine vanity ("She shrank . . . from the flight of youth, the ravage she saw in her looking glass"), but given the timing—shortly after Cather's depressed letter telling Jewett that her literary abilities had not kept pace with her age—it seems more likely that her failure to progress as a writer prompted this futile deception, which was more directed toward herself than to the anonymous readers of *Who's Who.*[16]

•

The likelihood that Cather's loyalty to her mentor as well as her work were eroding creative energies is even more strongly suggested by "The Willing Muse" (1907). Although in 1918 Cather would paint a straightforward portrait of McClure as O'Malley, the dynamic and erratic publisher of a "red-hot magazine of protest" in "Ardessa," the psychological and emotional currents linking her to the Chief are present covertly in "The Willing Muse" (*UV*, p. 101).

As in her other 1907 stories, here Cather employs Henry James' stylistic mannerisms and sketches familiar Jamesian types—the pure artist, the vulgar and successful commercial writer, and the reflective commentator. The story's subject is a parasitic, vampirelike relationship in which one partner thrives at the other's expense. This was also one of James' recurring interests, having just received its fullest treatment in *The Sacred Fount* (1901). But since symbiotic and parasitic relationships in which energy and identity are exchanged or absorbed also preoccupied Willa Cather, this seemingly derivative story expresses urgent personal concerns.

Parallels to Cather's situation abound. Kenneth Gray is a follower of her religion of art, pursuing the writer's holy "calling" in a sheltered Midwestern Parnassus, "Olympia, Ohio." Like hers, his was a long apprenticeship; his friends expected great things, but he did not produce his first book until he was thirty-five. His friends fear that "there ha[s] been a distressing leakage of power." Gray's second book, however, is more promising, and his friends believe that his marriage to Bertha Torrance will unleash more productivity.

But his marriage brings an even greater "leakage of power." Gray writes no fiction. Even his letters are "astonishing in their aridity." Meanwhile, his wife, a writer of popular historical fiction, undergoes a creative explosion that surpasses all "legendary accounts of phenomenal productiveness." Bertha achieves popular and commercial success by drawing on the historical past, the same material that inspired Gray's two aesthetically pure books, and thus she also draws on him ("There was every evidence that she had absorbed from Kenneth like a water plant"). As in *The Sacred Fount* or " 'A Death in the Desert,' " physical appearance mirrors the exchange of energy. When the narrator sees Gray after his four-year dry spell, he seems much older while Bertha is "taller, straighter, younger than I had left her—positively childlike in her freshness and candor." Bertha is not really to blame for Gray's creative aridity, we learn, for her husband conspires in causing his own silence. A "willing" muse and a willing secretary, he busies himself reading the manuscripts aspiring writers send her because he fears committing himself to his own work.

The story ends abruptly and unconvincingly. Gray mysteriously disappears, and we finally discover that he has left for China without informing wife or friends. The Jamesian observer tells us why: "His future was chalked out for him, and whichever way he turned he was confronted by his inescapable destiny"—being drained by his productive wife. Since Bertha's absorption of his will and energy was too subtle a phenomenon to be described, and since he partially conspired in the process, he couldn't be "churlish" and confront her. Gray's abrupt departure is a desperate grab at self-preservation (*CSF*, pp. 113–24).

That Cather projected her own fear of losing creative energy to Kenneth Gray is suggested by the many links between author and character—the Midwestern origins, the early promise, the two books, the supposedly promising change in life (his marriage, her job)—as well as by the letter to Jewett in which she echoes the story's themes and its imagery. Bertha in one sense represents all the people and the tasks absorbing Cather's energy and taking over her identity, all the writers for whom she was playing the muse's helping role. Yet there is a specific connection between the flourishing Bertha and the rejuvenated McClure, whose popular magazine was thriving, as Cather saw it, on her ebbing vitality. Certainly Cather's transforming this important personal and professional alliance into a marriage makes emotional sense. Given her loyalty to McClure, leaving her job would have been an act of betrayal, equivalent to deserting a spouse—or separating from a parent. In Kenneth Gray's unlikely flight to China we may then be seeing Cather's inability to imagine a satisfactory resolution both to the character's dilemma and to her own conflict between her commitment to *McClure's* and to her writing. Like Gray, Cather was in part tied to the uncongenial work of which she complained to Jewett not only because she needed to earn her own living, but also because she feared risking herself as a creative writer. Journalism might be draining, but it was safe.

Of course "The Willing Muse" and the letter to Jewett do not tell the only story of her relationship to McClure and the magazine. The excitment and gratitude she expressed in her letters to McClure, the pride and power she conveyed to Sergeant, were not false. In part, Cather was thrilled to be caught up in this competitive, fast-paced, male-dominated world.[17] Hence she might have been reluctant to leave McClure and the magazine if external forces had not intervened. As it happened, *McClure's* ran into financial difficulty and was taken over by new owners who eventually forced McClure out, so Cather could quit her job without having to abandon the Chief. She eased the separation process by taking a leave of absence in the fall of 1911, while McClure was still in command.

Cather remained unwilling to contradict or disappoint McClure, however, as their ensuing correspondence reveals. Even after he left his post, McClure kept hoping that Cather would collaborate with him on one journalistic project or another, and she characteristically put him off with excuses and delays rather than direct refusals. The two times when she could not avoid saying no, she used Mrs. McClure as intermediary, asking her to inform her husband that she was unavailable. Evidently this subterfuge kept the peace, for the two kept up a cordial relationship over the years although they saw each other infrequently. During the 1930s, when the elderly McClure was nearly destitute, Cather joined other ex-staffers (including Ida Tarbell) in establishing a fund for his support, and they had one last meeting in 1944 when both were honored by the National Institute of Arts and Letters. The only reference she made to difficulties on the job is an indirect one. Writing Ferris Greenslet in 1921, trying to console him for her departure from Houghton Mifflin to Knopf, Cather said that they might be better off once she had left; she often told Mr. McClure, she wrote, that they had not become the best possible friends until they were no longer business associates.[18]

Cather did gladly take on one of McClure's literary projects, however—ghost-writing his autobiography in 1912. By then she was feeling far more confident in her literary future than she had been in 1908; she had written *Alexander's Bridge*, "The Bohemian Girl," and part of *O Pioneers!* and felt that she had finally unleashed her creative powers. Cather received his letters requesting her help during her rejuvenating sojourn in the Southwest, and in the same letter in which she agreed to help with the autobiography she told McClure how rested, strong, and good-humored she felt. Although she hadn't written a line since she had left New York, she had such a headful of stories that she dreamed about them at night. So when she returned to New York, her head filled with her own stories, Cather could enjoy writing McClure's story.[19]

In one sense, Cather's ghostwriting dramatizes her continued subordination to her charismatic leader, duplicating the editor's loss of self she had experienced in rewriting Georgine Milmine's series. The book was published as *My Autobiography* by S. S. McClure, Cather's substantial contribution seemingly invisible. McClure acknowledged her role in its completion ("I am indebted to the cooperation of Miss Willa Sibert Cather for the very existence of this book"), but his name was on the title page.[20]

Yet from another perspective, one which I think more accurately captures the meaning of this collaboration to Cather, she was reversing their roles, taking over McClure's identity and voice. "He used to come down to 5 Bank Street," remembers Edith Lewis, and "walk up and down the room, talking it to her. I do not think she made many notes; when he had gone, she would write down what he had said" (*WCL*, p. 71). So McClure told Cather stories from his past, but she created the written text, retaining final control over the way in which he would be presented to, and remembered by, the public. Quite likely she was flattered by his dependence on her; the great crusading journalist, the dictator who commandeered so many writers, was acknowledging that she was the better writer, relying on Cather to tell the one story he should have been able to write himself.[21]

Later Cather described writing McClure's autobiography as a training ground that helped her to develop the novelist's sympathetic imagination. In a 1919 letter to Will Owen Jones, she defended her use of a male narrator in *My Ántonia* (a risky step for a woman writer to take, she contended), by saying that she had practiced assuming a male persona in McClure's autobiography. At first she had found it hampering to be Mr. McClure all the time, she went on, but finally she found it fascinating to work within the limits and color of a personality she knew so well. Finding a strengthening reversal in taking on an identity rather than being taken over by one, Cather controlled language and voice so effectively that the book impressed others as revealing McClure's characteristic turns of phrase. She was proud of this literary ventriloquism: she had managed to duplicate McClure's language so effectively, she boasted to Jones, that Mrs. McClure and Mr. Phillips (McClure's business partner) had told her that her sentences conveyed the man's characteristic abruptness and suddenness.[22] Here is Cather's description of McClure's unlikely beginnings, a good example of the book's plain style:

> During the first month I lived on bread and grapes varying this with soda crackers and grapes. I could get three pounds of delicious Concord grapes for ten cents. I found an unrented room in the dormitory building, perfectly empty except for a box. In this room I studied that first month, and there I ate my meals, keeping my supplies hidden under the box. I've forgotten how I arranged for a place to sleep. (p. 65)

•

In taking on the voice of this no-nonsense, straight-talking Midwestern man of business, Cather may have accomplished another feat: exorcising the literary mentor whose involuted, cultured, Eastern voice was absorbing hers during the *McClure's* years. If "The Willing Muse" suggests that Cather was muse to McClure, it simultaneously demonstrates that she was the willing disciple of Henry James, whose influence pervaded her *Troll Garden* stories.

In the 1920s the established author somewhat overstated the apprentice writer's debt to James. Perhaps Cather found it pleasurable to exaggerate the extent of her early servitude once her independence was secure: "In those days," she reminisced, "no one seemed so wonderful as Henry James; for me, he was the perfect writer." As she acknowledged, her early indebtedness to James extended beyond admiration to imitation. "All students imitate," she commented in 1925, and "I began by imitating Henry James. He was the most interesting American who was writing at the time, and I strove laboriously to pattern after him." This was, she continued, "a perfectly right form of education."[23] She described her younger self as "dazzled" by James; his was the "foremost mind that ever applied itself to literature in America," Cather told an interviewer, and she wanted to emulate his "best style."[24]

But Cather did not begin her literary career, as she later claimed, by imitating James. In the 1920s she misrepresented the timing of the apprenticeship as well as its extent. She had begun her literary career by writing stories of Nebraska farmers—"Peter," "Lou the Prophet," "The Clemency of the Court." After her college years, Cather's dependence on James did not steadily increase but fluctuated, waning in the best of the *Troll Garden* stories, waxing in the four stories she published in 1907—'The Willing Muse," "The Profile," "Eleanor's House," and "The Namesake"—vanishing in "The Bohemian Girl," and returning in *Alexander's Bridge* before being exorcised in *O Pioneers!* Since we cannot take Cather's later statements as an accurate guide to the young writer's experience of literary influence, we need to look at her literary relationship with James as it developed in the 1890s and early 1900s.

In the 1890s Cather was not always "dazzled" by James. She characterized him as "our only master of pure prose" in a college essay, yet in reviews of his work her tone is not worshipful but respectful and, at times, critical (*KA*, p. 382). Commenting on *The Lesson of the Master* (1892) and *Terminations* (1895), she found the short fiction "perfect" but "sometimes a little hard, always calculating and dispassionate" (*KA*, p. 361). Like most critics and theater-goers, she dismissed his plays, calling them as "laughable as plays by the property man" (*KA*, pp. 238–39).

It is understandable why Cather the *reader* would not have been strongly drawn to James during her college years. Then she was most entranced by those romantic artists who evoked "volcanoes, and earthquakes, and great, unsystematized forces"—not the imagery we would associate with Henry James. She identified with powerful, autonomous heroes who possessed the "boundless freedom" to which she aspired (*KA*, p. 317). Viewing the romantic exaltation of the self as naive, James was challenging the ideology of individualism from which Cather was deriving her narrative of self-creation. His characters inevitably face the limitations Cather wanted to transcend. In chastening Isabel Archer's presumptuousness and romanticism in *The Portrait of a Lady,* for example, James furnished his heroine with a female destiny Cather would not have found appealing. She wanted more than an expanded, enriched consciousness reached through suffering—she demanded the heroic accomplishments she found in romance.

As a beginning *writer,* however, Cather began to take James more seriously. Shortly after writing Mariel Gere in 1896, proclaiming that art was her religion, Cather wrote the first story in which James' influence is obvious—"The Count of Crow's Nest" (1896). "At that time," she observed in "Miss Jewett," James "was the commanding figure in American letters," and no beginning writer could ignore him (*NUF,* p. 91). Setting out in the 1890s to devote herself to the artist's holy calling, pledged not to use fiction as a means for exposing social or economic injustice, Cather could not avoid taking James as a literary and vocational model.

Although Cather later described her Jamesian phase as a "perfectly right form of education," the likelihood that she felt some discomfort with her growing allegiance to the "mighty master of language" is suggested by her 1899 review of Gertrude Atherton's *The Daughter of the Vine.* Like Kate Chopin and Ouida, Atherton roused Cather's anger because she disliked the distorted reflection she glimpsed in another woman's writing. Atherton had the temerity to avow "openly" that her "desire is to emulate Henry James," observed Cather as she began a self-revealing diatribe against a fellow Jamesian. The master's "trophies" would not let Atherton sleep, Cather imagined,and so she decided she must

> produce "literature" at any cost. Dear! Dear! How the ladies do fall in. . . . Now that the ladies have left us, whither shall we turn for the old-fashioned love story in which they always married and flouted art? Who will now supply the pages of Munsey's and Lippincott's?

Although the "old-fashioned love story" was one of the female plots Cather disliked, she went on to attribute, somewhat grudgingly, narrative power to one of Mrs. Atherton's pre-Jamesian novels, *The Doomswoman:*

> I remember a good while ago, before Mrs. Atherton went to England, before she ever dreamed of art or James, she wrote a story called *The Doomswoman*, which was fairly good of its kind. True, it was a pretty bad kind, the kind that summer hotel ladies term "light literature" and that schoolgirls keep hidden in their desks and weep over. Nevertheless it had . . . some sort of spirit because it was done without an ulterior purpose and was not an imitation. . . . It was a story that, in spite of its faults and unsubstantialness, held your attention. Its exaggerations were a part of its author, and one can put up with them. I shall be greatly surprised if Mrs. Atherton ever does so well again. . . . Art has done for Mrs. Atherton. (*WP* 2, pp. 694–96)

If an uncomfortable sense of resemblance contributed to her satiric dismissal of Atherton and her ilk, Cather's criticism raised a dilemma that she could not so easily banish: in imitating James she also might be "done for." Before Atherton devoted her self to James and art, her fiction—although limited—struck Cather as "vivid" and self-expressive ("part of its author") because she was engaged by her material, not by another writer. After she sought literary elevation by imitating James, her fiction was poor James rather than good Atherton.

If Cather knew in 1899 that imitating James could weaken rather than strengthen a woman's writing, why did she continue to "fall in" with the other Jamesians? Much of the fiction James was publishing during this period was concerned with the same issues that were preoccupying Cather: the nature of the artist, the mysteries of the creative process, the ambiguous connection between art and life, the conflict between fealty to art and the desire for popular success. She simply could not ignore a writer who addressed these mutual concerns.

Even more compelling than James' themes was his command of style and form. When she termed him the "mighty master of language," Cather acknowledged his control over fiction's formal elements, a point she frequently made. The "framework" of his fiction was "perfect," she observed in 1895, and

> the polish is absolutely without flaw. . . . He never lets his phrases run away with him. . . . He subjects them to the general tone of his sentence and has his whole paragraphs partake of the same predominating color. You are never startled, never surprised, never thrilled or never enraptured; always delighted by that masterly prose that is as correct, as classical, as calm and as subtle as the music of Mozart. (*KA*, p. 361)

The classically correct James contrasts sharply with Cather's romantic Whitman, who in drawing on life's "primitive elemental force" gave up the powers of control and selection, "enjoying everything with the unreasoning enthusiasm of a boy" (*KA*, pp. 351–52). Cather's "calm and subtle" James likewise is opposed to women writers like Ouida who did not control either their "power of loving" or their adjectives to "advance the effectiveness of a story" (*WP* 2, p. 696).

Although she found some of James' fiction cool and dispassionate, when he tapped profound emotions and wrote a story of passion, as Cather thought he did in *The Other House* (1896), he created a "Shakespearean" work that united "great art with great emotions.'' Previously, she admitted in her 1896 review, she had viewed James as the quiet producer of ''perfect masterpieces of style, [who] eschew[ed] the strange emotion[s] of humanity as if he were a little afraid of them or considered them somewhat vulgar.'' But in *The Other House,* a tale of sexual passion and violence which Cather likened to Greek tragedy, James had, she thought, thrown aside his ''graceful studies in repose'' to make a tragedy ''throbbing with the aching pulse of life.'' At his best, then, James placed passion within the four walls of form as did the Greek dramatists and Sappho, and at these moments the priest of art reached a wide audience without trying to cater to the public. ''It is not every day that a book by a really great artist is capable of making a sensation,'' Cather observed of *The Other House,* and yet James—the last writer one would have thought could ''arouse any excited admiration'' in the public mind—had done so, drawing on ''great elemental, human emotions'' to produce ''the sort of book that keeps one up until three o'clock in the morning'' (*WP* 2, pp. 551–53).

•

Cather thus saw in James the resolution to her own creative and psychic conflicts. She gravitated to a writer who possessed what she feared she lacked: the ability to shape powerful and disturbing feeling into art. James' Apollonian perfection could balance Dionysian self-indulgence; the male writer's order, balance, and command of language could counter the heightened emotionalism and lack of discipline Cather saw in her fervent and florid early style as well as in women's writing. Cather was in part drawn to James, then, because his craftsmanship balanced her excesses.

Since Cather's attraction to James was so overdetermined, it was hard to dispel. By 1905, however, she had written several stories in which she controlled the emotional sources of her material and the tools of her craft without the master's help; ''Paul's Case,'' ''The Sculptor's Funeral,'' and ''A Wagner Matinee'' were not Jamesian. And yet during her years at *McClure's* Cather's devotion to James increased. At this time James was, Elizabeth Sergeant remembers, the ''god'' of her literary life (*WC:AM,* p. 67). The four stories Cather published in 1907—''The Willing Muse,'' ''The Profile,'' ''Eleanor's House,'' and ''The Namesake''—reveal a thoroughgoing Jamesian. The author of ''Paul's Case'' seems to have vanished. Of course a writer's literary maturation resembles any human being's growth from childhood to adulthood in that it is not an unbroken progress but a complex evolution involving regressions as well as advancements. So it is not remarkable that Cather did not move immediately from ''Paul's Case'' to *O Pioneers!* But the length of the retreat is intriguing; James' presence is more insistent in the 1907 stories than in any she wrote previously. Why?

Like her original attraction to James, Cather's renewed subservience to her ''master'' during the *McClure's* years can be connected with several extraliterary patterns. Her move from Pittsburgh to New York pointed her in James' direction. As a writer and editor for *McClure's,* Cather traveled in the elite literary circles of New York, Boston, and London. On the surface, this new life offered her greater autonomy and power, but it also deepened the provincial's reverence for the Eastern literary establishment. In moving eastward Cather also came closer to a literary culture that praised James' work and rewarded young writers who wrote in his manner, as Cather had first discovered when *Cosmopolitan* took ''The Count of Crow's Nest.'' To be sure, critical dissent was growing in the late 1890s and early 1900s as the style of James' major phase struck some American readers as unnecessarily convoluted and his subjects excessively rarified. But James still had many admirers, and they, rather than his detractors, struck Cather as the literary trend-setters in the 1900s. As she observed in ''My First Novels,'' at that time ''Henry James and Mrs. Wharton were our most interesting novelists, and most of the younger writers followed their manner, without having their qualifications'' (*OW*, p. 93).[25]

In ''My First Novels,'' Cather quotes a New York critic who voiced a ''very general opinion'' when he declared ''I simply don't care a damn what happens in Nebraska, no matter who writes about it'' (*OW*, p. 94). The literary opinion-makers Cather sought to please (not the same editors, evidently, who published Crane, Dreiser, and Norris) had rules governing proper setting, characterization, and subject matter: the ''drawing-room was considered the proper setting for a novel, and the only characters worth reading about were smart people or clever people'' (*OW*, p. 93).

For Edith Wharton to write of well-born characters in genteel settings was one thing; for Willa Cather to do so was quite another. When she wrote ''The Willing Muse'' or ''Eleanor's House,'' she was not only imitating James; she was also conforming to the social and literary expectations she attributed to readers who didn't give a damn about Nebraska. Overlooking the immigrant farmers and prairie landscapes of her past—now in her view both unsophisticated and unsuitable subjects—Cather retreated from ''A Wagner Matinee'' and ''The Sculptor's Funeral'' (which hadn't pleased Nebraska folks anyway) to portray clever, upper-class, articulate Anglo-Saxon and European characters who trade psychological and aesthetic insights in drawing rooms, studios, and formal gardens.

Internal as well as external pressures drove Willa Cather closer to Henry James during these years, pressures deriving from the conflicts that marked her evolution as woman and writer. Most of Cather's Jamesian stories are concerned with the weakening or annihilation of the boundaries that separate self and other. Centering on questions of power, autonomy, and identity, the stories Cather wrote in New York, like her Jamesian *Troll Garden* stories, feature varieties of parasitic or symbiotic relationships in which one character feeds off another; characters who fuse or confuse identities; and power relationships in which one character is dominated or absorbed by another. These patterns suggest the mother–daughter issues that informed Cather's development and that she found replayed in the creative process. But these patterns can characterize the father–daughter as well as the mother–daughter bond, and for Cather they resurfaced in her act of imitating Henry James. Emulating her ''mighty master'' inevitably raised the issues of connection and separation, sameness and difference, identity and originality that Cather was also exploring in the stories; the creative process both mirrored, and produced, the content. Striving to duplicate James' style, setting, and subject, Cather was facing a paradoxical dilemma: the more she succeeded in creating Jamesian stories, the more she failed to create a distinctive literary voice.

Although learning from Henry James may have been a ''perfectly right form of education,'' the well-schooled author considered the fiction she wrote during this phase of her apprenticeship artificial and inauthentic. ''I never abandoned trying to make a compromise between the kind of matter that my experience had given me and the manner of writing which I admired, until I began my second novel, *O Pioneers!,*'' she told an interviewer in 1921. Her Jamesian phase seemed one of identity confusion: ''I could not decide which was the real and which the fake 'me.' ''[26] Cather's arrangement of Houghton Mifflin's Library Edition of her collected works reflects her continued desire to repudiate her Jamesian self: she placed *Alexander's Bridge,* the only novel she thought still written under the master's influence, third; she included none of her Jamesian short stories; and she began the edition with *O Pioneers!,* the first book she felt the ''real'' and not the ''fake'' Cather had written.

Although Cather's opposition between the ''fake'' and the ''real'' writer is too stark, masking the continuity between earlier and later styles and selves, her later connection of her Jamesian phase with a false identity suggests that her growing indebtedness to James during her years at *McClure's* was a very complicated matter. In part she was drawn to his work because his psychological themes—in particular, his interest in threats to the integrity of the self—reflected her own concerns with literary autonomy and identity. In imitating her master of form and language, she was attempting to borrow control over psychological and emotional material she had not yet assimilated or integrated. But the very act of imitating James, which required her to duplicate his settings and speak in his voice, intensified her fear that she was not attaining the literary authority she was seeking; this fear, in turn, reinforced her attachment to James as a model; and so the cycle continued. She was trapped by her indebtedness.

This cyclical pattern was also linked to her subservience to McClure. Because her relationship with McClure and her editorial work intensified Cather's preoccupation with questions of self-annihilation and identity diffusion, she was drawn to James' fiction, where these themes are explored. And yet her imitations of James further intensified her conviction that she had no literary self or voice, and so reinforced her conviction that McClure must be right: she didn't have what it took to be a fiction writer. Lacking confidence in her own creative abilities, she relied on James. And so forth.

•

The various forms of ''psychic transmission,'' possession, and symbiosis evident in Cather's Jamesian *Troll Garden* stories are even more insistent in the 1907 fictions, which explore the various ways in which self–other distinctions can be annihilated (*CSF,* p. 165). Although these stories are imitative, they are not ''fake''; as ''The Willing Muse'' demonstrates, Cather's imitative work often betrays deeply troubling conflicts. While ''The Willing Muse'' portrays one writer's absorption by another, ''Eleanor's House'' examines the same phenomenon in a marriage. Harold Forscythe is obsessed with the memory of his dead wife, the ghost occupying his imagination, just as Adriance Hilgarde absorbed Katharine Gaylord's. Harold's marriage to Eleanor had been a union of the sort that disturbed Cather—a fusion that obliterated separate identities. When he first met Eleanor, we discover, he "lost himself," and so when she dies he loses himself as well ("They had so grown together that when she died there was nothing in him left whole"). A "fragment" on his own, he seeks wholeness by recreating his wife in memory, but this compensatory process proves a deadly investment in the past that further obliterates selfhood ("He hadn't grown older, or wiser, or, in himself, better. He had simply grown more and more to be Eleanor").

On the surface the story is concerned with the power of sexual and romantic relationships to absorb and usurp the imagination, self, and will. Like the *Troll Garden* stories, however, "Eleanor's House" can also be read as an expression of Cather's anxiety of influence and imitation. From this perspective, the absent presence is Henry James, whose powerful ghost dominates the story thematically and stylistically. "Eleanor's House"—the structure which dominates and confines the character's imaginations, as did Poynton Mrs. Gereth's in *The Spoils of Poynton*—is also James' house of fiction. The story's conclusion, an unconvincing turn from death to life as Harold escapes the grip of the past and turns from Eleanor to his second wife, does not arise from the character's development. As in the ending of the "The Willing Muse," Cather's desire to escape James' confining architecture and invisible presence seems responsible for her character's last-minute decision to move on (*CSF,* pp. 95–112).

"The Profile" is another story concerned with similarity and difference, confusion and fusion of identities, in which Cather also creates parallels between the aesthetic and the sexual. On one level, the story is concerned with mother–daughter conflicts as seen in its doubling of characters (the daughter Eleanor and the mother-figure Eleanor, the scarred first wife and the scarred second wife) and its contrast between the "good" mother and the "bad." The ways in which these issues are worked out, however, reflect Cather's contemporary interest in the creative process. So, on another level, the story is concerned with questions of representation, authenticity, and truthfulness—aesthetic versions of the psychological issues that underlie the bonds among the characters.[27]

Cather explores the relationship between art and life by examining similarity and difference, representation and reality. The portrait Dunlap paints of his future wife—a profile of her unscarred side—seems to "resemble" her, and yet it differs from her because it suggests to the viewer that the part stands for the whole, and thus that her other profile is similarly perfect. Hence Dunlap's seemingly accurate representation of his subject is in fact deceptive and false, an aesthetic problem Cather foreshadows in her portrayal of his creative process. Dunlap is an imitator, not an original creator. As a painter of women's portraits he strives to duplicate his source—the model—in the finished painting, possessing the "faculty of transferring personalities to his canvas, rather than of putting conceptions there." The painter's subservience to his model, which ultimately results in his deceptive work, is dramatized in the scene where Dunlap begins to paint his future wife's portrait. Shocked by her disfiguring scar, he is immobilized; she then takes control, suggesting that he paint a profile, and turns her good side to him—in effect becoming the artist herself, creating an illusion that the relieved painter then copies.

The implications this story holds for a writer who imitates another artist—thereby reproducing a literary "model" in prose—are disturbing. To imitate, the story suggests, is to distort the connection between life and art; in this sense the "grinning distortion" in the story is not Virginia's scarred cheek but the beautiful painting that conceals it. Because Dunlap allows the model to determine the pose (or the point of view), he does not produce his own vision but reproduces a partial and deceptive surface. Because in writing the story Cather had allowed James to be the model, through Dunlap she comments on —in fact, undermines—her own process of composition and representation (*CSF*, pp. 125–36).

The fourth story Cather published in 1907—"The Namesake"—is an important one in charting her literary development, and I will return to it in the next chapter. Inspired by Cather's imagined bond with her uncle, the source also of the poem "The Namesake," as the title suggests the story is concerned with resemblance and identification; as in "The Profile," the subject is the creative process, which in this story is linked with the discovery of identity within a relationship. The unhealthy, threatening aspects of fusion and mutual absorption Cather explored in her other Jamesian stories are here resolved, as the sculptor Hartwell achieves creative power through an imagined attachment to his dead uncle. The story's opening presents both an unhealthy father–son bond and a debilitating example of artistic influence: Hartwell's sculptor father expects his son to carry on his work after his death. The sculptor's rescue from this artistic and parental entrapment—paradoxically achieved through the bond he creates to his own family heritage—would also characterize Cather's eventual escape from her "mighty master" Henry James when she took the "road home" in *O Pioneers!* (*KA*, p. 361).

Both imitative and deeply personal, derivative and self-reading, Cather's 1907 stories, despite her later protestations, are not "fake." Like "The Burglar's Christmas," they fail to equal the power of "Paul's Case" for two seemingly contradictory but related reasons: in them Cather conforms to external patterns and does not possess sufficient control over the psychological and emotional forces inspiring the stories. If, as Norman Holland has argued, literary form provides a defense for the reader, distracting attention from the unconscious fantasies informing the text, then imitation—whether of a popular literary formula or of another writer's work—can similarly be a defense for a writer who is not ready to draw on, or to shape, unconscious material.[28] Ultimately, Cather was able to feel that her books were writing themselves as she allowed repressed and forgotten material to surface. Meanwhile, during the early 1900s, imitating the master allowed her to defend against disturbing conflicts, many of which concerned the creative process and so were raised by the act of writing itself. Modeling herself after a writer whose "masterly prose" was as "correct, as classical, as calm and subtle" as Mozart's symphonies at least allowed her to keep writing, unlike the silenced narrators and protagonists of her college stories. But Cather could not depend on emulating James to conceal, reveal, and control unconscious conflicts if she were to create her own voice and subject matter.

•

When Cather called James the master of language rather than of plot, or narrative, or characterization, she suggested that hers was indeed a profound discipleship. Language is the writer's artistic medium, and until Cather felt that she commanded language, she did not have confidence in herself as a writer. The importance Cather later attached to language is evident in her 1925 "Preface" to *The Country of the Pointed Firs,* where she praised Jewett's ability to duplicate in prose the rhythms and idioms of colloquial American speech:

> She had not only the eye, she had the ear. From childhood she must have treasured up those pithy bits of local speech, of native idiom, which enrich and enliven her pages. The language her people speak to each other is a native tongue. No writer can invent it. It is made in the hard school of experience, in communities where language has been undisturbed long enough to take on color and character from the nature and experiences of the people. . . . Such an idiom makes the finest language any writer can have; and he can never get it with a notebook. He himself must be able to think and feel in that speech—it is a gift from heart to heart. (*CSF,* p. 10)

Paradoxically, perhaps, it was Jewett's ability to convey this gift she did not invent—the language of her Maine countryfolk—that contributed to her distinctive literary voice. Her stories, in Cather's view, had that "inherent, individual beauty" one finds when a "beautiful song is sung by a beautiful voice that is exactly suited to the song," the "cadence," the "quality of voice that is exclusively the writer's own, individual, unique" (*CSF,* p. 7).

Cather later attributed mastery of her own cadence to her receptivity to rhythms of American speech. "The American language works on my mind like light on a photographic plate," she told an interviewer in 1925. That was why, she explained, she could not work abroad; she had to be immersed in "the American idiom," she said, the language which touched "springs of memory, awaking past experience and knowledge necessary to my work."[29] When she was writing Jamesian stories, however, Cather was not working in her native idiom but in something like a foreign language. James' thematic concerns corresponded to her own, but his style and voice were not hers.

In some of her earliest stories Cather had tried, not always successfully, to capture a "native idiom" by conveying the polyglot voices of rural Americans, both immigrant and native-born. In "The Clemency of the Court" she tries to duplicate the foreigner's halting, broken speech ("I could make some broom, I think") while in "The Elopement of Allen Poole" she attempts a Southern dialect ("Take keer, darlin', yo' goin' to git yo' dress all bloody, yo' nice new frock what yo' goin' to wear to the Bethel picnic"). She does better at conveying the colloquial talk of small-town, native-born Midwesterners. The conversation of the watchers over Harvey Merrick's body in "The Sculptor's Funeral" suggests that, like Sarah Orne Jewett, Cather had "treasured up" some "pithy bits of local speech": "He killed a cow of mine that-a-way onct," a mourner reflects; "a pure Jersey and the best milker I had, an' the old man had to put up for her."

In her least imitative *Troll Garden* stories—those in which the accomplished writer would later discern her own cadence—Cather's mastery of language goes beyond her command of colloquial talk. She also knows when to let her characters speak, when to let them be silent, and how to connect speech and silence to her themes. "The Sculptor's Funeral" is a case in point. The extended coffin-watching conversation, which involves several of the town's judgmental elders, effectively evokes Sand City's "native idiom" and furthers Cather's contrast of West and East, materialism and art, conventional and unconventional visions. The easy flow of ungrammatical, complacent talk in which one speaker adds to a previous speaker's dismissal of Harvey Merrick is the sociolinguistic embodiment of the conflict Cather wants to sketch between a repressive society where ignorant, materialistic values are shared and the isolated, eventually silenced artist.

Cather ends the group conversation by allowing Jim Laird, the drunken lawyer who shares Merrick's values although not his commitment to practicing them, to interrupt. Laird reflects Cather's anger at small-town Nebraska, big-city Presbyteria, and America's Philistine society; she wants to interrupt too. But because the lawyer's monologue is consistent with his character psychologically as well as linguistically, it does not seem an authorial intrusion. The evidence he marshalls to support his attack, his use of rhetorical questions ("Why was Mrs. Thomas's son, here, shot in a gambling house? Why did young Adams burn his mill to beat the insurance companies and go to the pen? . . . I'll tell you why. Because you drummed nothing but money and knavery into their ears from the time they wore knickerbockers"), the length of his speech—all suggest the prosecuting attorney's final summation. Cather makes it seem likely that when a small-town lawyer finally confronts the neighbors he despises, releasing years of rage and self-disgust, he *would* talk like that, using his oratorical skills to attack the people who have controlled his voice as well as his legal services. And because she is sure of her intention, she does not end with Laird's self-dramatizing speech since he too is silenced by the community he will never be able to leave. The story concludes with the narrator's laconic voice telling us that Laird "got the cold he died of driving across the Colorado mountains to defend one of Phelps's sons who had got into trouble out there" (*CSF*, p. 186).

Whereas the lengthy group conversation in "The Sculptor's Funeral" conveys the narrow community's shared assumptions, the sparsity of spoken language in "A Wagner Matinee" and "Paul's Case" reflects the silence that envelops Aunt Georgianna and Paul. Georgianna speaks only three times during her Boston visit, each time becoming more assertive and self-expressive as her buried soul reawakens; her last outcry, however ("I don't want to go, Clark, I don't want to go!"), reminds us that she will return to the "silence of the plains" from which she has only briefly escaped (*CSF*, p. 242). In "Paul's Case" there is even less dialogue, fittingly since language is a social construction in which Paul cannot find self-expression. Asked by his high school principal "whether he thought that a courteous speech to have made a woman," he answers, "I don't know . . . I didn't mean to be polite or impolite, either. I guess it's a sort of way I have of saying things regardless" (*CSF*, p. 244). Paul's response reveals that he finds no correspondence, not even the inexact one language offers other speakers, between emotion and verbal expression. The restricted polarities he sees in speech (one has to be either "polite" or "impolite") cannot embody his inchoate feelings just as the polarized social roles and gender identities he confronts do not correspond to his inchoate desires; Paul does not ''fit'' into the male role just as his feelings do not ''fit'' into language. Hence Cather makes the correct decision to employ third-person narration; the narrator, who possesses the command of language Paul lacks, can continue to tell his story when he relapses into silence after the interview with the principal. He does not speak for the rest of the story, his silence reflecting both his muting by an oppressive culture and his increasing delight in nonverbal stimuli—music, perfumes, colors, food—as he regresses to the stage of human development prior to the acquisition of language.

In those *Troll Garden* stories where James' presence is least evident, then, Cather displays an impressive command of language. In deciding what voices to give her characters, she not only evokes their native idiom but she also demonstrates her skill in matching speech with character and theme, and—perhaps most impressive of all—she knows when to use dialogue and when to use third-person narration to speak for characters who are, for one reason or another, unable to speak for themselves. All these forms of mastering language contribute to the sense these stories give of Cather's own voice, a sense she shared when she allowed them to be reprinted in the Library Edition.

In the four Jamesian stories she published in 1907, however, Cather's control over language and speech wanes as her voice is absorbed by James' style. Here are two excerpts, the first from ''The Willing Muse'' and the second from ''Eleanor's House'':

> I don't believe he even knew where he stood; the thing had gone so, seemed to answer the purpose so wonderfully well, and there was never anything that one could really put one's finger on—except all of it. (*CSF*, p. 123)
>
> ''She is certainly going to do something,'' Harriet declared. ''But whatever can she hope to do now? What weapons has she left? How is she, after she's poured herself out so, ever to gather herself up again? What she'll do is the horror.'' (*CSF*, p. 105)

''There was nothing left for him to say,'' observes Harrison of the drained Kenneth Gray in ''The Willing Muse'' (*CSF*, p. 123). One cannot make the same statement about most of the characters in Cather's Jamesian fiction. The most striking feature of these stories is their awkward and excessive use of dialogue, which she had employed so effectively—and sparingly—in her best *Troll Garden* stories. Capturing the ebb and flow of conversation would never be one of Cather's literary strengths, however. When she achieved her own cadence in the major fiction, the most powerful speaking voice is generally that of the narrator, sometimes that of a character who momentarily becomes a narrator or storyteller. But the 1907 stories, like the Jamesian *Troll Garden* stories, are crammed with conversation. Trying to follow the literary fashion and set her fiction in the drawing room, faced with ''smart or clever'' characters, Cather had to let them talk. And when they speak, as they often do at great length, her characters use a formal, stilted prose that Cather evidently thought the well-bred and well-modulated residents of a Jamesian house of fiction should possess (''Shall you then, Harriet ventured, go to Fortuney?'' [''Eleanor's House'']). Unable to grant distinctive voices to her characters, Cather makes most of her upper-crust speakers sound identical—an effect which was most likely unintentional rather than satiric.

Cather would later associate the artificiality of her Jamesian phase with enforced, strained conversation. Writing *Alexander's Bridge,* her Jamesian novel, was like "riding in a park," she wrote in "My First Novels," "with someone not altogether congenial, to whom you had to be talking all the time" (*OW,* pp. 92–93). Having to talk "all the time" in her formal park, an ordered and confined space, like the characters she was writing about, Cather felt she had to speak according to external rules and expectations. This enforced speech was required by her unnamed companion in the park, whom it is tempting to imagine as Henry James, the author with whom she was politely conversing as late as 1912 (*KA,* p. 448).

Cather would later say that a young writer "must care vitally, fiercely, absurdly about the trickery and the arrangement of words, the beauty and power of phrases" (*KA,* p. 451). And yet in 1908, when she wrote Jewett her despairing letter about her failure to advance as a writer, she said that she was beginning to hate words and everything made from them. In the same letter she confessed that she was feeling more and more bereft of herself. Cather was referring specifically to the alienation from self and language she experienced as the editor required to subordinate her creative vision to other people's words. But the four stories Cather had published the previous year—instead of providing an outlet for a more personal voice—duplicated and reinforced the conflicts she was undergoing as an editor. Imitating James, she was losing her command of language as well as her literary identity, and becoming increasingly sure that McClure must be right—she lacked the originality and power to be a really original literary talent. Small wonder, then, that her 1907 stories are concerned with relationships where one partner is absorbed by another.

•

Cather never met her master of language, although in "148 Charles Street" she reports a near-miss in 1910:

> The next summer I was visiting Mrs. Fields at Manchester in a season of intense heat. We were daily expecting the arrival of Henry James, Jr., himself. One morning came a spluttery letter from the awaited friend, containing bitter references to the "Great American Summer," and saying that he was "lying at Nahant," prostrated by the weather. I was very much disappointed, but Mrs. Fields said wisely: "My dear, it is just as well. Mr. James is always greatly put about by the heat, and at Nahant there is the chance of a breeze." (*NUF,* p. 69)

Mrs. Fields may have been wiser than she knew in thinking it "just as well" that James did not visit: think of the irritable victim of the "Great American Summer," marooned in Manchester without a breeze, confronted by a worshipful female admirer. With the exception of Edith Wharton, James had a low opinion of women writers—young American ones in particular—and it is hard to imagine him offering Cather any encouragement.

Cather and James had another failed meeting which may also have been just as well. Her *McClure's* colleague Witter Bynner, one of the young men whom James enjoyed encouraging, sent *The Troll Garden* along to his mentor, asking James to look it over. Cather had the dubious pleasure of reading James' condescending reply to Bynner:

> I have your graceful letter about "The Troll Garden"; which duly reached me some time ago (as many appealing works of fiction duly reach me); and if I brazenly confess that I not only haven't yet read it, but haven't even been meaning to (till your words about it thus arrive), I do no more than register the sacred truth. That sacred truth is that, being now almost in my 100th year, with a long and weary experience of such matters behind me, promiscuous fiction has become abhorrent to me, and I find it the hardest thing in the world to read almost *any* new novel. Any is hard enough, but the hardest from the innocent hands of young females, young American females perhaps above all. . . . I've only time now to say that I *will* then (in spite of these professions) do my best for Miss Cather—so as not to be shamed by your so doing yours.
>
> Believe me,
> Yours ever,
> HENRY JAMES (*WC:AM*, pp. 68–69)

But Cather was impressed by this offhanded reply. Desperate to belong to the men's club that James had allowed Bynner to join, she took no offense at the master's dismissal of women writers. To her, James was the oracle for sacred truth, and she told Bynner that his reply had given her much satisfaction. She had assumed that he would feel so, but it was comforting for her to see his opinions in black and white. She particularly liked his condemnation of "promiscuous fiction"—this made his letter a bracing moral stimulant, she told Bynner. Of course, it was unsettling for one's work to be assessed by someone with such exalted standards, but Cather declared that she would be able to stand up to whatever punishment James might mete out in a second letter. She asked for Bynner's sympathy, however, for while she was waiting for James to read her book she confessed she felt as if she were about to undergo a searching physical examination that might undermine her confidence in the dependableness of her organs, horribly confirming her own doubts. As soon as Bynner received James' diagnosis of *The Troll Garden,* she asked that he write her.[30]

The medical imagery Cather uses to describe the act of reading and evaluation is telling: The text is identified with the female body being probed and entered by the male physician hunting for signs of illness. The former vivisectionist and aspiring surgeon now grants James the powers of dissection and analysis; she is the patient who can only await the doctor's diagnosis. The sexual power of entrance Cather imputes to the male reader with this imagery further suggests her anxiety that she might, like the "innocent" young females James mentioned, be found guilty of producing "promiscuous" fiction, fiction that was not selective about its readers. But James never did do his best for Miss Cather, evidently never reading *The Troll Garden;* in any case, he never sent Bynner his diagnosis. Thus Cather never had to confront the worst: the possibility that he might have grouped *The Troll Garden* with the promiscuous fiction produced by prolific American females.

Willa Cather did not always grant James the power to define or dismiss her. Once she achieved her own literary authority, like a daughter sure of her difference, she regarded her literary parent with a mixture of affection and detachment, seeing his flaws as well as his strengths. She could be "impatiently amused," Sergeant remembers, by his "elaborate, subtle phrases" (*WC:AM*, p. 139); reading *Notes of a Son and Brother* in 1914, Cather found the sentences "so full of after-thoughts and parentheses and recoil that he ended with everything cancelled out. Knut Hamsun's *Shallow Soil* was more rewarding" (*WC:AM*, p. 126). Even if his style began to seem comically involuted, James' critical ideals continued to stimulate Cather's thinking about literary theory, however, and she would not be afraid to acknowledge her indebtedness, entitling a short essay "On the Art of Fiction" and endorsing, in "The Novel Démeublé," James' principle of selection.[31]

Although imitating James' convoluted style and genteel settings was a false start for a writer whose strengths would lie in the evocative simplicity of her prose and her ability to convey a sense of American places, Cather did learn technical skills from the master, as she later acknowledged when she termed her apprenticeship a "perfectly right form of education." A particularly impressive acquisition is the way she made the Jamesian observer/reflector character her own. Convinced, like James, that drama took place in the human consciousness rather than in external events, Cather tried an observer figure as early as "The Count of Crow's Nest," and returned again with her choric commentators in "Eleanor's House" and "The Willing Muse." Yet these characters are stiff and undeveloped. Once Cather had assimilated James' influence, however, she could accommodate his techniques for her own purposes, as she did in *My Ántonia, A Lost Lady,* and *My Mortal Enemy,* where the observer's acknowledged and unacknowledged investment in the subject is the story itself.

This transition from the imitative to the assured use of James first required Cather to declare her independence from her literary father, however, which she would accomplish in *O Pioneers!* and *The Song of the Lark*. These are the literary equivalents of McClure's *My Autobiography*—texts in which Cather reverses the power dynamics of the father–daughter relationship, asserting her independence from a male mentor. Both novels are pointed rejections of James. Declaring her connection to Whitman in the first, Cather aligns herself with the American romantic, not the expatriated classicist, as she leaves James' drawing rooms for her own cornfields. In *The Song of the Lark,* in part a celebration of her artistic emergence in *O Pioneers!*, Cather even more pointedly rejects James by rewriting one of his novels. Her title refers to a painting by Jules Breton in which a peasant woman looks up from the fields at a singing lark. Concealing a portrait of a woman—not a lady—in her title, in her story Cather also subtly revises James' plot in order to reject the muted view of a gifted woman's possibilities he expressed in *Portrait.* In each novel a romantic, imaginative heroine dreams of achievement and adventure; she has three suitors or mentors; she profits from a legacy; and she finds that one suitor has a concealed sexual or marital relationship.

Yet whereas Isabel Archer must face constriction and disillusionment, Thea Kronborg gains the expansive life of her dreams, an outcome Cather associates with her choice of art rather than marriage. Granting her heroine the heroic fulfillment James denied his, in *The Song of the Lark* Cather also makes the female body—in her letter to Bynner associated with the disease-ridden text James was sure to dislike—the healthy and powerful source of Thea's art.

•

In *The Song of the Lark,* the five-hundred-dollar inheritance Thea receives from one of her suitors is a useful, but not ultimately a sufficient, bequest. Before she discovers her singer's voice, Thea is granted another inheritance—the clay vessels left by the ancient women potters of the Southwest—that helps her to envision her own artistry. At the same time that she was finding McClure's and James' influences inhibiting, Cather seemed to be finding such a maternal legacy through her friendship with Mrs. Fields. Hers was a compelling, authoritative voice that apparently countered the voices of the fathers.

NOTES

1. Eva Mahoney, "How Willa Cather Found Herself," *Omaha World-Herald,* November 27, 1921, p. 7.

2. Willa Cather to Will Owen Jones, May 7, 1903, Barrett-Cather Collection, University of Virginia Library, Charlottesville, Va.

3. As she told him in 1933, he was a success with young people because he always expected so much of them, and those who had anything in them tried to be as good as McClure thought they were. Willa Cather to S. S. McClure, May 26, [1933], Lilly Library, University of Indiana, Bloomington, Ind.

4. Cather to McClure, January 17, 1907, and n.d. Writing Zoë Akins in 1912, Cather referred to the illnesses and the exhaustion caused by the job's pressures (Willa Cather to Zoë Akins, n.d., Akins Collection, Huntington Library, San Marino, Calif.). Lewis also mentions Cather's suffering nervous exhaustion from the "necessity of talking to and being talked to by so many people" in her editorial role (*WCL,* p. 70).

5. Ellery Sedgwick, *The Happy Profession* (Boston: Little, Brown, 1946), p. 139.

6. In addition to Sedgwick, see Lincoln Steffens, *The Autobiography of Lincoln Steffens* (New York: Harcourt, Brace, 1931), pp. 363–64; Robert Stinson, "S. S. McClure's *My Autobiography:* The Progressive as Self-made Man," *American Quarterly* 22 (Summer 1970): 203–12; Ellen Moers, "Introduction" to Peter Lyons, *Success Story: The Life and Times of S. S. McClure* (Deland, Fla.: Everett/Edwards, 1967), pp. vii-xv.

7. Moers, p. xiii.

8. Sedgwick, p. 145.

9. S. S. McClure to Willa Cather, June 18 and 19, 1909, Lilly Library, University of Indiana, Bloomington, Ind.

10. Cather to McClure, January 17, 1907.

11. Since Cather disliked muckraking writers and scorned socially conscious fiction, she doubtless wanted to envision her "Chief" as placing aesthetic goals first, although most observers associate him with the investigative reporting he introduced to American journalism. As Ellen Moers points out, McClure's major fostering of literary talent oc-

curred during the 1890s, not during Cather's years at the magazine, when he was publishing writers like Stevenson, Norris, and Crane (Moers, pp. vii–x).

12. Willa Cather to Sarah Orne Jewett, December 17, 1908, Houghton Library, Harvard University, Cambridge, Mass.

13. McClure to Cather, October 14, 1911.

14. Mahoney, p. 7.

15. Cather to Jewett, December 17, 1908.

16. Leon Edel, "Homage to Cather," in *The Art of Willa Cather*, ed. Bernice Slote and Virginia Faulkner (Lincoln: University of Nebraska Press [Bison ed.], 1974), p. 194. Cather may have made the 1919 change for the same reason. Although relatively more well established then, she was going through an anxious period, wondering whether she should change publishers since Houghton Mifflin did not seem to be promoting her work with enough energy. Cather thought that she should be better known than she was, so once again her age and her accomplishments did not seem to coincide.

17. Her association with McClure gave her power over other writers, some far better known than she; given her low opinion of "Amygism," it must have given Cather pleasure to turn down Amy Lowell by telling her that her poem was too slight for their magazine. Willa Cather to Amy Lowell, n.d., Houghton Library, Harvard University, Cambridge, Mass.

18. Willa Cather to Mrs. McClure, October 2, 1919, and n.d., Lilly Library, University of Indiana, Bloomington, Ind.; Willa Cather to Ferris Greenslet, January 21, [1921], Houghton Library, Harvard University, Cambridge, Mass.

19. Cather to McClure, April 22, June 9, and June 12, [1912].

20. S. S. McClure, *My Autobiography* (New York: Frederic Stokes, 1914), p. v.

21. The likelihood that in writing the autobiography Cather regained some of the power she felt she had lost is supported by their correspondence. By 1912, McClure was in financial trouble, ill and anxious about his future, while Cather was an up-and-coming novelist with one book out and another on the way. Evidently he had suggested payment, which she refused; she wanted to return some of the favors he had done for her, she wrote, and to help him when he needed her. As I told you in New York, Cather continued, I have never felt the power to do things in you as strongly as now. She ended the letter by twice using the word "willing" to describe her eagerness to undertake the project; at this point, when she was the parent and McClure the weakened child, it was not threatening to imagine herself in the role of Kenneth Gray (Cather to McClure, June 12, 1912).

22. Cather to Jones, May 20, 1919.

23. Flora Merrill, "A Short Story Course Can Only Delay, It Cannot Kill an Artist, Says Willa Cather," *New York World*, April 19, 1925, Section 3, p. 1; Latrobe Carroll, "Willa Sibert Cather," *The Bookman* 53 (May 1921): 214.

24. Mahoney, p. 7.

25. See Millicent Bell, *Edith Wharton and Henry James: The Story of Their Friendship* (New York: George Braziller, 1965) for an informative account of Edith Wharton's Jamesian apprenticeship as well as of her divergence from the master.

26. Mahoney, p. 7.

27. In a letter to Elizabeth Sergeant, Witter Bynner stated that "The Profile" was based on a friend of Cather's who possessed a scar, and despite the protest of mutual friends, she refused to withhold the story from publication. If so, the guilt and distortion marking the relationship between life and art may be connected with this real-life source. Bynner was no fan of Cather's, however, and his possible bias should be taken into account (Witter Bynner to Elsie Sergeant, May 8, 1951, Houghton Library, Harvard University, Cambridge, Mass.).

28. Norman Holland, *The Dynamics of Literary Response* (New York: Norton, 1975), pp. 104–33.

29. Walter Tittle, "Glimpses of Interesting Americans," *The Century Magazine*, 110 (July 1925): 312.

30. Willa Cather to Witter Bynner, February 24, [1906], Houghton LIbrary, Harvard University, Cambridge, Mass.

31. For a discussion of the similarities between James' and Cather's aesthetic theory, see James E. Miller, Jr., "Willa Cather and the Art of Fiction," *The Art of Willa Cather*, ed. Bernice Slote and Virginia Faulkner (Lincoln: University of Nebraska Press, 1973), pp. 121–48. Miller sees the influence of James' house of fiction in Cather's use of architectural metaphors (as in "The Novel Démeublé"), but given Cather's persistent interest in space and landscape and her early linking of writing and architecture (as in "A Tale of the White Pyramid"), it seems likely that she could have arrived at these metaphors independently.

14

Maternal Voices

> You know, my dear, I think we sometimes forget how much we owe to Dryden's prefaces.
> Mrs. Fields, "148 Charles Street"

> But have you reserved your places? No? then I would advise you to do so at once.
> Madame Grout, "A Chance Meeting"

> What a beautiful voice, bright and gay and carelessly kind—but she continued to hold her head up haughtily. . . . I suppose I stared, for she said suddenly, "Does this necklace annoy you? I'll take it off if it does."
> I was utterly speechless.
> *My Mortal Enemy*

Cather's friendship with Annie Adams Fields flowered during the years from 1908 to 1912, the same period when she was entangled with McClure and James. It endured until Mrs. Fields' death in 1915. While serving her journalistic and literary fathers, Cather felt dispossessed, weakened, and drained, her energy, words, and self absorbed by others. But sipping tea in Mrs. Fields' maternal presence, as she later recalled in letters and her reminiscence "148 Charles Street," she felt as if she were receiving an inheritance and blooming in a temperate, sustaining climate, a Boston version of Isabelle McClung's "garden fair." In New York, Cather sensed vital energies dissipating; at 148 Charles Street, Mrs. Fields's Beacon Hill home, she felt enriched rather than depleted as she accepted the gifts of friendship, culture, and inclusion in the literary past.

After reading "148 Charles Street," Cather's honorific memoir of Annie Fields published in *Not Under Forty,* we might assume that she was a more benevolent parent than McClure or James because she was a mother who offered her new friend a woman's care and sympathy. Certainly theirs resembled a mother–daughter bond. But in some ways Cather benefitted from apprenticeship to McClure and James, and in some ways she found Mrs. Fields an inhibiting parent. The mother–daughter relationship is not based on equality, and Mrs. Fields' gifts had strings attached. So in exploring this important friendship we will have to be alert to a disturbing pattern that is hard to see in the flattering and retrospective "148 Charles Street": the dutiful and docile daughter seeking to please a matriarch whom she both loves and fears by modifying and modulating her voice and self.

•

As the young wife of James T. Fields—one of the founders of the Boston publishing house of Ticknor and Fields and editor of the *Atlantic Monthly* from 1861 to 1871—Annie Fields occupied a powerful position in nineteenth-century literary New England. Her husband and his partner George Ticknor had taken the lead in promoting and protecting American writers. While other nineteenth-century publishers were pirating cheap editions of British fiction, making quick profits before copyright laws were passed, James Fields and his partner were publishing Emerson, Hawthorne, Longfellow, Whittier, and Lowell. "Few were our native authors," recalled Henry James in his 1915 memorial essay "Mr. and Mrs. James T. Fields," and "the friendly Boston house had gathered them in almost all." Most American publishers viewed their profession solely as a trade, but Fields and Ticknor dedicated themselves to encouraging and supporting their authors, without, of course, losing money. As James put it, the Boston publishers managed to "link the upper half of the title-page with the lower."[1]

Annie Fields had an important role in her husband's enterprise. She read manuscripts, corresponded with authors, and promoted the work of women writers like Harriet Beecher Stowe, Rebecca Harding Davis, and Elizabeth Stuart Phelps, many of whom published in the *Atlantic Monthly;* under her husband's editorship, the magazine was friendly to local-color writing in the 1870s and 1880s, most of which was produced by women. After James Fields' death and the formation of her "Boston marriage" with Sarah Orne Jewett, who became the center of her emotional life, Annie Fields continued to be a literary power in New England, weaving a network of personal and professional bonds with women writers. Her circle included Celia Thaxter, Louise Imogen Guiney, Alice French, Rose Terry Cooke, Mary Wilkins Freeman, and Harriet Prescott Spofford. Early in 1908 it expanded to include Willa Cather. Writer and editor herself, Annie Fields was devoted to preserving her friends' memories and literary reputations, and published after their deaths the *Life and Letters of Harriet Beecher Stowe* (1897) and the *Letters of Sarah Orne Jewett* (1911).[2]

Although her creation and maintenance of a female literary network have intrigued feminist critics and biographers, Mrs. Fields was not associated in her contemporaries' minds primarily with women writers. Better known as a hostess than as a literary figure in her own right, during her husband's lifetime she made 148 Charles Street the symbolic center of American literary culture. There Mrs. Fields presided over a salon where writers and artists, European as well as American, visited and mingled. As Cather observed in "148 Charles Street," Mrs. Fields was known for her "hospitality to the aristocracy of letters and art," and Cather's description reveals that most members of the "aristocracy" were male.

> It was not only men of letters, Dickens, Thackeray, and Matthew Arnold, who met Mrs. Fields' friends there; Salvini and Modjeska and Edwin Booth and Christine Nilsson and Joseph Jefferson and Ole Bull, Winslow Homer and Sargent, came and went against the background of closely united friends who were a part of the very Charles Street scene. Longfellow, Emerson, Whittier, Hawthorne,

> Lowell, Sumner, Norton, Oliver Wendell Holmes—the list sounds like something in a school-book; but in Mrs. Fields' house one came to believe that they had been very living people—to feel that they had not been long absent from the rooms so full of their thoughts, their talk, their remembrances sent at Christmas to the hostess, or brought to her from foreign lands. (*NUF*, pp. 56–57)

Cather met Mrs. Fields in 1908 when she was temporarily quartered in Boston on the Mary Baker Eddy assignment. Pauline Goldmark, a New York friend, introduced Cather to her sister, who was married to Louis Brandeis, and Mrs. Brandeis in turn brought her to 148 Charles Street. At that time Cather had not yet resolved the conflicts marking her European trip and reflected in her first two books. As a woman, she was separated by gender from the single-named authors who dominated the monuments and the literary tradition; as an American, she could not claim a direct inheritance from European writers; as a Nebraskan, she suffered the same Midwestern insecurities and class anxieties that plagued her male colleagues Garland, Anderson, Hemingway, and Fitzgerald when they migrated eastward to prestigious cultural and social worlds like Boston and Princeton where they never felt securely rooted.

As we have seen, Cather's desire to attach herself to a richer tradition than the one she received from her family and community took several forms in the 1890s and 1900s: her imagined connection with Stephen Crane, her pilgrimage to A. E. Housman, her attraction to cultured, aristocratic, or European families—the Weiners, the Geres, the Westermanns, the McClungs. She wanted to be someone else's daughter, the heir of artistic sensibilities and cultural traditions that had, as she put it in "The Namesake," "not begun with [her]" (*CSF*, p. 146). To Elizabeth Sergeant, Cather expressed her regret that she had not "been born into an inheritance of musical scholarship and linguistic gifts like the Viennese." She had struggled to wrest that inheritance from the past by her own efforts, but there were "gaps," she told Sergeant, that "youthful temerity and native flair and assiduity could never fill" (*WC:AM*, p. 48).

The danger was that in filling the "gaps" with male-defined views of the artist, Cather was still excluded from the company to which she wanted to belong. When she met Mrs. Fields, however, Cather felt that she was at last included in Western culture. "Sometimes entering a new door can make a great change in one's life," she wrote in "148 Charles Street," and that image of entry epitomizes the importance of Annie Fields to Willa Cather. In Mrs. Fields' house Cather found the male literary tradition brought to life as her hostess, who appears in "148 Charles Street" as both a priestess and a medium, evoked the vanished ghosts for her new friend.

Other young writers might not have been as impressed by Boston's dusty greatness as Willa Cather was, for in the early 1900s literary prestige and activity had shifted from Boston to New York, by then the center of the publishing industry.[3] For Willa Cather, however, Boston was still the symbolic center of American literary culture. She had grown up admiring the blue-bound Ticknor and Fields editions of classic American authors in her father's library, and in stories like "The Sculptor's Funeral" and "A Wagner Matinee" she made Boston the oasis of culture. So even though she was living in New York, the magnet for other aspiring writers, she was prepared to invest Beacon Hill and its presiding deity, Mrs. Fields, with the power the city no longer possessed. Mrs. Fields was an "exquisite survivor of the Golden Age of American Literature," Cather told Elizabeth Sergeant exultantly. Lacking Cather's pious reverence for Boston's nineteenth-century men of letters, the more modish Sergeant, who preferred Amy to James Russell Lowell, wanted to exclaim "Come now, this is 1910!" But as a Red Cloud emigré Cather could not scoff at the cultural legacy that Sergeant, a native New Englander and Bryn Mawr graduate, took for granted and hence could repudiate ("New Englanders in their twenties, like me, did not bow to the ancient and honorable idols" [*WC:AM*, p. 41]).

Cather's first glimpse of Mrs. Fields and Sarah Orne Jewett in the long drawing room overlooking the Charles River remained imprinted on her memory. In "148 Charles Street" she tells us what she saw after walking through the open door:

> That room ran the depth of the house, its back windows looking down on a deep garden. Directly above the garden wall lay the Charles River, and, beyond, the Cambridge shore. At five o'clock in the afternoon the river was silvery from a half-hidden sun; over the great open space of water the western sky was dove-coloured with little ripples of rose. The air was full of soft moisture and the hint of approaching spring. Against this screen of pale winter light were the two ladies: Mrs. Fields reclining on a green sofa, directly under the youthful portrait of Charles Dickens (now in the Boston Art Museum), Miss Jewett seated, the low tea-table between them. . . . I do not at all remember what we talked about. . . . I was too intent upon the ladies. (*NUF*, pp. 53–55)

It is like a Mary Cassatt painting: the impressionist skyline, the soft air, the two ladies grouped around the tea table. In her novels, Cather would devote herself to the open spaces with which this scene-painting begins. But in "148 Charles Street" she is "intent upon the ladies" as she draws the reader's eye downward from the horizon to the tea table; the "western sky" is just a backdrop (a "screen") for the two women.

Although Fields and Jewett are seated under Dickens' portrait, their positioning does not reflect their subservience to male writers in Cather's imagination. In the context of both the memoir and the friendship it recalls, the detail suggests Mrs. Fields' possession of the cultural past. She owned the painting which that respected institution of Brahmin culture, the Boston Museum of Fine Arts, was grateful to receive after her death. Cather's inclusion of this detail—like her observation that the painting was of the "youthful" Dickens—stresses Mrs. Fields' power over the representatives of male literary culture whose memorabilia were housed at 148 Charles Street. (Informed by Elizabeth Sergeant that Boston assumed that Mrs. Fields was in love with Dickens, Cather "laughed and said it was more likely the other way round" [*WC:AM*, p. 64]).

Mrs. Fields also had the power, granted her by memory and long acquaintance, of transforming Henry James into a young writer who had once been as inexperienced as Cather herself. One morning at Manchester-by-the-Sea, her summer home north of Boston, Mrs. Fields revealed that the younger James' first essays and stories were "thought self-conscious, artificial, shallow," and that literary Boston had even "feared the young man had mistaken his calling." Cather's friendship with a woman who, far from being awed by James, remembered him as a "very young man" with "execrable" handwriting gave James human proportions. Throughout "148 Charles Street" Cather refers to James as "Henry James, Jr."; the "Jr," which does not appear in her other commentaries on James, suggests that in Mrs. Fields' presence, Cather's literary father became just another son. Later Cather's friendship with Mrs. Fields gave her the authority to criticize James directly; as she told her Houghton Mifflin editor Ferris Greenslet in 1915, his memoir "Mr. and Mrs. James T. Fields" was somewhat patronizing.[4]

Because Cather thought her Boston hostess possessed the cultural past, she gave Mrs. Fields the power to bestow it upon a "Midwest grandchild" (*WC:AM*, p. 65). At last Cather began to feel that she had received the legacy she had been seeking; in Mrs. Fields' presence "an American of the Apache period and territory could come to inherit a Colonial past" (*NUF*, p. 57). Whereas her European trip intensified her sense of exclusion and dispossession, her entrance into Mrs. Fields' drawing room gave her an exultant feeling of acquisition, as she told Sarah Orne Jewett in a joyous letter written several months after this first meeting. How beautifully Mrs. Fields evoked a vanished time, Cather wrote. Seeing her was the first time in her life that she felt Americans had any past of their own. That first meeting had been something so long awaited.[5]

In "148 Charles Street" Cather's cultural acquisition depends on Mrs. Fields' ability to connect past and present, European and American culture. A living link to the heroic times Cather venerated, Mrs. Fields seemed to "reach back to Waterloo," reviving at Beacon Hill the era of Napoleon and Keats, Shelley and Byron, as well as that of Longfellow and Hawthorne:

> As Mr. Howe reminds us, she had talked to Leigh Hunt about Shelley and his starlike beauty of face—and it is now more than a century since Shelley was drowned. She had known Severn well, and it was he who gave her a lock of Keats' hair, which, under glass with a drawing of Keats by the same artist, was one of the innumerable treasures of that house.

Henry James described the Fields' house as a "waterside museum," but to Cather it was a shrine.[6] The "treasures" she saw were holy relics as well as literary memorabilia: "When one was staying at that house the past lay in wait for one in all the corners; it exuded form the furniture, from the pictures, the rare editions and the cabinets of manuscript—the beautiful, clear manuscripts of a typewriter-less age, which even the printers had respected and kept clean" (*NUF*, p. 61).

When Henry James called Mrs. Fields the "literary and social executor, so to speak, of a hundred ghosts," he implied her subservience to the past. To Cather, however, she was a magician who controlled the ghosts she conjured up, possessing the past rather than being possessed by it.[7] "In casual conversation, at breakfast or tea," Cather recalled, "you might at any time unconsciously press a spring which liberated recollection, and one of the great shades seemed quietly to enter the room and to take the chair or the corner he had preferred in life'' (*NUF,* pp. 61–62). Having lost the powerful medium of language to Mrs. Fields and her stories, the ''great shade'' of some vanished and silenced writer sat ''quietly,'' owing his existence to her ''liberated'' memories.

In coming into her inheritance of American and European culture, then, Cather received the male literary tradition through a woman's mediation. In ''148 Charles Street'' Cather uses her favorite metaphor—voice—to suggest the ways in which the male tradition was made accessible to her by a female intermediary. ''Mrs. Fields read aloud beautifully, especially Shakespeare and Milton, for whom she had, even in age, a wonderful depth of voice.'' Her voice had the power, in passing on the literature of the past, to leave it tinged with her presence in Cather's imagination: ''Many of those lines I can only remember with the colour, the slight unsteadiness, of the fine old voice'' (*NUF,* pp. 63–64).

In ''The Novel Démeublé'' Cather praises Tolstoi's ability to invest material objects with emotional significance; in his work, she writes, ''they seem to exist, not so much in the author's mind, as in the emotional penumbra of the characters themselves'' (*NUF,* p. 48). In ''148 Charles Street,'' the ''emotional penumbra'' emanating from Mrs. Fields and her home is nurturant and maternal, suggesting that in the bastion of chilly male literary authority Cather found a life-giving feminine warmth, a ''great power'' to nourish and protect:

> It was a power so sufficient that one seldom felt it as one lived in the harmonious atmosphere it created—an atmosphere in which one seemed absolutely safe from everything ugly. Nobody can cherish the flower of social intercourse, can give it sun and sustenance and a tempered clime, without also being able very completely to dispose of anything that threatens it—not only the slug, but even the cold draught that ruffles its petals. (*NUF,* p. 58)

The hint of spring glimpsed in the opening portrait is here fulfilled as Cather turns Mrs. Fields into a Beacon Hill Demeter, a female deity who creates a warm, sun-filled garden—a protected aesthetic and emotional space in which the guest could feel ''safe from everything ugly.'' Throughout ''148 Charles Street'' Cather invests Mrs. Fields with these maternal powers; she imagines her house as a place where the past could be ''protected and cherished,'' given ''sanctuary from the noisy push of the present,'' and her drawing room as a magical enclosure where ''all possibility of wrenches and jars and wounding contacts'' seemed ''securely shut out'' (*NUF,* pp. 61–63).

This description in ''148 Charles Street'' is echoed in a comment Cather made to Elizabeth Sergeant in 1910: 148 Charles Street was a ''literary Elysium'' which Cather hoped would stay protected from the ''flimsiness of the present age,'' sure to distress Mrs. Fields if allowed to enter (*WC:AM,* p. 41).

Cather may have wanted this literary Elysium protected more for herself than for Mrs. Fields. When Cather took Elizabeth Sergeant to visit 148 Charles Street, the older woman revealed, in Sergeant's opinion, a ''real curiosity about the contemporaneous in English literary circles.'' She wanted to talk about the people whom Cather had just met on her London mission for *McClure's,* but Cather was "refractory," Sergeant recalls, and "turned the talk to the manuscripts, the portraits and photographs that crowded the walls and tables. When the atmosphere thickened with ghostly presences, murmuring their wit and wisdom, Willa was happy and things were as they should be." So when Mrs. Fields tried to step out of the role Cather wanted her to play—the genteel, high-born lady, engrossed with the past—Cather needed to rewrite the script. Before their visit, Sergeant asked her friend if it was all right to mention the controversial Amy Lowell in Mrs. Fields' presence. Cather's reply suggested that she may have wanted to shield Mrs. Fields from "new forms of art" like Amygism for her own purposes: "Willa shuddered—new poetic forms were not welcome where we were bound—nor would Willa want them to be!" (*WC:AM,* pp. 64–65).

Cather was not the only younger person who wanted to make Mrs. Fields the voice of the past. Henry James could have been speaking of Cather when he referred to people who surrounded Mrs. Fields, "admiringly and tenderly, only to do in their interest all the reminding." [8] Although Mrs. Fields gladly filled the reminiscer's role when required, James saw in her compliance a gesture to the modern age: it was her younger reminders, James observed, who "preferred her possibilities of allusion to any aspect of the current jostle, and her sweetness under their pressure made her consentingly modern even while the very sound of the consent was as the voice of a time so much less strident." Coaching Sergeant for their visit, Cather laid out a conversational rule that, as James might have noted, may have been more in her own interest than in Mrs. Fields': one got along with her, she warned Sergeant, "by asking a tactful question that would lead her to one of her yesterdays" (*WC:AM,* p. 64).

Cather's attempt to stage-manage this visit to Mrs. Fields suggests that she was in part acting from a script that had already been written. The maternal imagery in "148 Charles Street" hints at—far behind the scenes—the presence of Cather's mother, the woman she loved, feared, and tried to propitiate; and, closer by, Isabelle McClung, who like Mrs. Fields, gave her friend a protected space that enriched her imagination. As a member of Boston's literary and social aristocracy, Mrs. Fields was also a lady, seated in a higher realm. Cather wanted to secure her place in this drawing room even if it meant censoring or transforming herself. And so Mrs. Fields' drawing room, like her mother's parlor at Willow Shade, could become a constricting social interior. After seeing Mrs. Fields and Cather together, Sergeant thought that the older woman regarded the younger as a "creature of zestful surprises who still needed a little toning down"; Cather for her part was eager to finish off her Nebraska rough edges with a little Boston polish (*WC:AM,* p. 65).

"Toned down" by Mrs. Fields, on Beacon Hill Cather was not the powerful, assertive woman who seemed expansive enough to fill her *McClure's* office. Sitting by the tea table at 148 Charles Street, offering "tender homage" to her hostess, anxious to accommodate Mrs. Fields' "gently superior manner" as she sipped her Souchong tea (imported, Mrs. Fields informed her guests, by a "cousin of Professor Barrett Wendell of Harvard"), Cather "stepped right out of her dominant personality," Sergeant recalled, "and melted into the warm light of our hostess' coal fire" (*WC:AM,* p. 65).

This ''melting'' and diminishment of personality marked Cather's attachment to Mrs. Fields in other ways. If Cather found her access to Western literary culture enhanced by the sound of Mrs. Fields' memorable voice telling stories of the ''great shades'' or reading from Shakespeare and Milton, the price she paid for this legacy was the regulation—in a metaphoric sense, the silencing—of her own voice. Eager to satisfy the expectations she attributed to Mrs. Fields, Cather set herself strict rules for letter-writing as well as for conversation. One had to be careful; communicating with Mrs. Fields was a perilous business. Cather later refused to allow Mark DeWolfe Howe either to reprint her letters to Mrs. Fields or to donate them to the Huntington Library on the grounds that they were artificial. When writing Mrs. Fields she never felt at ease, she told Howe, because she was afraid that she might touch one of Mrs. Fields' prejudices or somehow let the noisy modern world intrude; she wrote letters from a sense of duty, she added, because Mrs. Fields liked to receive mail. Cather took no pleasure in this correspondence; she would write long meaningless sentences, filling pages with writing yet saying nothing at all.[9]

In ''148 Charles Street'' Cather associates Mrs. Fields' power to silence with maternal disapproval. The lady could withdraw the warmth that made the drawing room, ideally, a place for flowering. Mrs. Fields had a beautiful, musical laugh, she recalled, with ''countless shades of relish and appreciation and kindness in it,'' but occasionally

> a short laugh from that same fragile source could positively do police duty! It could put an end to a conversation that had taken an unfortunate turn, absolutely dismiss and silence impertinence or presumption. (*NUF,* p. 58)

Cather did not want to be the silenced recipient of Mrs. Fields' laughing dismissal, a politely punishing version of her own mother's whippings, which did the ''police duty'' in the Cather family. That Cather viewed this ''slight and fragile'' elderly lady, clad in black velvet and Venetian lace, as a potent goddess who could destroy as well as protect is suggested by the associations she made to ''Thunderbolt Hill,'' Mrs. Fields' Manchester summer home. ''When I went a-calling with Mrs. Fields and left her card with Thunderbolt Hill engraved in the corner, I felt that I was paying calls with the lady Juno herself. Why shouldn't such a name befit a hill of high decisions and judgements?'' (*NUF,* p. 69).

Perhaps because she associated them with her own imperious mother, Cather was attracted in life and fiction to autocratic, at times coldly domineering women. The hint of steel behind the black velvet, the thunderbolt hovering over the tea table, were compelling dangers. In Cather's reminiscence ''A Chance Meeting,'' also published in *Not Under Forty,* Madame Franklin Grout is such a woman. As Ellen Moers observes, the ''frightening'' Grout has ''a rather fearsome relationship to the younger Willa Cather.'' [10] The many parallels between ''A Chance Meeting'' and ''148 Charles Street'' and between Madame Grout and Mrs. Fields make overt the power relationship that is only hinted at in her recollections of Mrs. Fields. By briefly considering ''A Chance Meeting,'' we can see more clearly how Cather granted Mrs. Fields the ability to silence her, creating a mother–daughter relationship that was inhibiting as well as enabling; we will also be better able to distinguish between the roles Mrs. Fields and Sarah Orne Jewett played in Cather's literary development.

•

Published in the *Atlantic Monthly* in February 1933—the magazine favored by 148 Charles Street—"A Chance Meeting" recalls Cather's visit to France in the summer of 1930. She had stayed at the Grand Hotel in Aix-les-Bains and there met Madame Franklin Grout, Flaubert's niece and the "Caro" of his letters. Madame Grout died the next winter, and Cather wrote "A Chance Meeting" as both memoir and elegy. Since she had met Madame Grout during her own mother's final illness and wrote "A Chance Meeting" after Virginia Cather's death in August 1931, it is likely that these painful experiences informed her portrait of the demanding matriarch.

Like "148 Charles Street," this supposedly nonfictional account is a shaped narrative. Inscribing Elizabeth Sergeant's copy of *Not Under Forty,* Cather wrote "These are true stories, told just as they happened," and for our purposes we can consider "148 Charles Street" and "A Chance Meeting" stories, as "true" to Cather's imagination as *My Ántonia* or *A Lost Lady* (*WC:AM,* p. 258). Like her memoir of Mrs. Fields, Cather's description of her meeting with Madame Grout is the daughter's story of inclusion in a powerful mother's world, although the price she pays for sharing the maternal calling card is higher in France than in Boston.

Cather begins her story with details that link Aix-les-Bains with Charles Street and Madame Grout with Mrs. Fields. She is staying at the "Grand-hotel d'Aix," a stately establishment built for nineteenth-century travelers who liked large rooms, large baths, and quiet; she shuns, she tells us, modern hotels that cater to the "fashionable trade; the noise and jazz and dancing." Cather has thus chosen the public equivalent of 148 Charles Street, a quiet, cultured realm, safe from the "noisy push" of the present. The woman Cather notices in the hotel's drawing room is, like the Mrs. Fields of "148 Charles Street," an elderly lady who seems to defy age, impatiently disregarding her lameness when she walks; she is also a powerful, regal woman whose head reminds Cather of the "portrait busts of Roman ladies."

The fearsomeness Ellen Moers observed in this relationship emerges long before Cather discovers that Madame Grout is Flaubert's niece. Cather's Madame Grout strikes me as opinionated, selfish, domineering, and even a bit cruel, although to the worshipful narrator, she is an inspiring and compelling—if demanding—feminine presence.[11] Cather merely presents the dialogue, offering no editorial comments; in doing so, she records a conversation about opera that evolves into a drama featuring a dominant mother and an acquiescent daughter. Their first interaction in the drawing room introduces the power dynamics that characterize the entire memoir: Madame Grout commands and Cather obeys, inconveniencing herself in the process. The evening is stifling, but the lady feels a draft and asks Cather to see that the doors are closed, which she does. Then the lady's attention shifts. "You attend the Grande-Cercle? You heard the performance of Tristran and Iseult last night?" inquires Madame Grout, who is also asking whether Cather shares her world.

> I had not heard it. I told her I had thought the evening too frightfully hot to sit in a theatre.
>
> "But it was not hotter there than anywhere else. I was not uncomfortable."
>
> There was a reprimand in her tone, and I added the further excuse that I had thought the principals would probably not be very good, and that I liked to hear that opera well sung.
>
> "They were well enough," she declared. "With Wagner I do not so much care about the voices. It is the orchestra I go to hear . . . ."
>
> I said I was sorry I had missed the opera.

The lady then gives Cather another chance to share her world and her calling card, asking her whether she plans to attend a performance of Ravel's *La Valse* the next day. The concert will, Madame Grout announces, be superb—"if you care for modern music." "I hastily said that I meant to go."

Although the following day was also "intensely hot" and she "dreaded the heat of a concert hall," Cather dutifully attended the performance. She was rewarded: mother knew best, and she "heard such a rendition of Ravel's *La Valse* as I do not expect to hear again; a small orchestra, wonderfully trained, and a masterly conductor." Deciding to leave the performance after the Ravel because the rest of program seems uninteresting, Cather is recognized by Madame Grout during the intermission:

> She beckoned to me and asked whether I had enjoyed the music. I told her that I had, very much indeed; but now my capacity for enjoying, or even listening, was quite spent, and I was going up to the Square for tea.
>
> "Oh, no," said she, "that is not necessary. You can have your tea here at the Maison des Fleurs quite well, and still have time to go back for the last group."

But this time Cather rebels, although she leaves Madame Grout with the impression that she will obey her orders ("I thanked her and went across the garden, but I did not mean to see the concert through"). The extent of her subservience to the imperious *grande dame* is conveyed in the almost comic image of the by-then renowned novelist, a commanding presence in her mid-fifties, stealing away from the Casino gardens through an inconspicuous grotto like a guilty schoolgirl who "had escaped from an exacting preceptress."

The two women meet again a few days later, after Cather takes a trip to the mountains, following Madame Grout's suggested itinerary (which "proved excellent, and I felt very grateful to her"). Once again she is impressed by the elderly lady's nobility:

> No one could fail to recognize her distinction and authority; it was in the carriage of her head, in her fine hands, in her voice, in every word she uttered in any language, in her brilliant, very piercing eyes. I had no curiosity about her name; that would be an accident and could scarcely matter.

But her name does matter. The fact that she is Flaubert's niece, as Cather then discovers, only adds to Madame Grout's authority: "so this must be the "Caro" of the *Lettres à sa Nièce Caroline*. There was nothing to say, certainly. The room was absolutely quiet, but there was nothing to say to this disclosure." Struck dumb by this confrontation with literary divinity ("There was no word with which one could greet such a revelation"), Cather expresses her reverence with the gesture traditionally accorded royalty: "I took one of her lovely hands and kissed it, in homage to a great period, to the names that made her voice tremble."

Like Mrs. Fields in "148 Charles Street," Madame Grout has the power to make the past accessible to Cather, giving the younger woman the connection to Flaubert she had not been able to achieve during her 1902 trip to France. And once again Cather offers her reverence to the regal woman, not the "great shade" she evokes; her imagination is stirred by female, not male power ("It was the Flaubert *in her mind and heart* that was to give me a beautiful memory" [italics mine]). Even Cather's reference to her favorite Flaubert novel—*Salammbô*—suggests female power. As the two women discuss the master's *oeuvre,* Madame Grout tells Cather that she may view his manuscripts " 'when you come to my place at Antibes. I call my place the Villa Tanit, *pour la déesse,*' she added with a smile." *Salammbô*'s "reconstructions of the remote and cruel past" dominate Cather's memory for the remainder of the evening as she thinks about the moon in *Salammbô,* which in turn evokes the "remote and cruel" goddess Tanit, one of the matriarchal deities who constitute the mythic Great Mother—the feared and adored mistress of life and death who, like her twentieth-century descendant Madame Grout, demanded propitiation and sacrifice from her children.

The resemblances among maternal goddesses, Madame Grout, and Mrs. Fields are dramatized by one seemingly extraneous detail which Cather mentions—the calling card, here the sign of the lady's lofty identity and emblem of her social power:

> Next morning the *valet de chambre* brought me a visiting card on which was engraved:
>
> MADAME FRANKLIN GROUT
> ANTIBES
>
> In one corner *Villa Tanit* was written in purple ink.

Thunderbolt Hill, Villa Tanit, ink the color of royalty, the lady Juno herself, the kiss on the lovely hand: the associations accumulate, linking Cather's two aristocratic matriarchs and recalling the autocratic women in her fiction—Marian Forrester, Myra Henshawe, Sapphire Colbert—who are also "remote and cruel" to adolescent admirers infatuated with their beauty and to slaves and daughters subjected to their power (*NUF,* pp. 3–42).

•

In an often-cited passage from *A Room of One's Own,* Virginia Woolf makes a seemingly anomalous reference to a female voice that can help us to understand the power relationships in "A Chance Meeting" and "148 Charles Street." Trying to explain why Jane Austen and Emily Brontë were able to "hold fast" to their own vision of the world in their writing, Woolf suggests that they were able to ignore the "perpetual admonitions of the eternal pedagogue—write this, think that."

> They alone were deaf to the persistent voice, now grumbling, now patronising, now domineering, now grieved, now shocked, now angry, now avuncular, that voice which cannot let women alone, but must be at them, like some too conscientious governess, admonishing them, if they would be good and win, as I suppose, some shiny prize, to keep within certain limits which the gentleman in question thinks suitable.[12]

At first glance, Woolf's "eternal pedagogue" seems to be the eternal patriarch, the voice of male authority telling women what they can be, think, and write. But the allusion to the "conscientious governess" allows a female voice to join the admonitory chorus, suggesting that a woman can be lectured by an eternal matriarch. She may be as oppressive as her male counterpart—perhaps even more so, since the mother, although denied social and economic power in patriarchy, possesses immense emotional and psychological power over the daughter. Certainly the opinionated, hectoring voice of Madame Grout dominates Cather's reconstruction of her chance meeting: "You can have your tea here at the Maison des Fleurs quite well. . . . If you will ask the boy to close the doors, we shall not feel the air. . . . But have you reserved your places? No? Then I would advise you to do so at once. . . . Speak idiomatically, please." Woolf's "too conscientious governess" is also Cather's "exacting preceptress" whose "persistent voice" cannot let her alone.

If both "148 Charles Street" and "A Chance Meeting" are "true stories" about Cather's encounter with the eternal matriarch, in both Cather awards herself the less powerful role of daughter, awed less by the difference in age than by the older woman's inherited social and cultural aristocracy. The extent of the homage she describes has unsettling implications for a woman writer, since in different ways—and to different extents—the mother's voice is paired with the daughter's silencing. As we have seen, talking with Mrs. Fields involved strict conversational rules, and writing letters required self-regulation and suppression. The verbal constraints Cather reveals in "A Chance Meeting" are even more severe, just as Madame Grout was a more demanding matriarch. These conversational rules required that Cather adopt a deferential, submissive posture ("I tried to let Madame Grout direct our conversations without suggestion from me, and never to question her"), and Cather portrays herself as intimidated by the language difference. Even before she knew who Madame Grout was, she felt insecure about speaking to her:

> As she passed my table she often gave me a keen look and a half-smile (her eyes were extremely bright and clear), as if she were about to speak. But I remained blank. I am a poor linguist, and there would be no point in uttering commonplaces to this old lady; one knew that much about her, at a glance. If one spoke to her at all, one must be at ease.

But when they did speak, Willa Cather was not "at ease." Madame Grout spoke "excellent English," while Cather's command of her mother tongue seemed to wane:

> In replying to her questions I fell into the stupid way one sometimes adopts when speaking to people of another language; tried to explain something in very simple words. She frowned and checked me with: "Speak idiomatically, please."

A "poor linguist" afraid to attempt French, speaking English in a "stupid way," Cather portrays herself as almost voiceless while Madame Grout displays "distinction and authority" in "every word she uttered in any language." "A Chance Meeting" is filled with Cather's praise of Madame Grout's linguistic power: she "spoke very good English, and spoke it easily"; she "seemed, indeed, to have a rather special feeling for language"; her "facility in languages was a matter of the greatest pride" to her uncle; her "speech, when she was explaining something, had the qualities of good Latin prose: economy, elegance, and exactness."

Madame Grout's lecturing voice particularly dominates "A Chance Meeting" because Cather recalls her conversation directly, using quotation marks to set off her remarks. Meanwhile she records her own responses as indirect discourse. Most likely her intention was to keep the reader's attention on her subject, and yet in "148 Charles Street," where Cather's purpose was similar, she gave herself a few lines and quotation marks too (" 'And who,' I brazenly asked, 'was Doctor Donne?' "). Cather's ability to record a "brazen" question is consistent with her portrayal of Mrs. Fields, whose quoted voice is far gentler and less dogmatic than Madame Grout's: "You know, my dear, I think we sometimes forget how much we owe to Dryden's prefaces," Mrs. Fields observes, including herself among those who fail to do sufficient homage to Dryden. Clearly the effect of Cather's decision to award the quotation marks only to Madame Grout is to mute the narrator's voice for the reader, reflecting the power balance in their conversation and relationship. Cather's response to the Frenchwoman's revelation of identity ("There was nothing to say, certainly") then becomes part of a disturbing pattern. Her inability to find the "word with which one could greet such a revelation" suggests not only her difficulty in matching profound emotion with verbal expression, but also her intimidation by a woman who seems to undermine Cather's ability to speak at all.

Cather's retrospective desire to reverse this linguistic and psychological dynamic and regain the writer's mastery of words may explain the curious ending of "A Chance Meeting." Cather concludes her memoir not in her own words, but by quoting a newspaper clipping sent to her by Parisian friends that announces Madame Grout's death:

> MORT DE MME. FRANKLIN-GROUT
>
> Nous apprenons avec tristesse la mort de Mme. Franklin-Grout, qui s'est éteinte à Antibes, à la suite d'une courte maladie.
>
> Nièce de Gustave Flaubert, Mme. Franklin-Grout a joué un rôle important dans la diffusion et le succes des oeuvres de son oncle. . . . Mme. Franklin-Grout était une personne charmante et distinguée, très attachée a ses amis et qui, jusqu'à la plus extrême vieillesse, avait conservé l'intelligence et la bonté souriante d'une spirituelle femme du monde.[*]

Like the ghostwritten autobiography of McClure, this is an ambiguous linguistic and literary act. In one sense Cather is ending the memoir as she began it, ''blank,'' muted, allowing a voice that speaks the language she cannot to override hers. But it seems significant that the last statement is an obituary, the public recognition of the matriarch's death. That this may be a death that frees the daughter linguistically, culturally, and psychologically is suggested by the French newspaper; its inclusion reminds the reader that although Cather may not speak the language fluently, she reads it—possessing some command over Madame Grout's verbal territory. In addition, by closing an essay directed toward an American audience with a French passage she neglects to translate, Cather leaves the reader—if unversed in French—in the same subordinate roles vis-á-vis Cather that the narrator assumes with her exacting preceptress. And the obituary finally reminds us that Madame Grout has, in fact, been silenced by death, and that in the act of writing ''A Chance Meeting'' and turning the preceptress into a character, Cather regains the verbal—and perhaps the emotional—power she did not experience in Madame Grout's presence. As language moves from the spoken to the written word, from experience to reconstruction in art, power passes from the mother to the daughter, from the literary figure to the woman writer.[13]

•

Both ''148 Charles Street'' and ''A Chance Meeting'' are stories of the daughter's subjugation by a mother who usurps her power of speech, a threatening possibility indeed for a woman writer. Yet the Mrs. Fields Cather describes in ''148 Charles Street'' and in her letters is a far less domineering maternal presence than Madame Grout; her tone is gentler, the climate cooler, the daughter more vocal. Yet Mrs. Fields' presence intensified Cather's class anxiety, her sense of Midwestern crudity, and her uncertain command of language. Describing herself as an American of the ''Apache period and territory'' suggests how a Nebraska background could seem like savagery to Beacon Hill, where Cather's ''Boeotian ignorance'' of Donne's poetry and Dryden's prefaces could not be hidden. Mrs. Fields had her opinions, however gently expressed, and Cather knew that her laugh could do ''police duty'' to offenders.

Cather's eagerness to please this maternal voice meant that the daughter might be silenced—not by ceasing to speak, but by speaking inauthentically as a writer. As long as Mrs. Fields would be reading her work by her tea table, Cather would have to confront the possibility that her stories—like her letters—might not pass muster. Many of Mrs. Fields' literary opinions were genteel and class conscious, as her introduction to a biographical sketch of Hawthorne reveals:

> In the year 1804, when Hawthorne was born, Salem, his birthplace, was a flourishing town. Cultivated persons were living there, and many of his contemporaries were men and women accustomed to the finest amenities of social life.[14]

Given Mrs. Fields' association of literary excellence with gentility, in her way she was as inhibiting to Cather's discovery of her native ''cadence'' and subject as was Henry James. In fact, Mrs. Fields did not like "The Bohemian Girl" (1912), a tale of passion and adultery set in Nebraska and featuring Scandinavian and Bohemian immigrants. A powerful story that anticipates *O Pioneers!,* "The Bohemian Girl" and its foreign-born characters did not offer Mrs. Fields the "cultivated persons" she saw in Hawthorne's Salem. She found it utterly charmless, Cather confided to Louise Imogen Guiney—she simply could not get past the bad manners of her characters.[15] The two also disagreed on *Alexander's Bridge:* Cather felt she was "through with Bartley" after the novel was published, ready for a new direction, while Mrs. Fields was "all praise and delight" (*WC:AM,* p. 76). One can see why Mrs. Fields would have been happier with *Alexander's Bridge* than with "The Bohemian Girl": set in Boston and London, its Beacon Hill setting—the Alexanders' quiet, harmonious drawing room—mirrored her world, and although the characters commit adultery they do so with good manners.

In speaking of the novel later, Cather stressed her slavish fealty to Henry James. But given the time at which it was written, it is likely that she was also speaking, in part, to Mrs. Fields and the genteel, upper-class literary and social world she represented and to which Cather wanted to belong. The link between these two literary parents is the drawing room, the space for formal conversation and constrained expression, for writing "sonnet[s]" or "sonata[s]," as Cather observed in "148 Charles Street," but not for writing American novels that did not follow formal rules (*NUF,* p. 60). Cather would leave both 148 Charles Street and James' house of fiction before she freed her country-bred voice from its Boston accent. Sarah Orne Jewett would be her guide.

•

Although Cather was impressed by Mrs. Fields' command of Charles Street's literary ghosts, simultaneously she was hoping to find a maternal inheritance that would allow her to write of her own past. This desire is revealed in an important story Cather published shortly before she met Mrs. Fields and Sarah Orne Jewett. Although "The Namesake" (1907) on the surface seems to be a Jamesian tale, on a deeper level it anticipates Cather's later creative breakthrough: It imagines the creative process as a yielding to inspiration, as a return to origins, and as the discovery of an American subject. In one sense "The Namesake" suggests Cather's continued association of creativity with masculinity: she imitates James and depicts artistic inheritance as patrilineal. But Cather also creates a feminine subtext that attributes the artist's discovery of his subject to his receiving a maternal legacy. Hence the story suggests that even before Cather entered 148 Charles Street in 1908, she was prepared to accept what Sarah Orne Jewett represented and could pass on, a female literary inheritance that directed her to her Western roots.

"The Namesake" begins with a Jamesian frame: seven American art students are gathered in the Paris studio of the great sculptor Lyon Hartwell, whose best-known works derive from and celebrate American experience: "We recalled his *Scout,* his *Pioneer,* his *Gold Seekers,* and those monuments in which he had invested one and another of the heroes of the Civil War with such convincing dignity and power." That evening they admire a new sculpture, called *The Color*

Sergeant; it is the figure of a young soldier running while "clutching the folds of a flag, the staff of which had been shot away." One of the students is depressed because he has been "called home," forced to leave Paris—the symbolic realm of art and culture—for America, seemingly a cultural desert. But when Hartwell tells the story of his birth as an artist, which resulted from the connection he establishes with his native land and with his "namesake" uncle killed in the Civil War, being "called home" is redefined. In the inset story—Hartwell's tale of his return to America—Cather gives the non-Jamesian view of creativity to which she was moving: art has its source in American soil, American history, and American lives.

Before his return to Pennsylvania, Hartwell was the apprentice serving several fathers who inhibit his originality: his own sculptor–father who designated the son to carry on his work, an Italian sculptor, and the "one master after another" he studies under in Paris. "Called home" to care for an ailing and senile aunt, he then spends two years in exile from the artistic world, sequestered at his family homestead in western Pennsylvania. At first Hartwell feels "no sense of kinship" in his ancestral home. But gradually he becomes curious about his namesake uncle who died in the Civil War; he hears the story of his brave death from village storytellers: his right arm "torn away" by shrapnel, his uncle had picked up the flag with his left hand, Hartwell discovers, and kept running until a "second shell carried away his left arm at the armpit" and he falls, the flag "settling around him."

Hartwell becomes increasingly preoccupied with his soldier–uncle, who remains an elusive ghost until he stumbles across a leather trunk in the attic engraved with his "own name"—also his uncle's. As he opens the trunk and examines its contents—clothes, letters, toys, a battered copy of the *Aeneid*—he achieves the "sense of kinship" he has been seeking ("I seemed, somehow, at last to have known him"). This moment of sympathetic union is followed by Hartwell's even more intense experience of merging with a force larger than the self—the force of connection to family, land, and his own creative power. This subsequent revelation takes place in the garden:

> The experience of that night, coming so overwhelmingly to a man so dead, almost rent me in pieces. It was the same feeling that artists know when we, rarely, achieve truth in our work; the feeling of union with some great force, of purpose and security, of being glad that we have lived. For the first time I felt the pull of race and blood and kindred, and felt beating within me things that had not begun with me. It was as if the earth under my feet had grasped and rooted me, and were pouring its essence into me. I sat there until the dawn of morning, and all night long my life seemed to be pouring out of me and running into the ground. (*CSF,* pp. 137–46)

In achieving this sense of union and wholeness, Hartwell reverses the pattern of his uncle's death; his mysterious experience in the garden destroys an old, false, "dead" self and creates a new one. This wrenching change "almost rent [him] in pieces." Yet because the metamorphosis leads to reunion and integration, Hartwell escapes the psychological equivalent of his uncle's death—being rent in pieces; he thus escapes the disease Cather told Jewett she was suffering because of her draining work—split personality. When he turns his creative energies to Civil War monuments, Hartwell transforms the nation's historical experience of fragmentation and conflict into a public representation of unity. At the same time he recaptures in the creative process the strengthening self-dissolution and attachment he experienced in the garden.

The story imagines the creative process as Cather wanted it to be (and as she may, occasionally, have experienced it). The loss of self the artist feels in uniting with something beyond the ego—here the maternal earth—is not the debilitating self-annihilation Cather feared (and was then experiencing in her editorial work), but the exhilarating dissolvement of the self into "something complete and great" that she describes in *My Ántonia* (p. 18). Nor is this the loss of self Cather feared when she told Sergeant she did not want to become another indistinguishable "cell in the family bloodstream"; it is a familial connection in which the self is enriched, not destroyed, similar to Thea's discovery of her ancestors in the anonymous Indian women who first made art from the necessities of life.

Written in the midst of her Jamesian apprenticeship and yet foreshadowing her own creative emergence, "The Namesake" is one of Cather's most autobiographical fictions. In describing Hartwell's artistic coming of age, she draws on her identification with her uncle William Seibert Boak, whose middle name she had adopted a few years previously. Connecting artistic self-discovery with a return from Europe to America, she rejects her literary subservience to the expatriate Henry James that she simultaneously reveals in the story's framing device. Although in her 1903 poem "The Namesake" Cather had openly announced her desire to take her fallen uncle's place, here she makes Lyon Hartwell a Northern soldier and sets the family homestead in Pennsylvania, suggesting that she was not yet ready to draw fully on her own roots for artistic inspiration. The story's ending also returns her to Europe and to her literary apprenticeship as she concludes with the students deferentially listening to Hartwell in his Paris studio. Hartwell may have escaped his artistic masters by returning imaginatively to his past, but Cather was not quite prepared to leave Henry James. Nevertheless, "The Namesake" is an almost uncanny prophecy of her eventual return to memory, to the past, and to Nebraska for fictional material.

On the surface, "The Namesake" identifies artistic achievement with the discovery of a masculine inheritance; Cather's bond with William Boak is mirrored in Hartwell's identification with his soldier–uncle. But "The Namesake" both reveals an ambivalence toward masculine enterprises and contains a subtext in which femininity and creativity are allied, just as Cather's attachment to William Boak connected her to her mother as well as to her mother's brother. By examining these contradictory patterns we can see how the creator of Lyon Hartwell could respond so strongly to Sarah Orne Jewett a year later.

Although the story appears to equate the manly arts of war and sculpture—apparently revealing the same associations among violence, force, and virility that dominated Cather's thinking about art in the 1890s—on a deeper level it condemns masculine energies and relationships grounded in power rather than in reciprocity. When Hartwell returns to his Pennsylvania homestead he confronts a landscape that Cather imbues with metaphoric and sexual meaning: a gentle, feminine, pastoral world is being ravaged and destroyed by industry and technology, masculine forces centered in nearby Pittsburgh. The artist is alienated, not inspired, by this ''relentless energy'' and ''shrieking force,'' which Cather connects with masculine pursuits that do not ''feed'' or ''warm'' but ''scorch and consume.'' So Hartwell finds refuge from the machine in the sanctuary of the garden, ''where the dropping of a leaf or the whistle of a bird was the only incident.'' It might seem odd that it is in the garden's feminine space that he achieves a mystic bond with his soldier–uncle, who appears at first to signify masculine force. Yet his uncle is a victim as well as a hero, and Cather connects his body's mutilation by shrapnel with the similarly brutal ravaging of nature's body by ''gas wells and mine shafts.''

Cather's criticism of masculine power, evident in her symbolic portrayal of landscape, is consistent with her revelation that Hartwell owes his discovery of creative force to women—and to expressing ''feminine'' attributes himself. Let us retrace the chain of events that leads him to discover his uncle's trunk in the attic, a chain that will lead him, and us, to a female creative inheritance.

Hartwell returns to Pennsylvania in the first place to take care of a female relative, his aging aunt; in doing so, he both abandons the professional world of art, associated with his father, and takes on a woman's nurturing role. His aunt then helps Hartwell to find his attachment to his soldier–uncle; she still observes the ritual of honoring the fallen soldier on Memorial Day, and it is at her insistence that the flag is raised. She sends Hartwell to the attic to search for the flag; there he discovers his uncle's trunk. And the trunk is there because his uncle's mother has preserved her son's belongings; they are ''still smelling faintly of lavender and rose leaves,'' the scent suggesting her continuing if invisible presence. The aunt and the mother, then, preserve the past from which the sculptor eventually derives inspiration and wholeness.

The spaces where Hartwell receives his revelation—the attic and the garden—are feminine realms connected in Cather's life and fiction with dream, reverie, creativity, and maternal care. Isabelle McClung—the ''garden fair''—had metaphorically granted her friend these two spaces, offering the attic room where Cather's writing could flower. Further in the biographical background is Cather's mother, who first offered her daughter an attic room and who preserved her dead brother's memory; in her psychological background is Cather's association of the creative process with regression. In contrast to ''The Burglar's Christmas,'' another story in which a ''namesake'' (Cather's alter ego William) is ''called home'' to a maternal presence, ''The Namesake'' presents the return to the past, to family, and to a female heritage as strengthening rather than debilitating. Later Cather would make these submerged patterns overt in *The Song of the Lark,* where the daughter/artist feels ''beating within [her] things that had not begun with [her]'' when she discovers her maternal inheritance.

•

So in ''The Namesake'' Cather links her character's creative emergence with the discovery of a maternal legacy. To see the fruition of this story in her own life, we need to return to the other lady at the Charles Street tea table. Later Cather may have recognized how "The Namesake" anticipated her literary and personal bond with Sarah Orne Jewett, because she echoed its feminine imagery in her "Preface" to the 1925 edition of *The Country of the Pointed Firs,* in particular using the metaphor of perfume to suggest the invisible maternal presence. Hartwell's belongings are still smelling of "lavender and rose leaves" when the younger Hartwell discovers them, and the garden is scented with the "woody odor" of roses the mother has planted. Similarly Cather attributed to Jewett's fiction a "quality of voice" that one could experience "without the volume at hand . . . as one can experience in memory a melody, or the summer perfume of a garden" (*CPF,* p. 7).

Sarah Orne Jewett was more an older sister than a mother, however, or, perhaps, a maternal figure who offered Cather love and support without the coerciveness of the mother–daughter bond we have considered in this chapter. Cather's imaginary memoir "When I Knew Stephen Crane" and her pilgrimage to Housman's London flat reveal her need, in the years preceding her meeting with Jewett, for a personal, intimate connection with a literary forebear; her visit to Mrs. Southworth's home showed her unable to imagine such a connection with a woman writer. Yet in Sarah Orne Jewett Cather found many people: a female literary precursor, a woman writer whom she could respect, a mentor, and an encouraging friend. Offering Cather an American and a female literary tradition as well as a friendship, Jewett helped the younger woman to unleash her creative powers so that, ultimately, Cather could write even of women who left her speechless: for she did have something to say.

NOTES

1. Henry James, "Mr. and Mrs. James T. Fields," *Atlantic Monthly* 116, no. 1 (July 1915): 21–31, esp. 23–24.

2. Mrs. Fields' social and literary milieu is described in M. A. DeWolfe Howe, *Memories of a Hostess: A Chronicle of Eminent Friendships* (Boston: Atlantic Monthly, 1922). For a discussion of the networks among nineteenth-century New England literary women, see Josephine Donovan, *New England Local Color Literature: A Women's Tradition* (New York: Frederick Ungar, 1983), pp. 38–49. For an account of the network by one of its members, see Harriet Prescott Spofford, *A Little Circle of Friends* (Boston: Little, Brown, 1916).

3. Larzer Ziff sees William Dean Howells' move from Boston to New York and his exchange of the editorship of the *Atlantic* for that of *Harper's* in the late 1880s as representative of this trend. See *The American 1890s: The Life and Times of a Lost Generation* (New York: Viking, 1966), p. 24. Howells went in the same direction when he changed publishers: he moved from Ticknor and Fields to Harper's.

4. Willa Cather to Ferris Greenslet, Houghton Library, Harvard University, Cambridge, Mass.

5. Willa Cather to Sarah Orne Jewett, October 24, 1908, Houghton Library, Harvard University, Cambridge, Mass.

6. James, p. 24.

7. James, p. 30.
8. James, p. 30.
9. Willa Cather to Mark DeWolfe Howe, November 11, 1931, Houghton Library, Harvard University, Cambridge, Mass.
10. Moers is quoted in *The Art of Willa Cather,* ed. Bernice Slote and Virginia Faulkner (Lincoln: University of Nebraska Press [Bison ed.], 1974), p. 83. As Moers perceptively notes, there are parallels between "A Chance Meeting" and the "tensions in her portraits of all mature women in the fiction."
11. Perhaps because her remarks were made in conversation, Ellen Moers is even more direct: Madame Grout was, in her opinion, perhaps the "worst bitch in nineteenth-century French literature" (Slote and Faulkner, p. 82).
12. *A Room of One's Own* (New York: Harcourt, Brace, 1929), p. 78.
13. Here is an English translation of the obituary:

> DEATH OF MME. FRANKLIN-GROUT
>
> We learn with sadness of the death of Mme. Franklin-Grout, who passed away at Antibes, after a brief illness.
>
> Niece of Gustave Flaubert, Mme. Franklin-Grout played an important role in the dissemination and the success of her uncle's work. . . . Mme. Franklin-Grout was a charming and distinguished person who was very attached to her friends; and she was someone who, until the last years of her life, displayed the intelligence and the good humor of a witty woman of the world.

14. Annie Fields, *Nathaniel Hawthorne* (London: Kegan Paul, 1899), p. 1.
15. Willa Cather to Louise Imogen Guiney, August 17, 1912, Holy Cross College Library, Worcester, Mass.

15

"A Gift from Heart to Heart"

> The language her people speak to each other is a native tongue. . . . Such an idiom makes the finest language any writer can have; and he can never get it with a notebook. He himself must be able to think and feel in that speech—it is a gift from heart to heart. ''Miss Jewett''

> You must find a quiet place near the best companions (not those who admire and wonder at everything one does, but those who know the good things with delight!). Sarah Orne Jewett to Willa Cather, December 1908

Encountering Sarah Orne Jewett in the Boston twilight was a moment of recognition as well as discovery: Willa Cather had seen the other lady at the tea table before. Jewett ''looked very like the youthful picture of herself in the game of 'Authors' I had played as a child,'' she later recalled in ''Miss Jewett,'' ''except that she was fuller in figure and a little grey'' (*NUF*, p. 54). Jewett was already established in Cather's mind, then, as an ''Author,'' a literary authority like the other male companions on the playing cards, who included some of the ghosts of Charles Street—Hawthorne, Longfellow, and Holmes.

Even if Cather had not recognized Jewett's face from her childhood game, she knew the name of the woman sitting next to Mrs. Fields. Cather had admired her work in the 1890s, and by 1908 Jewett was a respected American writer whose *The Country of the Pointed Firs* (1897) was widely considered the high-water mark of local-color fiction. Jewett's stories and sketches had consistently been appearing in the *Atlantic Monthly* and other periodicals since the 1870s; she had published nineteen books, beginning with *Deephaven* (1877); and her fiction was admired by writers like Kipling, Howells, and James, who later praised her ''beautiful little quantum of achievement'' in ''Mr. and Mrs. James T. Fields.'' [1] If the ''little'' in James' temperate accolade anticipates the dismissals of Jewett's fiction common in the 1930s and 1940s when ''regional'' fiction became equated with ''female'' and ''minor,'' the ''beautiful . . . achievement'' more accurately reflects the substantial literary reputation Jewett enjoyed when Cather met her in the late winter of 1908.

Cather would later observe of Jewett that ''friendships occupied perhaps the first place in her life'' (*NUF*, p. 85). After reading Jewett's letters, I would instead say that she happily did not have to choose whether to let her writing or her friends come second: she found her professional and personal commitments complementary. Cather may have assumed that Jewett put her friends first because during their brief friendship (which lasted until Jewett's death in 1909) she was too ill to write fiction. Jewett never fully recovered from the injuries she suffered in a 1902 carriage accident and devoted her remaining years to the offices of friendship—conversation, visiting, letter-writing. Her last important attachment was to Willa Cather.

Both women gained from their short association. Cather offered the ailing Jewett the chance to be a mentor and to find a literary inheritor, while Jewett gave Cather the loving support and sound literary advice the younger woman needed during a time of professional conflict and self-doubt. In offering Cather the right advice at the right time—encouraging her to find her own material and to speak in her own voice—Jewett aided her new friend in leaving *McClure's,* discarding her Jamesian persona, and fashioning the literary self she would later call the "real me."[2] Catalyst to Cather's emerging creativity, the friendship—although brief—had a lasting impact on the younger writer.

Introducing her edition of Jewett's letters, Annie Fields told her readers they would find in them "the portrait of a friend and the power that lies in friendship to sustain the giver as well as the receiver."[3] This image of mutuality captures the essence of Jewett's friendship with Cather. Almost twenty-five years older, an accomplished writer, and the daughter of New England gentry, Jewett outranked Cather in age, literary achievements, and social class. Yet their relationship was characterized by reciprocity, not hierarchy. Unlike Mrs. Fields, Jewett was a mother and mentor who used her powers of persuasion and affection to encourage Cather to find literary independence, as we will see in looking at her influence on Cather's fiction in the next chapter.

Theirs may have been a chance meeting, but the growth of their friendship and Jewett's contribution to Cather's creative metamorphosis were not accidental. When Cather told Jewett that their first meeting had been something she had long awaited, she implied that she had already imagined Jewett—or someone like her—well before she entered that Charles Street door.[4] And so in meeting Jewett, Cather also met part of herself.

•

Although she disliked the quaintness and sentimentality she saw in most local-color writing, Cather admired Jewett's fiction. Even in the 1890s when she was attacking realism in the Lincoln and Pittsburgh papers, she made an exception of the Maine writer. She praised *A Country Doctor* to her Pittsburgh friend and literary confidante George Seibel and gave him a copy of *The Country of the Pointed Firs;* since Seibel and Cather shared literary tastes, the gift was a sign that the judgmental Cather considered Jewett an exceptional writer: America's "best woman writer," she informed Elizabeth Sergeant in 1910 (*WC:AM,* p. 40).

When she met Jewett, Cather was impressed by the woman as well as the writer. Jewett was a "lady, in the old high sense," she observes in "Miss Jewett" (*NUF,* p. 83). She represented a female identity that Cather associated with power rather than oppression—the cultured, self-assured woman born to a heritage of wealth, breeding, and refinement. Although the modest Jewett (who liked to think of herself as "one o' the Doctor's girls") would not have described herself in such aristocratic terms, she did come from a family higher in the social scale than Willa Cather's.[5] Her grandfather Captain Theodore Jewett, a prosperous shipowner and merchant, had left his descendants financially independent; unlike Cather, Jewett never had to struggle to support herself.

Captain Jewett also left a tangible bequest, the spacious and elegant colonial home in the center of South Berwick where Jewett was born and where she lived, at least for part of each year, until her death. She grew up surrounded by comfort and grace, accustomed to "mahogany four-post bedsteads, Adam mirrors, Chippendale chairs and tables, Wedgewood candlestick holders, a beautifully turned spinet, and swiss hand-embroidered curtains." [6] Cather loved Jewett's home—she loved its "summer coolness, the sweet breath of the garden flowers in the shuttered lower hall," the "patina of the Chippendale and Hepplewhite, the old portraits and prints" (*WC:AM*, p. 58). She frequently visited South Berwick after Jewett's death, staying with her sister Mary. The "composed and quiet atmosphere of the old Jewett house" made her realize how her "starvation" for a richer world than Red Cloud had distorted her portrayals of Nebraska in stories like "The Sculptor's Funeral" and "A Wagner Matinee." Jewett, she imagined, had been at "peace with her environment" (*WC:AM*, p. 67).

Jewett's genteel background and gracious home attracted Cather but they did not intimidate. Unlike Madame Grout or Annie Fields, she was not a lady who made Cather feel awkward, unrefined, or speechless, perhaps because Jewett's egalitarian ideal of a relationship, whether between a writer and a reader or between two friends, was based on a model of conversation. As Cather later observed, Jewett never made "pronouncements." Her opinions she always "voiced lightly, half-humorously; any expression of them was spontaneous, the outgrowth of the immediate conversation" (*NUF*, p. 88).

One can see Jewett's refusal to make pronouncements in her letters to Cather. She invariably presents her opinions delicately and considerately, as in her critique of "On the Gull's Road" (1908), where Jewett thought Cather had made a mistake in using a male narrator:

> And now I wish to tell you—the first of this letter being but a preface—with what deep happiness and recognition I have read the "McClure" story,—night before last I found it with surprise and delight. You have drawn your two figures of the wife and her husband with unerring touches and wonderful tenderness for her. It makes me the more sure that you are far on your road toward a fine and long story of very high class. The lover is as well done as he could be when a woman writes in the man's character,—it must always, I believe, be something of a masquerade. I think it is safer to write about him as you did about the others, and not try to be he! And you could almost have done it as yourself—a woman could love her in that same protecting way—a woman could even care enough to wish to take her away from such a life, by some means or other. But oh, how close—how tender—how true the feeling is! the sea air blows through the very letters on the page. Do not hurry too fast in these early winter days,—a quiet hour is worth more to you than anything you can do in it. (pp. 246–47)

How different this caring voice is from the hectoring, lecturing one of Madame Grout! Jewett makes her literary advice seem gentle—even tentative ("you could almost have done it as yourself"). These are suggestions, not commands ("Speak idiomatically!"). She mutes her criticism, weaving it in among so much praise that the recipient of this letter would have to be encouraged rather than disheartened. And she assures Cather that reading the story has strengthened her affection for her and concludes by returning to their personal bond, so that Cather's last impression would be of Jewett's tender concern ("Do not hurry too fast in these early winter days").

Exhausted and depleted by her *McClure's* work, Cather was drawn to the more peaceful and privileged life Jewett enjoyed, but similarities also contributed to her immediate sense of kinship. Jewett's "patina" might have been Beacon Hill and Chippendale, but her core was Maine's rocky soil. "I am a country person," she would tell guests at 148 Charles Street, and the fact that the two women shared rural origins helped Cather to appreciate the literary possibilities of her Nebraska past.[7] Jewett was also a writer whose creative imagination was sparked by memory; like Cather, she frequently attributed the writer's gift to the child's first observations, her earliest "[interest] in the lives of people about me." Jewett's early memories were of rural storytellers and their "graphic country talk"—upcountry farmers, elderly sea captains, Berwick townsfolk, Maine's equivalent of the Nebraska farm women who entranced the young Cather with their tales of Europe. "I listened as eagerly as any one," Jewett remembered, and from this material, like Cather, she later derived her fiction—the gift that she both received from and returned to her beloved homeland.

Even though as a child the writer had "no consciousness of watching or listening, or indeed of any special interest in the country interiors," Jewett later realized how much experience of others' lives she had absorbed in lonely inland farmhouses or seacoast cottages as she waited for her father, a country doctor. Once she realized that "everything that an imaginative child could ask" she had possessed in South Berwick, Jewett told aspiring writers not to yearn for different surroundings or greater opportunities; the materials for art, she knew, were in the "simple scenes close at hand." [8]

Elizabeth Sergeant uses Cather's comment that "Miss Jewett, too, had turned away from marriage" to explain Cather's choice of art over human relationships. But since both Cather and Jewett formed their primary intimacies with other women, Cather's remark suggests that Jewett may have been a model for her personal as well as her professional life (*WC:AM*, p. 116). Although Jewett, having come of age in the 1870s, was spared the conflicts the younger woman experienced when she realized that feminine friendship was "unnatural" in a society that was yoking lesbianism with deviance, Cather must have seen the parallels between their emotional choices. During the period of their friendship, Cather was making an important personal decision—to live with Edith Lewis; the Jewett–Fields ménage, however innocently Victorian, gave her the example of two women harmoniously sharing their lives and their work in a marriage-like bond. Assuming her approval, Cather wrote Jewett of the pleasure she and Lewis were taking in their new apartment.[9]

As Elaine Showalter has observed, because women writers—even those who draw on a distinct "women's culture"—inhabit a male-dominated society, their writing simultaneously exists inside two literary traditions. "If a man's text, as Harold Bloom and Edward Said have maintained, is fathered, then a woman's text is not only mothered but parented; it confronts both paternal and maternal precursors and must deal with the problems and advantages of both lines of inheritance." A woman writer thinks back through her fathers as well as her mothers, then, and is the product of both the "dominant" and the "muted" traditions.[10] Given the social assumptions of women's literary as well as social inferiority, the woman writer, like Willa Cather, may want for a time to ignore the maternal heritage entirely and view herself as her father's favorite daughter—or as his son. In either case, she ignores the muted tradition because she wants her own voice to be heard.

Sarah Orne Jewett, however, considered herself equally the daughter of the male and female literary traditions: unlike Cather, during her youth she neither scorned women writers nor viewed the ones she admired as "anything but women." Hence Jewett offered her protégée two lines of literary inheritance. Fittingly, Jewett's parents introduced her to her paternal and maternal precursors; her father urged her to read Sterne, Fielding, Smollett, and Cervantes, while her discovery of "the pleasant ways" of Austen, Eliot, and Mrs. Oliphant was a maternal legacy.[11]

Jewett's most treasured inheritance from the male literary tradition was her debt to Flaubert's theory of fiction. She took two of his dicta as "constant reminders": *Écrire la vie ordinaire comme on écrit l'histoire*" and "*Ce n'est pas de faire rire, ni de faire pleurer, ni de vous mettre à fureur, mais d'agir à la façon de la nature, c'est à dire de faire rêver*" (p. 165). These sentiments contributed to her characterization of her own work as "imaginative realism"; she wanted both to be faithful to the ordinary, probable course of social life—to write fiction like history—and to write allusive, suggestive, resonant prose that would leave circles of meaning to expand in the reader's mind.[12] Jewett's aesthetic theory was doubly a paternal inheritance, since both aspects of Flaubert's advice were echoed by her own father. "Don't try to write *about* people and things," he would say, "tell them just as they are." [13] And to tell things as they were, Dr. Jewett, like Flaubert, thought that the writer should make the reader dream:

> Father said this one day "A story should be managed so that it should *suggest* interesting things to the *reader* instead of the author's doing all the thinking for him, and setting it before him in black and white." [14]

Although she took Flaubert as her aesthetic conscience, Jewett was a discriminating reader who did not feel compelled to admire the nineteenth-century's great men unless their work affected her, and she once shrewdly suggested the debt at least one writer may have owed to a woman: "How much that we call Wordsworth himself was Dorothy to begin with," she reflected to Mrs. Fields (p. 77). Jewett was not blind to literary politics, realizing that a writer's worth might fail to be recognized if she were a woman. She considered Ouida's *A Village Com-*

mune to be as ''powerful a story'' as any of Tolstoi's. Ouida was a ''great writer,'' she wrote Annie Fields, ''when she is at her best, there is no getting over that fact. If she didn't lose her head, and—perhaps—were she not a woman, we should hear much more of Ouida!'' (p. 209).

Conscious of the social and literary assumptions that kept readers from hearing ''much more'' about women writers, Jewett did not receive her literary inheritance solely from the dominant literary tradition. ''Muted'' female voices spoke as loudly to her as did those of Flaubert, Howells, or James. Harriet Beecher Stowe, the most prominent literary woman in the preceding generation, was her first model; Flaubert may have given her a literary aesthetic, but in Stowe's fiction Jewett discovered her subject: after reading *The Pearl of Orr's Island,* where Stowe wrote ''about people of rustic life just as they were,'' Jewett was inspired to attempt her own sketches of Maine life.[15] Stowe was an enabling mentor because her work showed Jewett the possibilities of local-color fiction without exhausting them; *Pearl* was an ''incomplete piece of work,'' Jewett thought, and so more New England stories could be written (p. 47). George Sand's stories of rural French life were another important early influence, and Jewett kept a portrait of Sand over the mantle in her South Berwick home.

As she developed this friendship, Cather made an important bond not only with an individual writer, but also with a woman nourished by nineteenth-century female literary culture and women's friendships.[16] So Cather became, by inheritance, a member of a female literary tradition as well. When she later placed a portrait of Sand over the mantle in her New York apartment, the gesture reveals her inclusion in this tradition since she was honoring Jewett as well as Sand. Members of Jewett's inner circle included Boston poet Louise Imogen Guiney; poet and naturalist Celia Thaxter; artist Sarah Wyman Whitman, who designed the covers of Jewett's *Strangers and Wayfarers* (which Jewett in turn dedicated to Whitman), *The Queen's Twin,* and *The King of Folly Island;* cellist Sara Norton, daughter of Harvard's Charles Eliot Norton; and New England writer Harriet Prescott Spofford. If we extend the circle we find among her correspondents most of the women writers of the day: Mary Wilkins Freeman, Alice French, Elizabeth Stuart Phelps, Mary Murfree, Rose Terry Cooke, Susan Coolidge, Alice Brown, Violet Paget (Vernon Lee), Mrs. Humphrey Ward, and Julia Ward Howe.[17] Many of these women were friends with each other, and through letters and visits they sustained each other personally and professionally. Pet names and endearments were the signs of special attachment and inclusion within the group. Jewett referred to Thaxter as ''Sandpiper,'' and Thaxter signed her letters to Jewett with a sketch of the bird; Louise Imogen Guiney was ''Linnet''; Jewett was ''Owlet'' to Celia Thaxter and ''Pinny'' to Annie Fields, who in turn was ''Fuff'' or ''Mouse'' to Jewett.[18]

•

Along with late nineteenth-century women writers like Mary Wilkins Freeman and Kate Chopin, Jewett belonged to the literary generation following the ''literary domestics'' of the mid-nineteenth century and preceding that of the major women writers of the early twentieth—Cather, Wharton, Glasgow, Stein. In this intervening position, Jewett escaped the conflicts surrounding gender and vocation that affected the domestic writers, who feared that writing might be unwomanly, and the young Willa Cather, who at first wanted writing to be unwomanly.[19] As Cather observed, Jewett's neighbors considered her writing a "ladylike accomplishment," and Jewett evidently did as well (*NUF,* p 86). Her letters to friends, editors, and publishers show none of the conflicts between gender and vocation evident in Cather's 1890s journalism and her early fiction.

Why was this so?

Working within the largely female-dominated mode of local-color writing and specializing in the short story, Jewett both inherited and helped create a school of fiction in which women's contributions were valued and indeed sanctioned by New England's literary men; thus she did not have to feel that she was straying too far from women's appropriate literary realm. The modesty of Jewett's temperament extended to her view of her writing. She did not aspire to individual greatness: her stories were "sketches" and their purpose was to enrich readers' understanding of Maine life, not to enhance her reputation. At the height of her fame, she was known to say, "You know, I seem impressive, but really I only come up to my own shoulder."[20]

Jewett did not have to struggle as fiercely as Cather did to find an acceptable female literary tradition. Not only did she draw on Stowe, whom Cather scorned as a moralist, but she also placed herself in a non-literary and non-elite tradition of female creativity that Cather did not discover until she was sure of her artistic power. Perhaps because Jewett had not been as captured by the romantic and masculine ideology of the heroic artist as Cather, she found it easier to imagine art as collaborative and domestic and to see the links among the stories Maine countrywomen told, the creativity they expressed in daily tasks like cooking, gardening, and herb-collecting, and the artist's imagination. Describing a visit to her South Berwick dressmaker to Annie Fields, Jewett commented that Miss Grant had been "in the full tide of successful narration" (p.37). She was always willing to acknowledge how the "successful narration[s]" of Miss Grant and other Maine storytellers contributed to her own fiction. After attending the dressmaker's funeral in 1905, Jewett wrote Annie Fields that she couldn't "take it in that I shall see that lively, friendly, quaint, busy creature no more. My stories are full of her here and there, as you know" (p. 208).

Able to see the literary powers displayed in women's daily conversation, Jewett creates in her fiction female characters who are also successful narrators. Her women demonstrate a flair for figurative language, derived from folk sayings and Maine's native idiom, that makes their conversation vivid and expressive. "I must say I like variety myself," remarks Mrs. Todd; "some folks washes Monday an' irons Tuesday even when the circus is goin' by!" (*CPF,* p. 37). But Jewett's women rise most fully to artistry not in casual remarks but in the stories they tell, either about themselves or, more frequently, about other women whose stories they want to pass on, as in "The Foreigner" and the "Poor Joanna" sequence in *The Country of the Pointed Firs.* The narrator who listens to these tales and in turn tells them to us is a writer, a woman like Jewett herself—and so Jewett dramatizes not only the connections between storytellers and listeners but also those between unconscious women artists and conscious ones.

Moreover, Jewett did not find artistry inconsistent with qualities and tasks conventionally considered feminine: Mrs. Todd's creativity is expressed in her knowledge of herbs, spices, and the ancient female art of healing. Jewett thought of herself as a healer; through her work she hoped to ease tensions and conflicts among groups who did not understand each other—country people and city people, in particular—by increasing the reader's sympathy for others. Explaining to her friend Mary Mulholland why she valued both writing and reading, Jewett wrote that "the people in books are apt to make us understand 'real' people better, and to know why they do things, and so we learn to have sympathy and patience and enthusiasm for those we live with, and can try to help them in what they are doing, instead of being half suspicious and finding fault."[21] Whereas Willa Cather chose the opera singer to represent her own literary evolution in *The Song of the Lark,* in *A Country Doctor* Jewett makes the physician the woman writer's alter ego and suggests that her desire to be a doctor is both natural and feminine; Nan Prince expresses the power of healing that Jewett ascribes metaphorically to the woman writer and literally to the herb-gatherer Mrs. Todd.

Associating writing and medicine with the traditionally female traits of nurturance and empathy, in granting herself and her heroines a nondomestic social role Jewett was not overtly challenging Victorian assumptions about woman's nature. Her vision of womanhood in fact is not too different from that held by Stowe and the other literary domestics whose fiction Cather scorned. How, then, could she have been an artistic model for Cather?

Despite her modesty, Jewett regarded her vocation with greater aesthetic seriousness than did the domestic and sentimental writers of the previous generation. Hence she offered Cather a more acceptable version of the woman writer than either Stowe, Southworth, or Ouida—without seeming "anything but" a woman, as had Sand and Eliot. Implicitly challenging the mutually exclusive categories and polarities Cather had used to define and dismiss women writers in the 1890s, embodying the "magic of contradictions" as would Cather's Marian Forrester, Jewett refuted the young Cather's charges against literary women (*LL,* p. 79).

Most women writers, Cather had complained in her 1890s journalism, were not artists. They did not shape or perfect their craft, but carelessly spewed forth whatever thoughts or emotions ran through their pens. Yet Jewett took Flaubert as her mentor because she wanted to affect her readers by a finely honed art. Whereas some members of the preceding generation of women writers disclaimed artistic responsibility for their works, Jewett self-consciously experimented with form and technique, willingly devoting time and effort to improving her craft. As her self-confidence increased along with her technical mastery, she increasingly wanted credit for her own work, so her modesty coexisted with a writer's pride in achievement. She thus rejected the anonymity and pesudonyms favored by the previous generation of women writers. Jewett published two early stories as "Alice Eliot" ("Mr. Bruce" and "The Shipwrecked Buttons"), but by 1876 she had adopted a more assertive attitude toward her editor, her name and her work: "My dear Mr. Scudder," she wrote. "My name is Sarah Orne Jewett, if you please; and when you are arranging the index, can you credit a story to me which was called *Mr. Bruce?*"[22]

After rereading *The Pearl of Orr's Island,* Jewett concluded that she and her mentor Stowe had different attitudes toward their vocation. Regretting that Stowe had not finished the book "in the same noble key of simplicity and harmony" which she began it, Jewett reflected on the "unconscious opposition" that confronts the writer, the demands of life that deflect the attention from art: "You must throw everything and everybody aside at times, but a woman made like Mrs. Stowe cannot bring herself to that cold selfishness of the moment for one's work's sake" (p. 47). While in her stories she explored the female capacity for attachment, for friendship, for sympathy, Jewett was capable of that seemingly unfeminine "cold selfishness of the moment" for the sake of her writing, the solitary devotion to art that Cather thought essential to creative endeavor. Even when she was not writing, she told an editor, she was "always thinking about [her] work." [23] Having made the commitment to art that Stowe could not, Jewett could then tell Cather that she must have time, quiet, and concentration in order to "keep and guard and mature" the creative force she was pouring into her editorial work (p. 248).

During her undergraduate years, Cather imagined that the woman artist who achieved such a single-minded commitment would, like Eleanora Duse, be a solitary priestess of art. Jewett, however, did not find the writer's momentary "selfishness" contradictory to the demands and delights of friendship, and so demonstrated that love and work might coexist, a hopeful sign to Cather who was developing her relationship with Edith Lewis at this time.[24] Her Boston marriage allowed Jewett to enjoy both intimacy and solitude; she would spend half of each year either at 148 Charles Street or traveling with Annie Fields, then back to South Berwick to reattach herself to her native soil. With a busy social and literary life of her own, Mrs. Fields could easily grant her friend the time she needed to write; Jewett in turn trusted her friend's support of her writing, and so could postpone visits to Boston without fearing that Mrs. Fields would be hurt by the delay.

Jewett escaped other limitations Cather had attributed to women writers in her 1890s journalism. She consciously tried to avoid the aesthetic pitfalls that claimed female authors who devoted themselves, as Cather phrased it, to "hobbies and missions"; knowing that she had to guard against her tendency to inject Christian truths into her fiction, Jewett strove for a suggestive rather than a didactic art. Although she drew on the "power of loving" which Cather thought the woman writer's particular strength, Jewett was determined to control sentimentality. Jewett's training in nineteeth-century Continental fiction taught her the virtues of simplicity, control, and understatement; she did not want to "lose her head" as Ouida had. As she warned Cather in 1908, one had to be careful not to let "sentiment" degenerate into "sentimentality"—a code word for the emotional excesses of the female pen (p. 249). Writing of Jewett's fiction in her 1925 "Preface" to *The Country of the Pointed Firs,* Cather paid her a high compliment: her best work was "tightly built and significant in design" (p. 6). Unlike the women writers Cather despised, then, Jewett could both draw on emotion and control it; she combined the woman writer's gift of emotion with the male writer's command of form.

A woman writer who drew on the European and American, the male and female literary traditions; a woman writer who wanted to acknowledge female precursors, yet sought to transcend their aesthetic limitations; a woman writer who was modest yet who took herself and her craft seriously; a woman writer who spent half the year talking with Boston literati on Beacon Hill and half the year listening to village dressmakers in South Berwick; a woman writer who combined work and friendship; a woman writer who considered herself both a woman and a writer, without struggling to connect those identities—Jewett could help Cather achieve ''integration and tranquillity'' because she had done so herself (*WC:AM*, p. 54). By expressing traits the culture had divided between ''male'' and ''female,'' uniting her commitments to art and to friendship, seriously pursuing a literary vocation while being a ''lady, in the old high sense,'' Jewett demonstrated to Cather that professional artistry need not be a male preserve. This possibility might have been lost on the college student, but Cather was ready for it by the time she met Jewett; she had been gradually reaching the same conclusion herself.

Even before meeting Jewett, Cather had encountered her indirectly. Witter Bynner, her *McClure's* colleague, was impressed by *The Troll Garden* and took it upon himself to send it to Jewett as well as to James. Unlike James, Jewett read the book, and she sent Bynner her assessment in May 1905; we can assume that Bynner showed the letter to Cather, as he did the reply from James. Jewett found the stories ''full of talent,'' but the first part of her letter expresses her discomfort with their bleakness:

> I cannot help wishing that a writer of such promise chose rather to show the hopeful, *constructive* yes—even the pleasant side of unpleasant things and disappointed lives! Is not this what we are bound to do in our own lives and still more bound to do as writers? I shrink more and more from anything that looks like giving up the game.

Jewett went on to say that the ''better stories'' in the collection most revealed the ''danger'' of pessimism, which she linked with the outmoded literary mode of naturalism (''we have outgrown the bleaker realism of twenty years ago''). But she went on to honor the power of Cather's stories. ''I was *very* glad to read The Sculptor's Funeral again—and again,'' she wrote. ''It has touched me profoundly.'' She concluded by asking Bynner to ''give my best messages to Miss Cather'' and speculating that ''We should have much to say if we could talk together.'' [25]

Eager for her own fiction to contribute to people's spiritual and moral welfare, Jewett evidently found stories like ''Paul's Case'' and ''The Sculptor's Funeral'' disturbing even though they ''touched [her] profoundly.'' When she and Cather did finally talk together, brought together by their chance meeting at 148 Charles Street in 1908, Jewett did not impose her views on the younger writer by urging her to express a softer, more benevolent view of humanity in her work. Indeed, she praised ''The Sculptor's Funeral'' as the high point of Cather's fiction, and this was one of grimmer stories in *The Troll Garden*. Once Jewett knew Cather personally, she became a more sympathetic and supportive critic than she shows herself to be in this letter because she was responding not so much to the work

Cather had already written as to the work she thought Cather would write—to Cather's literary promise. And since Jewett viewed Cather as a writer of great talent she became, for a time, both the mirror of Cather's artistic self and the maternal friend who helped her to construct that self. To see how Jewett's friendship was enabling rather than oppressive, we need to examine the letters the two women exchanged in 1908. Like Jewett's Maine stories, these were a "gift from heart to heart."

•

The first letter from Jewett to Cather reprinted in Annie Fields' edition is dated August 17, 1908, several months after their meeting in the Charles Street drawing room. In it Jewett strikes a note of mutuality and reciprocity that characterizes the rest of the correspondence. Evidently she had promised Cather that she would write a story for *McClure's* if health permitted, for she tells Cather that

> You will find that I sent a verse that I found among my papers to "McClure's,"—and I did it as a sort of sign and warrant of my promise to you. No story yet, but I do not despair; I begin to dare to think that if I could get a quiet week or two, I could really get something done for you, and it should be for you who gave me a "Hand up" in the spring! (p. 234)

In her next letter Jewett describes a friendship in which both women are givers and receivers. She was "sorry to miss the drive to the station and last talk about the story and other things," she wrote, but "I was too tired—'spent quite bankrupt!' " In missing the chance to talk with Cather about her story—a conversation that would have benefitted the younger writer—Jewett keenly felt the loss herself: "And I knew that I was disappointing you, besides disappointing and robbing myself, which made it all the harder. It would have been such a good piece of a half hour!" (p. 245). She then went on to offer heartfelt praise and gentle criticism of "On the Gull's Road."

In December 1908 Jewett sent perhaps the most important letter Cather ever received. "I think it became a permanent inhabitant of her thoughts," Edith Lewis observes (*WCL,* p. 67). It is a wonderful letter, the sign of the first mentoring relationship between two major American women writers. Like Jewett, I wonder whether we might not have heard "much more" of this letter were the sender and the recipient not women, perhaps even as much as we have heard of Emerson's congratulatory letter to Whitman. At once encouraging and admonitory, Jewett reminds Cather of her great talent and warns that magazine work can only impede the development of her literary "gifts":

> My dear Willa,—I have been thinking about you and hoping that things are going well. I cannot help saying what I think about your writing and its being hindered by such incessant, important, responsible work as you have in your hands now. I do think that it is impossible for you to work so hard and yet have your gifts mature as they should. . . . I do wish in my heart that the force of this year could have gone into three or four stories. In the "Troll-Garden" the Sculptor's Funeral stands alone a head higher than the rest, and it is to that level you must

> hold and take for a starting-point. You are older now than that book in general; you have been living and reading and knowing new types; but if you don't keep and guard and mature your force, and above all, have time and quiet to perfect your work, you will be writing things not much better than you did five years ago. This you are anxiously saying to yourself! but I am wondering how to get at the right conditions. (pp. 247–48)

As Jewett realized, the "right conditions" for writing fiction were psychological and emotional as well as vocational. To be sure, Cather's magazine work was taking time and energy away from her writing, but Jewett also felt that Cather did not possess sufficient detachment from—and control over—her material. What Cather needed to do, Jewett thought, was paradoxical. She needed both to be "surer" of her "backgrounds," the subject matter she had inherited (Nebraska, Virginia, newspaper life); and she needed to see them more "from the outside,—you stand right in the middle of each of them when you write, without having the standpoint of the looker-on who takes them each in their relations to letters, to the world" (p. 248).

This is the advice that struck home. Cather gives a similar formulation in the often-quoted 1922 "Preface" to *Alexander's Bridge:*

> One of the few really helpful words I ever heard from an older writer, I had from Sarah Orne Jewett when she said to me: "Of course, one day you will write about your own country. In the meantime, get all you can. One must know the world *so well* before one can know the parish." (*AB*, [1922], p. vii)

In suggesting that one must know the world before one can see the world in the parish, Jewett was passing on to Cather the goal Emerson thought the American writer should pursue: to find transcendent meaning in the everyday materials of American life, as Whitman did in *Leaves of Grass*. Jewett was thus not giving uncommon advice; drawing on strains in both American romanticism and realism, she was telling Cather that if she found her "own quiet centre of life"—her own deepest self—she would be able to "write life itself," to create fiction so realistic that it would not seem like art. Cather would find her writer's voice, Jewett thought, in the vernacular rather than the genteel tradition.

In later references to Jewett's influence Cather always connected her literary development to the self-discovery she felt Jewett had aided. The language Cather uses suggests how closely Jewett's advice was derived from romanticism and American individualism, the intertwined narratives of self-creation that Cather had been using all along. Shortly after *O Pioneers!* appeared, Cather attributed her return to Nebraska material to

> Sarah Orne Jewett, who had read all of my early stories and had very clear and definite opinions about them and about where my work fell short. She said, "Write it as it is, don't try to make it like this or that. You can't do it in anybody else's way—you will have to make a way of your own. If the way happens to be new, don't let that frighten you. Don't try to write the kind of short story that this or that magazine wants—write the truth, and let them take it or leave it." (*KA*, p. 449)

This advice seems even more Emersonian than Jewett's 1908 letter; Cather remembers Jewett echoing "Self-Reliance" 's injunction to express the self, even if the way be "wholly strange and new." [26] But the difference between Jewett's and Emerson's or Thoreau's formulation of romantic self-discovery also emerges in her letter to Cather, where she suggests that the literary self is not found in isolation but in relationship. The writer can only transcend convention, she hints to Cather, if she speaks to an audience that encourages self-expression:

> Your vivid, exciting companionship in the office must not be your audience, you must find your own quiet centre of life, and write from that to the world that holds offices, and all society, all Bohemia; the city, the country—in short, you must write to the human heart. (p. 249)

Jewett did not leave the question of audience as abstract as the "human heart." While she stressed Cather's need for solitude and autonomy ("To work in silence and with all one's heart, that is the writer's lot; he is the only artist who must be a solitary, and yet needs the widest outlook upon the world"), she also encouraged her to "find a quiet place near the best companions (not those who admire and wonder at everything one does, but those who know the good things with delight!)." Writers must be "ourselves," she went on to observe—indeed, "our best selves"—but her letter implies that one cannot discover or express one's "best self" by oneself; the "best companions" to whom one speaks, and who expect the best, must help to elicit this self.[27] The creative process, as Jewett described it to Cather, is an extension of the domestic art of letter-writing; the writer creates with an audience always in mind, and like the letter the text is a bridge connecting the writer and the reader.

Jewett ended the letter by reassuring the younger woman that she had been "growing" even when she felt "most hindered" and reminding Cather that she was not alone: "I have been full of thought about you." Asking for a reply to her letter, Jewett assumed the role of the "best companion" to whom Cather might speak honestly: "You will let me hear again from you before long?" she queried (p. 250).

Unlike Mrs. Fields and Madame Grout, Jewett engaged Cather in an unstrained conversation. Shortly after receiving this letter, Cather sent off the long, self-revealing reply we looked at in Chapter 13, in which she openly confessed her dissatisfaction with magazine work and her fears of literary inadequacy. Jewett had become her literary conscience as well as her mentor. In 1911, two years after Jewett's death, Cather would tell Elizabeth Sergeant that her spirit "separated her in a benign way from her worst temptations . . . the lure of big money rewards in the crude sense—for cheap work—letting all the people who wanted you steal your energy and time" (*WC:AM*, p. 62).

•

Jewett was Cather's guide out of Henry James' and Annie Fields' drawing rooms to the "road home," and not only because she was a woman writer whose art grew from her native soil. In Jewett, Cather also saw dramatized a pattern of creativity similar to that she had discerned in Shakespeare and Carlyle: artistic inspiration arising not from power or force but from love and sympathy, the same emotions of attachment that flowered in friendship. In her 1891 essay "Shakespeare and Hamlet" Cather imagined Shakespeare so moved by his sympathy for Hamlet's legend that he wrote his play to express this feeling; in the process, he gave back to Hamlet the "legacy of his love" which, in turn, he passed on to the reader or audience (*KA*, p. 433). In her 1925 "Preface" to *The Country of the Pointed Firs* Cather twice used the word "gift" to describe Jewett's artistry, a metaphor recalling the "legacy" she had imagined Shakespeare receiving and passing on. The "gift from heart to heart" was the fresh, colloquial, idiomatic language Jewett inherited from her Maine countryfolk, the foundation for her own distinctive voice and cadence, while her "gift of sympathy" was her ability to give herself "absolutely to [her] material" (*CPF*, pp. 10, 7).

In describing the creative process as she imagined Jewett experiencing it, in the "Preface" Cather describes a self-effacement that may seem to reveal the Victorian equation of femininity with self-denial and submissiveness: "this gift of sympathy is his *[sic]* great gift; is the fine thing in him that alone can make his work fine. He fades away into the land and people of his heart, and dies of love only to be born again. The artist spends a life-time in loving the things that haunt him" (*CPF*, p. 7). But this is not the abnegation of selfhood Victorians thought a female vocation. Cather grants Jewett the artist's sympathetic imagination that the undergraduate critic had attributed to Shakespeare. This is the alternative pattern of creativity Cather mentions in her 1890s journalism, opposed to the athletic force Kipling represented; this is the pattern she found in Jewett. Jewett thus helped Cather to envision Shakespeare's legacy—the self-transcending imaginative power that was not, if conventional definitions of gender were applied, "manly"—being passed on to a woman writer.

In using the word "gift" to describe the source and quality of Jewett's art, Cather implies self-expression as well as inheritance and connection. On the one hand, a writer's "gift" is something individual, unaccountable, innate—a mysterious presence with no clear human source or explanation. Jewett used "gift" in this sense when she warned Cather that her "gifts" were not maturing as they should and when she told editor Horace Scudder that she did not "wish to ignore such a great gift as this God has given me. I have not the slightest conceit on account of it—indeed I believe it frightens me" (p. 29). But a gift is also something received from a person or a place, like the "gift from heart to heart," or something the writer gives back, like the "gift of sympathy." So in Cather's writing, as in Jewett's, the gift suggests the magic of contradictions, being simultaneously a metaphor of selfhood and of relationship. The gift metaphor appears frequently in Jewett's fiction; after their meeting it begins to appear in Cather's fiction as well, where it signifies both creativity and attachment. By imagining women as the gift-givers, both writers revise the traditional associations between "woman" and "gift" in patriarchal culture, in which women are the gifts which men give each other.

The parallels between friendship and the creative process as Jewett, and later Cather, experienced them can best be illustrated by considering a Jewett story that develops the gift metaphor: ''Martha's Lady,'' which Cather treasured and included in her 1925 edition of Jewett's fiction. The story of a friendship between women that persists and deepens despite differences in social class, decades of separation, and marriage, ''Martha's Lady'' is also the story of the growth of an artist. Like the village dressmaker whose storytelling flair Jewett admired, Martha is a creator whose raw material is the texture of everyday life; her birth as an artist—like Willa Cather's—is connected with her discovery of self in relationship with another.

In a letter to Jewett, Cather confessed that she had read ''Martha's Lady'' over and over, and it is possible that her intense response to the story arose from her identification with Martha, the clumsy maid who learns housekeeping skills from Helena.[28] Awed by Harriet Pyne's gracious home, Martha is at first an uncouth outsider unfamiliar with the rituals of the tea table. But Helena, Harriet Pyne's cousin, instructs Martha in the skills she will need. Passing on the traditions of flower-arranging and food preparation that she has inherited from her own mother, Helena gives Martha a ''gift from heart to heart.''

Loving Helena and eager to emulate and please her, Martha becomes an expert housekeeper during her friend's absence. Because Helena is both the model and the imagined audience for Martha's creativity—the reader of the texts she creates in table settings and flower arrangements—maintaining their relationship does not demand the friend's physical presence. Just as Jewett wanted Cather to write to the ''best companions'' and to the ''human heart,'' so Martha envisions her absent friend approving of her variations on the ''fine art of housekeeping.''

Jewett signifies the self-expression this reciprocal friendship allows by the gifts Helena leaves Martha: a little box filled with a few ''trifles from her dressing-table.'' The most important keepsake is the mirror, into which Martha gazes always hoping to ''see some faint reflection of Helena.'' But this memento—which suggests the daughter's recognition of identity and difference in relation to her mother, her first mirror—does not give her a ''reflection of Helena'' but of her own self: ''there was only her own brown old New England face to look back at her wonderingly.'' Her memory of Helena is thus a mirror in which Martha finds the gift of herself. In seeking to please the ''perfect friend'' through her housekeeping arts, Martha begins to express herself—just as Cather created her own novelist's voice in carrying on a literary conversation with Jewett.

Elizabeth Abel suggests that the ''distinctive dynamics of fusion and differentiation'' characterizing female friendship—the formation of self in relation to another first structured in the daughter's preoedipal bond with the mother—may characterize women writers' experience of literary influence.[29] Cather's personal and literary relationship with Jewett supports this speculation: neither defining herself as identical to Jewett nor as completely different, Cather looked into Jewett's mirror and saw herself. In a letter Cather expressed her hope that Jewett would like ''The Enchanted Bluff,'' a slight story which, she admitted, McClure had sniffed at, saying it was all introduction. But Jewett might like it, Cather continued, because it was different from the things she had known as a child in Maine. Growing up in the West, she had been surrounded by a Latin influence, Cather wrote, just as Jewett had been by an English one; Nebraskans like Cather were exposed to many Spanish words, just as Jewett had heard words left over from Chaucer's day. So she was going to be bold, Cather declared, and send along the story McClure had spurned.[30]

Just as Jewett allowed for both similarity and difference in her letter of literary advice to Cather (one day she would write about her "own country," as Jewett had, but she would do it in her own way), so Cather notes both points of contrast and comparison in her letter. Her country had Spanish roots and words, Jewett's had British—this was a difference she thought Jewett would appreciate. And yet in returning to the landscape and legends of her Nebraska childhood, as she did in "The Enchanted Bluff," Cather was following Jewett's example.

•

Jewett's death in 1909 was a terrible blow to Cather. She heard of her loss while in London, on a scouting trip for *McClure's*, and immediately wrote Mrs. Fields. Numbness and inertia were descending, she wrote. Now that Jewett was dead, Cather realized that her London activities only had meaning because she thought they might interest her. Life seemed dark and purposeless.[31]

Cather's depression, however, did not endure. Although the friendship was brief, the conversation the two women began was enduring. After her death Jewett became a permanent resident of Cather's imagination and memory, just as Helena did in Martha's. During 1910 and 1911, after Jewett's death and before *O Pioneers!*, Cather made several pilgrimages to South Berwick. Staying in Jewett's house and surrounded by her possessions, lingering in the garden where she had walked, she regained an emotional and spiritual connection with her absent friend. After one visit she felt a particular urgency to commit herself full-time to writing. "The old house had stirred something up—she felt goaded," she told Sergeant. "It was as if Miss Jewett's spirit, which filled the place, had warned her that time was flying" (*WC:AM*, p. 60).

In 1911 Annie Fields published her edition of Jewett's letters, which included the 1908 letter of literary advice and warning to Cather. Re-encountering Jewett's words in print may have influenced Cather's decision, made in that year, to leave the magazine; she was in a sense receiving the letter again, but now as a public statement. In deciding to "give myself up to the pleasure of recapturing in memory people and places I had believed forgotten," as she did in writing *O Pioneers!*, Cather was continuing the conversation with Jewett she began in 1908, inventing a literary voice while talking to her "best companion":

> I dedicated my novel *O Pioneers!* to Miss Jewett because I had talked over some of the characters in it with her one day at Manchester, and in this book I tried to tell the story of the people as truthfully and simply as if I were telling it to her by word of mouth. (*KA*, p. 448)

Writing *O Pioneers!* was then a reciprocal act of communication: a letter, a gift received and returned, a conversation with an absent friend.

The relationship Cather enjoyed with Jewett in memory—when the older writer took the roles of ideal reader, maternal friend, and literary authority—was spared the vicissitudes and conflicts of actual human relationships as well as the particular tensions that characterize mother–daughter relationships in life. These tensions might have arisen had the two enjoyed a longer friendship, as a comment Cather made about "Martha's Lady" suggests. Every time she read "Martha's Lady," Cather confided to Jewett, she felt the way she used to after a childhood whipping—as if she must try very hard to be good.[32] Reading the story evidently recalled her desire to please her own mother, who administered the family whippings. Even though the relationship Jewett created between Helena and Martha was marked by mutual affection, for Cather it evoked the coercive aspects of the mother–daughter relationship

Cather mourned Jewett's death in 1909, but this loss may have prevented her from turning Jewett into a silencing or an inhibiting literary mother, someone who would have required her to be "good." For it is possible that if Jewett had read more of Cather's work—which did not become as optimistic as she might have hoped—the concerns she expressed to Bynner might have returned and Cather might have had to confront Jewett's gentle disappointment. Thus Jewett may have played an even more important role in Cather's memory than she could have in life. Jewett probably would have liked *O Pioneers!* had she lived, but in Cather's imagination she unquestionably did.

•

As Cather's literary precursor and encouraging friend, Jewett was a parental figure who did not dominate or silence not only because she encouraged Cather to speak in her own voice but also because her own achievement, although substantial, was still modest. Jewett's medium was the short story, not the novel, and she did not have a lengthy *oeuvre* in this genre; *The Country of the Pointed Firs* and a dozen short stories constituted the best of her work, as Cather decided when she edited the Houghton Mifflin collection of Jewett's fiction in 1925. This was a less intimidating body of fiction to contemplate than Henry James' numerous and lengthy novels. So the daughter could surpass the mother without having to free herself from an inhibiting predecessor. In fact, once Cather had committed herself to novel-writing she wanted to acknowledge her debt to Jewett and to grant her mentor renewed life. She first reciprocated Jewett's creative gift when she wrote the dedication to *O Pioneers!*:

TO THE MEMORY OF
SARAH ORNE JEWETT
IN WHOSE BEAUTIFUL AND DELICATE WORK
THERE IS THE PERFECTION THAT ENDURES

When Cather prepared her edition of Jewett's fiction in 1925, the roles of mother and daughter, established writer and obscure writer, had been reversed. By then one of America's foremost novelists, the winner of the Pulitzer Prize for *One of Ours,* praised by such shapers of literary taste as H. L. Mencken and Carl Van Doren, Cather could take the role of Demeter, trying to rescue her Persephone from the literary darkness of the 1920s when Jewett's quiet voice—seeming old-fashioned and quaint to writers and readers listening to Joyce, Eliot, and Lawrence—was not being heard. In preparing the edition and writing the preface,

Cather was hoping to give Jewett a new generation of readers and to ensure that the perfection she found in her work would, in fact, endure.

Cather's letters to Houghton Mifflin editor Ferris Greenslet concerning this edition show her desire to award Jewett both literally and figuratively the stature she deserved. Writing Greenslet in February 1924, Cather insisted that the edition be issued in standard-sized volumes. Some people, she told Greenslet, would ask for "The White Heron" and then give it back to librarians or booksellers because its diminutive size made it look like a child's book, and she did not want that to happen with the new edition. Greenslet was thinking of prefacing the new edition with a poetic tribute by a friend of Jewett's which Cather considered "old-lady-poetry," and she demanded that it be removed: she did not want Jewett associated with second-rate women writers.[33]

Because Cather wanted to control the reader's encounter with Jewett and to distinguish her work from children's and women's fiction, she further insisted on exercising the editor's prerogative to write a preface. In doing so, she had a literary purpose: she wanted to suggest the "range and character" of Jewett's achievement by defining the spirit of her fiction—her sympathy for her subject matter, her sensitivity to place and landscape, her command of colloquial speech, her mastery of her craft.

But Cather also had a political purpose: she wanted to place Jewett firmly within the American literary canon, which she accomplished by grouping her with the nation's foremost male writers: "If I were asked to name three American books which have the possibility of a long, long life, I would say at once, 'The Scarlet Letter,' 'Huckleberry Finn,' and 'The Country of the Pointed Firs.' I can think of no others that confront time and change so serenely." While it is possible to imagine formal or thematic reasons why Cather would have associated *Country* with these two novels, given her major purpose in preparing the edition—to rescue Jewett from obscurity and belittlement—it is most likely that Cather simply wanted to associate Jewett's masterpiece with two novels of unquestioned stature. By juxtaposing *Country* with American classics, Cather was hoping to increase Jewett's size.[34]

Certainly Cather wanted to ensure that Jewett's work would last. No writer can achieve literary immortality without a continuing readership, and Cather predicted that Jewett would have this (in part, of course, because of the edition she was editing):

> The latter book [*Country*] seems to me fairly to shine with the reflection of its long, joyous future. It is so tightly yet so lightly built, so little encumbered with heavy materialism that deteriorates and grows old-fashioned. I like to think with what pleasure, with what a sense of rich discovery, the young student of American literature in far distant years to come will take up this book and say, "A masterpiece!" as proudly as if he himself had made it. It will be a message to the future, a message in a universal language, like the tuft of flowers in Robert Frost's fine poem, which the mower abroad in the early morning left standing, just skirted by the scythe, for the mower of the afternoon to gaze upon and wonder at—the one message that even the scythe of Time spares. (*CPF*, p. 11)

Concerned in 1925 with cultural continuity, distressed by personal as well as social change, Cather then wanted to believe that in speaking the dialect of the parish she and Jewett were still sending a "message in a universal language" to the world. In placing Jewett in the male-dominated literary canon from which she had once felt excluded, as her literary inheritor Cather was also including herself. So even though in some ways the world divided in 1922, the year Cather later chose as her symbolic moment of discontinuity, in 1925 she imagined that Jewett's work—and thus her own—would span the gap, finding readers even in the age when male writers like Fitzgerald, Faulkner, and Hemingway were struggling to supplant James, Hawthorne, and Twain. In March of the same year, attending a dinner in honor of Robert Frost's fiftieth birthday, she gave the poet a copy of her edition of *The Country of the Pointed Firs* as a present, perhaps a symbolic way of ensuring Jewett's continuity and immortality as well as her own.

•

Jewett's advice to Cather was not only literary. When she told her to "find a quiet place near the best companions," Jewett was recommending a creative life similar to her own. But an artist's life filled with friends as well as work was not what Cather had imagined in the 1890s when she had assumed that the artist must be solitary as well as masculine. Literary achievements, she then thought, required the creator to reject both family and friends for a lonely quest. "If an artist does any good work he must do it alone," she wrote in "Shakespeare and Hamlet." Indeed, "no number of encouraging or admiring friends can assist him, they retard him rather" (KA, pp. 434–35). And in *The Song of the Lark* she seems to echo this view when she has Thea Kronborg proclaim that "Your work becomes your personal life. You are not much good until it does" (*SL*, p. 456).

But in reality Cather's creative life was closer to Sarah Orne Jewett's than to Thea Kronborg's. Like Jewett, Cather did not have to choose between writing and companionship. She found several close friends who, in different and complementary ways, nourished her creativity. Recognizing the "good things with delight," these friends—all women—provided emotional and professional support during the crucial period when Cather was deciding to commit herself full-time to writing.[35]

Several women composed Cather's outer circle of friends during the *McClure's* years: Boston poet Louise Imogen Guiney; Pennsylvania-born novelist Margaret Deland, by then an adopted citizen of literary Boston; New York social reformers Josephine and Pauline Goldmark; *McClure's* manuscript reader Viola Roseboro'; and Ida Tarbell, who maintained cordial ties with her replacement. But it was the inner circle of friends who contributed the most to Cather's literary emergence, as Jewett had predicted. In addition to Isabelle McClung, Edith Lewis, Elizabeth Sergeant, and Zoë Akins remained Cather's friends and readers for a lifetime.

Even after Cather's departure from Pittsburgh, Isabelle McClung remained her primary love, her best "best companion." The mother of her writing, the solicitous Isabelle provided love and companionship while not intruding on her friend's solitude. She kept the attic study always ready, and Cather completed *O Pioneers!* and a good part of *The Song of the Lark* in Pittsburgh.

Although Edith Lewis was not as important a figure in Cather's life during this time of transition, she was beginning to take over some of Isabelle's supportive functions. In 1911 Cather observed to Elizabeth Sergeant that she needed to work "in a corner protected by someone who knew what it was all about," and she mentioned both Edith Lewis and Isabelle McClung as her protectors (*WC:AM*, p. 61). Lewis could not offer Cather Murray Hill Avenue's spacious, gracious living, but she knew how to create the emotional and psychological sanctuary Cather needed to write.

Lewis' account of her first meeting with Cather reveals the younger woman's awed deference to Cather's literary and conversational powers. Having just graduated from Smith, Lewis was spending the summer of 1903 with her family in Lincoln. She met Cather at the home of Sara Harris, editor of the *Courier*. As a girl Lewis had read Cather's book and drama reviews in her hometown papers. She found them the work of a "daring, provocative, original, imaginative" mind and had long wondered "who Willa Cather was." Prepared to be impressed, Lewis remembers how silent she became in the presence of a "genius" with "extraordinary eyes" (*WCL*, pp. ix–xii). Cather and Harris did all the talking while Lewis remained a spectator, as overwhelmed by Cather as Cather would later be by Madame Grout.

Lewis' memoir records her initial and continued fascination with Cather's literary and personal power, but she does not suggest why the older woman pursued the friendship. Quite likely, Cather found Lewis' unqualified admiration attractive: in 1903 she had not yet published *The Troll Garden*, and yet here was someone who regarded a Pittsburgh schoolteacher as a gifted writer. In any case Cather asked Lewis to visit her at the McClungs', and Lewis spent a week with her in New York the following summer. When Cather took the job with McClure, she first moved into Lewis' apartment house at 60 South Washington Square. In 1908 the two women set up their own apartment at 82 Washington Place, and in 1913 they began housekeeping at 5 Bank Street.

A spacious, sunny apartment where she worked on most of her major novels, Bank Street was an ordered, harmonious, almost sacred space to Cather. Although she frequently left New York for rest, travel, and inspiration—writing in her summer retreats in Jaffrey, New Hampshire, and Grand Manan Island, New Brunswick—the Bank Street apartment was, as Sergeant astutely observes, the "walled stronghold of her very self" (*WC:AM*, p. 112). After years of living in other people's houses—her cramped family home in Red Cloud, Pittsburgh boardinghouses, the McClungs' mansion—in this setting Cather established her own "quiet centre of life." For a woman whose emotional and creative equilibrium could be so disrupted by change and dislocation, this was a significant act, even more important than her entry into Murray Hill Avenue since this was a home of her own.

At Bank Street Cather literally created her own "garden fair," a sustaining, fertile environment where her imagination could bloom. She liked to write surrounded by flowers—orange blossoms and camellias in the winter, jonquils and lilacs in the spring. Entering her apartment, Sergeant recalls, was like walking into a greenhouse—the air was sensuous, at any season filled with the "heavy-spicy perfume" of flowers (*WC:AM*, p. 112). Here Cather was "at home" to friends on Friday afternoons, "eager as a countrywoman," Sergeant remembers, "to feed and warm her guests" (*WC:AM,* p. 130). The importance of this home to Cather's creative and emotional well-being is evident in her response to its loss: forced to relocate in 1927, she felt "exposed and miserable," like a turtle without its shell (*WC:AM,* p. 227).

Cather, however, did not create this sheltering home alone. This was a shared space, and it is possible that without Edith Lewis (or someone like her) Cather might not have been able to fashion such a nourishing and harmonious domestic environment. The two women's tastes were similar. After they had decorated the apartment with mahogany chests and a dining room table, oriental rugs, comfortable chairs, books, etchings, and photographs, they agreed to give "no more thought to acquiring new things, or getting better ones than those we had. What money we had we preferred to spend on flowers, music, and entertaining our friends" (*WCL,* p. 89).

Cather and Lewis were close professional colleagues as well as friends and housemates. After Lewis took a job with *McClure's* (on Cather's suggestion), she read proof for the Mary Baker Eddy series, and this marked the start of a lifelong collaboration. Beginning with *The Song of the Lark,* Lewis began to help Cather proofread her fiction, an activity which became "one of our greatest pleasures" (*WCL,* p. xviii). But Lewis' description of her first meeting with Cather suggests that she offered more than technical assistance. "Whatever she said had an evocative quality," Lewis remembers, "a quality of creating much more than her words actually stated, of summoning up images, suggestions, overtones and undertones of feeling that opened long vistas to one's imagination" (*WCL,* p. xvi). Lewis presents herself here as the ideal reader Cather imagined in "The Novel Démeublé"—sensitive to the "inexplicable presence of the thing not named"—and indeed her memoir shows her to be a sensitive, although frequently an uncritical, reader of Cather's fiction.

Lewis remained employed in publishing and advertising (eventually working for J. Walter Thompson), but she would take long vacations to accompany Cather on trips and help her establish summer residences in Jaffrey and Grand Manan. Clear to both whose career and needs took precedence, the two women lived together, sharing life and work, for almost forty years. When subway construction uprooted them from Bank Street in 1927, Cather and Lewis found temporary quarters at the Grosvenor Hotel (which became a five-year-residence). They moved to their final home, a Park Avenue apartment, in 1932.

One doesn't learn much about this relationship from Lewis' cautious memoir, *Willa Cather Living.* Despite the title, Cather never comes to life—she is Genius observed from a respectful distance. In her zeal to protect her friend's privacy, Lewis gives no sense of the rhythms of their daily life, the reasons for their mutual affection, the domestic and emotional roles each played. It is possible to reconstruct some aspects of the friendship by drawing on Cather's letters and the views of contemporary observers, but we will probably never be able to glimpse its deeper emotional currents.

Although their bond was intimate, it is unclear whether or not it was sexual. Cather and Lewis had separate bedrooms in the Bank Street apartment as they did in their Grand Manan cottage, and they always reserved two rooms when traveling together. The evidence of separate bedrooms is ambiguous, however—it may signify two friends who have an emotional and nonsexual bond; or two lovers who enjoy their privacy; or two lovers who want visitors and outsiders to assume that they are only friends. If the relationship did have a sexual component, it was not a turbulent, disruptive one: Lewis offered Cather tranquillity and protection, not passionate intensity.

In Elizabeth Sergeant's view, Cather was unquestionably the dominant partner in the ménage. Yet every ship's captain has a first officer, she writes, who "does a lot the captain never knows about to steer the boat through rocks and reefs." As Cather's first mate, Lewis "gracefully helped and seconded" when the captain prevailed at social gatherings, Sergeant remembers, "unquestioned in her primacy" (*WC:AM*, pp. 202, 130). Lewis helped and seconded in many ways besides proofreading galleys and passing the canapés. She steered the boat into smooth waters by taking care of those messy, intruding details of daily life that we all wish someone else would handle for us; Lewis hunted for apartments, made hotel reservations, took care of the baggage, ran errands, fended off unwanted visitors, and generally served as a buffer between Cather and the outside world. When Cather entered the hospital for a gallbladder operation in 1942, she checked in under Lewis' name; the deception, intended to ensure privacy, suggests how important Lewis eventually became as a shield and a protector.[36]

In addition to removing these burdens, Lewis refrained from making excessive demands on her friend's time. She was available when needed, absent when not. Cather could devote herself to writing knowing that she had a companion for dinner, opera, and the theater as well as for her vacations, and yet she knew that she did not have to include Lewis if she preferred not to. During the 1920s and 1930s, for example, Cather frequently dined with Alfred and Blanche Knopf, often accompanying them to their box at the Metropolitan. Their invitations were directed to her alone, however, and Cather's acceptances do not mention Lewis. Similarly, Lewis would accompany Cather on their summer vacations but return to New York when Cather was ready to stop off in Red Cloud to see her parents and childhood friends.

Although Lewis was Cather's traveling companion during the 1920s and 1930s, at an earlier time she was second choice: in 1915 Cather and Isabelle were planning a European trip when Judge McClung forbade his daughter to go. Lewis then filled in, accompanying Cather to the Southwest. Second in line as a friend as well as a vacationer, Lewis may never have supplanted Isabelle in Cather's deepest affections, and yet her willingness to accommodate herself to a demanding writer's wishes gave her an essential role in the relationship, which in my view was not as one-sided as Sergeant describes it. Cather doubtless dominated Lewis, but she was also dependent on her—and that means that she invested Lewis with a certain amount of power. And Cather was also capable of making at least one sacrifice for Lewis; after they were forced to leave Bank Street, Cather toyed with the notion of moving out of New York—she was finding the city increasingly oppressive—but since Lewis did not wish to leave her job, she agreed to stay (*WC:AM*, p. 237).

After Cather's death, Lewis retained her devotion to the genius she had discovered in 1903 until her death in 1972. As if tending a shrine, she kept their Park Avenue apartment exactly as it had been when her friend died. She seemed to move into a "state of curious exaltation," Leon Edel thought, as she became absorbed in protecting Cather's memory and tending her literary reputation.[37] Committed to preserving and perpetuating the identity Cather had fashioned for herself, Lewis marked the birth date as 1876 on the Jaffrey gravestone and tried, unsuccessfully, to keep the hated "Wilella" out of the official biography. As Cather's literary executor along with Alfred Knopf, Lewis carried out her friend's wishes, protecting her novels from film adaptation or dramatization; she agreed to paperback publication of Cather's Houghton Mifflin novels only in 1960, and that concession was in return for the publisher's agreement not to sell the motion picture rights. Equally dedicated to keeping Cather's personal life out of the public gaze, Lewis bought up photographs and letters—including several to Louise Pound—and evidently destroyed them. "No one could have been more unselfishly devoted to her memory," Alfred Knopf recalls; Lewis answered all literary questions exactly as she thought "her friend would have done had she still been there to speak for herself."[38]

Lewis' devotion to Cather's memory and reputation may not have been quite as selfless as Knopf imagined. She had been required to share her friend in life but in death Cather was hers, and the exaltation Edel noticed may have been connected with this sense of exclusive possession. The writer could indeed no longer "speak for herself," but Lewis could, thus assuming more power than she had enjoyed when her friend was alive. Courted by biographers, sought after by publishers, as the arbiter of Cather's literary and biographical destiny, in a sense Lewis took over the writer's as well as the celebrity's role, becoming both guardian and author of her friend's life. She made the latter identity a literal one when she offered to complete E. K. Brown's biography after his death and, later, when she published *Willa Cather Living,* which appeared the same year as the Brown–Edel biography.[39]

Most readers of Cather's life conclude that Lewis played a less important emotional role in Cather's life than did Isabelle McClung, and in one sense this is true. Isabelle was her grand romance, her muse, her ideal reader, and it is likely that their painful separation not only provided Cather with creative inspiration but also kept her feelings for Isabelle quite intense: the two women never had to confront the sometimes unromantic reality of a shared life. In writing close friends Cather rarely refers to Edith Lewis, and when she does, the context is practical (Lewis is performing some function or task for her) rather than emotional. By contrast, Cather always mentions Isabelle with deep feeling, whether joy or grief; her letters to Zoë Akins and Dorothy Canfield after Isabelle's death, for example, are heartfelt cries of pain, loss, and abandonment. (Of course, she never had to cope with Lewis' loss, which might have been equally, if not more, devastating.) Cather imagined herself writing all her books for Isabelle, whom she saw infrequently after 1923, and yet she dedicated no books to the woman with whom she was living, wrote no prefatory poems celebrating Lewis' protective presence.

But Cather wanted to be buried beside Edith Lewis. During the 1940s she made it clear to family members that she was not going to be buried in Red Cloud but in Jaffrey, the retreat associated both with her writing and with Lewis, who accompanied her to the Shattuck Inn for many summers.[40] Those who seek the grave site may find further evidence of Lewis' subordination to Cather. Her grave is smaller, humbler. But the juxtaposition of the graves tells a story besides that of Lewis' self-effacement: that of her importance to Cather. The two women had entertained together, traveled together, and lived together for almost forty years. Even if she never placed Lewis' name before a title page, Cather was willing to leave a public sign of this intimacy that could be read after their deaths. What this suggests, I think, is that Lewis' place in Cather's emotional and creative life was not subordinate to Isabelle's but different, and quite likely equally important. She gave Cather, to use Blanche Cook's phrase, "a living environment in which to work creatively and independently," and that is not an insignificant gift.[41]

•

Although Cather treated Edith Lewis like a fellow writer when they first met (she had published a few articles in the Lincoln *Courier*), had Lewis actually been a professional journalist or novelist it is doubtful that they could have created such a harmonious household (*WCL*, p. xii). Like Thomas Carlyle, Cather needed a partner willing to be a "secondary consideration," not a literary competitor or a strong-willed colleague with equal demands for deference and attention (*KA*, p. 423). Elizabeth Sergeant, the "shy New Englander, full of self-doubts" who timidly brought her story on sweated tenement workers to the managing editor of *McClure's* in 1910, eventually became such a colleague, and when she did their friendship waned (*WC:AM*, p. 34). Sergeant soon wanted to be a captain herself, not someone else's first mate. But during Cather's first years in New York, Sergeant was one of the "best companions" Jewett had envisioned. She played an important role in Cather's return from the world to the parish.

A Bostonian brought up in the privileged New England world Cather venerated, Sergeant attended Bryn Mawr during the years of M. Carey Thomas' reign and emerged in 1910 dedicated to the Progressive Era's cause of reform. A "Bryn Mawrter" of that generation, Sergeant writes, "must try to right these terrible social wrongs that blistered and festered under the shiny urban surface of Manhattan Island," and she brought her first muckraking article to Willa Cather assuming that the *McClure's* editor shared her reformer's zeal (*WC:AM*, pp. 35–36). During their first interview Sergeant discovered her mistake, but the two women were immediately drawn together by their love for France and French culture and by their shared literary enthusiasms. Within minutes, Sergeant remembers, they were far from the tenements and deep into books: "Flaubert, Balzac, Tolstoy, Henry James, Edith Wharton, Sarah Orne Jewett." Sergeant's appreciation of Jewett's writing was a particular bond; once Cather found that she had given a copy of *The Country of the Pointed Firs* to a French critic who had compared it to Turgenev, "her eyes sparkled" with pleasure (*WC:AM*, p. 41). Sergeant was one of the few people with whom Cather felt she could "talk about books," and Sergeant in turn was captivated by the untapped literary energies she saw in Cather ("My young heart was hungry to discover—to keep up with her, to follow her, to tag along and find out where she was going'' [*WC:AM,* p. 421]).

Sergeant may have felt that she was ''tag[ging] along,'' but Cather found her an important friend and literary confidante during her period of transition from magazine editor to full-time writer: Cather told Sergeant of her desire to write serious fiction; she revealed how important Jewett had been (and still was) to her, she took her to meet Mrs. Fields, thus including her in the sacred world of 148 Charles Street. (Since Sergeant was a Boston native, she must have struck Cather as deserving a place by the tea table.) Sergeant became her traveling companion on her journey into creativity; Cather wanted her ''to come along,'' Sergeant recalls, ''to be patient with her, to understand her complexities and her goal, without words,'' and she remembers how, after confessing her desire to write novels, Cather took her arm and said ''Now it will always be easier to catch step'' (*WC:AM,* p. 53).

Understanding Cather's literary goal, Sergeant kept step. A valuable reader during her friend's time of uncertainty, she responded to her work as Jewett might have, encouraging Cather to develop her Western, not her Jamesian, voice. Shortly after they met, Sergeant went to the Boston Public Library to look up the stories Cather had published in 1907 and saw immediately that they had been written under the influence of James and Wharton; she also recognized the ''Jamesian touch'' in *Alexander's Bridge* and regretted that she did not find there the strength and power she knew Cather possessed. As Jewett had hoped, Sergeant was the ''best companion'' who did not ''admire and wonder'' at everything Cather wrote, and she delicately expressed this disappointment to Cather, who ''more or less agreed with it'' (*WC:AM,* p. 76).

Sergeant conveyed her approval of Cather's work even more strongly: a perceptive reader of *The Troll Garden,* she found three of the stories ''brilliant''—''The Sculptor's Funeral,'' ''Paul's Case,'' and ''A Wagner Matinee''—the same stories that Cather would later decide to republish in the Library Edition. She immediately wrote Cather of her delight. So when Sergeant read ''The Bohemian Girl'' and felt that ''this was it,'' her response had weight. Sergeant was particularly important as a reader of *O Pioneers!* Reading the novel in manuscript, Sergeant reassured Cather that this was the right literary direction; here was the power and originality she had been hoping for. When Cather finally gave her a copy of *O Pioneers!* she signed it ''To Elsie Sergeant, the first friend of this book'' (*WC:AM,* p. 113). During this period of transition, Sergeant was thus the ''best companion'' who helped Cather to express her ''best self,'' just as Jewett had hoped. In giving Sergeant an inscribed copy of *O Pioneers!,* Cather reciprocated another ''gift from heart to heart.''

The letters Cather wrote to Sergeant during the first years of their friendship are likewise self-revealing gifts. Frank, open, intimate, they reveal a range of emotions: Cather was able to share with Sergeant exhilaration and depression, confidence and anxiety. Viewing her new friend as a kindred spirit, Cather did not have to censor herself as she did in writing Mrs. Fields. But in the 1920s and 1930s the letters become both less frequent and more impersonal as the friendship waned. The differences between the two women's social and political opinions intensified. Cather's gradual disaffection with postwar culture as the world seemed to "break in two" and Sergeant's transition from timid apprentice to accomplished journalist strained the relationship. Their contrasting responses to World War I are representative of this process. A staff writer for the liberal *New Republic,* Sergeant went to France during World War I, was injured on a supposedly safe battlefield, and later wrote a war memoir *(Shadow-Shapes)* to convey the horrors of war. Cather (who spent the war years writing *My Ántonia*) didn't much care for the book, and Sergeant in turn did not like *One of Ours,* which struck her as romanticized ("War was a *story* to Willa" [*WC:AM,* p. 163]).

Nor did the two friends agree on national politics. Sergeant maintained her reformer's conscience, becoming an ardent New Deal supporter, while Cather detested both Eleanor and Franklin Roosevelt. They also differed on Freud and psychoanalysis (Sergeant was a believer, Cather a skeptic), on new currents in the theater (Sergeant loved O'Neill's expressionist drama, Cather preferred Galsworthy), on socially conscious fiction (Sergeant yes, Cather no). Of course such disagreements need not separate friends, but Sergeant felt Cather, always a forceful person, became rigid, opinionated, and domineering over the years:

> Willa did seem not to enjoy talk as difference; as argument and ferment. When she had made her mind up, she wanted to prevail. I had been young in experience, and very much a learner and tyro in writing when I first met her. Now I was at least on my own, and at least must stand my ground—or be silent. (*WC:AM,* p. 163)

In her memoir Sergeant gives us several examples of her friend's imperiousness: we see Cather commanding waiters, Cather intimidating customs officials, Cather dismissing Amy Lowell, Cather defining the correct salad dressing: "light French olive oil . . . the richest wine vinegar, with a dash of tarragon. She insisted on the tarragon" (*WC:AM,* p. 163).

Although the friendship diminished as Cather kept insisting on the tarragon, it was never broken. The two kept in touch with occasional letters and phone calls, and after Cather's death Sergeant decided to write her own reminiscence *(Willa Cather: A Memoir),* a far richer and more complex portrait than Lewis' *Willa Cather Living.* The reason Sergeant gives for writing her book suggests why the early years of the friendship were so important to Cather. She had been present at "the critical turning point" when Cather dedicated herself to the "art of fiction," Sergeant explains, and wanted to share what she had witnessed with other readers (*WC:AM,* p. 9). Catalyst as well as witness to Cather's literary turning point, Sergeant's responsiveness to the "more fertile process" that engendered "The Bohemian Girl" and *O Pioneers!* gave her an important role in Cather's creative emergence. Her friend ended a letter, Sergeant remembers, "thanking me for caring for her book . . . , but even more for caring about the whole thing—that is, the art of creative writing" (*WC:AM,* p. 99).

•

Another friend who cared both about Willa Cather and the art of creative writing was playwright Zoë Akins. Like Sergeant, Akins was a younger writer who met Willa Cather after submitting some poems to *McClure's*. Writing Akins in 1909, Cather was sorry to reject her submissions but added that she thought her real gift was in playwriting. She should settle into her work, Cather advised, and write what she wanted to say rather than what she thought an audience might like—thus passing on Jewett's literary advice to another beginner.[42] In other letters written during her *McClure's* tenure Cather praises some of Akins' poems and criticizes others, looking for the same freshness and authenticity she was trying to capture in her own writing.[43] In some ways Cather was Akins' mentor, but the younger woman was also important to her as a critic. Like Sergeant, Akins applauded Cather's return to her Nebraska material in "The Bohemian Girl" and *O Pioneers!*

Their friendship endured over the years with no evident rivalries even when Akins became a well-known playwright after the success of *Déclassé* in 1919. Cather always felt that Akins' good fortune was hers as well.[44] She praised Akins for writing plays like *The Texas Nightingale* (1927), which she thought brilliant, and chided her when Akins seemed to desert her own material, as in *Starvation on Red River* (1940). (She could not detect Akins' voice in the play—always Cather's most important criterion for determining literary authenticity.)[45] But she became enraged when she felt that critics and reviewers were attacking her friend unfairly, and several of the letters Cather wrote her in the 1930s express her distress with the negative reviews Akins was getting, particularly from leftist critics who disliked her romanticism. Attacked herself during the decade, Cather identified with Akins; as she confided, she had her critics, too.[46] Akins acknowledged Cather's support of her creative life when she dedicated her 1937 volume of poetry, *The Hills Grow Smaller,* to Cather. And the correspondence between them continually refers to gifts. Knowing how to please her friend, for birthdays and holidays Akins would invariably send fruit, flowers, or potted trees, and many of Cather's replies begin with a thank you.

Once she moved into her major fiction Cather had other best companions and readers: childhood friends Carrie Miner Sherwood and Irene Miner Weisz, opera singer Olive Fremstad, novelists Sigrid Undset and Zona Gale, and—particularly important in the 1920s—her regained college friend and fellow writer Dorothy Canfield. Thus, even though the act of writing was solitary, it was sustained and supported by women friends who played many roles—protectors, catalysts, critics, supporters. Knowing that she had a sympathetic female audience may have been even more important to Cather while she was working on a manuscript than after she completed a novel. By imagining a female reader Cather could at least alleviate, if not dissipate, the pressure to please a male audience or critical establishment.

•

When Jewett told Cather that she needed to add to her "recognitions," she suggested that her friend's literary power would come from knowing again—from the mature writer's return to previous experience, at last understood from an enriched perspective (p. 250). That process of return and recognition was already under way when Cather met Jewett, but as their friendship developed Cather gradually found herself ''called home'' in her fiction, encouraged by Jewett to continue a literary journey she had already begun. And after her death Jewett left Cather more than a trunkful of memories: in her mentor's fiction Cather could reencounter the literary voice that helped her to speak in her own.

NOTES

1. ''Mr. and Mrs. James T. Fields,'' *Atlantic Monthly* 116, no. 1 (July 1915): 21–31, esp. 30.

2. Eva Mahoney, ''How Willa Cather Found Herself,'' *Omaha World-Herald,* November 27, 1921, p. 7.

3. Annie Fields, ed., *Letters of Sarah Orne Jewett* (Boston: Houghton Mifflin, 1911), p. 11. Further references to this edition included in the text.

4. Willa Cather to Sarah Orne Jewett, October 24, 1908, Houghton Library, Harvard University, Cambridge, Mass.

5. In her higher social origins, Jewett resembled Louise Pound and Isabelle McClung—the other ''ladies'' to whom Cather was attracted.

6. Richard Cary, ed., *Sarah Orne Jewett Letters* (Waterville, Me.: Colby College Press, 1956), p. 17.

7. Cary, *Letters,* p. 11.

8. ''Looking Back on Girlhood,'' in *Uncollected Short Stories of Sarah Orne Jewett,* ed. Richard Cary (Waterville, Me.: Colby College Press, 1971), pp. 3–7.

9. Willa Cather to Sarah Orne Jewett, October 24, 1908.

10. Elaine Showalter, ''Feminist Criticism in the Wilderness,'' *Critical Inquiry* 8, no. 2 (Winter 1981): 179–205; esp. 203.

11. Cary, ''Looking Back on Girlhood,'' p. 6.

12. For a fuller discussion of ''imaginative realism,'' see Cary, *Letters,* p. 91. The two Flaubert quotations can be translated: ''To write about ordinary life as one writes history''; and ''It is not to make you laugh, nor to make you cry, nor to make you angry, but to act as nature does, that is to say—to make you dream.''

13. Cary, ''Looking Back on Girlhood,'' p. 6.

14. Quoted in Josephine Donovan, *Sarah Orne Jewett* (New York: Frederick Ungar, 1980), p. 4. Donovan also notes the similarities between Flaubert's and Dr. Jewett's advice (p. 5).

15. Cary, *Letters,* p. 65.

16. For a discussion of Jewett's literary context, see Josephine Donovan, *New England Local Color Literature: A Women's Tradition* (New York: Frederick Ungar, 1983), chapters 1–4.

17. See Donovan, *Jewett,* pp. 6–9 and Harriet Prescott Spofford, *A Little Book of Friends* (Boston: Little, Brown, 1916).

18. Cary, *Letters,* p. 74.

19. For a study of the domestic writers' conflicts, see Mary Kelley, *Private Woman, Public Stage: Literary Domesticity in Nineteenth-Century America* (New York: Oxford University Press, 1984).

20. Cary, *Letters,* p. 11.

21. Cary, *Letters,* p. 90.

22. Cary, *Letters,* p. 30.

23. Cary, *Letters,* p. 64.

24. Unlike Stowe, Jewett did not have to integrate writing with the demands of marriage and family. For a discussion of Stowe's difficulties in combining motherhood and writing, see Kelley, pp. 247–48.

25. Sarah Orne Jewett to Witter Bynner, May 3, 1905, Houghton Library, Harvard University, Cambridge, Mass.

26. Stephen Whicher, ed., *Selections from Ralph Waldo Emerson* (Boston: Houghton Mifflin [Riverside ed.], 1957), p. 158. The conjunction between the American ideology of romantic individualism and female self-expression evident here was not unusual in nineteenth- and early twentieth-century America. Feminists and writers like Elizabeth Cady Stanton, Margaret Fuller, and Kate Chopin derived a vocabulary for redefining femininity from the discourse of romanticism; the notion of the individualistic, autonomous self, emancipated from social forms and conventions, could be adapted to feminist purposes.

27. Jewett presented Cather with a model of creativity remarkably similar to the models of feminine development presented by Nancy Chodorow and Carol Gilligan, in which the female self is created through relationship rather than separation. See Nancy Chodorow, *The Reproduction of Mothering: Psychoanalysis and the Sociology of Gender* (Berkeley: University of California Press, 1978), p. 167, and Carol Gilligan, *In a Different Voice: Psychological Theory and Women's Development* (Cambridge: Harvard University Press, 1982), chapter 2.

28. Cather to Jewett, December 17, 1908.

29. Elizabeth Abel, "(E)merging Identities: The Dynamics of Female Friendship in Contemporary Fiction by Women," *Signs: Journal of Women in Culture and Society* 6, no. 3 (Spring 1981): 413–35, esp. 432. As Abel observes, the woman writer's "willingness to absorb literary influence instead of defending the poetic self from it is consistent with the flexible ego boundaries and relational self-definition described in theories and novels of women's relationships" (p. 433). The same "relational self-definition" characterizes the fictional friendship between Martha and Helena, the real-life bond between Cather and Jewett, Jewett's literary influence on Cather, and both writers' experience of the creative process.

30. Cather to Jewett, October 24, 1908.

31. Willa Cather to Mrs. Fields, June 27, 1909, Houghton Library, Harvard University, Cambridge, Mass.

32. Cather to Jewett, October 24, 1908.

33. Willa Cather to Ferris Greenslet, February 17, [1924] and May 10, [1924], Houghton Library, Harvard University, Cambridge, Mass.

34. In 1913 Cather told an interviewer that the "three great" American writers were Twain, James, and Jewett (*KA*, p. 446). Her replacing of James with Hawthorne in 1925 may reflect her desire to place Jewett in an unquestionably American tradition as well as, perhaps, her wish to underplay James' influence on her own early fiction.

35. James Woodress discusses the importance of women friends to Cather's creative and emotional life in "Cather and Her Friends," *Critical Essays on Willa Cather*, ed. John J. Murphy (Boston: G. K. Hall, 1984), pp. 181–95. Men like Charles Gere, Will Owen Jones, S. S. McClure, and in particular Alfred Knopf were also important to Cather's professional and literary life, but she always maintained a certain formality in these relationships.

36. In *Only One Point of the Compass: Willa Cather in the Northeast* (Danbury, Conn.: Archer Editions Press, 1980), Marion Marsh Brown and Ruth Crone give a portrait of the Cather–Lewis relationship as remembered by innkeepers, waitresses, guests, and neighbors in their three New England retreats: the Shattuck Inn in Jaffrey, Whale Cove Cottage on Grand Manan Island, and the Asticou Inn in Northeast Harbor, Maine. Lewis is portrayed

as self-effacing and subordinate, fierce only in protecting Cather's privacy; it was she who went to the post office and the bank, did the necessary shopping, supervised construction of the Grand Manan cottage while Cather worked, and performed secretarial duties when necessary.

37. Leon Edel, ''Homage to Willa Cather,'' in *The Art of Willa Cather,* ed. Bernice Slote and Virginia Faulkner (Lincoln: University of Nebraska Press, 1974), p. 190.

38. Alfred Knopf, ''Miss Cather'' in Slote and Faulkner, p. 221.

39. For an account of the delicate negotiations between Leon Edel and Edith Lewis when E. K. Brown died before completing his biography, see Edel, pp. 190–91. As Edel points out, Lewis originally did not intend to publish a memoir, but wrote an account of her life with Cather for Brown to use in preparing the official biography. When she was not asked to complete Brown's biography, Lewis decided to publish her memories as a book—perhaps, Edel speculates, a ladylike act of vengeance (p. 191). Whether or not Lewis' motives included a desire for revenge, her publication of *Willa Cather Living* shows how, after Cather's death, she assumed the role which, during their forty-year relationship, was associated with power: authorship.

40. Elsie Cather to E. K. Brown, September 23, 1949, Beineke Library, Yale University, New Haven, Conn.

41. Blanche W. Cook, '' 'Women Alone Stir My Imagination': Lesbianism and the Cultural Tradition,'' *Signs: Journal of Women in Culture and Society* 4, no. 4 (Summer 1979): 719–39, esp. 738.

42. Willa Cather to Zoë Akins, January 27, 1909, Akins Collection, Huntington Library, San Marino, Calif.

43. Cather to Akins, [n.d.], February 6, 1912, and March 14, 1912.

44. Cather to Akins, March 20, 1932.

45. Cather to Akins, February 15, 1940.

46. Cather to Akins, October 28, 1937.

16

Literary Inheritance 1908–1911

I dedicated my novel *O Pioneers!* to Miss Jewett because I had talked over some of the characters in it with her one day at Manchester, and in this book I tried to tell the story of the people as truthfully and simply as if I were telling it to her by word of mouth. *The Kingdom of Art*

Then I had the good fortune to meet Sarah Orne Jewett, who had read all of my early stories and had very clear and definite opinions. . . . She said, "Write it as it is, don't try to make it like this or that. You can't do it in anybody else's way. You will have to make a way of your own."
The Kingdom of Art

Viewing her literary development through the romantic ideology of individualism as well as the classic American narrative of religious conversion, in later years Willa Cather retrospectively created a "false" and a "true" literary self. The "false" Cather had written *Alexander's Bridge,* the "true" Cather *O Pioneers!* She attributed the birth of the second Cather to her transformative months in the Southwest—her Walden—where she "recovered from the conventional editorial point of view" (*NUF,* p. 92). While *O Pioneers! was* a creative breakthrough—recalling Whitman's artistic emergence in *Leaves of Grass*—it was not an anomalous, mysteriously new work. Nor did she create this work in isolation.

The Puritans used the phrase the "prepared heart" to describe the readiness of a soul for conversion, and we can say that Cather possessed the prepared heart when she met Sarah Orne Jewett, the other actor in her drama of literary genesis. If she were mother to herself in writing *O Pioneers!* she was also daughter to Jewett, inheritor as well as originator. Cather may have felt that she first spoke in her native accents in *O Pioneers!,* but in fact her voice was emerging during the preceding years. Intermingled with the strained tones of "The Willing Muse" and "Eleanor's House" are a few stories that carry on an invigorating and unaffected conversation with Sarah Orne Jewett, beginning with "The Sentimentality of William Tavener" in 1900. In contrast to the Freudian pattern of struggle and competition that Harold Bloom sees characterizing male poets' relationships with their forebears, Cather's literary response to Jewett's fiction and friendship suggests a less competitive paradigm: the woman writer's willingness to define herself in connection with, rather than in opposition to, her female precursor.[1]

The first three stories Cather wrote after meeting Jewett—"On the Gull's Road" (1908), "The Enchanted Bluff" (1909), and "The Joy of Nelly Deane" (1911)—show the immediate impact of Jewett's influence; evidently, Cather's receptivity to her mentor's literary example grew along with their developing friendship. In different ways, these stories are all concerned with the creative process, and they all portray creativity not as an isolated achievement—won, as Cather had once thought, by a solitary manly artist who struggles with his subject—but as a gift or an inheritance. Just as Martha's bond with the absent Helena in "Martha's Lady" contributed to her growing artistry, and just as Cather's friendship with Jewett aided her literary emergence, so in these stories human relationships stimulate self-expression. Since the alliance between creativity and friendship was one of Jewett's major subjects, before we turn to Cather's stories we need to look briefly at her mentor's fiction.

•

When she described the origins of *O Pioneers!* to an interviewer in 1913, Cather gave herself the roles of both storyteller and listener: she had told the story to Jewett as "simply as if I were telling it to her by word of mouth"; yet this was also a novel inherited, in a sense, from the old farm women who told her a "great many stories about the old country" (KA, pp. 448–49). This reciprocal pattern of creativity—the "gift from heart to heart"—is also central to Jewett's fiction. In the years immediately preceding *O Pioneers!,* when Cather was more and more intrigued by models of the artist and the creative process that were not exclusively masculine, she found a congenial presentation of female creativity in Jewett's fiction. In Jewett's Maine, storytelling is a collaborative act. Like all good raconteurs, her women are inspired to create only when they have receptive listeners who contribute to the creative process. An encouraging, concerned female audience is particularly important because Jewett's storytellers are proud, reticent countrywomen who guard their emotions and frequently find verbal expression difficult. Because emotion is the source of art in Jewett's fiction (as it was in her life), a sympathetic listener can help a normally reserved woman to speak.

Jewett suggests the importance of the female listener—at times a questioner—in creating the tale in "The Only Rose," where Mrs. Bickford is inspired to tell the story of her three husbands because of Miss Pendexter's kindly attention (and occasional prompting). Mrs. Bickford is usually "speechless," not from "poverty of thought," Jewett writes, but from an "over-burdening sense of the inexpressible." But Miss Pendexter lightens the burden with a thoughtful question in a "sympathetic tone," which gets her hostess "well started in conversation." A good listener, Miss Pendexter is so "quiet and sympathetic" that Mrs. Bickford is "no more embarrassed than if she had been talking only to herself" (*CPF*, pp. 218–32).

In "The Flight of Betsey Lane" Jewett similarly connects self-expression and friendship, storytellers and listeners. Betsey Lane, Miss Peggy Bond, and Mrs. Lavinia Dow are friends and fellow-residents in the By-fleet poorhouse. Despite their impoverished surroundings they pass the time pleasantly, conversing and storytelling; Mrs. Dow is the prime narrator, offering "inexpressible delight to Betsey Lane" with her stories of "events prior to the Revolution." When Betsey Lane suddenly embarks on a heroic, solitary journey to the Philadelphia Centennial, the stories cease: with her best listener gone, Mrs. Dow's creative fires are quenched. When Betsey returns, however, we learn that her individualistic venture was sparked as much by a desire to have stories to tell her friends as to have her own adventures: "I guess I've got enough to think of and tell ye for the rest o' my days," she exclaims, and to her credit Mrs. Dow does not try to hold on to the spotlight, gladly assuming the role of receptive listener along with Peggy Bond. "I want to know," the two women exclaim together, validating Betsey Lane's conclusion about collaborative art: "What's for the good o' one's for the good of all. You just wait till we're setting together up in the old shed chamber!" (*CPF*, pp. 172–93).

Female storytelling reaches its fullest expression in *The Country of the Pointed Firs* where the connection between the inset stories and the narrator's written art structurally embodies the lineage of creativity and affection linking inheritors and originators, unconscious artists and conscious ones, mothers and daughters. The women storytellers in *Country* are the narrators and creators of other women's lives as well as of their own. In a realm where men control public forms of representation (like the sermon preached by the pompous minister Mr. Dimmick), stories are an essential means of preserving female identity and continuity. (It is no accident that Mrs. Todd is both the principal storyteller and a geneologist.) The inset narrative that best illustrates the nature and function of collaborative female creativity in *Country* is the story of Shell-heap Island's female hermit, Joanna Burden, which Mrs. Todd and Mrs. Fosdick jointly tell to the narrator. This shared narrative requires four women's participation: Joanna, whose mysterious retreat provides the original "text"; Mrs. Todd and Mrs. Fosdick, who both read and tell Joanna's story; and the narrator, whose interest and curiosity prompt the two women to tell the story—and who ultimately retells the story to the reader.[2]

Joanna Burden's continued existence depends on these inheritors and creators of her story. The conventional social narrative of a woman's life does not represent her since she did not marry, and Joanna leaves no enduring record after her death—no child or book—that could preserve her memory. So Mrs. Todd and Mrs. Fosdick are the recorders who give her story to the narrator, the woman writer. The oral narrative is a "gift from heart to heart" and is signified by the coral pin that Mrs. Todd brings to Joanna, who then gives it back ("I want you to have it, Almiry"), just as she gave her story to her sympathetic observer (*CPF*, p. 70). When Mrs. Todd gives the departing narrator presents to take back to the city—"a quaint West Indian basket which I knew its owner had valued" and "a little old leather box which held the coral pin" (*CPF*, p. 159)—the gifts link the narrator both to Mrs. Todd and to Joanna. They are, Jewett implies, mementoes that will inspire both memory and imagination—the sources of her art as well as of Cather's.

•

Having already written of artistic power as the product of a maternal inheritance in "The Namesake," Cather was ready to respond to the model of female creativity she saw in Jewett and her fiction, as the short stories she wrote between 1908 and 1911 demonstrate. "On the Gull's Road," on the surface the least innovative of the three stories, on a deeper level shows Cather attributing the creative process to the child's yearning to return to a maternal presence, although here, as in "The Burglar's Christmas," Cather unconvincingly casts the child as male.

Like *My Ántonia,* "On the Gull's Road" is retrospectively narrated by a man who recalls a lost lady. During an Atlantic crossing twenty years ago, he confides, he had fallen in love with a beautiful woman named Alexandra Ebbling. Unsatisfied in a loveless marriage to a brutal, domineering man, she returns the narrator's devotion but leaves him at the voyage's end. He does not then know why, but discovers later that she has a fatal illness and wants to spare him the pain of seeing her die—an outcome which Cather signals with her name (Ebbling/ebbing). When the narrator learns that she is going to leave him, he exclaims "You will give me nothing." He is the abandoned and dispossessed child: where is his gift, his inheritance? He will have a rich one, she tells him: "I shall have given you all my life." And this she does by giving him his memory of her, incarnated in the gift she leaves—a box of mementoes containing a faded magnolia blossom, two seashells, and a lock of her reddish-gold hair.

The story ends in the present. Sitting by the fire, our narrator savors his memories and keepsakes, including the sensuous lock of hair, an almost living symbol of his lost lady. "How it gleamed, how it still gleams in the firelight!" he rhapsodizes. The imagery of the gifts suggests Cather's intention to imply a sexual attachment, as does her association of Alexandra with the sea, an overwhelming erotic and maternal presence in the story, allied with creative and destructive powers. An Aphrodite linked with seashells, warm Mediterranean waters, and gleaming hair that "curls and undulates with the tide," Alexandra is a compelling sexual force, and the sea imagery the narrator uses to describe her influence becomes one with the inexorable waves of desire. She is also a maternal power; her life-giving force is visible in her healthy daughter, as "round and red as a little pumpkin," and the narrator wants to bask in the warmth she emanates.

"On the Gull's Road" has generally been considered one of Cather's Jamesian imitations, and indeed there are many Jamesian elements—the international setting; the stiff, formal dialogue; the reflector/narrator who keeps his passions at a safe distance. Moreover, the story's subtitle ("The Ambassador's Story") directly alludes to *The Ambassadors* and associates the narrator with James' cautious emissary Lambert Strether. But when Cather described the story in a letter to Jewett she disparaged it by saying that it fairly screamed, the scent of the tube-rose emanated from its pages.[3] As this image suggests, she considered "On the Gull's Road" excessively emotional and overwritten, flaws she associated with women's writing and with her own early style rather than with the master's classic restraint. So the story's overt reference to James, along with its allusion to "The Raven" ("This, . . . nothing more. And it was all twenty years ago!"), shows Cather trying to compensate for her story's preoccupation with the "feminine" realm of the emotions by associating it with male-authored texts.

Cather was too hard on the story. "On the Gull's Road" does betray the signs of a tearjerker—the shipboard romance, the dying woman, the treasured keepsakes—that link it with the conventions of women's magazine fiction. But it is not merely made-to-order romance, as Jewett herself noted when she told Cather that it made her "the more sure" that Cather was "far on [her] way toward a fine and long story of very high class." She had read the story, she assured Cather, with "deep happiness and recognition."[4] Jewett may have read the story with "recognition" because she sensed its similarities to "Martha's Lady." Presenting a love that endures despite time and separation (caused here by the woman's death rather than by her physical absence), "On the Gull's Road" gives the narrator the role of Martha; he preserves his lover's memory by contemplating the box of mementoes she gives him; when he opens the box, he is inspired to dream and to speak, to become a narrator. These gifts, like the gifts in Jewett's stories, are signs of her continuing presence in his memory and imagination.

The gifts are also signs of the connections among love, absence, and creativity. In remembering his lost lady, the narrator momentarily becomes a storyteller—like the women in Jewett's fiction, where gift-giving represents both intimacy and storytelling. "On the Gull's Road" is of interest for this reason; like much of Cather's apprenticeship fiction, the story comments on its own creative process. Alexandra is muse as well as love-object, and it is her absence that sparks the narrator's storytelling art, where he momentarily recaptures the intimacy denied him in life. So, in a sense, the gifts resemble D. W. Winnicott's transitional objects: they are symbols that allow the child/artist to maintain his attachment to a maternal figure he has lost. The retrospective narrative allows the speaker to preserve what Cather terms "the precious, the incommunicable past" in *My Ántonia;* it also lets him return to—while keeping a safe distance from—a maternal/erotic presence who possesses the power to obliterate the self as well as to nourish it (*MA*, p. 372).

Thus the story shows Cather addressing the regressive aspects of the creative process without the fear of annihilation we saw in "The Burglar's Christmas." Intermingling the desire for the mother, regression, and creativity, "On the Gull's Road" suggests that by 1908 Cather was able to imagine surviving the journey to what she later called the "bottom of . . . consciousness" (*AB* [1922], p. vi). She could confront the possibility of self-annihilation the daughter/artist risks in relaxing the ego's vigilance and in merging with another self (here suggested by the maternal but threatening sea which "satisfied the soul like sleep"). The story's literal journey reflects the child/artist's symbolic one: from the regressive desire for fusion and gratification (represented by the warm, sensuous waters of the Mediterranean), to the adult's separation from the mother (the chilly Atlantic, the sea of "reality"), and finally to the artist's regaining of the lost connection through narrative (the storyteller's "voyage" into the past). The retrospective narration provides distance—simultaneously offering the satisfactions of separation and connection—as does the male narrator, and it seems likely here, as elsewhere in her fiction, that Cather employed a male consciousness in part as a defense against fusion with a maternal presence.

In contrast to "Martha's Lady" or *The Country of the Pointed Firs,* then, the artist-figure here is male. Or is he? When Jewett gently chided Cather for her "masquerade" as a male narrator, commenting that a "woman could have loved her just as well,'' she was responding to an ambiguity in the story.[5] Although cast as male, the narrator's gender is left so indeterminate that it seems quite possible a woman could have loved her just as well, as Jewett thought, because a woman *was* loving her; unlike Jewett, who could write directly of passionate female friendship in ''Martha's Lady,'' Cather was forced to disguise the love inspiring the story. The literary techniques she employs here anticipate those she uses to express the ''thing not named'' in her later fiction. Sexuality is displaced into the landscape and into feminine images such as the box, the hair, and the seashells; mythic allusions provide a lesbian subtext (Alexandra is identified with Aphrodite, whom Cather in turn associated with Sappho and her hymn to the goddess); and the male mask allows Cather to explore a woman's passion for another woman.

The male narrators and perceivers Cather uses throughout her fiction to confront sexually compelling women are not invariably, or solely, masks for lesbian desire, but in this case the narrator's assigned gender is not convincing. The story's framing device in which a bachelor sits by the fire in his study and reflects on a lost lover recalls another storyteller whose name revealed that he had something to hide: Miles Coverdale, the untrustworthy narrator of *The Blithedale Romance*. That Cather's Coverdale has something to hide is signified by the absence rather than by the presence of a name, however; despite their intimacy, Alexandra never mentions his name in conversation, and the one time we would expect her to use it—in her farewell letter—she addresses him as ''My Friend.'' His namelessness both reinforces his indeterminate gender and suggests Cather's inability to name the love inspiring the story, the emotional ''aura'' unnamed on the page (*NUF*, p. 50).

Although Jewett was disappointed by Cather's decision to write ''in the man's character,'' since her narrative impulse was connected with a form of desire deemed unnatural by her society, Cather had to evolve different narrative strategies to tell her stories, as ''On the Gull's Road'' demonstrates. And yet the metaphor of the gift reveals the story's even stronger resemblance to Jewett's fiction. On the one hand, Cather was receiving her subject matter from her own deepest concerns, from the things that ''teased the mind''; on the other hand, Cather was also receiving the story from her literary and personal relationship with Sarah Orne Jewett, a connection she may not have consciously acknowledged. The importance of this maternal inheritance can be seen in the changed relationship we see between the artist and subject in ''The Burglar's Christmas'' and ''On the Gull's Road,'' conveyed through the metaphor of the box. In the latter story, written after her meeting with Jewett, Cather imagines the mother granting the child the gift: he is an inheritor, not a thief.

Cather's friendship with Jewett may have contributed to the portrait of the fatally-ill Alexandra in an even more direct way. Having just entered the charmed world of 148 Charles Street, Cather was delighted by her sense of cultural acquisition; and yet her literary gift-giver was ill, possibly dying. In fact, in the same letter praising ''On the Gull's Road,'' Jewett refers to her failing health: ''It takes but little care about affairs, and almost less true pleasure, to make me feel overdone,'' she wrote, ''and I have to be careful—it is only stupid and disappointing, but there it is, as an old friend of mine often says dolefully. . . . I only deplore [ill-health] and often think it a tiresome sort of mortification.''[6] Like Alexandra's, Jewett's illness might require the mother to abandon the daughter; even so, the story suggests, she would leave behind her enduring presence in memory and in imagination.

•

''The Enchanted Bluff'' (1909), published a year later in *Harper's*, demonstrates the connection between Jewett's influence and Cather's artistic self-discovery even more strongly than ''On the Gull's Road.'' Leaving her Jamesian drawing rooms and formal gardens, Cather follows the implicit prescription of ''The Namesake'' as well as Jewett's explicit advice, returning to the landscapes of her childhood—the island and the river where she played with her brothers and dreamed ''Of the conquest of the world together'' (*AT*, p. 3). Edith Lewis felt that her friend was experiencing the creative process in a new way when she was writing ''The Enchanted Bluff'': ''It was as if she had stopped trying to make a story, and had let it make itself out of instinctive memories, deep-rooted, forgotten things'' (*WCL*, p. 70). Even before *O Pioneers!* came along, Cather sensed that she was taking the right road in her fiction; critical of all her *Troll Garden* stories, in 1911 she sent ''The Enchanted Bluff'' to Elizabeth Sergeant for her examination. Like Lewis, Sergeant thought that it ''seemed to come from the heart'' (*WC:AM*, p. 70).

The tale is a simple one. The adult narrator tells us the story of the last evening of his childhood, twenty years ago, when he and the other Sandtown boys camped out on their sandbar island and shared adventurous tales. The most gripping one was Tip's legend of the Enchanted Bluff. An immense mesa in the midst of the New Mexico desert, the Bluff was once the shelter for a peaceful Indian tribe that made cloth and pottery. Although they ''went up there to get out of the wars,'' the Indians cannot long escape the violence they want to elude. One day, while the braves are out hunting, they come upon a tribe of warriors and are massacred before they can make their way back to safety. Their link to the outside world severed, the old people, women, and children left on the Bluff starve to death, and the tribe becomes extinct. ''Nobody has ever been up there since,'' Tip concludes dramatically. All the boys are immediately captured by this tale-within-the-tale and endorse Tip's plan of climbing the Bluff together. But, the narrator tells us regretfully, none of them have done this; separated by the responsibilities of adult life, the former playmates have abandoned their dreams of shared adventure.

As well as showing Cather's growing ability to draw on her Nebraska memories, ''The Enchanted Bluff'' reveals her disenchantment with the masculine values she once prized. The story contrasts two views of creative achievement: one masculine, the other feminine; one solitary, the other communal; one recalling Cather's past, the other anticipating her future. A space for dreaming, play, and possibility, the Sandtown boys' island—like Jewett's Maine—is a pastoral world separated from the competitiveness and aggressiveness of American life. On their last night together, however, the boys look up at the stars in search of ideals and stories and conjure up the heroes that had gripped Cather's adolescent imagination—Columbus, Napoleon, Coronado. These explorers and conquerors were her first models for the artist, the virile, Kiplingesque figures likely to produce manly battle yarns and stout sea tales. Yet Cather then replaces this heroic model of the artist with a vision of communal, collaborative creativity more in harmony with the island setting and closer to that offered by Jewett in *The Country of the Pointed Firs*. Both the content of the inset story of the Bluff and its method of narration contradict the image of the artist as a heroic, solitary conqueror.

It's easy to see why the listeners are more captivated by Tip's narrative than by the individualistic tales of Columbus, Napoleon, and Coronado: the story of the Bluff is a version of their story. The sheltering mesa is the spatial and social equivalent of the island; the gentle tribe that seeks refuge there, intending to cultivate the cooperative, creative arts of cloth-making and pottery rather than the aggressive arts of war, are the ancestors of the six boys "sworn to the spirit of the stream" who seek a shared imaginative world in the midst of a competitive and materialistic society. The method of narration reflects the Indians' and the boys' commitment to communal dreams. Tip, who tells the story to his assembled playmates, has heard it from his uncle, a wandering miner; he in turn had come across the tale in New Mexico, where the Indians had been recounting the legend of their ancestors for hundreds of years. Not only is the tale of the Bluff passed on, from the Indians to Tip's uncle to Tip to the listeners, but as in *The Country of the Pointed Firs* and other Jewett stories, the listeners help create the story. Tip can speak because he knows his friends are sympathetic, and their eager questions help him begin, continue, and amplify his narrative.

Twenty years later, when the narrator reflects upon his failure to climb the Bluff with his friends, it seems as if only the compromises of adult life have eroded the boys' dreams of romantic adventure. Their failure to perpetuate the island's values is foreshadowed, however, by the listeners' reaction to the story, which contradicts the communal manner of its telling and parallels the inset story's tragic outcome. At first eager to plan a joint pilgrimage to the Bluff, the boys soon become preoccupied with getting there "first." The narrator thinks he "must have dreamed about a race for the Bluff, for I awoke in a kind of fear that other people were getting ahead of me and that I was losing my chance." Their dreams become competitive rather than cooperative; they carry the values of the shore within them. But the result of adopting individual goals, Cather suggests, is that no one wins. Without the support of his friends, no one has the courage to act. No one climbs the Bluff (*CSF*, pp. 69–78).

In "The Enchanted Bluff," neither the communal nor the individualistic models of creative endeavor succeed. The Indian potters and cloth-makers are destroyed and the boys do not climb the Bluff together, and yet no one climbs the Bluff separately. But Cather's story succeeds because she was integrating these methods, finding both connection and self-expression through her literary relationship with Sarah Orne Jewett. Even though S. S. McClure had dismissed the story as inconsequential, Cather had sent it to Jewett because she thought she would like it. In explaining why Jewett might approve, Cather stressed the difference which she thought would interest her mentor: in her Midwestern childhood, she had been exposed to a Spanish influence instead of Jewett's English one.[7] Cather could not have been so certain that Jewett would appreciate these nuances if she had not sensed the similarities between "The Enchanted Bluff" and Jewett's Maine sketches.

The language Cather uses to describe the Sandtown boys' river haunts reveals how Jewett's example was helping her discard her anxious need to align herself with the male literary tradition through imitation and allusion. In the passage describing the origin of Nebraska sandbars—impossible for Cather to have written when her mind was on Henry James—a voice emerges which, retrospectively, we can identify as hers:

> The turbulence of the river in springtime discouraged milling, and, beyond keeping the old red bridge in repair, the busy farmers did not concern themselves with the stream; so the Sandtown boys were left in undisputed possession. . . . The channel was never the same for two successive seasons. Every spring the swollen stream undermined a bluff to the east, or bit out a few acres of corn field to the west and whirled the soil away to deposit it in spumy mud banks somewhere else. When the water fell low in midsummer, new sand bars were thus exposed to dry and whiten in the august sun. . . . Here and there a cottonwood soon glittered among them, quivering in the low current of air that, even on breathless days when the dust hung like smoke above the wagon road, trembled along the face of the water.
>
> It was on such an island, in the third summer of its yellow-green, that we built our watch fire . . . on the level terrace of fine sand which had been added that spring; a little new bit of world, beautifully ridged with ripple marks. . . . We had been careful not to mar the freshness of the place, although we often swam to it on summer evenings and lay on the sand to rest.

In this simple but resonant prose, which unobtrusively fuses the descriptive and the metaphoric, the two streams of Willa Cather's literary inheritance invisibly join. The female and male literary traditions underlie this description but they do not dominate it. Like Jewett's sketches of Maine landscape and life, which Cather had taken as models, "The Enchanted Bluff" and its cottonwoods grew "out of a thin new soil" (as Cather observed of "Marsh Rosemary" [*CPF*, p. 7]). And like *Huckleberry Finn* and *Walden* (the classic American stories of innocence and regeneration), "The Enchanted Bluff" infuses an Edenic landscape with spiritual and imaginative meaning, making Nebraska's soil less thin. Cather was probably not thinking consciously of Twain or Thoreau when she wrote "The Enchanted Bluff"; by contrast, "The Treasure of Far Island," the earlier story drawing on Cather's childhood memories of the island and river, was overladen with intrusive references to male writers—Stevenson, Emerson, Shakespeare, Homer, Housman. Like the Jamesian subtitle to "On the Gull's Road," such allusions demonstrate the woman writer's need to declare herself a descendant of the male literary tradition to which she fears she does not belong. But in "The Enchanted Bluff," the one story she thought Jewett would like, Cather does not have to flash her author's credentials in the reader's face.

•

In writing of collaborative art in "On the Gull's Road" and "The Enchanted Bluff," Cather does not explicitly connect creative power with feminine characteristics, although she anticipates her discovery of a native female tradition in the Southwest with her portrayal of the Indian tribe's skill at pottery-making. But the storyteller in "On the Gull's Road" is ostensibly male, and no female version of "Willie" Cather can be detected among the Sandtown boys.

In the first short story Cather wrote after Jewett's death, however, her artist-figure is female. Although not as confident of female creative power as Jewett's fiction, "The Joy of Nelly Deane" (1911) is the first story in which Cather openly addresses Jewett's major subject, the love between women; it is also the last until *My Mortal Enemy* (1926).[8] "Nelly Deane" is also the first story in which Cather portrays a nurturing female community like that in *The Country of the Pointed Firs,* one in which the mother-daughter bond is the strongest affectional tie.

Jewett's loss was a potent creative source for "Nelly Deane." She remained alive in Cather's imagination, and her invisible presence became increasingly important as Cather struggled toward her second birth, the creation of her artistic self. Traveling to South Berwick often in 1910 and 1911, Cather needed to reattach herself to Jewett as well as to pay her respects to Mary Jewett. After one visit—in March 1911—Cather felt a particular urgency to commit herself full-time to writing. "The old house had stirred something up—she felt goaded," Cather told Elizabeth Sergeant. "It was as if Miss Jewett's spirit, which filled the place, had warned her that time was flying" (*WC:AM,* p. 60). Cather was despairing when she told Sergeant of this experience since she was still working at McClure's and did not have the time to write. But somehow she was able to produce "Nelly Deane," which appeared in the October 1911 issue of the *Century*. Her first literary response to the warning she imagined Jewett giving her, the story simultaneously reveals Cather's interest in women's power—based in affection rather than in domination—and her fear that such power is ineffective in a world ruled by men.

As in "The Enchanted Bluff," Cather returns to Nebraska for her setting. Here the small town rather than the island suggests woman's more complete entrapment by social structures and conventions. Nelly is a preliminary sketch for romantic heroines like Marie Shabata and Lucy Gayheart; like them she is vital, gay, intense, a figure always in motion—running, coasting, skating, against a stationary (and ultimately confining) background. Like many of Cather's protagonists, Nelly dreams of an exciting life beyond the constraints of small-town society. A gifted singer, she imagines the adventures she will have when she marries Guy Franklin, goes to Chicago, and takes singing lessons. But Franklin deserts her for a richer bride. Nelly then marries a cold, stingy, small-town merchant and dies shortly after the birth of her second child.

Despite this rather bleak plot, "The Joy of Nelly Deane" is not a despairing story, as its title suggests, although it is sharply critical of romantic love and patriarchal marriage. Perhaps because Jewett was so much on her mind at this time, Cather depicts a world of supportive female friendships which, as in "Martha's Lady," endure despite separation and loss. The female narrator, a woman who loved Nelly Deane when they were girlhood friends, retains their bond in memory because of her continuing friendship with the three motherly women who also loved Nelly: Mrs. Dow, Mrs. Freeze, and Mrs. Spinny, the ladies of the Baptist sewing circle who adopt Nelly as their favorite daughter. Cather's portrait of these kindly women recalls Jewett's work and its focus on female friendship, signaled by the borrowing of the name ''Mrs. Dow'' from ''The Flight of Betsey Lane.''

Cather had originally entitled the story ''Flower in the Grass,'' and the three Baptist ladies, despite their chilly names, are associated with the maternal powers of warmth, nurturance, and fertility, linked with the sun, crackling fires, flowers and plants. They are also hometown Fates; the Clotho, Lachesis, and Atropos of the sewing circle watch over their favorite child, plotting glorious futures for Nelly as they sew ''pretty things'' for her. Conventional themselves, they love Nelly even though—or because—she is irrepressible and exuberant, a girl who does not seem to be contained by social molds or expectations. ''I think they loved her for her unquenchable joy,'' the narrator reflects, and the ladies grant her freedom along with affection. Of course they want her to join the Baptist church, but these are ''artless plans,'' the narrator reveals. Their ''watchful affection'' does not demand that Nelly be like them; these are mothers who can love a daughter for her differences and wish her greater freedom than they possess themselves.

The ladies of the sewing circle would like to grant a powerful role to Nelly, as they seem to when they attend her as dressers for the cantata performance of *Queen Esther* in which she shines as Esther, the royal heroine. But the clothes they make and adjust are only the female options available within patriarchy, as Cather suggests by casting Guy Franklin as a dry-goods salesman (they must buy his material). In the story's opening, Mrs. Spinny divides her time between ''pulling up and tucking down the 'illusion' that filled in the square neck of Nelly's dress'': she isn't sure whether she should conceal the girl's body or reveal it, grant her modesty or sexuality as socially defined. Hence these Fates have limited powers; they can only spin clothes, stories, and futures within the limits of woman's traditional role; they can clothe the female body, but they cannot redefine its meaning or change its fate.

Nelly's childhood friend Margaret, who narrates the story, is as much her adorer as the boys whose photographs litter Nelly's dressing table. Like the sewing ladies, Margaret loves to see her friend ''bloom and glow,'' but she knows their friendship, unlike the mother–daughter bond Nelly enjoys with the three ladies, must be temporary: her photograph is not on the dressing table, and Nelly will inevitably marry one of her male admirers. Yet Margaret's love is all for Nelly. It is as if Cather had just reread Jewett's criticism of the male persona in ''On the Gull's Road,'' where she observed that ''a woman could have loved her in that same protecting way—a woman could even care enough to wish to take her away from such a life, by some means or other,'' and had decided to cast aside the male mask.[9]

Cather could allow this dangerous step because she makes the relationship between Nelly and Margaret tender and caring but not passionate—in contrast to the romantic love portrayed in ''Martha's Lady'' or ''On the Gull's Road.'' Cather's careful attempt to remove any lesbian overtones from the friendship is evident in her description of the night the two spend together (Margaret shares Nelly's bed after their performance of *Queen Esther*). Cather defuses the scene's erotic potential by having Nelly blushingly confide that she is in love with Guy Franklin before they go to sleep. As if following Jewett's prescription, Cather then gives Margaret a maternal response: "I was somehow afraid for Nelly when I heard her breathing so quickly beside me, and I put my arm about her protectingly as we drifted toward sleep."

Unlike Jewett's narrator in *Country,* Margaret is not a writer, although she does leave Riverbend for an enriched destiny and a profession Cather does not specify. She sees the world in terms of narrative, however: fifteen years after her departure from Riverbend, after she hears of Nelly's death, Margaret reflects that even in childhood "it was all arranged, written out like a story, that at this moment I should be sitting among the crumbling bricks and drying grass, and she should be lying in the place I knew so well, on that green hill far away." Nelly's life and death have indeed been "arranged" and "written out," though not by the narrator or the three loving Fates of the sewing circle, who would have given her a different destiny; hers is a male-authored story, Cather suggests, but one to which women give their consent. As in a fairy tale where the godmother's charm only works under certain conditions, the ladies' wishes cannot counteract the more powerful spell of male dominance. Nelly's three mothers and her friend do not have the power to rewrite her story, to change the woman's traditional narrative of marriage and motherhood—which here, since Cather does not finish her story with the expected happy ending of women's fiction, leads to death.

Nelly needed a different story; her exuberant desire for an expanded life does not fit the traditional female plot. Cather gives her the gift she was searching for as a writer—a beautiful, individual voice—and allows her briefly to escape the limits of male authority in her performance of *Queen Esther,* where she outshines the tenor—the schoolmaster who unsuccessfully tries to "eclipse" her moment in the sun.[10] Nelly, however, invests her imagination in the romantic plot, the "self-limited" story of female desire Cather attacked in her review of *The Awakening.* She hopes that Guy Franklin will take her to Chicago: unlike Thea Kronborg (whose story revises Nelly Deane's), she never dreams of going there herself.

For the first time in her fiction Cather sharply contrasts female and male realms and clearly opposes herself to the second, which is portrayed as dark, cold, and life-destroying. The men in the story either fail Nelly, desert her, or oppress her: first her loving but irresponsible father loses his money; as a result Guy Franklin leaves her for a wealthier bride; then she must marry cold, grasping Scott Spinny, who darkens her life and indirectly causes her death by alienating the town's two doctors; finally the inexperienced doctor who tends her in her final illness (connected with the birth of her son) botches her case. That Nelly's death is connected with her subservience to men—represented here by Cather's recurring paradigm of male power and female powerlessness, the doctor–patient relationship—is demonstrated by her first deathlike "immersion," the baptism in which she "disappeared under the dark water." She becomes a Baptist not to please the ladies but because she is to marry Scott, who has also told her how to dress.

The parallels to Jewett's fiction in "Nelly Deane" are extensive—the community of women, the centrality of the mother–daughter bond, the criticism of male dominance. And yet Cather's story does not revise the traditional female plot as fully as do some of Jewett's stories. Jewett frequently alters conventional narrative structures in order to offer her heroines stories and lives outside of patriarchal institutions and definitions. "A White Heron," for example, subverts the conventions of the fairy tale: the heroine refuses the handsome stranger's offer, choosing to protect the white heron and to maintain identity and autonomy. And in *The Country of the Pointed Firs,* Jewett counters the American and masculine narrative of the individualistic, isolated self with a vision of female identity and creativity formed within a women's community.[11]

The female artist-figures in "Nelly Deane," however, accept the scripts that have already been written for women. For Nelly, love and marriage—or even marriage without love—are woman's only imaginable destiny, and the sewing ladies can only hope that her life with Scott Spinny will not be too onerous. When the narrator learns of Nelly's death because of the doctor's bungling, she asks Mrs. Dow whether "it needn't have been," but male authority is not, finally, to be defied, since the ladies will never question Christianity: "We must just feel that our Lord wanted her then, and took her to Himself," Mrs. Dow tells Margaret. The Baptist sewing ladies are loving but ineffective mother–goddesses, Cather suggests, because they accept God the Father as the sole Author of a woman's life story.

Yet the author who creates these limited female artists ends her story recalling the conclusion of *The Country of the Pointed Firs* and foreshadowing her own transformation of inherited male and female narratives in *O Pioneers!* Although not achieved within the story's frame, the possibility exists that women may be able to transform masculine values and discover an independent source of female power. When the narrator returns to Riverbend after Nelly's death, she sees a sign of hope, continuity, and maternal inheritance in Nelly's daughter Margaret (named after her) who romps in the snow, wearing her mother's "old-fashioned" brown fur cap. The mother's gift which keeps the daughter warm in the wintry climate suggests the matrilineal heritage of creativity that Cather received when she wrote a story "goaded" by Jewett's presence; simultaneously it refers to the gift-giving scene at the end of *Country* (*CSF,* pp. 55–68).[12]

•

The differences between "Nelly Deane" (1911) and "Tommy, the Unsentimental" (1896), Cather's earlier version of the young woman who stands out in a small, conventional Nebraska town, succinctly reveal how much her views of gender had changed since the 1890s. In "Tommy" the gifted woman is masculine; her supportive circle of friends, the town's elder statesmen; and her sworn enemies, the silly, conventional women who care only for "babies and salads." But in "Nelly Deane" the vibrant heroine is no longer boyish; her supportive friends are older women; and her enemies are the men who control and dominate her.

Yet even as it demonstrates Cather's moving from male to female identification, "Nelly Deane" reflects her continuing difficulty in envisioning a success story for a gifted woman. This failure is understandable: despite her journalistic and publishing triumphs, despite her literary inheritance from Jewett, Cather had not yet achieved her highest goal. Although she had been writing fiction for twenty years, she had not yet produced a novel.

NOTES

1. Cather's literary relationship with Jewett thus parallels the psychodynamics of their personal bond, which allowed both for attachment and for difference; it further suggests Nancy Chodorow's theory of female development, in which daughters define themselves always in relation to mothers. In "Dickinson and Rich: Toward a Theory of Female Poetic Influence," Betsey Erkkila presents a model of female literary influence based on Adrienne Rich's response to Dickinson, which strongly resembles the patterns I see linking Jewett and Cather (*American Literature* 56, no. 4 [December 1984], pp. 541–59). As Erkkila observes, the feelings of kinship women poets establish with each other become "a source of personal identity, self-confirmation, and creative power" (p. 544).

2. As Marjorie Pryse observes, in "The Foreigner," in which the narrator of *The Country of the Pointed Firs* is told the story of a mother–daughter reunion by Mrs. Todd, the "relationship that exists between storyteller and listener resembles the ties that link mother and daughter." The listener is thus an inheritor who receives the story from previous generations of storytellers and passes it on to the reader, who in turn becomes another "daughter" in this matrilineal pattern of inheritance. See Marjorie Pryse, "Introduction" to *The Country of the Pointed Firs* (New York: Norton, 1981), pp. x–xi.

3. Willa Cather to Sarah Orne Jewett, October 24, 1908, Houghton Library, Harvard University, Cambridge, Mass.

4. Sarah Orne Jewett to Willa Cather, November 27, 1908, in *Letters of Sarah Orne Jewett,* ed. Annie Fields (Boston: Houghton Mifflin, 1911), p. 246.

5. Jewett to Cather, November 27, 1908, in Fields, p. 246.

6. "My Dear Willa," in Fields, p. 246.

7. Cather to Jewett, October 24, 1908.

8. It might be argued that the friendship between Alexandra and Marie in *O Pioneers!* is an exception. But it seems to me that the friendship that Cather openly describes is not central to either woman's life; it is Alexandra's unconscious attraction to Marie, the subject that Cather could not explore openly, that provides the intensity in this bond.

9. Fields, pp. 246–47.

10. The story of Esther recorded in the oratorio echoes Cather's view of women's power in the story: they have some, but not enough. Esther is a heroine who saves her people from the evil Haman, and yet to do so she must appeal to the power of another male ruler, Ahasuerus. For an analysis of the musical references in the story, see Richard Giannone, *Music in Willa Cather's Fiction* (Lincoln: University of Nebraska Press, 1968), pp. 47–50.

11. For a discussion of the ways in which "A White Heron" inverts fairy-tale motifs, see Annis Pratt, "Women and Nature in Modern Fiction," *Contemporary Literature* 13, no. 4 (Fall 1972): 479. Marjorie Pryse observes that *Country* depicts "a world in which women were once united with their mothers and inherited their mothers' powers" ("Introduction," p. xiii). Elizabeth Ammons analyzes the connections between Jewett's narrative techniques and her portrayal of the female self as created in human relationships, drawing on Carol Gilligan's theory of female moral development in "Going in Circles: The Female Geography of Jewett's *Country of the Pointed Firs,*" *Studies in the Literary Imagination* (Fall 1983): 83–92.

12. The story's last scene, in which the narrator cradles Nelly's infant son surrounded by the three sewing-women, further demonstrates Cather's interest in revising male-authored narratives—here the nativity story. Cather juxtaposes the picture of the three Wise Men on the wall to the three women bending over the baby, substituting her female goddesses for the Bible's patriarchal kings. Yet Cather is still writing within the limits of Christian ideology; even though she imagines the child as inheriting Nelly's joyous nature—this is the mother's son rather than the Father's—she still uses the vocabulary of Christianity and casts her Christ child as male.

III

EMERGENCE

17

Building Stories: *Alexander's Bridge* and "The Bohemian Girl"

> There is a time in a writer's development when his ''life line'' and the line of his personal endeavor meet. This may come early or late, but after it occurs his work is never quite the same. After he has once or twice done a story that formed itself, inevitably, in his mind, he will not often turn back to the building of external stories again.
>
> ''Preface'' to *Alexander's Bridge* (1922 ed.)

> The young writer must have his affair with the external material he covets; must imitate and strive to follow the masters he most admires, until he finds he is starving for reality and cannot make this go any longer. Then he learns that it is not the adventure he sought, but the adventure that sought him, which has made the enduring mark upon him.
>
> ''Preface'' to *Alexander's Bridge* (1922 ed.)

In later statements, Willa Cather liked to portray her transformation from the author of *Alexander's Bridge* (1912) to the author of *O Pioneers!* (1913) as miraculous. Her first novel was the apprentice writer's laborious construction of an ''external'' story, her second novel the mature writer's effortless expression of an internal story (*AB* [1922], p. vi). In between was her watershed experience, the liberating weeks she spent in the Southwest early in 1912. According to Cather's retrospective account, the year 1911—when she wrote *Alexander's Bridge*—was thus the final year of her apprenticeship, during which she was ''follow[ing] the [master]'' Henry James rather than speaking in her own voice (*AB* [1922], p. viii).

Cather's dramatic version of her literary evolution is oversimplified. What actually happened during 1911 is far more complex and more interesting than she later acknowledged. This was a confusing, fertile, transitional year, filled with strains and discontinuities as Cather struggled to become a novelist. She was redefining herself in two important ways: moving from magazine work to full-time authorship, she was changing her professional identity; moving from the short story to the novel, she was changing her literary identity. The fiction she wrote during this period itself is transitional. *Alexander's Bridge*, which Cather later described as a shallow, conventional work, in fact is a divided, Janus-like text: in part it looks back to her Jamesian apprenticeship, in part it looks ahead to her creative breakthrough in *O Pioneers!* Moreover, at the same time that Cather was working on *Alexander's Bridge* she was writing two Nebraska stories, "The Bohemian Girl" and "Alexandra"; the latter story eventually became part of *O Pioneers!* So her personal, Western cadences were emerging even as she was speaking in self-consciously cultured tones.

The chronology of this year shows how intertwined were Cather's two voices. She completed a draft of *Alexander's Bridge* in the spring and summer of 1911 while she was still working at the magazine. Then, in the fall of 1911, she took a leave of absence and went to Cherry Valley, New York, where she wrote "The Bohemian Girl," revised *Alexander's Bridge,* and worked on "Alexandra." With all this accomplished, early in 1912 she left for the Southwest. Thus Cather's romantic portrait of herself as the "superficial" author of *Alexander's Bridge* who "recovered from the conventional editorial point of view" in the Southwest's bracing climate is not quite accurate (*OW,* p. 92). Not only is *Alexander's Bridge* less conventional than she later admitted, but she had also begun to express her unconventional literary self in "The Bohemian Girl" and "Alexandra" as well as in earlier short stories like "The Joy of Nelly Deane," "The Enchanted Bluff," "On the Gull's Road," and "The Namesake."

In exploring the two major works of fiction Willa Cather completed during this transitional year, *Alexander's Bridge* and "The Bohemian Girl," we then have two tasks. We need to see in what ways *Alexander's Bridge* represents a return to Cather's Jamesian mode; and we need to see how *Alexander's Bridge* and "The Bohemian Girl" represent an advance that led her in *O Pioneers!*

•

To understand why Cather would have in part abandoned the advances of "The Enchanted Bluff" and "The Joy of Nelly Deane" and reverted to James' tutelage in *Alexander's Bridge,* we must first consider what writing a novel, rather than a short story, meant to her.

In 1911 Cather was thirty-seven. Growing more and more impatient with her editorial duties and feeling compelled to answer Jewett's summons to write a "fine and long story," she feared that time was running out—and she had yet to leave the magazine, devote herself to a literary vocation, and complete the novel that Jewett wanted her to write.[1] Yet even if Cather had been blessed with free time, writing a novel would have been a conflict-ridden endeavor. Moving from the short story to the novel meant expanding her range, the same assertive act that she later saw dramatized in opera singer Olive Fremstad's transformation from contralto to soprano. Moreover, since the short story was the local-color genre, moving to the novel also meant leaving the domain of women's writing—and thus leaving Sarah Orne Jewett.

Believing that plot construction was not her strength, Jewett always felt more at home with the short story or sketch than the novel. Writing *Atlantic Monthly* editor Horace Scudder in 1873, she explained that she could not write the "long story" he had requested. "In the first place I have no dramatic talent," she confessed. ''The story would have no plot. . . . It seems to me I can furnish the theatre, show you the actors, the scenery, the audience, but there never is any play!''[2] Thus when Jewett told Cather in 1908 that the young woman was ''far on [her] road'' to a novel ''of very high class,'' she was acknowledging a potentiality in the younger woman she did not feel she possessed herself.[3] And Cather understood her mentor's limitations. As she later observed, Jewett was ''content to be slight, if she could be true.'' She always spoke of ''the Pointed Fir papers'' or the ''Pointed Fir sketches,'' Cather recalled in ''Miss Jewett'': ''I never heard her call them stories'' (*NUF*, p. 89).

In granting Cather permission to surpass her own ''sketches,'' Jewett was also sending her protégée into literary territory where she herself had rarely ventured and where she did not feel she had succeeded. Thus when Cather prepared to embark on the ''solemn and terrible'' act of novel-writing in which she had once thought women writers were exposed as liars and sentimental fools, she could not look to Jewett's work as a model (*KA*, p. 409). Nor could she regard this ambitious endeavor with Jewett's ''fine flame of modesty'' (*NUF*, p. 89). The genre Cather had once conceded to male writers in 1911 (''If I see a novel by a woman, I confess I take one by a man instead; I prefer to take no chances when I read''), the novel still seemed a masculine preserve to her. And so in writing *Alexander's Bridge,* she was entering the domain where James, rather than Jewett, reigned. Even though she made several visits to South Berwick and 148 Charles Street during 1911, hoping to find Jewett's encouraging spirit, Cather could not be sure that her literary mother had the power to guide her into the male writer's province.[4]

The anxieties Cather experienced in contemplating the longer, more ambitious, and male-dominated genre surface in ''Behind the Singer Tower,'' a short story written in 1911—the same year as *Alexander's Bridge*—and published in the May 1912 *Collier's.* Based on a real-life disaster, the Windsor Hotel fire of 1899, ''Behind the Singer Tower'' has been viewed as one of Cather's rare works of social criticism. The story does expose the human costs of American progress: Cather contrasts two tragedies, a fire in a new skyscraper-hotel that stuns New York because hundreds of wealthy and powerful people perish, and an earlier accident during the hotel's construction that killed dozens of Italian immigrant workers but never made the news. New York's skyscrapers are her metaphors of Americans' destructive commitment to commercial and technological progress. Visual signs of upward mobility, ambitious buildings like the hotel embody the ''New York idea in architecture,'' outscaling ''everything in the known world.'' The American scheme of ''life and progress and profit was perpendicular,'' observes the narrator, who describes American life as a ''race,'' the same image of competitive individualism that Cather had used in ''The Enchanted Bluff'' to account for the destruction of the communal ideal.[5]

''Behind the Singer Tower'' is more than social criticism, however; it also reveals Cather's own ''perpendicular'' desire to rise as a novelist, going beyond both her editorial work and the short story. It is easy to see why her literary aspirations could be imagined as vertical: writing a novel meant going ''up'' as a writer, constructing a work of both greater length and height. The vertical imagery in "Singer Tower" reminds us that such ambitions are male, as do the many references to Flaubert's *Salammbô,* one of Cather's treasured novels. Flaubert's subject was the siege of Carthage—the war of mercenaries against the Carthaginian masters—and as Evelyn Haller points out, *Salammbô* "bristles with verticality," featuring towers, "engines of war, terraced buildings, aqueducts."[6]

In writing *Alexander's Bridge,* like New York's architects and Flaubert's warriors, Cather was entering a public realm of display and competition (her novel would be reviewed). Leaving the modest, and female, genre of the short story for the individualistic and male genre of the novel, she was also scaling heights from which she might fall. She includes a gruesome punishment in "Singer Tower" for the woman novelist's literary crime: after the hotel fire, rescuers discover on a window ledge "a man's hand snapped off at the wrist as cleanly as if it had been taken off by a cutlass." The hand belongs to "Graziani, the tenor"; the identification is possible because of the "little-finger ring," and the narrator remembers seeing the "same hand" when the singer "placed it so confidently over his chest as he began his 'Celeste Aida,' or when he lifted . . . his little glasses of white arrack." Cather thus associates this image of mutilation and dismemberment with the artist's expansive, ascending gestures.

"Lop away so much as a finger" Cather had written in "The Profile" (1907), "and you have wounded the creature beyond reparation" (*CSF,* p. 125). As we have seen, Cather's preoccupation with injuries to the hand reflects her literary anxieties. The hand which holds the pen is the physical agent of creation, as Cather noted in speaking of the creative process as the "voyage perilous" from the "brain to the hand" (*KA,* p. 76). But the severed hand in "Singer Tower" suggests a more serious concern than Cather's fear of literary inadequacy: it recalls the fear of retribution for writing shown in her early stories through images of mutilation, castration, and disfiguration. In 1911, however, the metaphor of wounding is less connected with the child's unlawful desire for the mother than with the woman writer's unfeminine ambition. Daring to claim novel-writing as her vocation, Cather merited punishment for her overreaching desire: destruction of her creative powers.

Although it is possible that Cather's fear of mutilation had a biographical source in the childhood trauma she recounted to Edith Lewis—the story of the half-witted boy who threatened to cut off her hand—images of mutilation and castration in her fiction and real-life problems with her right hand and arm frequently occurred at times of professional and personal stress when she was attempting to "rise" as a writer. We see this pattern when she first began to write fiction ("The Clemency of the Court"), when she took Henry James most directly as her mentor ("The Namesake" and "The Profile"), when she moved from the short story to the novel ("Singer Tower"), and when she wrote her first and only war novel, *One of Ours* (1922), a text filled with images of dismemberment, most often injuries to the hand and arm. After completing *One of Ours* and suffering the critical attacks she half expected for venturing into such masculine territory, Cather faced the punishment she had imaginatively anticipated in her novel: she experienced a painful attack of neuritis in her right arm and found herself unable to write.

"Singer Tower" suggests that the beginning novelist faced internal risks as well as external retribution in writing a "long story." Injury, mutilation, and death afflict both the wealthy residents in the hotel and the Italian laborers in the "hole," those above and those below the ground. If we consider only the story's social message, we see that both the masters and the workers are vulnerable to catastrophe in an unjust capitalist society. Yet if we consider the story's covert concern with creativity and literary aspiration, the detail reflects the risks Cather associated with "digging" into the self. The taller the building, the deeper the foundation; the longer the story, the deeper the descent into what Cather called the "bottom . . . of consciousness," and the greater the possibility of self-exposure (*AB* [1922], p. vi). In "A Tale of the White Pyramid," Cather resolved the dangers of self-disclosure by sealing the structure and ending the story. But if she were to build novels, "Singer Tower" implies, then she would have to descend further into the self than she had as a writer of short fiction (*CSF,* pp. 43–54).

•

In March 1911 Cather told Elizabeth Sergeant that she was contemplating a story about a "bridge-builder, with a double nature" that might become her first novel (*WC:AM,* p. 67). Using the conflicts she was trying to heal in her own life as the source of her protagonist's self-division, somehow Cather reserved enough time from her magazine work over the next few months to write *Alexander's Bridge.* The novel first appeared in three issues of *McClure's,* beginning in February 1912, under the title *Alexander's Masquerade;* it was accepted by Houghton Mifflin, the Boston inheritors of Ticknor and Fields, who brought the novel out in April 1912.

Although Cather's first title refers to the false, destructive self that absorbs Bartley Alexander when he deceives his wife and tries to regain his youth through an affair with Hilda, his former lover, Cather must have been aware that she also was indulging in a "masquerade" similar to the one Jewett had detected in "On the Gull's Road"—adopting a male persona. Since Cather was attempting a subject Jewett generally avoided—romantic and sexual passion—she had to cast the character who is entranced by female erotic power as male. Moreover, Cather may have thought that a male protagonist would distinguish her book from "women's fiction." Certainly she did not want her first novel compared to stories of female desire like *The Awakening* and dismissed by critics like herself.[7]

Constrained from following Jewett's advice to cast away male masks, Cather was not ready to follow her mentor's example and return from the world to the parish. If we look only at its formal properties, her first novel takes the road of "The Willing Muse" rather than that of "The Enchanted Bluff." Less derivative than the 1907 Jamesian stories, *Alexander's Bridge* still reveals Cather's allegiance to Henry James, 148 Charles Street, and the genteel Eastern literary establishment. The novel features elegant international settings (Boston and London), sophisticated characters, a Jamesian observer, and unconvincing drawing-room talk. As Cather later acknowledged, the "people and the places of the story interested me intensely at the time when it was written, because they were new to me and were in themselves attractive." It is not "always easy," she conceded, for the "inexperienced writer to distinguish between his own material and that which he would like to make his own'' *(AB* [1922], p. v). But she had written her first novel at a time when the ''drawing room was considered the proper setting'' and ''Henry James and Mrs. Wharton were our most interesting novelists'' *(OW,* p. 93).

The novel's frontispiece (also an illustration in the *McClure's* serialization) visually presents the upper-class, aristocratic world Cather was trying to make her own. *(See photograph section.)* Two slim, elegant people in evening dress (Bartley and Hilda) are embracing in a drawing room, the woman uttering one of the more unfortunate lines from the novel: ''Are you going to let me love you a little, Bartley?'' Filled with such stilted dialogue, *Alexander's Bridge* matches Cather's later description of its composition: it was as if she were forced to talk with ''someone not altogether congenial'' while riding in a park *(OW,* p. 93). This uncongenial companion also demanded that she not write of that unpromising literary site, Nebraska. ''As everyone knows,'' she observed in ''My First Novels,'' Nebraska then was ''distinctly déclassé as a literary background; its very name [threw] the delicately attuned critic into a clammy shiver of embarrassment'' *(OW,* p. 94).

Even before it was published, *Alexander's Bridge* was a novel Cather had her doubts about. When she submitted the manuscript to *McClure's,* she reverted to her earlier use of camouflage and employed a pseudonym, ''Miss Fanny Cadwallader''; she even had the manuscript mailed from St. Louis.[8] Several motives can be seen in this rather elaborate deception: unsure of her talents, Cather must have wanted to know if *McClure's* would accept the manuscript on its own merits; and having been told by McClure himself that she should abandon fiction, she may have wanted to force him to acknowledge her literary skill. The use of this particular pseudonym also suggests her desire to present her novel as the work of a more aristocratic, and Jamesian, self. With its aura of ladylike gentility, the name she adopted parallels her literary ''masquerade'' in *Alexander's Bridge:* Miss Fanny Cadwallader was more likely to have grown up in Boston or London than in Red Cloud, Nebraska. The name is a bit stagy and overdone, though, and hints that Cather's Jamesian self might not be her real identity.

When Edith Lewis reread the novel after Cather's death, she saw signs of Willa Cather as well as of Fanny Cadwallader. Her friend had unfairly disparaged *Alexander's Bridge* as an unrelievedly imitative work, she thought. ''It is true that it is a contrived novel,'' Lewis allowed, ''and one feels a slightness and an artificiality of structure in the early parts.'' But when it

> at last moves into its true theme, the mortal division in a man's nature, it gathers an intensity and power which comes from some deeper level of feeling. . . . It is as if her true voice, submerged before in conventional speech, had broken through, and were speaking in irrepressible accents of passion and authority. *(WCL,* p. 78)

Lewis has a point. *Alexander's Bridge* is not solely, as Cather later wanted to think, a conventional or imitative novel. In turning to Henry James and the language of the drawing room, Cather was not merely demonstrating her allegiance to literary fashions. In a profound way, she was both concealing and revealing the anxieties accompanying the woman writer's ambitious, risky decision to write and to publish a "long story." The novel's imitative style functions as psychological defense as well as literary deference, concealing volatile and troubling content. To understand the role *Alexander's Bridge* played in Cather's literary development, we need to take the novel more seriously than the author later did, focusing on the key issues of gender and creativity.

•

Cather's relationship to masculine values and enterprises in *Alexander's Bridge* is complex and ambivalent, displaying both respect and rejection. Her momentary return to Henry James for style and setting shows the woman novelist's anxiety of authorship: Cather suspected that she might not be equal to this demanding and public genre. In fact she had tried, and failed, to write a novel previously. In 1905 she completed a novel with a Pittsburgh setting that she intended to publish with McClure; whether McClure or Cather decided that the manuscript should not appear, her first novel-writing venture was not successful.[9] And so in writing *Alexander's Bridge,* again tackling the genre she had not mastered, Cather sought reassurance by apprenticing herself, once again, to her "mighty master." Her uncertainty of success, as well as her anticipation of punishment, can be seen in her decision to include a disaster in her story: the destruction of Bartley Alexander's bridge, which recalls the hotel fire in "Singer Tower."

That the untried novelist still feared that the novelist's vocation and genre were male can be seen in her portrayal of Bartley Alexander—engineer, bridge-builder, and the novel's artist-figure. The novel ends with the collapse of Bartley's most ambitious project, the Moorlock Bridge, and his drowning in the St. Lawrence. Cather based this event on a real-life tragedy, the Quebec Bridge disaster of 1907, and in part she borrows from the Quebec findings to explain why Bartley's bridge cracks: cramped by a "niggardly commission," he had used "lighter structural material than he thought proper" (p. 37).[10] Cather gives us a more self-revealing reason for the bridge's failure, however, that suggests her projection of authorial anxiety into the structure's collapse and the engineer's drowning.

In taking on the Moorlock Bridge project, Bartley is moving from the area of his expertise—bridges of "ordinary" length—to the "longest cantilever in existence" (p. 37). His Moorlock Bridge falls not only because he neglects to use adequate building materials: there is, we discover, a flaw in his original design. The bridge collapses because Bartley applies the methods he derived from designing bridges of average length to this oversized construction. Eventually Bartley acknowledges his complicity in the flawed design: "we were never justified in assuming that a scale that was perfectly safe for an ordinary bridge would work with anything of such length," he tells his assistant (p. 122). Fearing that her novel might turn out to be only a long short story, Cather evidently sensed that novel-writing might require more than an expansion of the literary techniques she had already learned in the woman writer's domain.

And yet—while *Alexander's Bridge* demonstrates Cather's insecurity in aspiring to a genre and a vocation she associated with men—the novel shows an even stronger rejection of male-defined views of the artist and the creative process. Cather not only condemns men's urge to dominate, as she had in "The Joy of Nelly Deane": she shows that this impulse, if expressed by the artist, paradoxically results in weak, flawed structures. In *Alexander's Bridge,* creative methods that employ force and subjugation are shown to be ineffective and self-destructive when Bartley's bridge plunges into the St. Lawrence.

In her 1922 "Preface" to *Alexander's Bridge,* Cather suggests why Bartley Alexander's seemingly powerful artistry is in fact fundamentally weak when she refers to the inexperienced writer's "building of external stories," an inferior means of construction which she contrasts to the mature writer's allowing a story to form itself. She develops this architectural metaphor at more length in her 1925 "Preface" to *The Country of the Pointed Firs* where she juxtaposes two modes of creativity, one "masculine," the other "feminine":

> It is a common fallacy that the writer, if he is talented enough, can achieve this poignant quality [a "unique" voice or cadence] by improving upon his subject-matter, by using his "imagination" upon it and twisting it to suit his purpose. The truth is that by such a process (which is not imaginative at all!) he can at best produce a brilliant sham, which, like a badly built and pretentious house, looks poor and shabby after a few years. If he achieves anything noble, anything enduring, it must be by giving himself absolutely to his material. And this gift of sympathy is his great gift; is the fine thing in him that alone can make his work fine. *(CFP,* p. 7)

Examining the first creative method in *Alexander's Bridge*— a method that resembles the "manly" model of the creative process she had thought superior in the 1890s—Cather exposes its inadequacies. The builder/artist's forceful attempt to impose a design upon his subject rather than to "[give] himself absolutely to his material"—evident in Bartley's desire to span the river with his bridge—results in a flawed, impermanent structure. Unlike Jewett's "tightly yet so lightly built" masterpiece *The Country of the Pointed Firs,* which Cather thought would endure forever, Bartley's construction does not "confront time and change" with strength and serenity *(CPF,* p. 11). Ponderous, badly designed, and shoddily built, his bridge falls of its own weight.

In sending Bartley's bridge into the St. Lawrence, Cather was in part expressing her fear that her own novel—an external structure that she willed into existence—would not be "noble" or "enduring." The woman novelist's work might thus be insufficient if judged by masculine standards and values. But since Cather was also questioning masculine standards and values, wondering whether the surface of strength might hide internal weakness, the bridge's collapse can be interpreted as a defiant and liberating gesture for a woman writer: Cather was also saying her last farewell to Kipling and the masculine aesthetic. In doing so she was clearing the path to *O Pioneers!* and her reconciliation of gender and vocation, for it was this excessively male-identified vision of the artist that kept the woman writer from recognizing that her own creative power lay in her "gift of sympathy," her ability to abandon the ego rather than to impose it upon her subject.

That Cather wanted to exorcize her youthful fascination with Kipling and masculine aesthetics is evident in the many similarities between Bartley Alexander and the Kipling who appears in her 1890s journalism. Signifying the power of war, imperialism, and technology to the younger Cather, Kipling was the bard of the machine, and she had praised his ability to write of the world's "gigantic industries." Cather had found Kipling's "worship of force" evident both in his sympathy for the "energy of great machines" and in his praise for "men in their hand-to-hand fight for a foothold in the world." She was particularly impressed by Kipling's story "The Bridge Builders," which also features an engineer–hero whose bridge is the emblem of his creative power. Unlike Bartley Alexander, Kipling's engineer is equal to his heroic task. After years of labor and hardship his bridge triumphantly subdues "Mother" Ganges: "we have bitted and bridled her," he exults.[11] Once Cather had liked this image of male power subduing feminine nature. "He [the engineer] had changed the face of the country for miles around," she wrote approvingly in her 1899 commentary on the story, "burrowed out pits and thrown up embankments, and seen a village of workmen grow up and about him" *(WP* 2, pp. 555–61).

In the 1890s, Cather considered technology's victory over nature analogous to the "virile" writer's praiseworthy triumph over recalcitrant subject matter in the creative process *(WP* 2, p. 556). But as she gradually began to distance herself from male identification, she wrote two stories in which she condemned this model of creativity: in "The Professor's Commencement" (1902) and "The Namesake" (1907), she criticizes an interrelated series of hierarchies and oppositions—male/female, technology/nature, artist/subject—in which the first element in the equation seeks to dominate and to subdue the second. In these stories she expresses her discomfort with male dominance through her description of pastoral landscapes ravaged by commercial and technological forces: gentle hills and valleys are turned into "blackened waste[s]" by industry; "virgin soil" is torn by the "chilled products" of the mills' "red forges"; timber is "ruthlessly torn away" from mountains, leaving "wounds in the earth" *(CSF* pp. 286, 141). In portraying these scarred landscapes, Cather was not merely lamenting technological and commercial threats to the natural world: she was also criticizing Kipling's "virile" model of creativity in which the artist tries to "[twist]" his subject matter "to suit his purpose."

Alexander's Bridge represents a continuance and a culmination of these attacks on masculine energies, not a return to the "worship of force" evident in her 1890s journalism. Bartley Alexander is a double for Kipling, and in exposing his weaknesses Cather rejects her earlier fascination with the virile artist, showing that the conqueror's urge to dominate, if adopted by the artist, produces a "brilliant sham."

Cather begins her deconstruction of masculine aesthetics by attributing heroic qualities to Bartley. The younger writer had associated her hero Kipling with Alexander the Great, and she gives the conqueror's name to her engineer–protagonist. "Glowing with strength and cordiality and rugged, blond good looks," linked with the power of technology and machinery, Bartley Alexander at first seems to be an unflawed hero (p. 9). Certainly Cather's readers would have initially viewed him positively because of the associations they brought to the novel:

embodying force and energy, the engineer was a popular symbol of American power and progress in the early twentieth century. The men who designed and built bridges, tunnels, and canals were embodiments of what Jackson Lears calls the "cult of strenuosity"—red-blooded men who united scientific, aesthetic, and economic power in making nature submit to the American will and imagination.[12]

Cather, however, challenges her contemporaries' romance with the engineer as well as her own former "worship of force" when she shows that Bartley's power is only superficial. She grants him only the appearance of strength:

> It was always Alexander's picture that the Sunday Supplement men wanted, because he looked as a tamer of rivers ought to look. Under his tumbled sandy hair his head seemed as hard and powerful as a catapult, and his shoulders looked strong enough in themselves to support a span of any one of his ten great bridges that cut the air above as many rivers. (p. 9)

A Paul Bunyan for the modern age, Bartley at first appears to be a Western folk hero who combines technological and natural powers. Yet Cather undermines these associations when she defines Bartley as "*only* [italics mine] a powerful machine" controlled by larger social and economic forces: he is a "mechanism useful to society," something that can be "bought in the market" (p. 39). The apparently independent and autonomous hero is thus the servant of the business interests that employ him, and Cather progressively rewrites the hero's plot as she brings Bartley's inner weaknesses to the surface. She laboriously foreshadows the correspondence between the flawed engineer and his shoddy bridge in the first chapter, when the Jamesian observer Lucius Wilson tells Bartley, and us, of his suspicions: "I always used to feel that there was a weak spot where some day strain would tell," Wilson confides. "The more dazzling the front you presented, the higher your facade rose, the more I expected to see a big crack zigzagging from top to bottom . . . then a crash and clouds of dust" (p. 12).

What is this "weak spot" that makes Bartley, despite his heroic demeanor, an inferior artist–builder?

If we read *Alexander's Bridge* only as a novel of adultery, then we would say that Bartley's weakness is his passion for Hilda; unable either to leave his wife or to give up his mistress, he vacillates between two women, and this self-division is the "crack" in his nature. And Cather offers some support for this hypothesis by having Bartley miss the warning telegram because he is with Hilda. Since Cather uses the language of adultery to describe the imitative artist who must have "his affair with the external material he covets," however, this surface plot directs us to a deeper level of significance. If we read *Alexander's Bridge* as a novel about the creative process, as I think we should, then Bartley's weakness is his adopting the inferior mode of creativity Cather describes in her "Preface" to *The Country of the Pointed Firs:* as the bridge-builder, Bartley imposes his designs upon nature, asserting the self rather than giving it up. Consequently the "crack" in his nature is his failure to integrate into the creative process the self-effacing qualities Cather eventually associated with the "gift of sympathy."

When Cather later described the act of writing, she portrayed a process that combined opposites: creating required both receptivity and activity, self-abandonment and self-assertion, emotion and control—both ''feminine'' and ''masculine'' qualities, if one were applying conventional categories of gender. But Bartley's method of construction—like that of the shoddy builder Cather portrays in the ''Preface'' to *Country*—reveals dominance and subordination, not the integration of opposites. His bridges seem to represent union since they link opposite shores, yet they span and subdue rivers—an image of form arching over passion, control precariously bridging the unconscious. Rivers—often underground rivers—are Cather's recurrent metaphors for unconscious energies; she would use the river again in *O Pioneers,* where the ''underground stream'' of Alexandra Bergson's ''subconscious existence'' nourishes her crops *(OP,* p. 203). In *Alexander's Bridge,* however, the engineer who forces the wrong design on his river has not ''[given] himself absolutely'' to his subject, as Alexandra Bergson gives herself to the land. Instead, Bartley strives to maintain control and dominance, and so his formal designs—imposed by the ego rather than arising from the unconscious—are weak.

As the novel progresses, it becomes increasingly clear that the artist–builder with a ''double nature'' is unable to integrate the passive, yielding, ''feminine'' aspects of creativity with the active, assertive, ''masculine'' ones; as a result, the unexpressed components become self-destructive when Bartley releases them with Hilda. If his bridge-building reveals Bartley's urge to dominate, then his affair with Hilda reveals an even stronger desire for self-abandonment; Cather uses this romance not to demonstrate the allure of passion but to explore deflections and distortions of the creative process. Unable to experience the artist's healthful ''regression in the service of the ego,'' Bartley pursues a self-destructive regression with Hilda, seeking to regain his lost childhood self. So finally he becomes not the child/artist who creates but the regressive child—like the silenced narrators and protagonists in Cather's first stories—who loses an adult sense of self.

Bartley's inability to integrate his regressive desires into a creative process becomes clear in Cather's description of his ocean voyage from New York to London. At the outset of his love affair, Bartley feels as if Hilda has revived the energy and vigor of his youth, but the shipboard crossing is characterized by ''enervating reveries'' rather than a sense of ''quickened life'' (p. 68). Passive and quiet in his deck chair, covered with steamer rugs and a fur-lined coat, Bartley lies in a ''blessed gray oblivion'' and enjoys feeling ''submerged in the vast grayness about him.'' This self-abandonment is seductively pleasurable; his adult worries dissipate and he feels ''released from everything that troubled and perplexed him.'' But his other adult attributes vanish also, and soon Bartley does not even dream or remember; sitting by the ocean, ''losing himself in the obliterating blackness and drowsing in the rush of the gale,'' he is overcome by ''numbness and torpor'' and finally thinks of ''nothing at all'' (pp. 72–74). Surrounding Bartley with imagery of self-annihilation and death, Cather foreshadows his drowning when the bridge collapses.

In ''The Namesake,'' Cather had described an enriching creative moment when the artist–hero Lyon Hartwell found an invigorating rebirth by merging with a ''great force.'' Unlike Lyon Hartwell, however, Bartley does not emerge from his "gray oblivion" into artistic endeavor. As Bartley's regressive ocean voyage suggests, he is drifting toward a deadly fusion—the undifferentiated state in which the infant is merged with the mother. Unlike the child whose transitional objects allow him to separate from the mother while symbolically returning to her, unlike the artist who travels into the "bottom of . . . consciousness" and returns to create, Bartley does not desire to connect with another by fashioning symbols or stories. Thus he does not create the bridges the artist builds between self and other, imaginative structures that simultaneously offer attachment and separation. And so the collapse of his flawed bridge and his drowning, far from being accidents, seem to represent and to offer the self-annihilation he unconsciously seeks.

Bartley's regressive desires distort the creative process in one other significant way. Later Cather described her creative turning point in *O Pioneers!* as a blessed moment of integration between personal and professional selves, childhood and adult identities, when she said that at some time a writer's "life line" and the "line of his personal endeavor" would meet *(AB* [1922], p. vi). Bartley is also seeking such psychic wholeness, hoping to find a "continuous identity" linking the adult man with the youth. But he does this in the same way he fashions his bridge: by imposing meaning, by building an "external" story. Having enjoyed a brief affair with Hilda before his marriage, Bartley makes her into a symbol of his lost self; he rekindles their affair, Cather tells us, not because he seeks to know her but because he wants to regain "some one vastly dearer to him than she had ever been—his own young self" (p. 40). Bartley hopes, then, to build a bridge between his past and his present by using Hilda. Since he has tried to force this integration between the child and the adult rather than letting it emerge—like the writer who "twists" his subject matter to fit his designs—Bartley finds self-division rather than self-unification. His psychological design, like his bridge, cracks with the strain.

Thus *Alexander's Bridge* is not simply the young writer's flirtation with glamorous but impersonal subjects and settings. Here Cather is concerned not only with masculine values she had already repudiated but also with a model of creativity she was exhibiting in the novel itself. Cather's fear that she, like Bartley, was imposing a design and building an external story paradoxically makes the novel an expression of her internal self; she projects her own anxieties about the possibly flawed structure she was creating into the collapse of the bridge. And in showing a character destroyed by the regressive impulses he cannot integrate into creative endeavor, Cather reveals her own recurring anxieties about writing.

Like the bridge that is its major symbol, Cather's first novel points in two directions, revealing signs both of her apprenticeship and of her emergence, of imitation and of authenticity. The awkward dialogue, drawing-room settings, and Jamesian touches place Cather's first novel in her apprenticeship, while the portrait of Bartley Alexander, the flawed artist, looks forward to *O Pioneers!* In her second novel Cather dramatically revises *Alexander's Bridge* when she turns Alexander into Alexandra: replacing her failed masculine hero with her triumphant heroine, she creates an artist–conqueror who achieves her creative designs by letting them emerge from the soil, not by seeking to subdue nature through force. And yet the similarity of the names also shows what close relatives these two characters are: Alexandra simply succeeds at integrating the "double nature" that stays divided in Bartley.

Retrospectively we can see the connections between *Alexander's Bridge* and *O Pioneers!,* but Cather never acknowledged that her first novel had anything in common with her later work. Elizabeth Sargeant found it conventional and Cather agreed, confessing that she had "tried to be literary" and "feared that she had come a cropper." Despite Mrs. Fields' "praise and delight," she knew that she was "through with Bartley" *(WC:AM,* p. 76). Writing Zoë Akins in March 1912, Cather told her that she was working on a Nebraska story ("The Bohemian Girl") which was much better than *Alexander's Bridge;* she knew that Akins would prefer the Western story, although probably most people would choose Bartley.[13] And once her literary career took its new direction with *O Pioneers!,* Cather distanced herself more and more from *Alexander's Bridge.* In her 1922 "Preface" to the second edition she portrayed it as the inexperienced writer's imitative venture, and in the 1930s she completed its demotion when she gave first place in the Library Edition to *O Pioneers!,* placing *Alexander's Bridge* in Volume Three along with *April Twilights,* the other apprentice work she wanted to disown.

And yet writing *Alexander's Bridge* was very important to her at the time. Dispensing with Kipling, literary virility, and the masculine model of the creative process, Cather was ready for the "adventure that sought [her]" in *O Pioneers!,* ready to integrate her "life line" with the "line of [her] personal endeavor." Not only was the novel more self-expressive than she later acknowledged, but finishing it—and having it accepted for publication by Houghton Mifflin—bolstered her self-confidence. She had published it, she admitted to H. L. Mencken in 1922, with pride.[14] Because she had completed her first novel successfully, Cather felt justified in taking a risk—leaving the magazine for a few months to devote herself to her fiction. Partly because of *Alexander's Bridge,* she was willing to define herself as writer.

•

In the fall of 1911 Cather took a leave of absence from *McClure's* and went with Isabelle McClung to Cherry Valley, a small village near Cooperstown in the Finger Lakes region of New York. Isabelle had selected the site—her mother had grown up there—and she presided over Cather's retreat. The two women rented a house; Isabelle took care of the amenities, and Cather spent her time sleeping, hiking, and writing. Elizabeth Sergeant recalls Cather's exuberance when she announced her plan to leave the city for a few months to "write a couple of stories that had been bothering her for some time" *(WC:AM,* p. 72). Mrs. Fields' edition of Jewett's letters had just been published, and Sergeant wondered whether seeing Jewett's advice in print had prompted Cather's decision to take a leave of absence. Certainly Cather was at last taking the step Jewett had called for, leaving the distractions and skyscrapers of New York City for a "quiet center of life."

This was a blissful period. Five years of high-pressure life at the magazine had left her ill-tempered, fragmented, and drained; she needed rest and recuperation. As she wrote to S. S. McClure, she had not enjoyed herself so much in years. The only thing that happened in Cherry Valley was the weather, Cather informed him, but that was fine with her—when one was resting that was enough excitement. And she was so good-humored that even bad weather did not bother her; it had been raining for four days, she wrote, but she and Isabelle went walking anyway. After a day in the wet woods, she could sleep nine hours without turning over.[15]

Cather could write freely to McClure of her joy in leaving New York and its pressures because she did not think that she was deserting her Chief. McClure had plunged his magazine into financial difficulty, and during a forced reorganization, lost both economic and editorial control. During the time Cather was at Cherry Valley he was still nominally editor-in-chief and she was planning to return after her leave, but McClure was soon dismissed by the new owners.[16] Although she wrote a few articles for *McClure's* between 1913 and 1915, Cather never went back to work there. S. S. McClure's professional calamity was, in a sense, her gain; she did not have to regard leaving journalism for full-time writing as a betrayal of the fatherly mentor she loved and respected. Simultaneously she was in the enviable position of exploring a new identity and vocation without being forced to abandon the old. Assuming that she could return from Cherry Valley to her secure job and salary, Cather could regard her writing as play rather than work.

McClure himself encouraged Cather to throw herself into her writing, and his enthusiasm contributed to her creative well-being at Cherry Valley; she must have assumed that Jewett would approve, and after she received a letter from McClure she knew she had her father's blessing as well. "You must dismiss the magazine from your mind entirely," McClure wrote her in October, "forget it exists, and when you come back I hope you will not let yourself be tied up in office machinery."[17] Several days later, en route to Battle Creek, Michigan, McClure stopped off to see her for a few hours and reaffirmed his support in person. His visit had made her even more cheerful, Cather wrote him a few days later; their talk had helped to straighten her out, and she was resolved not to think about the magazine at all.[18]

McClure's encouragement of Cather's writing, so different from his previous belittlement of her literary abilities, may not have been quite the reversal it seems to be. After telling her to "dismiss the magazine" from her mind and devote herself to fiction, he mentioned that he hoped that when she returned to the magazine she would "work out a series of articles that will give us distinction."[19] Evidently McClure hoped that Cather—once she had played at being a fiction writer for a few months—would return to the magazine with renewed dedication. But Cather did not question the Chief's motivations, assuming that he was solidly behind her experiment.

Whereas McClure's demands had formerly contributed to Cather's loss of creative energy, now his support aided her regeneration. During this restorative period, the imaginative powers she feared dissipated by her editorial duties were revived. Within three months she completed her revisions for *Alexander's Bridge,* finished "Alexandra" (which she had started previously), and wrote "The Bohemian Girl," a short story in the same vein as *O Pioneers!* This story was "different from anything she'd ever written," Cather told Elizabeth Sergeant. "She was convinced that no magazine would consider it. But she liked it herself" *(WC:AM,* p. 76). What was so different about "The Bohemian Girl"?

•

The story opens as a character who has escaped Nebraska's confinement returns to his homeland. Nils Ericson pays a visit to Sand River Valley after spending twelve years in Europe, where he has made his fortune working for a Norwegian shipping line. His goal is not to revive childhood memories or to resurrect his youthful self; unlike Bartley Alexander, Nils is satisfied with adult life. His return is temporary, his purpose to elope with the fiery Bohemian Clara Vavrika, who has married his stolid, taciturn brother Olaf during Nils' absence. He manages to rekindle her passionate spirit, subdued but not crushed by her loveless marriage. And in contrast to *Alexander's Bridge,* Cather gives this tale of adulterous love a happy ending: Nils and Clara flee Nebraska, leaving Olaf Ericson and his petty concerns—money, new barns, and the Bohemian vote—far behind.

Although Cather had occasionally drawn on her Nebraska past before, "The Bohemian Girl" does differ from her earlier work, as she had thought; her use of Nebraska materials is both more compassionate and more thickly-textured than in her previous fiction. A closer relative of *O Pioneers!* or *My Ántonia* than of "A Wagner Matinee" or "The Enchanted Bluff," the story is about the immigrant groups who settled the Divide (here Scandinavian and Bohemian); it places individuals against the backdrop of family and community rituals, here the barn-raising festival and dance at Olaf Ericson's; and it suggests the profound interrelationship between people and the land. Cather now returns to the Divide in a sympathetic spirit. More removed in time and space from her past than she had been in *The Troll Garden,* she writes from understanding rather than anger. And in contrast to "The Enchanted Bluff," the memories that are awakened here are not of herself but of others—the immigrants whose stories she had received as a child, stories she was now ready to retell. In telling their stories she was also telling her own, creating her characters by drawing, in part, upon herself; and yet because she was thinking of her subject rather than of herself, she creates full-bodied characters in Nils and Clara, not pale stand-ins for the author. As a child, Cather remembered the joy of "[getting] inside another person's skin" when she listened to the pioneer women tell stories; in "The Bohemian Girl" she does the same herself, getting inside the "skin" of her characters *(KA,* p. 409).

Cather had not been able to do this very effectively in *Alexander's Bridge:* Bartley and Hilda never really come to life. Cather's ability to create realistic, vital characters in "The Bohemian Girl" suggests that she was becoming the second writer she described in her 1925 "Preface" to *The Country of the Pointed Firs,* the writer whose "gift of sympathy" allows her to "[fade] away into the land and people of [her] heart" *(CPF,* p. 7). As Jewett had hoped, she was now able to look back on her native material with both detachment and empathy. And it is possible also that Cather could create convincing characters because she had resolved many of the conflicts that had made writing a difficult process; her attention could now be given to her subject.

Although "The Bohemian Girl" is not as preoccupied with creativity as *Alex-*

ander's Bridge—in which Cather's focus on gender and art forms a scarcely-concealed subtext—it cannot be classified solely as a love story. "The Bohemian Girl" is also a story about risk and self-transformation, the possibilities facing Willa Cather at Cherry Valley. Clara's story is, in part, the author's; Clara's response to Nils reflects Cather's response to the creative power she sensed was within her, a power she was releasing even more strongly in "The Bohemian Girl" than in *Alexander's Bridge*.

At first Clara fears expressing the emotions she has buried. The passion Nils awakens seems to threaten an "erasure of personality"—the state that attracted and terrified Cather. Imagery of violation, drowning, and death surrounds the lovers' first embrace:

> [Nils] drew the horse under the shadow of the straw stack. Clara felt him take her foot out of the stirrup and she slid softly down into his arms. He kissed her slowly. He was a deliberate man, but his nerves were steel when he wanted anything. Something flashed out from him like a knife out of a sheath. Clara felt everything slipping away from her; she was flooded by the summer night.

This enigmatic passage suggests the dangers of releasing either sexual or creative desire. Clara must face the potentially wounding power of male sexuality and aggression, conveyed in the knife image; then she must risk the return to the infant's merging with the mother, suggested by the sensation of being "flooded." Similarly, the writer who relaxes the ego's control risks "the knife," a metaphor conveying her vulnerability: in giving herself up to the possibly wounding experiences of life from which art arises, she may be, as Cather later commented to an interviewer, "cut . . . deep" *(KA,* p. 413); and if she draws deeply upon the self in the creative process, she may be "cut" or mutilated by hostile critics.[20] Since the creative process replays the dynamics of the mother–daughter bond, the writer, like the lover, risks being "flooded," particularly if she abandons her ego-ridden attempt to build a story and lets the story write itself, releasing the regressive components of the creative process.

Clara confronts these sexually potent forces, just as Cather was confronting the powerful creative forces she was beginning to unleash in her writing. "You have to plunge," Nils tells her. "There's nothing so dangerous as sitting still." At first, however, Clara does not know whether she will be able to leave with him; the desire he awakens threatens identity and autonomy, and she does not know if she can abandon her old self:

> She felt as if she could not bear separation from her old sorrows, from her old discontent. They were dear to her, they had kept her alive, they were a part of her. There would be nothing left of her if she were wrenched away from them. Never could she pass beyond that skyline against which her restlessness had beat so many times. She felt as if her soul had built itself a nest there on that horizon at which she looked every morning and every evening, and it was dear to her, inexpressibly dear.

Cather beautifully captures the ambivalence with which most human beings regard change. Old patterns may be restrictive, but at least they are familiar: Clara fears that "there would be nothing left of her" if she abandons her confining past. The landscape mirrors these ambiguities: the skyline is both the encircling boundary that keeps Clara locked within an old, incomplete self and an "inexpressibly dear" maternal embrace within which her soul builds a nest. Take away the boundaries that restrict and define the self, and one confronts the possibility of self-annihilation. Writing her own fears of literary risk-taking into Clara, Cather also allies herself with Nils, the voice of change: "One has to tear loose."

He helps Clara to decide by lifting her "lightly to her saddle," reversing the motion of dismounting which began the scene and represented her initial loss of volition ("Clara felt him take her foot out of the stirrup"). This second gesture does not lead to his domination or absorption of her, however, but rather to her attainment of autonomy and power. Clara leaves riding her own horse, just as she entered the story—giving herself up to a force and motion she can also guide. Although Clara is not as transparent an artist-figure as was Bartley Alexander, the metaphor Cather later used in "My First Novels" to convey the exhilaration she felt when writing *O Pioneers!*—"taking a ride through a familiar country on a horse that knew the way" *(OW,* pp. 92–93)—recalls the horsewoman Clara riding with a "free, rhythmical gallop" on a horse "under tight rein." Like Clara, in writing "The Bohemian Girl" and, later, *O Pioneers!,* Cather found a way to give up and yet maintain control in the creative process, combining inspiration and technique, passion and form, self-abandonment and self-expression.

"The Bohemian Girl" does not end with the lovers' flight, however, but a year later as Nils' brother Eric is about to join them in Sweden—a conclusion that has struck some readers as a problematic diversion from the main plot.[21] Nils has sent Eric money to escape a repressive home. Since Nils' departure, Mrs. Ericson has kept a close watch on her son and forbids him to go into town by himself. Eric wants to leave, but since he is tied to the mother whom he both loves and fears, he cannot break away from his childhood home and his child's identity. Cather's final scene is his ambiguous return. Sitting by his mother, Eric is rewarded for his sacrifice of independence by the one gesture of affection she grants him in the story: "her fingers twined themselves in his soft, pale hair. His tears splashed down on the boards; happiness filled his heart."

Yet this ending is a digression only if one reads "The Bohemian Girl" as merely a love story instead of a story about risk, change, and self-renewal—issues that confront writers as well as lovers. The characters' choices comment on each other: Clara leaves home even though part of her wants to stay, Eric stays home even though part of him wants to leave. She risks the future, he finds safety in the past. David Stouck feels that Eric's choice reflects Cather's deepest wish—to return home to a "protective" mother. "We cannot help feeling that his choice involves the author's deepest sympathies and emotional preference."[22] Although Cather feels the pull of Eric's choice, in my view she allies herself more strongly with the character who dares to leave home—with Clara. Cather shows us that Eric returns to a problematic maternal embrace when he reenters the "nest" Clara has left: his mother offers affection at the price of a tangled, symbiotic bond, conveyed by the almost frightening image of her fingers "[twining] themselves" in his hair.

Happily for Willa Cather, writing the story allowed her to make Clara's and

Eric's choices while avoiding the negative consequences of both. Returning in memory to her homeland, Cather created a new literary identity without rejecting her past; returning in imagination, she could reattach herself to her Nebraska roots without being bound by them. Eric's return is regressive, Cather's creative; for her, the "road home" was also the road into her literary future.

"The Bohemian Girl" also anticipates Willa Cather's literary direction in *O Pioneers!* by hinting that femininity and artistry could be united. We see this liberating potential not in Clara but in the old farm women. Cather's portrayal of their strength and dignity shows her beginning to accept her own inheritance of female power and creativity—from the Nebraska farm wives, from the storytellers of Willow Shade, and, although this would be the last gift she would acknowledge, from the women in her own family.

> The older women, having assured themselves that there were twenty kinds of cake, not counting cookies, and three dozen fat pies, repaired to the corner behind the pile of watermelons, put on their white aprons, and fell to their knitting and fancywork. They were a fine company of old women, and a Dutch painter would have loved to find them together, where the sun made bright patches on the floor and sent long, quivering shafts of gold through the dusky shade up to the rafters. There were fat, rosy women with brown, dark-veined hands; and several of almost heroic frame, not less massive than old Mrs. Ericson herself. . . . [Nils] fell into amazement when he thought of the Herculean labors those fifteen pairs of hands had performed: of the cows they had milked, the butter they had made, the gardens they had planted, the children and grandchildren they had tended, the brooms they had worn out, the mountains of food they had cooked. It made him dizzy. *(CSF,* pp. 3–42)

When Cather wrote "A Wagner Matinee" (1903), she painted a bitter portrait of a farm woman warped and stunted by the demands of serving others; at that point, only high culture signified creativity. But here, surer of her literary abilities, she looks back at the immigrant women from her childhood and makes their domestic labors seem heroic. Memory becomes a source of empowerment for the writer who recalls "old neighbors, once very dear, whom I had almost forgotten in the hurry and excitement of growing up" *(OW,* p. 93). Cather was not quite ready to create a living, vital link between such women and a daughter/artist—that she would do in *O Pioneers!,* where Alexandra Bergson learns how to cultivate the land from her mother. But she shows us that some forms of maternal power are liberating rather than oppressive.

•

Cather felt both confident and insecure about the new direction her work was taking in "The Bohemian Girl." She told Zoë Akins that she was sure her friend would prefer her Nebraska piece to *Alexander's Bridge*—it was the better story[23]—and yet it took Elizabeth Sergeant two days of fierce arguing to convince Cather that "this was it": she simply must submit "The Bohemian Girl" to *McClure's.* The story revealed the "potential I had divined in her," Sergeant recalls, "and had not found in the bridge-builder's story, where her pen had been guided by a fashionable pattern. Here, at last, was a Willa Cather story that was true to the person" *(WC:AM,* p. 76).

Encouraged by Sergeant's response, early in 1912 Cather sent "The Bohemian Girl" to Cameron Mackenzie, McClure's son-in-law, who was then making editorial decisions at the magazine. She was overwhelmed and somewhat surprised by Mackenzie's enthusiasm. She wrote Sergeant immediately: Mackenzie had taken her to tea at the Brevoort and offered her $750 for "The Bohemian Girl"! Yet she knew that *McClure's* never paid more than $500 for a story, and told him that she could not accept so much. Mackenzie agree to let Cather take $500 but warned her that she would have to accept more money the next time *(WC:AM,* p. 78). This unusual scene—a writer insisting that she be paid less than an editor proposed—reflects Cather's amazement that sophisticated Easterners would appreciate fiction about Nebraska immigrants. "What did they see in it?" she asked Sergeant *(WC:AM,* p. 78). By the time "The Bohemian Girl" appeared in the August 1912 issue of *McClure's,* however, Cather was a more self-assured writer. The story did not pass Mrs. Fields' inspection, Cather told Louise Imogen Guiney, but *she* knew that "The Bohemian Girl" was better than *Alexander's Bridge.*[24] Writing Zoë Akins later in the year, Cather thanked her for liking "The Bohemian Girl" and confessed that she thought it pretty good herself.[25]

•

Cather worked on one other story at Cherry Valley—"Alexandra," which she later incorporated in *O Pioneers!* She had written a draft of the story previously, even before *Alexander's Bridge,* although exactly when she did so is uncertain. She had talked it over with Sarah Orne Jewett, so Alexandra Bergson's story had been on her mind for a long time.[26] Although Cather completed "Alexandra" at Cherry Valley, she was even more unsure of it than "The Bohemian Girl"; according to Edith Lewis, she "doubted whether anyone but herself would find it interesting." Although no copy of this draft remains, Cather read the story to Lewis before leaving for the Southwest, so we at least have Lewis' description of "Alexandra."

The story included Part I of *O Pioneers!* ("The Wild Land"), Lewis recalls, and ended with Alexandra's dream of being lifted by a strong, godlike male figure (which now concludes "Winter Memories," Part III of *O Pioneers!).* Cather "had this story very much at heart," Lewis remembers, but was "dissatisfied" and did not try to get it published. "I think she felt there was something there which she did not wish to waste by inadequate presentation," Lewis writes; "if she held it for a while, she might get some new light on it" *(WCL,* p. 83). Evidently Cather also still feared that fiction which did not follow the conventional literary patterns might be inferior. Writing H. L. Mencken in 1922, she said that at first she feared that "Alexandra" seemed real to her only because she had a romantic and lyrical attachment to her homeland.[27] So Cather held back her "chill and grey" Swedish story: it was, she told Sergeant, just not "ready to show" *(WC:AM,* p. 76).

The story may not have been "ready to show" because Cather was not quite ready to envision Alexandra—or herself—as an artist. If Lewis' recollections are accurate, then the draft of "Alexandra" did not include "Neighboring Fields," the long chapter in which we see the flowering of the wild land, proof of Alexandra's creative power. It was not until her return from the Southwest that Cather could finally integrate the woman and the artist and clearly see literary possibilities in Nebraska's soil.

•

Cather was pleased with her accomplishments at Cherry Valley. Along with Clara, she had chosen "to plunge," and the risk of leaving the magazine paid off. She returned to New York early in 1912 and soon had the pleasure of having "The Bohemian Girl" praised by friends and accepted for publication. Knowing that she had a sympathetic audience was particularly important to Cather at this time because she was discarding conventional literary patterns and using her own memories more directly in her fiction than she ever had before. Whereas "Fanny Cadwallader" had sent *Alexander's Bridge* to *McClure's,* Willa Cather took "The Bohemian Girl" directly to Cameron Mackenzie. Using her own voice and her own name, she needed to be reassured that her literary self-expression could be valued by others.

And so Willa Cather was well on her way to *O Pioneers!* even before her trip to the Southwest in the spring of 1912. And yet the accomplished novelist liked to attribute her artistic birth to this transformative experience.

Although Cather's later accounts of the profound differences between the fiction she wrote before and after this trip are somewhat exaggerated, nevertheless something did happen to her during this vacation that had a profound impact on her writing. Cather was in the process of discovering herself as woman and artist before she left: she had been doing so since the 1890s. Yet after her return, as Ellen Moers observes, we date "her serious beginnings as a novelist."[28] Within a few months she had completed *O Pioneers!,* and after that her novels kept appearing, one after another, with scarcely a break. It was as if a dam had been broken and the stories she had been storing up came pouring forth. The Southwest trip and its immediate aftermath—the writing of *O Pioneers!*—constitute the turning point in Willa Cather's life as a writer.

NOTES

1. *Letters of Sarah Orne Jewett,* ed. Annie Fields (Boston: Houghton Mifflin, 1911), p. 246.

2. *Sarah Orne Jewett,* ed. Richard Cary (Waterville, Me.: Colby College Press, 1967), p. 29.

3. Fields, p. 246.

4. For descriptions of these visits, see *WC:AM,* pp. 59–67, and Willa Cather to Louise Imogen Guiney, May 25, 1911, Holy Cross College Library, Worcester, Mass.

5. In the February 1911 issue of *McClure's* a poem ("From a Skyscraper" by Allan Updegraff) reveals similar themes and images: four people are "reading" the New York skyline, the first three seeing fame, wealth, and power, while the fourth sees "shame" *(McClure's* 36: 438). Cather obviously had read the poem, and it may have inspired her short story.

6. Evelyn Haller, " 'Behind the Singer Tower': Cather's Visual Commentary on New York in the Carthaginian Mode," paper presented at the annual conference of the Northeast Modern Language Association, Spring 1980. Haller also connects the story with Cather's decision to seek the "perilous prize" of artistic success (p. 8).

7. It is possible, as E. K. Brown speculates, that the title *Alexander's Masquerade* was "invented in the office of the magazine" *(Willa Cather: A Critical Biography,* completed by Leon Edel [New York: Knopf, 1953], p. 153). Given Cather's role as managing editor, however, it is likely that she had something to do with its invention; at the very least, she must have agreed to its use. The novel was published in England as *Alexander's Bridges.*

8. Bernice Slote, "Introduction" to *Alexander's Bridge* (Lincoln: University of Nebraska Press, 1977), p. vi.

9. Slote, p. v.

10. For a discussion of the Quebec bridge disaster, see David Plowden, *Bridges: The Spans of North America* (New York: Viking, 1974), pp. 172–75.

11. Rudyard Kipling, *The Day's Work (Part I),* from *The Writings in Prose and Verse of Rudyard Kipling* (New York: Scribner's, 1899), p. 13.

12. T. J. Jackson Lears, *No Place of Grace: Antimodernism and the Transformation of American Culture, 1880–1920* (New York: Pantheon, 1981), p. 100. For an analysis of the meaning of bridges in the late-nineteenth-century American imagination, see Alan Trachtenburg, *Brooklyn Bridge: Fact and Symbol* (New York: Oxford University Press, 1965); for a discussion of Cather's demystification of the engineer–hero in *Alexander's Bridge,* see Elizabeth Ammons, "The Engineer in the Garden: Willa Cather's First Novel and America's 'Soldier of Fortune,' " *American Quarterly* (forthcoming).

13. Willa Cather to Zoë Akins, March 14, 1912, Akins Collection, Huntington Library, San Marino, Calif.

14. Willa Cather to H. L. Mencken, February 6, [1922], Enoch Pratt Free Library, Baltimore, Md.

15. Willa Cather to S. S. McClure, October 21, 1911, November 5, 1911, and November 17, 1911, Lilly Library, University of Indiana, Bloomington, Ind.

16. For an account of McClure's loss of power, see Harold S. Wilson, *McClure's Magazine and the Muckrakers* (Princeton, N.J.: Princeton University Press, 1970), pp. 320–21.

17. S. S. McClure to Willa Cather, October 14, 1911, Lilly Library, University of Indiana, Bloomington, Ind.

18. Cather to McClure, November 5, 1911.

19. McClure to Cather, October 14, 1911.

20. In her ground-breaking article "The Forgotten Reaping-Hook: Sex in *My Ántonia,*" Blanche H. Gelfant was the first critic to note the connection between Cather's use of "sharp instruments and painful cutting" and the novel's discomfort with sexuality *(American Literature* 43 [March 1971]: 60–72; reprinted in *Critical Essays on Willa Cather,* ed. John J. Murphy [Boston: G. K. Hall, 1984], p. 158. Knives and sharp instruments have multiple meanings in Cather's fiction, however—they can be tools of creation as well as punishment. And while Gelfant is right in arguing that Cather often portrays sexuality as a threat to the self's autonomy, some of her characters—like Clara Vavrika—do not flee from the threat of obliteration that sexuality poses, but face it and find themselves enriched rather than depleted.

21. E. K. Brown finds the ending "unexpected and unsatisfactory" as well as "sentimental"; the moment of affection between Eric and his mother may have been added, he speculates, because Cather decided that her story was "too severe in its estimate of the Divide" (pp. 165–66). In Richard Giannone's view, the final scene is "irrelevant to the

action'' *(Music in Willa Cather's Fiction* [Lincoln: University of Nebraska Press, 1968], p. 58).

22. David Stouck, *Willa Cather's Imagination* (Lincoln: University of Nebraska Press, 1975), p. 22.

23. Cather to Akins, March 14, 1912.

24. Cather to Guiney, August 17, 1912.

25. Cather to Akins, October 31, 1912.

26. In a letter to H. L. Mencken, Cather explained that she had written a draft of *O Pioneers!* even before *Alexander's Bridge,* evidently referring to ''Alexandra'' (Cather to Mencken, February 6, [1922].

27. Cather to Mencken, February 6, [1922].

28. Ellen Moers, *Literary Women* (Garden City, N.Y.: Doubleday, 1976, p. 259).

18

Every Artist Makes Herself Born

> Every artist makes himself born. Your mother did not bring anything into the world to play piano. You must bring that into the world yourself.
>
> *The Song of the Lark*

> And when I see you again, I shall not see you, but your daughter.
>
> *The Song of the Lark*

Before Willa Cather traveled to the Southwest in 1912, Elizabeth Sergeant observes, she had suffered a "truly gruelling inner pull" between Eastern literary culture and her Western background. After she returned and began to write of Nebraska in her novels, however, Cather found a new "integration and tranquillity," seeming to "be all of a piece" *(WC:AM,* p. 54). Cather's weeks in the Southwest had something to do with this transition and integration. In this new land—which affected her emotionally, spiritually, imaginatively, erotically—polarities seemed momentarily reconciled. The Southwest offered her a magical, harmonious balancing of earth and sky, nature and culture, life and art, body and soul, feminine and masculine. This integration of opposites was mirrored in the region's unusual contours: instead of Nebraska's unrelieved flatness, here Cather found indentations in the earth—gulleys, ravines, crevices—matched by mesas, bluffs, cliffs. The sky and air thus entered the earth while the earth rose into the sky, an interpenetration she would associate with sexuality and fertility in the Whitmanesque passage describing spring plowing in *O Pioneers!:* "The air and the earth are curiously mated and intermingled, as if the one were the breath of the other" *(OP,* p. 77).

Cather envisioned the trip as a restorative vacation. She had been ill for several weeks early in 1912, so like many Americans who migrated to the Southwest for its dry climate, she associated the region with a return to health. Cather was already feeling better when she left New York; for the first time in years she was not thinking about work. "The Bohemian Girl" and *Alexander's Bridge* were safely with the publisher, so she could leave with a "clear conscience," ready to explore a new country with Douglass, her brother and childhood playmate *(WC:AM,* p. 78). Even after the Southwest had exerted its powerful imaginative influence on her in *O Pioneers!* and *The Song of the Lark,* Cather did not go back consciously to look for literary material. This was her land of escape, freedom, and play, not of work. "She loved the Southwest for its own sake," recalls Edith Lewis, who accompanied her friend there on several return trips. "She did not keep any notebook or diary," and "never spoke of its literary possibilities" *(WCL,* p. 101).

Cather's visit with Douglass, who was working on the Santa Fe railroad and stationed in Winslow, Arizona, differed from the literary pilgrimage she had taken to Europe in 1902 when the traveler not only sought out the tombs and monuments of great artists but also immediately transferred her impressions into writing. In Winslow she stayed in her brother's house—a rough-and-ready structure—along with Douglass' housemates, a brakeman named Tooker and their drunken Cockney cook. When Cather first arrived there in April, she planned to do some writing and told Elizabeth Sergeant that she might have to leave to find a quieter place to work.[1] But she soon forgot about both leaving and writing. She and Douglass explored canyons and cliff-dwellings, saw an Indian ritual (the Hopi snake dance), rode, hiked, and camped out. It was wonderful seeing her brother again, she told S. S. McClure. They were having wonderful adventures together, just as they had as children; she had not been so happy since childhood. She loved being with her brother and his wild pals.[2]

Cather did not tell McClure about one of her vacation companions, however, possibly because she thought he might not approve. When her brother and his railroad pals were at work, Cather spent a good deal of time with a young Mexican named Julio who was living in Winslow's Spanish settlement. Julio sang Mexican love songs to her, took her to the Painted Desert, and escorted her to a Mexican dance where Cather was the only Anglo woman. He also told her local legends and myths, including the grimly erotic story of an Aztec Cleopatra which Cather included in "Coming, Aphrodite!" as "The Forty Lovers of the Queen." She was entranced, delighted, infatuated. Having met Julio she was not sure that she could ever leave, but Cather finally tore herself away by the middle of June, even though she kept thinking about going back to spirit him away to New York.[3] Cather did not forget the appeal of the Southwest's non-Anglo-Saxon men; years later, when asked why Mabel Dodge Luhan had married an Indian, she reportedly replied "How could she help it?" *(WC:AM,* p. 206).

Cather immediately felt that she had been reborn in the Southwest. Writing S. S. McClure shortly after leaving, she told him that he would never recognize her; she was a new person, dark-skinned and good-humored. She was ready for a different life, she thought; she felt confident in herself, in touch with fundamentals, sure that she would never be bothered by little things again.[4] Drawing on her Southwest experience in *The Song of the Lark,* Cather gave Thea a similar transformation:

> Here everything was simple and definite, as things had been in childhood. Her mind was like a ragbag into which she had been frantically thrusting whatever she could grab. And here she must throw this lumber away. The things that were really hers separated themselves from the rest. Her ideas were simplified, became sharper and clearer. She felt united and strong. (p. 306)

In the Southwest—without ever singing a note—Thea Kronborg finds herself as an artist. Similarly, after her return to New York, Cather seemed "suddenly in control of her inner creative forces," Elizabeth Sergeant recalls. Somehow the Southwest had made available a "new artistic method" as well as her Nebraska memories *(WC:AM,* p. 85).

In her preface to the second edition of *The Song of the Lark,* Cather observed that her heroine was the sort of person to whom "fortunate accidents" always happen. She could have made the same observation of herself. Her trip to the Southwest was a "fortunate accident" that was not really all that accidental; Cather was ready for this turning point, and had even imagined it in stories like "Eric Hermannson's Soul," "The Namesake," and "The Bohemian Girl." Cather did not discover herself as woman and artist there: her ecstatic weeks in the Southwest were the culmination of a process of self-creation that had been happening for many years. But her weeks there were empowering, guiding her to the fictional material that was "really hers" and aiding her reconciliation of the once-opposed identities of woman and artist. Cather returned from the Southwest at peace with herself, with her femaleness, and with her literary imagination.

In *The Professor's House* (1925) "Tom Outland" is the name Cather gives to the character who discovers the ruins of the cliff-dwellers, and his surname embodies the significance the Southwest held for Cather: it was Outland, a magical place remote in time and space, far from New York's ascending skyscrapers and perpendicular aspirations. Even more than the fishing village of Lavandou—another compelling landscape off the map of conventional social structures and expectations—the Southwest corresponded, in ways Cather could never fully articulate, to unexpressed and perhaps unconscious dreams and desires. Speaking both to imagination and to memory, the landscape encouraged her to answer in her own voice. Why did she want to reply to this country?

•

Cather had already traveled to the Southwest in her mind, and so, like Lavandou, the region was emotionally and psychologically mapped before she ever arrived there. As children, she and her brothers had been fascinated by the story of Mesa Verde, the mesa in southwestern Colorado where the ruins of an Indian civilization had been discovered by Richard Weatherill in 1888. "The cliff-dwellers were one of the native myths of the American West," Edith Lewis observes, and "children knew about them before they were conscious of knowing about them" *(WCL,* p. 81). Cather and her brothers continued to think and speculate about the cliff-dwellings in later years, and "The Enchanted Bluff" reflects her continuing interest in the story of Mesa Verde. In her letter to Sarah Orne Jewett describing "The Enchanted Bluff," Cather had referred to the Spanish influence informing the imagination of anyone who had grown up in the West: Spanish words, legends, and history were part of her cultural heritage.[5]

Cather had also written versions of the passionate and creative release she found in the Southwest in two stories where a change of landscape unleashes buried energies: "Eric Hermannson's Soul" (1900) and "The Namesake" (1907), fictional dress rehearsals for her 1912 trip. "Eric Hermannson's Soul" almost uncannily predicts her Southwest experience—with the important exception that Cather, unlike the story's protagonist, did not flee the West's passionate challenge. Margaret, her intellectual, sophisticated Eastern heroine, travels West to visit her brother, a childhood playmate with whom she had read fairy tales and "dreamed the dreams that never came true." Consciously she wants to see the "wild country of which her brother had told her so much," and unconsciously she wants to find her own wildness by running her "whole soul's length out to the wind—just once." Eric—silent, sensual, primitive—is the agent of this momentary release, sparking her "gypsy blood" and unlocking the passion that "allures and terrifies" before she escapes back to the East and safety *(CSF,* pp. 359–79).

Whereas "Eric" anticipates Cather's response to Julio, "The Namesake" foreshadows the impact the Southwest would have on her writing. In this story Lyon Hartwell also travels westward, returning from Europe to his American roots and a mystic union with his past, his uncle, and his artistic power. His ecstatic discovery of a "great force" in the spiritual legacy left by his uncle becomes a reservoir of wholeness and strength he later channels into his art—a series of sculptures depicting the story of the American West, including one of a "Pioneer" *(CSF,* pp. 137–46).

Having written two stories in which Western journeys spark sexual and creative release, Cather may have been prepared to invest her trip to Winslow with similar significance and to live out both plots—the lover's and the artist's—at the same time. If so, her willingness to let life imitate art was encouraged by her preexisting views of the Southwest: she identified the region's landscape and climate with the expression of energies and desires that were regulated or forbidden elsewhere. While at times Cather portrayed desert landscapes as psychic and creative wastelands, as in stories like " 'A Death in the Desert' " and "The Sculptor's Funeral" (set in "Sand City"), she also viewed arid countries as places for sensual liberation or spiritual revelation. Examining the latter associations, which she derived from literary and religious traditions, will help us to understand the assumptions Cather brought to the Southwest.

During the 1890s Cather considered desert landscapes to be sultry, foreign regions imbued with the "oriental feeling" she found in the French literature she loved—Flaubert's *Salammbô* and Gautier's *One of Cleopatra's Nights.* For Cather, Gautier's oriental sketches "palpitate[d] with heat" like sand hills that "danc[d] and vibrate[d] in the yellow glare of noon" *(KA,* p. 138). That she connected this palpitating desert landscape with the "great passions" never to be "wholly conventionalized" is apparent in her 1895 review of Pierre Loti's *The Romance of a Spahi.* In Loti's novel a French soldier is taken from his mountain village to Africa, where he confronts the "glaring lights and eternal flatness of the desert." This new land both represents and evokes the passions he has controlled in France, and the soldier becomes entrapped by a native woman whom he both desires and despises: "She had charmed him with her amulets," Cather notes, and "thrown a spell over him by her savage chants." Cather liked Loti's portrayal of the "savage emotions" that flourished in the tropics and the deserts. "We see too much of civilization," she concluded. "We sometimes need solitude and the desert, which Balzac said was 'God without mankind.' Loti is . . . a knight-errant who gives . . . poor cold-bound, sense-dwarfed dwellers in the North the scent of sandalwood and the glitter of southern stars'' *(KA,* pp. 365–67).

Cather did not view the desert solely as the place where ''sense-dwarfed'' Anglo-Saxons could release primitive passions. As early as 1891—thinking of Moses, Christ, St. John the Evangelist, St. Anthony, as well as *Pilgrim's Progress*—she made the desert an important stopover in her artist/pilgrim's journey to the kingdom of art. It was only after Moses had left the luxury of the Egyptian court, she noted in ''Shakespeare and Hamlet,'' and had fled into the desert ''and dreamed for years in the sand hills, that the bush burned before him and was not consumed.'' She envisioned Carlyle drawing artistic strength from the barren heaths and finally rushing ''off into the desert to suffer alone,'' while Shakespeare entered the metaphoric ''desert and . . . waste places'' of trial and testing before he emerged, like Moses, with his vision *(KA* pp. 425, 435). For Cather, the Southwest thus became America's place for relevation: shortly after arriving in Winslow, she wrote Sergeant that the country seemed to ''forecast some great spiritual event—something like a Crusade, perhaps'' *(WC:AM,* p. 82).

So Cather approached the Southwest's arid climate and desert landscape with contrasting associations drawn from nineteenth-century French fiction, the Bible, and church history: the desert was the locus both for primitive passion and for celibate withdrawal, for sensual indulgence and for spiritual and aesthetic revelation. From one point of view these were contradictory possibilities, but in the Southwest Cather found spiritual, sensual, and creative experience unified and indistinguishable. Had she traveled to the Southwest a few years earlier—during the period when she doubted her artistic power and imagined women artists like Katherine Gaylord withering in a ''wilderness of sand''—she might have associated the region with the barrenness attributed to Nebraska in her childhood memories and to Western landscapes in her *Troll Garden* stories *(CSF,* p. 199). But in 1912 she knew she was on her way to becoming a writer. Indeed, she had already imagined the desert transformed into a garden in her Southwest story ''The Enchanted Bluff.'' There, in the sandbar's desert landscape, the willow seedlings root themselves and break into ''spring leaf''—an image of growth and flowering mirrored in Cather's own literary emergence from her Nebraska past *(CSF,* p. 69). Thus even before Cather visited the Southwest, the desert had become the locus for creative transformation.

•

And yet the religious and literary associations Cather brought with her to Winslow did not adequately represent this landscape: the Southwest eluded and transcended her inherited assumptions. Its ''sharp contours, brutal contrasts, glorious color and blinding light'' affected her powerfully.[6] This country demanded new responses, a new vocabulary of feeling. Later she expressed the land's expansive impact on her spirit in *Death Comes for the Archbishop* when Bishop Latour associates his adopted desert country with something ''soft and wild and free, something that . . . softly picked the lock, slid the bolts, and released the prisoned spirit of man into the wind, into the blue and gold, into the morning, into the morning!'' *(DCA,* p. 276).

Ellen Moers connects Cather's experience of ''earthbound ecstasy'' in the

Southwest with the "oceanic feelings" enjoyed by Emily Brontë and Isak Dinesen, who also found a joyous sense of "physical dissolution" in landscapes without topographical limits—for Brontë, Yorkshire's moors, and for Dinesen, Africa's highlands.[7] Moers assumes that Cather enjoyed similar feelings of release in Nebraska, her other "oceanic" landscape. But when Cather was physically in Nebraska, she did not enjoy the happiness of "dissolv[ing] into something complete and great"; this experience she later granted to Jim Burden, but not herself *(MA,* p. 18). Although the Southwest's topography does resemble that of Nebraska, it is not the same, and Cather found its distinctive features—the mesas and deserts, canyons and mountains—more congenial to release, expansion, and self-abandonment than she did Nebraska's prairies.

Right before leaving for Winslow Cather had in fact stopped in Red Cloud, where she experienced a recurrence of her childhood fear of drowning in a boundless landscape. Writing Elizabeth Sergeant in April, Cather spoke of the bad attack of fright she had just suffered in Nebraska. Being in the West always paralyzed her a little, she said, made her afraid of losing something, although she did not know what it was. She had felt the same way when she was a child, but then it was even worse. It was as if she could not let herself go with the current: she always fought the West a little, the way people who can't swim fight to survive. In the West there were just so many miles between you and anything, and then the everlasting wind might put you to sleep. If she were a fit person to write about this country, should she feel this way? It did not seem right.[8]

In Arizona and New Mexico, however, Cather did not suffer from the terror of engulfment that overcame her in Nebraska, where the "something" she feared losing was her identity. Perhaps this was in part because the Southwest was not the country of her childhood and adolescence. The anxieties she at times felt in Nebraska owed something to the vastness of the landscape, but they also suggest the association Cather made in her fiction between the return to the mother and death. Cather uses the same imagery of drowning and sleeping to describe both her fear of the prairies and the child's desire to reunite with the mother in stories like "The Clemency of the Court," "The Burglar's Christmas," and "On the Gull's Road." Evidently her vacation in the Southwest allowed her to reexperience the happiness she had known as a child without the threatening aspects of the mother-daughter bond.

In the Southwest, Cather could let herself go with the current. The new landscape was similar to Nebraska in its immensity of scale, but its mesas, plateaus, and mountains intervened between the self and the horizon, forming the boundaries and barriers Nebraska lacked. And there were cracks and gaps in the desert and mesas that could shelter life instead of threaten it, spaces where the cliff-dwellers had made their homes. The Southwest, unlike Nebraska's cornfields (where, in Cather's words, there was "no place to hide"), offered many refuges. This was an entire landscape of "intimate immensity"—the union of shelter and exposure, enclosure and expansion that, in Gaston Bachelard's view, nourishes the artist's imagination.[9]

Cather's portrayals of the Southwest in *The Song of the Lark* and *The Professor's House* are filled with such paradoxical spaces. While Thea Kronborg is sheltered in her rocky "nest" in the cliff-dwellers' settlement, she looks up into the "blazing blue arch over the rim of the canyon" and sees an eagle ascending into the limitless sky *(SL,* pp. 298, 320). She is concealed and enclosed in her cave/room yet simultaneously open to the sun, just as the canyon is open to the sky. In describing the cliff-dwellers' settlement in "Tom Outland's Story" in *The Professor's House,* a sequence based on her own trip to Mesa Verde in 1915, Cather describes a landscape offering the same combination of spatial and psychological opposites. The cliff-dwellers' refuge is a hidden, protected space, concealed inside the supposedly inaccessible Blue Mesa. Yet this rock shelters a civilization, as Tom Outland discovers when he uncovers its secret: "a little city of stone, asleep . . . guarded by the cliffs and the river and the desert" (p. 201). What is most hidden to outsiders is the expansiveness of this enclosure: once Outland enters the Indians' guarded city he finds that the inhabitants could look out on limitless spaces—their dwellings look off "into the box canyon below, and beyond into the wide valley we called Cow Canyon, facing an ocean of clear air" (p. 213). Hidden spaces that are not isolated, enclosed spaces that are not closed—like the Southwest itself, Thea's "nest" in the cliff and the Indians' stone city integrate opposites.

Encountering such a landscape was important to Cather because its topography externalized the psychological state central to her creative process. As Bachelard observes, spaces of intimate immensity are the psychological catalysts for reverie, a state in which the ego relaxes its guard over consciousness, a state which Willa Cather found too frightening on the prairies. The shapes of the Southwest both symbolized and promoted the relaxing of ego boundaries—in Cather's phrase, "letting go with the current"—while assuring their protection. Combining the intimacy of Virginia with the immensity of Nebraska, the Southwest presented on a grand scale the spaces that fostered Cather's creativity—attic rooms and islands, sheltered realms with views or windows connecting to vaster worlds beyond.

Allowing the self to enter and be entered by the external world while maintaining its boundaries, such spaces also mediated between the creative methods the younger writer had thought opposed: disclosure and concealment, the tensions she could not resolve in her first short stories. The conflict between disclosure and concealment was particularly intense for a lesbian writer who knew she could not write directly of her love for women; as we have seen, Cather's fear of exposure and expression was reflected in the spatial metaphors of containment (the pyramid, the cell) that dominate her early stories. Then as Cather gradually drew more deeply on memory, imagination, and passion—the sources of her art—the containers she describes change as well: the closed structure in "A Tale of the White Pyramid" becomes the opened box in "On the Gull's Road," which in turn becomes the vase/vessel in *The Song of the Lark,* the receptacle that is always open. Since the Southwest's landscape aided this process of metaphoric transformation and creative unveiling, we should consider more fully the psychological and creative significance of these spaces.

The closed box represents the human need for secrecy, Bachelard observes, the desire for an emotional hiding place: "It is not merely a matter of keeping a possession well guarded. . . . A lock is a psychological threshold." [10] In Cather's early fiction closed boxes signify both the female body with its sexual secrets and the creative or hidden self. In "The Burglar's Christmas," for example, when the son breaks into his mother's jewel box, he is revealing not only his desire to possess the mother but also his desire to possess himself and his own creative power. Inside the box he seems to discover the sign of that power—the open receptacle, the silver cup with his name on it—but in the context of this early story, the cup signifies regression rather than a creative return to origins. In addition, the cup fails to be an adequate symbol of adult creativity because it is neither a gift nor a possession. To obtain it the thief must force open the lock. The box is not ready to be opened: its contents belong to the mother, not to the child.

As Bachelard points out, while the closed box connotes secrecy, separating the inside from the outside, when the box is opened the "outside is effaced with one stroke . . . a new dimension—the dimension of intimacy—has just opened up."[11] In later stories such as "The Treasure of Far Island" and "The Namesake," Cather lets some protagonists open a box which either belongs to them or is a gift or legacy from someone else; when the box is opened and the "dimension of intimacy" revealed, dreaming and creativity are inspired. So when Cather discovered the Southwest's enclosed yet continually open spaces—its canyons and cliff-dwellings, mirrored in the Indians' pots and vases—she confronted structures she had already, in part, imagined. The continual flow they represented—between inside and outside, structure and air, container and contained—reflected and helped to create the release, freedom, and expansion she enjoyed in this landscape; this was a place where the hidden could be revealed, where unconscious and conscious powers could be integrated. The Southwest's spaces helped Cather to cross Bachelard's "psychological threshold," to "[pick] the lock," and to open the container that held her creative energies.

•

In Bachelard's account, the space of "intimate immensity" is genderless. In her descriptions of the Southwest, however, Cather imagined the land as female. When she drew on her 1912 trip for the Panther Canyon sequence in *The Song of the Lark,* she presented her readers with what Ellen Moers calls the "most thoroughly elaborated female landscape in literature." As Moers notes, the canyon topography is marked by "unguarded sexuality."[12] Thea's birth as an artist, like Cather's, takes place in a landscape that resembles the female body:

> Panther Canyon was like a thousand others,—one of those abrupt fissures with which the earth in the Southwest is riddled. . . . The canyon walls, for the first two hundred feet below the surface, were perpendicular cliffs, striped with even-running strata of rock. From there on to the bottom the sides were less abrupt, were shelving, and lightly fringed with *piñons* and dwarf cedars. The effect was that of a gentler canyon within a wilder one. The dead city lay at the point where the perpendicular outer wall ceased and the V-shaped inner gorge began. There a stratum of rock, softer than those above, had been hollowed out by the action of time until it was like a deep groove running along the sides of the canyon. In this hollow (like a great fold in the rock) the Ancient People had built their houses of yellowish stone and mortar. *(SL,* p. 297)

Thea explores the "long horizontal groove" and finds among the Cliff-dweller ruins another feminine space, a "cave," a womblike "rock-room" where she can touch the stone roof with her finger tips. "This was her old idea: a nest in the high cliff; full of sun" *(SL,* pp. 298–99).

It may at first seem contradictory that Cather's guide to this erotic and maternal landscape was male, but this was "Outland"—a marginal realm where conventional social rules and categories did not apply. Writing Elizabeth Sergeant, Cather said that she had not been as happy since childhood—which to her was an androgynous, Edenic time of play and play-acting before gender limited the identities men and women could assume.[13] Certainly Cather could momentarily elude some of her adult selves: away from New York and her demanding job, she could cast off the professional woman's identity; away from Isabelle McClung and Edith Lewis, she also was free to be different from the woman she was in those relationships. Released in a world of play, Cather could try out the script she had already written in "Eric Hermannson's Soul" and let herself be enraptured by a man who incarnated the Southwest's spirit of place.

Cather described her love affair with Julio in a series of letters to Elizabeth Sergeant. Among available letters, only those to Louise Pound match the romantic intensity of this correspondence. Yet while the letters to Pound are melodramatic and despairing, these are joyous and exuberant. The letters do not tell us whether she and Julio actually become lovers, or even how much attention he really gave her. What is clear is that Cather experienced a lover's intense emotions—rapture, wonder, self-abandonment, self-renewal. Writing Sergeant in May, she told her friend she should visit the Southwest and also find a Mexican lover. Cather went on to describe the romance she imagined Sergeant might have in terms that obviously echoed her own: she would not be able to help falling in love, and neither would he; and he would take as much time and strength from her as she would give, and she would find him so attractive that she would not be tightfisted.[14] As this scenario suggests, with Julio Cather found the involuntary aspects of romantic love pleasurable. Whereas self-abandonment meant a frightening self-annihilation in "Eric Hermannson's Soul," in this real-life drama giving up her time and strength to another was invigorating rather than draining.

Cather's letters to Sergeant reveal a double consciousness. On the one hand she seems like an adolescent girl rhapsodizing about her first love; on the other hand she is a thirty-eight-year-old woman looking a little wryly and ironically at her infatuation with a much younger man. Both voices emerge clearly in her letter of May 21. She began by telling Sergeant that Julio was too beautiful to be true, different from anyone else in the world. After resolving not to write too much about him—she did not want to be a bore—Cather went on to devote the whole letter to Julio, humorously acknowledging her inability to write about anything else. She apologized for her preoccupation, admitting that *she* hated to get letters from women raving about the beauty of untutored youths of Latin extraction—the kind of mail one received after one's friends went to Italy. "Nothing bored *her* more than to hear ecstatic accounts . . . about the charm of Venetian gondoliers and Sicilian donkey boys" *(WC:AM,* p. 80). But after grouping herself with slightly silly older women, Cather then distinguished herself from such enraptured females and Julio from their gondoliers and donkey boys: "But this one was different. His words seemed to come from the Breviary, they were so full of simple piety and directness'' (*WC:AM,* p. 81).

Cather even imagined bringing Julio back with her to New York, where she knew he could make a living as an artist's model. He would be lovely at Mrs. Fields', she told Sergeant; but if she took him back East she would be sure to lose him: sophisticated women would fight to possess this beautiful primitive, and she imagined Isabelle Stuart Gardener trying to snap him up and take him back to Boston as a trophy. Julio would doubtless prefer Mrs. Gardener's mansion to her humble Bank Street apartment.[15]

Why was Cather so enraptured by this ''young Antinous of a singer'' with a ''few simple thoughts and feelings'' (*WC:AM,* p. 80)?

Much of Julio's appeal must have been his otherness: younger, male, from another culture, he was an emissary from the sensual Latin races who were at home in the desert, Cather's land of passionate relevation. In her letters she refers to him as silent and opaque; perhaps because of the language and cultural difference, he did not speak enough to intrude upon her as a real human presence. Because he was different and in a sense unknowable, Cather could create him imaginatively, and her letters to Sergeant are full of interpretations of Julio: like the desert, he was timeless, without beginning and without end; he had classic beauty, reminding her both of Aztec and Roman sculpture; he reminded her of old gold and old races.[16]

Doubtless, too, Cather responded to Julio with the intensity she conveys in her letters because she knew their intimacy could not last. On vacation in a special place for a limited time, she could play with real feelings that she assumed would not have real consequences. Even though she told Sergeant that she was thinking about bringing Julio back to New York, it was precisely because he would never be integrated into her life that she could let herself experience emotions which, in another context, might have been too threatening to allow to surface.

Julio's appeal also had to do with the fact that he was male and yet not manly, neither a virile, Kiplingesque soldier of fortune nor a patriarchal, dominating father. Because he was much younger than Cather and from a sensuous, non-Anglo-Saxon culture that loved music and dance, he combined traits Victorian America divided between ''male'' and ''female.'' Falling in love with this mysterious, androgynous being, Cather did not experience the submission to male power and authority she associated with women's subservience in romantic love. In some ways Julio even reminded her of her lost childhood self (''he would go anywhere to find wild flowers, or hunt a spring of water,'' she told Sergeant, ''as she would do as a child in Nebraska'' [*WC:AM,* p. 81]), so he may well have struck her as a psychic twin, Siegmund to her Sieglinde. As the letters to Sergeant suggest, Julio played many roles in Cather's imagination—lover, brother, twin, other self, and son. When she drew on this experience obliquely in *O Pioneers!,* she gave Emil some of these roles, although he performs them with different women; he is son and brother to Alexandra, lover and other self to Marie.

The one role Julio did not play was mother to Cather's daughter. His gender marked him as different, and given the age difference, Cather was the parental figure. The absence of maternal power was perhaps a relief. Whatever or whoever

Julio was, he was unlike her mother (and the mother-figures in her fiction), and so self-abandonment with him did not mean the ''erasure of personality'' a daughter might fear with a same-sex lover. Because Julio was clearly other, Cather did not have to fear absorption and fusion.

It is difficult to assess with great certainty Julio's role either in Cather's entrancement with the Southwest or in her creative emergence in *O Pioneers!* In my view, however, he was a key figure in both dramas: among the hundreds of letters I have read, there are none as exuberant and lighthearted as those she wrote to Sergeant about him. What we can say with certainty, though, is that her romance with Julio (like her life and fiction in general) exposes the inadequacy of the categories we generally use—male, female, heterosexual, homosexual, sexual, nonsexual—to describe human experience. It was important that Julio was male, because the gender difference meant that Cather could release powerful feelings without fear of engulfment by a maternal presence, and this was a helpful dress rehearsal for her subsequent experience of self-abandonment in the creative process. But it was equally important that he was not *conventionally* male. This was a love affair with a man, but it was not a heterosexual relationship as defined and arranged in patriarchal society.

Finding Julio and the Southwest pivotal in Cather's subsequent creative breakthrough (already under way) is not to say that an incomplete woman boarded the train for Winslow who was mysteriously made whole through this relationship. Nor is it to say—with benefit of hindsight—that if Cather had not met this man (or someone like him) she would have continued to write imitative Jamesian novels for the rest of her life. She was already on the verge of writing *O Pioneers!* shortly before she left the East. Her romance with Julio, however, added a kind of emotional experience she had not, as far as we know, enjoyed before. Over the preceding twenty years Cather had grown to experience and to value herself as a woman in relation to other women; with Julio, she experienced herself as a woman in relation to a man, and this newness appears to have been both exciting and energizing. In a letter to Elizabeth Sergeant, Cather later paid indirect tribute to Julio when she said how glad she was that Sergeant had liked the Mexican ball scene in *The Song of the Lark.* She had meant to use the scene to portray the flowering of whatever was feminine in Thea.[17]

It is also possible that Cather's response to Julio was as strong as it was because she was simultaneously regaining her close childhood attachment with her brother Douglass. We have seen how Cather found the Siegmund–Sieglinde story a powerful narrative to describe both male–female love and the woman artist's self-discovery; in the Southwest she was living a version of that story and yet avoiding the dangers of incestuous love by dividing the brother/lover into two people. Later, writing of Katherine Mansfield, Cather observed that the person who had brought Mansfield to ''her realest self'' was ''not one of her literary friends but, quite simply, her own brother.'' Cather's understanding of the role played by the brother/muse in Mansfield's literary development may have owed something to the happy weeks she spent with Douglass and Julio shortly before completing *O Pioneers!*

If Julio contributed indirectly to the composition of *O Pioneers!* and directly to that novel's Mexican references, he also may have given Cather a more enduring legacy: an ability to portray woman's heterosexual desire more convincingly than she had before. The two novels she wrote after returning from the Southwest, *O Pioneers!* and *The Song of the Lark,* contain the most passionate male–female relationships in her fiction. In describing the love affairs between Emil Bergson and Marie Shabata and Fred Ottenburg and Thea Kronborg, Cather was not only the lesbian writer donning a socially acceptable heterosexual mask; she was drawing, in part, on her own experience of desire. And the "unguarded sexuality" Cather attributed to Panther Canyon may also owe something to this romance, since she was seeing both the Southwest's and her body's landscape through her lover's eyes as well as her own.

•

Julio was also Cather's guide to the non-Anglo peoples and cultures of the Southwest. While she was delighting in Mexican songs, dances, and traditions, Cather was also being introduced to a native American cultural inheritance preserved in the cliff-dwellers' ruins and living in the dances, rituals, and folk art of the Hopis. She experienced the same exultant feeling of acquisition she had felt after entering 148 Charles Street, yet this was a different inheritance, the art of an indigenous culture with roots in her Western past. The Southwest thus expanded her boundaries in time as well as space.

Like other writers and artists who gravitated to the Southwest in the first decades of this century—Mary Austin, Laura Gilpin, Mabel Dodge Luhan, D. H. Lawrence—Cather was enthralled by Pueblo civilization. Communal, ritualistic, mystic, the Southwest Indians' culture seemed a healthy counterpoint to the aggressive individualism, spiritual emptiness, and corrupt materialism of modern American life, where getting and spending were the only sacred enterprises. "Their lives were so full of ritual and symbolism that all their common actions were ceremonial," she wrote later of the cliff-dwellers, "planting, harvesting, hunting, feasting, fasting." [18] In this society where myths were part of people's lives—not narratives to be read about in books—Cather found the dichotomy between "art" and "life" erased. In Indian culture, both past and present, everyday life was imbued with order and meaning:

> When you see those ancient pyramidal pueblos once more brought nearer by the sunset light that beats on them like gold-beaters' hammers, when the aromatic pinon smoke begins to curl up in the still air and the boys bring in the cattle and the old Indians come out in their white burnouses and take their accustomed grave positions upon the housetops, you begin to feel that custom, ritual, integrity of tradition have a reality that goes deeper than the bustling business of the world.[19]

In the harmonious relationship the Indians enjoyed with the land, Cather found a metaphoric representation of the creative process to which she was moving. As we have seen, she condemned the human desire to master the environment in stories like "The Professor's Commencement," "The Namesake," and *Alexander's Bridge,* where masculine forces seek to control and subdue a feminine nature. In the Southwest, however, human beings had marked the land by adapting to its requirements rather than by imposing their wills upon it. The Indians "seem not to have struggled to overcome their environment," she noted approvingly:

> They accommodated themselves to it, interpreted it and made it personal; lived in a dignified relation with it. In more senses than one they built themselves into it. . . . House building, in those great natural arches of stone, was but carrying out a suggestion that stared them in the face; often they used a great rock that had fallen down into the archway as a cornerstone and anchor for their own lighter masonry. When they felled cedars with stone axes they were but accelerating a natural process.[20]

These native artisans and architects built their houses the way Cather was beginning to fashion her stories: not by imposing their designs upon subject matter, as did the "manly" artist she had just repudiated in "Behind the Singer Tower" and *Alexander's Bridge,* but by accommodating to the forms and shapes that were already present in nature; not by importing foreign materials (like the writer who imitated others' models) but by working with the material that had been given to them.

So in the Southwest Cather felt herself the recipient of a mode of creativity marked by receptivity rather than self-assertion. As representatives of a native American artistic tradition, the cliff-dwellers and their present-day descendents were artistic forebears and relatives she wanted to claim. Unlike the famous men whose single-named graves she contemplated in France, these artists were communal and anonymous, so Cather did not feel excluded or intimidated by their achievements. As Tom Outland observes of the cliff-dwellers, "They belonged to boys like you and me, that have no other ancestors to inherit from" *(PH,* p. 242).

In imagining herself the inheritor of the cliff-dwellers' art, however, Cather was not a "boy" looking back to his fathers but a daughter looking, and thinking, back to her mothers. Her imagination was most captured by the women potters who crafted their beautiful, yet functional, vessels from the earth. Seeing the shards of pottery still scattered among the cliff-dwellings, Cather felt herself the heir of a female artistic tradition; having been granted this gift by the Indian women, she took one of the clay fragments back to New York—a visible sign, like Joanna's coral pin in *The Country of the Pointed Firs,* of female inheritance and attachment.

After her return to the East, Cather remained fascinated by these anonymous women artists. In May 1914 she visited the Museum of Natural History with Elizabeth Sergeant to see an exhibit of cliff-dweller pottery. Viewing the "great black and red pots with complex geometrical patterns" displayed behind glass, Cather thought about the vessels' creators and reflected on "the women who, under conditions of incredible difficulty and fear of enemies, had still designed and molded them, 'dreamed' the fine geometry of the designs, and made beautiful objects for daily use out of river-bottom clay" *(WC:AM,* p. 123). In the cliff-dwellers' civilization, unlike her own, "woman" and "artist" were not conflicting identities. The Indian women made pots to hold grain and water because of their roles as food preparers and preservers, but they decorated the pots for creative, not functional, reasons, thus finding in their domestic work the materials with which to express creative desire. This unnecessary decoration captured Cather's imagination: "This care, expended upon vessels that could not hold food or water any better for the additional labor put upon them, made her heart go out to those ancient potters," she wrote in *The Song of the Lark.* "They had not only expressed their desire, but they had expressed it as beautifully as they could" *(SL,* p. 305).

Yet Cather found the decorations as important as the shapes of the pots. In forming and decorating their vessels, the women potters integrated womanhood and creativity, nature and culture, body and spirit, as the singer's voice also did for the younger writer. Like the opera singer, the Indian women drew on the primary materials of nature—the clay and earth under their feet—to make containers whose form echoed the shape of the female body. So these were not merely women who happened to be artists: these were women artists whose art was profoundly female, integrating procreative and creative powers.

As Adrienne Rich reminds us in *Of Woman Born,* in tribal cultures pottery-making was a "deeply reverenced art" practiced only by women. In such cultures, the vessel or container—far from being a symbol of female passivity—signified women's primordial transformative and creative force:

> It does not seem unlikely that the women potter molded, not simply vessels, but images of herself, the vessel of life, the transformer of blood into life and milk—that in so doing she was expressing, celebrating, and giving concrete form to her experience as a creative being possessed of indispensable powers. Without her biological endowment the child—the future and sustainer of the tribe—could not be born; without her invention and skill the pot or vessel—the most sacred of handmade objects—would not exist.[21]

In *The Song of the Lark* Cather makes these equivalences among the female body, the pot's shape, and the artist's creativity explicit as Thea Kronborg, nestled in her womblike "nest" in the cliffs, comes to understand that in singing "one made a vessel of one's throat and nostrils" *(SL,* p. 304). Unlike the singer, however, the Indian women potters had expressed their creative desire in enduring forms, and so in their work Cather could see a stronger analogue to her own: the writer also had to place her creative passion in a shape, or text, that would exist apart from, and outlive, the creator's body.

Following so soon after her literary inheritance from Sarah Orne Jewett, Cather's discovery of the Indian women pottery-makers strengthened her association of femaleness with creativity. Connecting to this lost tradition, she gained a vision of what was not written in her culture; receiving a message from female artists in the past, she could better envision her own artistic future. In doing so, Cather was simultaneously daughter and mother—daughter and mother to remote and forgotten women artists, daughter and mother to herself as artist. In *The Song of the Lark* Cather grants Thea the same dual identity: after she "makes [herself] born" as an artist, Fred Ottenburg knows that he will see not her but ''[her] daughter'' (p. 376). In describing Thea's artistic ''birth'' in *The Song of the Lark,* which mirrors her own, Cather describes her heroine's creative emergence as a mythic return to maternal power. Entering into the earth's womb, sheltered in a ''crack in the world'' where she finds evidence of female creativity, Thea emerges from her ''rock-room'' reborn as an artist—daughter to the earth and the women potters, mother to herself. This return to maternal origins—unlike the failed quests of the silenced protagonists in Cather's first short stories—is as empowering to Thea as Cather's sojourn in the Southwest was for her.

Cather's recognition of the resemblances between writer and potter had enduring impact on her creative process, an impact evident in the dramatic moment Elizabeth Sergeant records in her memoir. Early in 1916, when Cather was preparing to write *My Ántonia,* she suddenly set an ''old Sicilian apothecary jar'' in the middle of a table and exclaimed

> ''I want my new heroine to be like this—like a rare object in the middle of a table, which one may examine from all sides.''
>
> She moved the lamp so that light streamed brightly down on my Taormina jar, with its glazed orange and blue design.
>
> Saying this her fervent, enthusiastic voice faltered and her eyes filled with tears. (*WC:AM,* p. 139)

Cather gives us many equivalences here: the story, the heroine, and the vessel; the storyteller, the creator, and the potter. And the shape of the pot, identified here with a woman, leads us back to the Southwest's topography—the canyons, mesas, and caves where Cather found women's creative power written into a female landscape. Receiving this strengthening message in a land whose cracks and fissures metaphorically granted her spaces outside of patriarchal definitions of gender and creativity, Cather was empowered to fashion vessels from the materials of a woman's life and to find artistic resources in Nebraska's soil.

Although the Southwest was only the catalyst for creative and psychological processes already under way, later Willa Cather viewed the region as her artistic birthplace. There, like Thea Kronborg, she thought she had found her voice. Her intense response to the landscape interestingly anticipates the experience of other women artists who, a few years later, also found their imaginations stimulated by the region: Alice Corbin Henderson and Mabel Dodge Luhan, who both believed they had found artistic voices in New Mexico; [22] Mary Austin, who first wrote in the *Land of Little Rain* (1903) of the desert country whose preservation and transcription would become her life's vocation; Laura Gilpin, who devoted her photographs to recording and preserving the desert country and its native cultures; and Georgia O'Keeffe, perhaps the best-known woman artist to be inspired by the region. Like Willa Cather, these women artists and writers found the Southwest's culture and topography a potent antidote to patriarchal culture; there they could reimagine nature and the feminine and define themselves as artists.[23]

Perhaps because she had come to understand the creative role women had played in preindustrial, agricultural societies during her visit to the Southwest, Cather returned to Back Creek Valley for a visit in 1913. She was eager to re-encounter her childhood home, where rural women had expressed their creative desires through the anonymous, female pursuits of gardening, quilting, and storytelling. Having seen female creativity expressed in the work of ancient women potters, she could now more easily discern artistry in the farm wife's work—and hence in her own, since she was daughter to such mothers. We see evidence of Cather's changed view of the farm wife's role in *O Pioneers!* The daily tasks she had portrayed as oppressive in "A Wagner Matinee" became self-expressive in her novel, where Mrs. Bergson—like the native Indian house-builders and artists of the Southwest—adapts to a harsh climate, making her preserves and growing her garden by accommodating to the land.

•

After two months Cather had had enough of Julio, the desert, and ecstasy. She needed to get away, back to civilization and tranquillity so she could savor the Southwest's "golden elixir" and integrate the new self into the old *(WC:AM,* p. 83). One day, sitting by the Rio Grande, she became panic-stricken, she told Sergeant. A voice seemed to say "the West is consuming you, make tracks for home" *(WC:AM,* p. 84). The country's vastness now triggered her old fear of self-annihilation, and leaving became an act of self-preservation. In explaining to Sergeant why she had to escape, Cather identified the desert country with Julio, saying that, although he was wonderful, he could not replace a whole civilization. One could play with the desert, she said, love it, be filled and intoxicated with it, but then the time came when one had to go—to go bleeding, but go.[24] Describing herself as "bleeding" as she left the Southwest, Cather suggests both the aftermath of sexual experience and her emotional vulnerability. In this case, however, the imagery of penetration does not signify violation or dismemberment, as it did in her early short stories; Cather was marked by this experience but not mutilated by it. She left the desert feeling whole and integrated.

As she sat by the Rio Grande, she heard the voice of culture as well as of nature. The "civilization" Julio and the desert could not replace appeared to her in a vision, which she described in a letter to Elizabeth Sergeant. For several weeks she had been reading only the signs and messages of the land and its Indian and Mexican inhabitants, but suddenly "written before her in the dust" she saw a "sentence from Balzac she had long forgotten" (*WC:AM,* p. 84):

> *Dans le desert, voyez-vous, il y a tout et il n'y a rien—Dieu, sans les hommes.*
>
> Everything and nothing—God without men![25]

The sentence, which is from Balzac's short story "A Passion in the Desert" (also quoted in her 1895 review of *The Romance of a Spahi)* sends a complex warning. The allusion probably occurred to Cather because of the similarities between her own experience and Balzac's desert story. His protagonist is a French soldier marooned in the African desert where the "infinite vastness in every direction weighed on his soul." Overwhelmed with solitude and diminished by immensity, he finds a natural grotto in a hillock, a cave where he can hide. But again he is terrified when his refuge is invaded by creature of the desert—a panther. Hoping to tame the beast, he embarks on what amounts to a love affair with the panther, a lithe, seductive female presence, an ''imperious courtesan''; their romance is complete with sensuous caresses, jealousy, and reconciliation. In loving the panther the soldier comes to love and to understand the desert as well. All its ''secrets'' are revealed to his receptive imagination as he thrills to the ''sublime beauties'' of the once frightening landscape. Yet this idyllic interlude ceases when he is rescued and returns to France; the civilized and the primitive cannot coexist for long, suggests Balzac, who ends his story with the quotation that Cather saw written in the sand.[26] Evidently sensing the parallels between herself and the soldier, Cather referred to his story in the name she later gave to Thea's place of regeneration in *The Song of the Lark:* Panther Canyon.

Of course the very fact that she remembered this line from Balzac—a fragment from the literary tradition she had left behind as well as a reminder of the writer's profession she longed to enter—reveals her inclusion in the world of civilization. Unlike Julio and the grave, contemplative Indians, Cather had read writers like Balzac and so could not make the Southwest her only culture. Even though she found important signs of creativity in the landscape and the Indians' artifacts, if she were going to be an artist she would have to take Balzac's way and use written language.

Cather's departure from the Southwest was only momentary, however; the region remained a permanent feature of her emotional and imaginative landscape, as her fiction attests. She strengthened her attachment to the country and its Indian and Spanish cultures on several return trips. In 1914, while she was in the midst of *The Song of the Lark,* she visited her brother Roscoe in Wyoming and then went on to Arizona and New Mexico; in the summer of 1915 she and Edith Lewis visited Mesa Verde and Taos, and the two returned the following summer to Santa Fe and Taos. Then in 1917 Cather discovered Jaffrey, New Hampshire, which became her favorite summer retreat until she and Lewis traveled to Grand Manan in August 1921. So it was not until 1925 that Cather went back to the Southwest. By then the region had been discovered by other artists and writers. Mabel Dodge Luhan and D. H. Lawrence were living in Taos, and for a short time Cather and Lewis stayed in Mabel's guest house and took long drives through the country with their hostess' Indian husband Tony Luhan. Coming across the story of Bishop Lamy and his friend and missionary colleague Father Machebeuf that summer, Cather went back to New Mexico the next year to gather more material for *Death Comes for the Archbishop.*

The Mexican background in *O Pioneers!* and the ''Ancient People'' section of *The Song of the Lark* testify to Cather's love for this land. After completing *Song,* Cather momentarily thought she might write a novel set in the Southwest which she would call ''The Blue Mesa'' and then perhaps a travel book.[27] Neither materialized, although she wrote an essay on Mesa Verde that appeared in *The Denver Times* early in 1916. Her 1915 trip to Mesa Verde found its fictional reworking ten years later in *The Professor's House,* and her 1925 rediscovery of the Southwest inspired *Death Comes for the Archbishop.*

Although Cather's 1912 trip to the Southwest opened a rich vein of fictional material, its greatest significance was its immediate impact on the creative process. Cather went home, as she told S. S. McClure, with her head so filled with stories that she dreamed about them at night. She could hardly wait to write them down.[28]

•

By mid-June 1912 Cather was back with her family in Red Cloud in time for the wheat harvest, a communal activity she had not witnessed in several years. Her perceptions sharpened and cleansed by years of separation as well as her weeks in the Southwest, Cather was eager to "[soak] herself in the scents, the sounds, the colours of Nebraska, the old memories" *(WCL,* p. 84). Her new impressions of her homeland inspired her to write "Prairie Spring," a poem which she thought good enough to send immediately to Elizabeth Sergeant:

> Evening and the flat land,
> Rich and sombre and always silent;
> The miles of fresh-plowed soil,
> Heavy and black, full of strength and harshness;
> The growing wheat, the growing weeds,
> The toiling horses, the tired men;
> The long empty roads,
> Sullen fires of sunset, fading,
> The eternal, unresponsive sky.
> Against all this, Youth,
> Flaming like the wild roses,
> Singing like the larks over the plowed fields,
> Flashing like a star out of the twilight;
> Youth with its insupportable sweetness,
> Its fierce necessity,
> Its sharp desire,
> Singing and singing,
> Out of the lips of silence,
> Out of the earthy dusk.
>
> (*WC:AM,* pp. 84–85)

Whereas Cather had selected quotations from Kingsley and Rossetti as epigraphs to *The Troll Garden,* she chose this poem to introduce *O Pioneers!,* using her own work to represent the first novel she considered hers. But "Prairie Spring" foreshadows the lovers' subplot more accurately than it does *O Pioneers!* as a whole. The "fierce necessity" and "sharp desire" of youth aptly describe Emil Bergson and Marie Shabata, but the natural world in *O Pioneers!* is not "silent" or unresponsive. Nor does the poem suggest Alexandra Bergson's central role in making the land speak; in fact, in the novel Cather gives the power of song and expression to her artist/farmer heroine (who returns from her trip to the river farms humming "an old Swedish hymn") and to the natural world that replies to her "love and yearning" *(OP,* p. 65). In *O Pioneers!* it is the lovers whose lips are silenced.

"Prairie Spring" thus did not anticipate the novel's celebration of Alexandra's creative and transformative power, perhaps because Cather had not yet fully imagined it, having at that point only written "Alexandra," or what amounted to Part I of *O Pioneers!*, "The Wild Land." Her mind was then preoccupied with youth's romantic desire rather than Alexandra's creative desire: she was about to write "The White Mulberry Tree," the story of adulterous love and death which she thought would "terrify" her circumspect Boston editor Ferris Greenslet (*WC:AM*, p. 84).

By October 1912 Cather was in Pittsburgh, happily settled in with Isabelle and ready to begin "The White Mulberry Tree," which would become the source of the lovers' subplot in *O Pioneers!* "The Bohemian Girl" had appeared in the August issue of *McClure's* and Cather had received some complimentary letters, so she had some encouragement for the new direction her fiction was taking. Work on "The White Mulberry Tree" was proceeding "amazingly well" until, all of sudden, Cather found that she had a novel on her hands (*WC:AM*, p. 86). She had experienced a "sudden inner explosion and enlightenment" when she realized that "Alexandra" and "The White Mulberry Tree" belonged together (*WC:AM*, p. 6). Somehow, she told Sergeant, the "cold Swedish story" had entwined itself with the Bohemian story, and she was faced with a "two-part pastoral: the most foolish endeavor imaginable" (*WC:AM*, p. 86). But soon the unplanned endeavor seemed less foolish; as she integrated and expanded the two stories, she knew that the novel was going to be "much better than 'The Bohemian Girl' " (*WC:AM*, p. 86).

Since her second novel had such an unplanned genesis, Cather was spared the anxiety of authorship that accompanied the composition of *Alexander's Bridge*. She had gone to Pittsburgh assuming that she would be writing some more short stories, not thinking about expanding her range with another "long story." The transformation of her two short stories into a novel had been dictated by a sudden, unbidden insight, not by the woman writer's perpendicular aspirations to scale this lofty male genre once again. So a few months after her return from the Southwest Cather found herself the pleased author of a novel she had not known she was going to write, having achieved a goal she was not, at that point, consciously seeking.

Cather felt ambivalent about the novel she had just finished. On the one hand, as she told Sergeant, it was the story she had always wanted to write, and she had written it without thinking about anybody else's opinion, without being afraid of anybody. She had let the country be the hero; the novel was about crops and cows, and yet she found it interesting. She knew it was good. But would anyone else? Probably no more than six people in the world: who would want to read about Nebraska farmers? On the whole, she concluded, summing up her uncertainties, it was either pretty good or an utter failure—it could be nothing in between.[29]

Despite her fear that a Nebraska tale might not please a sophisticated Eastern audience, Cather still exhibited some pride in her new novel by connecting it with the work of Walt Whitman and Sarah Orne Jewett. She took her title from Whitman's hearty paean to Western emigration, "Pioneers! O Pioneers!" Although the novel's relationship to Whitman's poetry is complex and contradictory, the title announces Cather's desire to represent her work as the product of indigenous American materials—literary equivalents of the Indians' clay and stone—and herself as a descendant of the author of *Leaves of Grass* rather than of *The Golden Bowl.* This author, the title declared, was singing a song of ordinary American people and speaking unaffectedly in a vernacular voice.

Dedicating the novel to Jewett, Cather reciprocated Jewett's gift of literary inheritance and proclaimed that, in her estimation, she had reached the point of literary maturity Jewett had hoped for. Returning to her "own country," as Jewett had prophesied, Cather now was able to link the world and the parish, to draw deeply on her own past while still taking the "looker-on['s]" standpoint.[30] As she later said, she had written the novel as if she were talking to Jewett, and we can assume that she also envisioned receiving her mentor's warm congratulations.

An artist must possess "two things," Cather once observed, which must be "strong enough to mate, without either killing the other"—"technique and a birthright to write."[31] She was referring to the union of formal control and an intuitive, passionate response to people and places that characterized her major fiction, but she might also have been describing the composition of *O Pioneers!,* born when two of her stories were "strong enough to mate." Taken together, the title and the dedication suggest another "mating" and integrating in *O Pioneers!:* the novel and the author united a double heritage, the male and the female literary traditions. By mating Whitman and Jewett, Cather showed that these traditions united in her and her work—and even more daring, that she and her work united these traditions.

This proud declaration of literary inheritance matched the boldness of her novel. *O Pioneers!* was something new in American literature as well as in Cather's own work. Woman and artist, Alexandra represents both the American pioneers' story of inhabitation and cultivation and the woman writer's attainment of authorship and authority. Whereas Bartley Alexander represented the failure of Cather's male identification to provide her with a satisfactory artistic identity and aesthetic method, her transformed conqueror Alexandra Bergson unites the two identities which by then coexisted harmoniously within the author, who by 1912 was willing to regard herself as a woman writer.

•

The "new artistic method" she found in *O Pioneers!* contributed to Cather's reconciliation of gender and vocation. In the two major statements she made on the novel's composition—the 1922 "Preface" to *Alexander's Bridge* and "My First Novels"—as well as in less formal commentaries, Cather later portrayed her newfound creative process as a paradoxical combination of psychic and literary opposites. The polarities she stresses the most are self-discovery and self-abandonment.

In returning to her "own country" for inspiration and a subject, she had, Cather insisted, discovered herself as a writer. Instead of producing what she thought others might want to read, as she had in *Alexander's Bridge,* in *O Pioneers!* she had written a novel only to please herself, responding to her profound love for the Divide and its people. In doing so she had found her own voice, becoming "another writer," different from the person who had written *Alexander's Bridge* just a few months earlier. Whereas the author of *Alexander's Bridge* was unsuccessfully trying to incorporate external material into her novel, the author of *O Pioneers!* drew on her internal "deepest experience," the things she "knew best" and had once "[taken] for granted." The material she had been living, not the patterns she was copying, was her "own material" (*AB* [1922], pp. v–ix). Sophisticated New York critics might not "care a damn" about Nebraska, but she knew that books like *Alexander's Bridge* were "unnecessary and superficial" because they were written with such critics in mind; *O Pioneers!*, by contrast, was "entirely for [her]self" (*OW*, pp. 91–97).

Cather offered a slightly different version of this romantic (and American) story of self-creation and self-reliance when she sent a copy of *O Pioneers!* to her childhood friend Carrie Miner, by then married but still living in Red Cloud. She put the following inscription on the flyleaf:

> This was the first time I walked off on my own feet—everything before was half real and half an imitation of writers whom I admired. In this one I hit the home pasture and found that I was Yance Sorgeson and not Henry James. (*WWC*, p. 200)

As well as celebrating her departure from James' circuitous stylistic paths, in saying that she found herself to be "Yance Sorgeson"—a Nebraska farmer—rather than "Willa Cather," Cather shows that she had found Jewett's "gift of sympathy." In *O Pioneers!*, as in this comment, Cather displays the artist's empathic power to dispense with self and ego by "giving" herself "absolutely to [her] material" and "fad[ing] away into the land and people of [her] heart" (*CPF*, p. 7).

Cather "faded away" in creating characters like Yance Sorgeson, but in her 1922 "Preface" and "My First Novels," this self-abnegation takes another form: the writer's humble submission to her creative impulse, a nonrational "wisdom of intuition" that could only be accepted, not sought. In composing *Alexander's Bridge* Cather had consciously shaped the story, but *O Pioneers!* seemed to write itself after she had let the narrative form itself spontaneously in her mind. Working with her own material, she found she had "less and less power of choice" in determining its shape and form: "It seem[ed] to be there of itself, already moulded" (*AB* [1922], pp. viii–ix). Cather could only use metaphor to describe the creative intuition to which she had submitted. In the 1922 "Preface" it was "the thing by which our feet find the road home on a dark night, accounting of themselves for roots and stones which we had never noticed by day," while in "My First Novels" it was "taking a ride through familiar country on a horse that knew the way, on a fine morning when you felt like riding" (*AB* [1922], p. ix; *OW*, pp. 92–93). The metaphors have two elements in common: in both the writer is moving through a familiar landscape, and in both her progress comes from surrendering to a power the ego cannot control.

At times Cather would call this power "memory," as when she told Elizabeth

Sergeant that her creative life began when she ceased to admire and began to remember. She used another phrase in the 1922 "Preface" that is more useful in understanding what happened when *O Pioneers!* seemed to write itself: the writer's true subjects, she said, her "deepest" experiences and emotions, did not lie at the surface of the mind but were submerged in the "bottom of . . . consciousness" (*AB* [1922], p. vi). Hence finding the "road home" required descending into the unconscious and allowing submerged, forgotten, or repressed material to surface. In writing *O Pioneers!* Cather allowed the self to "fade away" not only by projecting herself into her characters, but also by permitting herself to experience those states of drowsing and dreaming she had found terrifying on the Nebraska prairies—states in which the ego momentarily relaxes its control. Of course there were always questions of "effective presentation" for the conscious artist to consider—point of view, for example—for the art of composition was not automatic writing (*AB* [1922], p. ix). Nevertheless, in retrospectively describing her literary breakthrough, she always stressed the unleashing of spontaneous forces and her unplanned discovery of organic form.

Cather experienced such moments of self-abandonment and release in the Southwest, and the connection she made between the region's topography and her own plunge into the "bottom of . . . consciousness" can be seen in a remark she made years later to Elizabeth Sergeant. Cather was working on *Shadows on the Rock* throughout 1929 and 1930, troubling years when her mother was paralyzed after a stroke. Being with her mother was painful, but she kept her new book in an "underground place," she told Sergeant, a refuge to which she could retreat for a few hours each day (*WC:AM*, p. 240). Like Alexandra Bergson's "underground river" (*OP*, p. 203) and the canyon into which Thea descends in *The Song of the Lark*, this "underground place" was Cather's psychic underworld, and the "road home" led her there as well as to Nebraska. In contrast to "The Burglar's Christmas," however, in which the child never emerges from the maternal embrace, this homeward journey led back out to adult creativity for Willa Cather as well as for her heroines. Like the "road of Destiny" Jim Burden imagines himself having traveled at the end of *My Ántonia*, Cather's creative road was circular, leading into and out of the self, connecting inner and outer worlds (*MA*, p. 372).

Hence another union of opposites was present in Cather's creative emergence: as for many artists and writers, her breakthrough was a return in order to go forward. Traveling back to early memories and impressions, she advanced as a writer. Cather acknowledged that the artist's road led to her own psychic and familial origins as well as to the Divide when she explained that the mature writer would discover—or rediscover—"homely truths which have been about him from his cradle" (*AB* [1922], p. vii). "Homely" here suggests "homelike" as well as "ordinary" or "humble," and in the "Preface" Cather uses the word "home" several times to describe her discovery of, and return to, her own material. The material with which Cather felt "at home" was not only childhood memories of Nebraska, however, but any experience—whether real or imagined—that marked her deeply. So while all the novels she would write after *O Pioneers!* are not set in the West—indeed, she has a wider range of setting and subject than she is generally given credit for—it is safe to say that after *O Pioneers!*, Cather never wrote an inauthentic or "external" novel.

•

Returning home to an "underground place" as well as to her past in *O Pioneers!*, Cather was no longer disturbed by the regressive aspects of the creative process. By 1912 she was sure enough of her identity to abandon the self without fear of annihilation, now enjoying the "regression in the service of the ego" that Ernst Kris describes.[32] She could also abandon herself to the "wisdom of intuition" in 1912 because by then she was a well-trained writer. Having mastered the tools of her craft through her long apprenticeship as both writer and editor, she could let herself follow her instincts. In her 1890s journalism she had savagely attacked women writers like Ouida and Marie Corelli because their breathless, uncontrolled, extravagant prose resembled her own overly emotional and undisciplined writing. After her apprenticeship to Henry James and Sarah Orne Jewett, however, Cather could draw on what she had once thought the woman writer's potent but perhaps uncontrollable creative source: her "power of loving."

Cather does not mention gender in her essays and comments on *O Pioneers!*; in fact, she always referred to the artist as "he" in making general literary statements. But the creative process she describes is far from masculine. In *Alexander's Bridge* Cather had firmly dispensed with male-defined artistic methods when she sent Bartley and his shoddy bridge crashing into the St. Lawrence, but in that novel she proposed no alternative model of creativity. In her essays on the composition of *O Pioneers!* she does, and it displays qualities socially defined as female—submission, receptivity, self-effacement—as well as the "power of loving" she had earlier associated with women poets. Although in most contexts such qualities would ensure a woman's subordination, they were the source of Cather's creative power.

Cather had found these qualities in Shakespeare, Carlyle, and Crane as well as in Jewett; and in her own fiction she would grant them to Godfrey St. Peter, Tom Outland, and Bishop Latour as well as to Alexandra Bergson and Thea Kronborg. As she came to believe, the artist possesses traits conventionally divided between the sexes: intellect, discipline, and control as well as intuition, passion, and self-abandonment. So in moving from male identification to female identification, from *Alexander's Bridge* to *O Pioneers!*, she did not conclude that artistry was solely a woman's possession or that men were inevitably inferior artists. Cather did, however, detach the artist's creative power and vocation from socially defined notions of masculinity and feminity; having done so, she was able to create artist-figures—both male and female—who combine "masculine" and "feminine" powers.

Women and men who cannot be defined by conventional categories of gender, Cather's artists are not disembodied, ungendered, or neutered human beings. In *O Pioneers!* the artist-figure is a woman who defies traditional definitions of womanhood, to be sure, but she is a woman nonetheless—not an imitation man, not a grown-up William Cather. This breakthrough novel reveals many advances. Cather challenges the social definitions of gender and creativity that had once entrapped her. She demonstrates her newfound connection to her prairie background and her female strain of blood by showing three identities harmoniously inhabiting one self: pioneer, artist, and woman. And she rewrites the male and female narratives she had inherited from literary and family history—the epic, the pastoral, the *Künstlerroman,* the romance, the story of westward migration—to make a woman her hero. It is a stunning achievement.

NOTES

1. Willa Cather to Elizabeth Sergeant, April 20, 1912, Barrett-Cather Collection, University of Virginia Library, Charlottesville, Va.

2. Willa Cather to S. S. McClure, April 22, [1912] and June 12, [1912], Lilly Library, University of Indiana, Bloomington, Ind.

3. Cather to Sergeant, June 15, 1912.

4. Cather to McClure, June 12, 1912.

5. Willa Cather to Sarah Orne Jewett, October 24, 1908, Houghton Library, Harvard University, Cambridge, Mass.

6. Susan J. Rosowski and Bernice Slote, ''Willa Cather's 1916 Mesa Verde Essay: The Genesis of *The Professor's House,'' Prairie Schooner* 58 (Winter 1984): 81–92, esp. 86.

7. Ellen Moers, *Literary Women* (Garden City, NY.: Doubleday, 1976), pp. 259–60.

8. Cather to Sergeant, April 20, 1912.

9. Gaston Bachelard, *The Poetics of Space,* translated by Marie Jolas (Boston: Beacon Press, 1969), pp. 183–210. Working independently from me, Judith Fryer has also applied Bachelard's phenomenological meditations on space to Willa Cather's fascination with the Southwest. See ''Cather's Felicitous Space,'' *Prairie Schooner* 55 (Spring/Summer 1981): 185–98.

10. Bachelard, p. 81.

11. Bachelard, pp. 217–18.

12. Moers, p. 258.

13. Cather to Sergeant, April 26, 1912.

14. Cather to Sergeant, May 21, 1912.

15. Cather to Sergeant, June 5, 1912.

16. Cather to Sergeant, May 21, 1912.

17. Cather to Sergeant, n.d. [1915].

18. Rosowski and Slote, p. 85.

19. Rosowski and Slote, p. 86.

20. Rosowski and Slote, p. 85.

21. Adrienne Rich, *Of Woman Born: Motherhood as Experience and Institution* (New York: Norton, 1976), pp. 96–97.

22. Lois Rudnick, ''The Female Imagination and the Southwestern Landscape: Alice Corbin Henderson, Mabel Dodge Luhan, and Mary Austin,'' unpublished ms. In her biography of Luhan, Rudnick associates Austin's self-creation as a ''new world'' heroine with her discovery of the Southwest and its Indian cultures. See *Mabel Dodge Luhan: New Woman, New Worlds* (Albuquerque: University of New Mexico Press, 1984).

23. On the significance of the Southwest as a place of regeneration and self-creation for other women artists, see Vera Norwood, ''The Photographer and the Naturalist: Laura Gilpin and Mary Austin in the Southwest,'' *Journal of American Culture* (Winter 1982):

1–27; Lois P. Rudnick, "Mabel Dodge Luhan and the Myth of the Southwest," *Southwest Review* 68 (Summer 1983): 205–21; and Rudnick, *Mabel Dodge Luhan,* pp. 143–90.

24. Cather to Sergeant, n.d. [August 1912].

25. Writing Louise Imogen Guiney shortly after her return to the East, Cather quoted the same line from Balzac and said that, although the country had been wonderful, after a while one became lonely in a place that had only a geological history (Willa Cather to Louise Imogen Guiney, August 17, 1912, Holy Cross College Library, Holy Cross College, Worcester, Mass.). Evidently Cather suddenly found her newly discovered cultural heritage to be alien and inaccessible, and had to place distance between herself and the Southwest to recapture it.

26. Honoré de Balzac, "A Passion in the Desert," from *Scenes of Military and Political Life* (Philadelphia: George Barre and Son, 1898), pp. 3–24.

27. Willa Cather to Ferris Greenslet, August 22, [1916], Houghton Library, Harvard University, Cambridge, Mass.

28. Cather to McClure, June 12, [1912].

29. Cather to Sergeant, n.d. [1913].

30. "My Dear Willa," in *Letters of Sarah Orne Jewett,* ed. Annie Fields (Boston: Houghton Mifflin, 1911), pp. 248–49.

31. Eva Mahoney, "How Willa Cather Found Herself," *Omaha World-Herald,* November 27, 1921, p. 7.

32. Ernst Kris, *Psychoanalytic Explorations in Art* (New York: International Universities Press, 1959), p. 65.

19

The Road Home: *O Pioneers!*

> She had never known before how much the country meant to her. The chirping of the insects down in the long grass had been like the sweetest music. She had felt as if her heart were hiding down there, somewhere, with the quail and the plover and all the little wild things that crooned or buzzed in the sun. Under the long shaggy ridges, she felt the future stirring.
>
> *O Pioneers!*

"Isn't it queer: there are only two or three human stories," Carl Linstrum remarks to Alexandra Bergson halfway through *O Pioneers!,* "and they go on repeating themselves as fiercely as if they had never happened before" (p. 119). But Carl's vision is more limited than Cather's. In chronicling Alexandra's transformation of "The Wild Land" into "Neighboring Fields"—the novel's first two parts—Cather was telling a new story in American women's writing as well as in her own work: the narrative of a creative woman whose life story follows neither the conventional female nor male plot.

In contrast to Marie Shabata, the vibrant, impulsive Bohemian woman whose spontaneity is channeled into conventional social and narrative structures—marriage, adultery, the *Liebestod*—Alexandra's imagination is imprinted by her readings of nature's text. Consequently her mind is filled not with romantic fantasies but with "clear writing about weather and beasts and growing things"; it is a "white book" which "very few people would have cared to read," Cather observes, "only a happy few" (p. 205). Cather includes her own readers among those "happy few" privileged to read a story that did not conform to the "one string" of female narrative—the often-repeated tale of "women and love" she disliked in nineteenth-century women's writing (*KA,* p. 409).

As Rachel Blau DuPlessis observes, twentieth-century women writers have tried to "replace the alternate endings in marriage and death that are their cultural legacy from nineteenth-century life and letters" by offering different choices to their female characters, a narrative and social act of revision which she terms "writing beyond the ending."[1] Refusing to give Alexandra the conventional romantic story that ends in marriage, Cather also declines to punish her with death, the ending she gives to Marie but not to her novel. In nineteenth-century fiction, death was the fate not only of the adulterous or sexually emancipated woman but also of the failed or potential woman artist who could not—within the limits of her restricted social world—find "the solution of art."[2] Writing beyond the endings of *Madame Bovary, The Mill on the Floss, The Awakening,* and *The House of Mirth,* as well as of *Jane Eyre* and *Emma,* in *O Pioneers!* Cather also rewrites her own stories of failed women artists— " 'A Death in the Desert,' " "A Wagner Matinee," and "The Joy of Nelly Deane."

•

In the 1890s Willa Cather had only one solution for the woman writer who wished to revise the "one string" of women's fiction: write men's fiction, using male plots and genres (*KA,* p. 449). But in *O Pioneers!* Cather does not conceive the pioneer adventure as a "manly battle yarn." A conqueror who wins by yielding, Cather's female Alexander tames the land not through force but through "love and yearning" (p. 65). Nor is Alexandra a male-identified woman, one of Cather's romantic heroes in female garb. Although Alexandra possesses the seemingly contradictory traits American society divides between men and women—strength and pragmatism as well as intuition and compassion—she is not a male in disguise. Cather represents her heroine's inner complexity by the outer signs of dress and physical appearance—signs which teach us how to read the novel as a whole.

In the first few pages we see Alexandra wearing a "round plush cap, tied down with a thick veil" and a "man's long ulster," which she carries comfortably, like a "young soldier." Yet if we now assume that she is mannish, Cather suddenly confuses us when this "young soldier" takes off her veil to reveal a sign of female beauty, a "shining mass of hair . . . two thick braids, pinned about her head in the German way, with a fringe of reddish-yellow curls" (pp. 6–8). Later Cather grants the mature woman a physical presence that is not conventionally feminine but is still female. "Sunnier" and "more vigorous" than she had been as a girl, Alexandra still has the same unruly, "fiery" hair that makes her resemble "one of the big double sunflowers" in her garden. Her face is "always tanned" in summer, because she neglects to wear her sunbonnet, but where "her collar falls away from her neck" or where "her sleeves are pushed back from her wrist," her skin has the "smoothness and whiteness as none but Swedish women ever possess; skin with the freshness of the snow itself" (pp. 87–88).

As these oppositions—tanned/white, sun/snow—suggest, Alexandra cannot be easily categorized. She is neither Tommy nor Miss Jessica, the contrasting female characters of "Tommy, the Unsentimental"; she is neither the lanky boy–girl nor the conventional heroine. Both descriptions of Alexandra begin with images that might be considered masculine by conventional viewers (the man's coat, the military bearing, the "calmness and deliberation of manner"), but end with unveilings of female beauty—the beautiful hair hidden under the veil, the smooth, white skin exposed when the collar or the sleeves are pushed back. So this is a woman, Cather tells us, who cannot be understood if we apply our polarized categories of gender. On the surface she may seem unfeminine, but if we wait to see what is unveiled we discover our mistake. This progression reflects the narrative structure of the novel as a whole: at first we see Alexandra solely as a strong, heroic, and unemotional figure; later we learn of her need for human attachment when she despairs of life after her brother's death.

In reading the novel we also have to wait to see that the story Cather is telling is unconventional. At first we appear to be in the same wasteland of deprivation that mirrored Katharine Gaylord's and Aunt Georgianna's defeats and assume we are about to encounter another failed woman artist. Settling the Nebraska Divide in the 1880s, the novel's Swedish and Bohemian farmers face an inhospitable landscape:

> A mist of fine snowflakes was curling and eddying about the cluster of low drab buildings huddled on the gray prairie, under a gray sky. The dwelling-houses were set about haphazard on the tough prairie sod; some of them looked as if they had been moved in overnight, and others as if they were straying off by themselves, headed straight for the open plain. None of them had any appearance of permanence, and the howling wind blew under them as well as over them. . . . The great fact was the land itself, which seemed to overwhelm the little beginnings of human society that struggled in its sombre wastes. (pp. 3, 15)

This ''stern frozen country'' is neither the bountiful mother nor the yielding virgin dear to the American imagination, but a genderless ''fact'' unresponsive to human desire or will. Recalling the harsh plains of her early stories as well as the grim land ''as bare as a piece of sheet iron'' Cather confronted as a young girl in 1883, this ''disheartening'' landscape without ''human landmarks'' is a world where human beings have not yet been able to write their story:

> The roads were but faint tracks in the grass, and the fields were scarcely noticeable. The record of the plow was insignificant, like the feeble scratches in stone left by prehistoric races, so indeterminate that they may, after all, be only the markings of glaciers, and not a record of human strivings. (p. 20)

The unwritten land is an ''enigma'' to men like Alexandra's father who can only interpret its resistance to cultivation as hostility: ''Its Genius was unfriendly to man'' (pp. 22, 20). After eleven years of homesteading, John Bergson has ''made but little impression upon the wild land he had come to tame'' (p. 20). Depressed and sick, defeated by years of crop failures, blizzards, and mischance, Bergson is dying. Meanwhile many of his neighbors have given up the fight and are moving to the city. Bergson does sense, however, that the land is not infertile; instead, he concludes, ''no one understood how to farm it properly'' (p. 22). Knowing that his daughter Alexandra understands more about farming than his unimaginative sons Lou and Oscar, Bergson leaves the land to her, and she pledges that she will never give it up.

Although the unmarked land can be disheartening to settlers who leave only ''indeterminate'' scratches in the soil, it can also be inspiring to a farmer who wants not to follow others' patterns but to create her own. ''A pioneer should have imagination,'' Cather observes, ''should be able to enjoy the idea of things more than the things themselves'' (p. 48). Alexandra is such a pioneer. Gifted with imagination, able to see possibilities in the soil that no one else has glimpsed, she triumphs because she combines a mystic faith in the Divide with a pragmatic willingness to experiment with new farming techniques After she inherits her father's land, she spends several days among the more fertile river farms, talking "to the men about their crops and to the women about their poultry," conferring with a young farmer who was experimenting with "a new kind of clover hay. She learned a great deal" (p. 64). Returning to the high plains of the Divide, she tells her brothers that they must buy the land others are deserting and mortgage their farm to expand their holdings. Although Alexandra has carefully thought over the financial complexities of this venture, ultimately she can only answer her brothers' question—How does she know that land will increase in crops and value?—with what Cather termed the "wisdom of intuition": "I *know,* that's all. When you drive about over the country you can feel it coming" (p. 67).

Cather ends "The Wild Land" with Alexandra's commitment to the future and opens "Neighboring Fields" after a sixteen-year gap with this description of the inscribed land:

> From the Norwegian graveyard one looks out over a vast checker-board, marked off in squares of wheat and corn; light and dark, dark and light. . . . From the graveyard gate one can count a dozen gayly painted farmhouses; the gilded weather-vanes on the big red barns wink at each other across the green and brown and yellow fields. . . . The Divide is now thickly populated. The rich soil yields heavy harvests; the dry, bracing climate and the smoothness of the land make labor easy for men and beasts. There are few scenes more gratifying than a spring plowing in that country, where . . . the brown earth, with such a strong, clean smell, and such a power of growth and fertility in it, yields itself eagerly to the plow. (pp. 75–76)

Cather then moves from these signs of human presence to the author of this profound transformation. The narrator is also our guide to Alexandra's farm, taking us through unfamiliar country and telling us that a "stranger"—or an Eastern reader, perhaps someone like us—could not help "noticing the beauty and fruitfulness of the outlying fields." As we get closer in this cinematic focusing, we learn that there is something "individual" about Alexandra's farm, something distinguishing; "a most unusual trimness and care for detail." Then we are allowed to see some of the details—osage orange hedges, their "glossy green" marking off a yellow field; a "low sheltered swale"; an orchard, its "fruit trees knee-deep in timothy grass." After this sumptuous introduction we can only agree with the narrator: "Any one thereabouts would have told you that this was one of the richest farms on the Divide, and that the farmer was a woman, Alexandra Bergson" (p. 83).

After passing through this Edenic landscape the narrator allows us to enter Alexandra's "big house," but immediately wants us to leave; the house is "curiously unfinished and uneven in comfort," some rooms overfurnished and others bare. This woman farmer is not, we learn, a conventional woman whose home is her sphere. And so, with a sense of relief, the narrator guides us back outside:

> When you go out of the house into the flower garden, there you feel again the order and fine arrangement manifest all over the great farm; in the fencing and

> hedging, in the windbreaks and sheds, in the symmetrical pasture ponds, planted with scrub willows to give shade to the cattle in fly-time. There is even a white row of beehives in the orchard, under the walnut trees. You feel that, properly, Alexandra's house is the big out-of-doors, and that it is in the soil that she expresses herself best. (pp. 83–94).

Drawing on the desert/garden opposition she inherited from her childhood, Cather gives Alexandra the power women lacked in her *Troll Garden* stories: the power to erase the opposition between the two spaces by turning the desert into a garden. In granting a female character the ability to grow her own fruit—and so to resist the goblin men's seductive cries—Cather also revises the myth of America, man's creation of an Edenic, gardenlike realm from an untamed wilderness. Simultaneously she rewrites the literary genres most often used to tell that story, pastoral and epic, by placing a woman at the center of the narrative.

As Gilbert and Gubar observe, most Western literary genres "are, after all, essentially male—devised by male authors to tell male stories about the world."[3] In telling their stories about American culture and the American self, male American writers have drawn most heavily on the themes and conventions of the pastoral and the epic. Both have seemed most adaptable to narrating what we think of as the American experience: the creation of a new culture in a new world and the formation of an Adamic self, cleansed of European imperfections and reborn in the land of spiritual and economic opportunity.[4] The hero of the new American story, as R. W. B. Lewis describes him in his classic study *The American Adam,* is

> an individual emancipated from history, happily bereft of ancestry, untouched and undefiled by the usual inheritances of family and race; an individual standing alone, self-reliant and self-propelling, ready to confront whatever awaited him with the aid of his own unique and inherent resources.

Given the religious background most nineteenth-century Americans shared, it is not surprising, notes Lewis, that the new hero became identified with "Adam before the Fall," who was both the "first, the archetypal man" and the poet invested with Adam's power of naming. Although, as Lewis observes, this new hero was supposed to embody a "new set of ideal *human* attributes" (emphasis mine), feminist critics have pointed out that this representative American self is male.[5]

In her articulation of the myth of America, Nina Baym argues that the "essential quality of America comes to reside in its unsettled wilderness" rather than in society and in "the opportunities that such a wilderness offers to the individual as the medium on which he may inscribe, unhindered, his own destiny and his own nature." As Baym observes, although there is no reason why a woman could not identify with this mythic protagonist, the fact that the other participants in his story—the "entrammeling society and the promising landscape"—are imagined as female makes it difficult for a woman to insert herself into this engendered narrative and to accept its "misogynist implications."[6]

In *O Pioneers!*, however, Cather both uses and transforms this central myth, suggesting that the "essential quality of America" resides not in the unsettled land but in the transformation from "The Wild Land" to "Neighboring Fields," the process of inhabitation that made writers like Cooper and Twain uneasy but that provided, in her view, a more accurate representation of the immigrants' desires and accomplishments. In doing so, she removes the misogyny from the Adamic myth as well as the valorization of the isolated, self-reliant individual; by inscribing herself in the land, Alexandra writes her community's story as well as her own.

Cather was able to challenge the implications of the American pastoral, in which the masculine self flourishes separate from society, by drawing on the epic, in which the hero's triumph is also that of his culture.[7] Wishing to portray Alexandra as a gardener on an epic scale, Cather begins both "The Wild Land" and "Neighboring Fields" with a description of the land and the community before she focuses on her protagonist, placing her in a larger cultural context. Throughout the novel Cather surrounds her heroine with communal events—the farmers buying supplies in town, the forty French boys riding to meet the Bishop, the church fair—that integrate her story with that of her society. Cather's decision to make Alexandra a representative figure is evident in the title change: "Alexandra" became *O Pioneers!*. In borrowing her novel's title from Whitman, Cather both connected Alexandra's story to the pioneer experience and declared that her novel was not an isolated text but part of a shared endeavor by American writers to understand American history and culture.

Recalling Alexander and his founding of empire, Alexandra's name reveals her epic role. More covertly the name refers to a failed epic poet, one who could have told the story of her culture had she not been deprived of authorship and authority by Apollo: Cassandra, the Greek heroine granted the power of prophecy but denied a sympathetic audience. Cassandra was also known by the name "Alexandra," as the classicist Willa Cather would have known. The parallels and the distinctions between the Swedish and the Greek Alexandra reveal how Cather was making her protagonist an artist/heroine who speaks for and to the culture. Like Cassandra, Alexandra is linked with the power of the sun; she also has the gift of prophecy, being able to read the future of her civilization in the land. But here the resemblance ends. Unlike Cassandra, Alexandra does not lose the sun's generative force; both seer and conqueror, she communicates and achieves her visions, turning prophesy into fulfillment as the land awakens. In her 1895 commentary on "Three Women Poets," Cather had linked Christina Rossetti—the woman gifted with the "power of loving" but condemned to the minor genre of the lyric—with Cassandra/Alexandra. In giving this name to her pioneer/artist, Cather thus revises the fate of Cassandra, Rossetti, and the marginal women poets, prophets, and artists they represented.[8]

In associating her novel with Whitman's poetry, Cather was both claiming an affiliation with America's epic and prophetic writer and associating herself with the heir of Emerson, the nineteenth-century progenitor of the Adamic myth. In doing so—like Alexandra—she made her house of fiction in the "big out-of-doors," taking as literary territory not the domestic space of women's writing but the external realm male writers had traditionally claimed as their own. And yet she did this without claiming to be writing a "manly" story. In saying that Alexandra's house *was* the out-of-doors—thus making the domestic realm coterminous with the world—Cather was collapsing the traditional nineteenth-century distinction between "public" and "private," male and female space. Simultaneously she was suggesting that a female hero and a female author could write stories that both sprang from and transcended female experience.

•

Alexandra's role as artist is evident in her visionary imagination, her combination of creative ability and technical skill, her discovery of self-expression in the soil, and in the emergence of the fertile, ordered landscape. Cather, however, wanted to be sure that her readers would not overlook the alliance of the pioneer and the artist, so she has Carl announce it directly. Returning from his sixteen-year absence and seeing the wilderness transformed into settled land, he acknowledges that his is the imitative and Alexandra's the creative sensibility: "I've been away engraving other men's pictures, and you've stayed at home and made your own" (p. 116). Carl's imagination is inspired by others' texts: as a boy he dreams over the illustrated paper, studies patterns for china-painting in the drug store, and collects slides for his magic lantern, preparing for his adult career as the engraver who copies others' designs.[9] Meanwhile Alexandra sees pictures no one has ever viewed before in the land.

In responding with passion and imagination to the Divide—which is her muse, her medium, her subject—Alexandra experiences the creative process in the way Cather later described the composition of *O Pioneers!:* as a yielding to inspiration, as an intermingling of self and other. Erasing subject–object dichotomies and hierarchies, Cather once again rejects the "manly" vision of creativity as a struggle to dominate nature; simultaneously in her "two-part pastoral" she criticizes the power relationship between a male protagonist and a feminine landscape that informs the traditional American pastoral, in which nature is an object—either the virgin land to be raped or the bountiful mother to be sought—against which the male self is defined.[10] In Cather's vision, both the female hero and the maternal land are subjects, and Alexandra defines herself in relation to—instead of against—the natural world. Challenging the traditional culture–nature dichotomy in which men are aligned with culture, women with nature, Cather portrays Alexandra both as connected to nature and as separate from it, a subject with will, imagination, and desire who shapes the land as she is shaped by it.[11] Hence it is she, rather than her father, who succeeds.

Although John Bergson is portrayed sympathetically, Cather makes him the heir of Bartley Alexander and the nameless men in "Behind the Singer Tower," "The Professor's Commencement," and "The Namesake" who try to subdue the "wild country" with their "relentless energy" (*CSF,* p. 141). Like other American Adams, John Bergson wants to make his mark on the soil by imposing his will upon it; he views the wild land as something he "had come to tame," as a horse not yet broken to harness (p. 20). Feeling no connection or similarity between himself and his environment, Bergson projects his own thwarted aggression onto the land, viewing it as "unfriendly to man" just as Ahab construes Moby Dick as evil (p. 20). His sons Lou and Oscar, concerned with their own power, wealth, and status, regard the land only as a source of profit. Like the mean-spirited farmers Thoreau attacks in "The Bean-Field" chapter of *Walden,* they do not recognize that "husbandry was once a sacred art" and define the soil "as property, or the means of acquiring property."[12] The violence submerged in John Bergson's struggle to dominate nature emerges in his sons: Lou and Oscar enjoy hunting wild birds and are disgruntled when they cannot bring their guns with them to Crazy Ivar's house.

With Frank Shabata and Emil Bergson Cather draws a clear relationship between man's violence to nature and to women. Emil foreshadows the lovers' deaths when he shoots the wild ducks, laughing "delightedly" as he transforms vivid wildlife into a "rumpled ball of feathers with the blood dripping slowly from its mouth" (pp. 127–28). Husband and lover are not so different, Cather suggests; Frank Shabata also uses a gun to kill both Emil and the wife who has betrayed him. Such violence is possible because the hunter objectifies what he destroys, separating himself from the object he wants to possess and control; but such possession, as Cather demonstrates, kills the object of desire.[13] Although throughout her fiction Cather would connect the human urge to dominate and to destroy with masculine forces (as in *A Lost Lady*), she creates many individual male characters who deviate from this pattern, rejecting the social definition of masculinity as aggressive and competitive. Cather knew that such traits were not inherently male: she had destroyed innocent life herself when she was dissecting Red Cloud's dogs and cats in pursuit of scientific truth.

In *O Pioneers!* two male characters differ from the masculine stereotype in their relationship to nature: Carl Linstrum and Crazy Ivar. A "thin, frail boy" whose mouth was "too sensitive for a boy's," Carl abandons the struggle with the land. But when he returns, he suddenly realizes that he has missed the beauty of the wild country; even though he had found farming uncongenial, memories of the "wild old beast" haunted him during his years in the city (p. 118). Ivar, the hermit and mystic who views nature as sacred, is even more opposed to the men who destroy life. He doctors sick animals, prohibits guns on his land, and unites himself so fully with nature that his home is indistinguishable from the landscape: "But for the piece of rusty stovepipe sticking up through the sod, you could have walked over the roof of Ivar's dwelling without dreaming that you were near a human habitation" (p. 36). Dissolving the boundaries between himself and the natural world, Ivar does not separate himself from the animals he cares for: "They say when horses have distemper he takes the medicine himself, and then prays over the horses" (p. 33). The identity Ivar experiences between self and world makes him protect life but prevents him from marking his environment: he does no harm, but he does not create.

Alexandra mediates between these polarities of dominance and submission; she views the external world neither as completely separate from nor as coextensive with the self. Her relationship with the land dissolves boundaries between self and other, but with the soil she can both erase the self and create it. Cather first describes the reciprocal bond Alexandra will have with nature when she is returning from her scouting trip to the river farms. Her face seems ''radiant'' to her brother Emil, and Cather tells us why:

> For the first time, perhaps, since that land emerged from the waters of geologic ages, a human face was set toward it with love and yearning. It seemed beautiful to her, rich and strong and glorious. Her eyes drank in the breadth of it, until her tears blinded her. Then the Genius of the Divide, the great, free spirit which breathes across it, must have bent lower than it ever bent to a human will before. The history of every country begins in the heart of a man or a woman. (p. 65)

This is an image of mutual submission, not of power and domination: the ''free spirit'' of the land bends low to Alexandra's will because her heart has been overwhelmed by its beauty, her eyes ''blinded'' by its breadth.

In a later passage, Cather develops the same paradox, linking Alexandra's discovery of self-expression in the soil with her abandonment of self: ''Her personal life, her own realization of herself, was almost a subconscious existence; like an underground river that came to the surface only here and there . . . and then sank to flow on under her own fields'' (p. 203). The metaphor of the underground stream conveys the fertilizing power of Alexandra's unconscious creativity, which nourishes her crops even if hidden from view.

Like an artist who draws on deep and hidden energies, Alexandra takes the external world into the self and returns the self to the world. The times when she was ''peculiarly happy'' are days when she feels ''close to the flat, fallow world about her'' and senses in ''her own body the joyous germination in the soil'' (p. 204). At other times it seems as if her own body extends into the land, as when she feels as if ''her heart were hiding'' along with the ''little wild things that crooned or buzzed in the sun.'' During these moments of connection she receives her vision of the land's transformation: ''Under the long shaggy ridges, she felt the future stirring'' (p. 71). In discovering the land's power, she releases her own creative force, which comes out of hiding along with the wheat and corn.

Unlike Ivar, however, Alexandra wants to shape the soil. Just as Cather knew that writing demanded technique, form, and hard work as well as the ''wisdom of intuition,'' so Alexandra studies crop rotation and irrigation, experiments with silos and pig corrals, and consults with other farmers. Cather does not attribute her success solely to mystic moments of communion with the Genius of the Divide: ''It was Alexandra who read the papers and followed the markets, and who learned by the mistakes of their neighbors'' (p. 23). And at times pragmatic concerns draw her away from romantic contemplation: one summer evening as she ''dreamily'' watches her brothers swimming in a ''shimmering pool,'' her eyes cannot help but drift ''back to the sorghum patch south of the barn, where she was planning to make her new pig corral'' (p. 46).

In her bond with the land—a female presence in the novel, although one filled with ''strength and resoluteness'' like Alexandra herself (p. 77)—she can both assert the self and give it up, playing the roles of both mother and daughter, creator and created. Cather offers an ideal model of feminine development: Alexandra discovers herself in relationship to a maternal presence that allows separation as well as connection. The land is a mirror within which Alexandra sees an image of her "best self," just as Helena was for Martha in "Martha's Lady" and Sarah Orne Jewett was for Cather. Simultaneously Alexandra's relationship to the land mirrors the creative process Cather was experiencing as she faded away into her subject, the "land and people of [her] heart," and so found her voice as a writer.

Yet Alexandra's daydream of the powerful male figure who lifts her up and carries her off raises the disturbing aspects of self-abandonment. A recurrent fantasy Alexandra finds alluring as well as distressing, it emerges when her conscious mind relinquishes control and she lies in bed in the morning in a trancelike state of reverie, hovering between sleep and wakefulness:

> Sometimes, as she lay thus luxuriously idle, her eyes closed, she used to have an illusion of being lifted up bodily and carried lightly by some one very strong. It was a man, certainly, who carried her, but he was like no man she knew; he was much larger and stronger and swifter, and he carried her as easily as if she were a sheaf of wheat. She never saw him, but, with eyes closed, she could feel that he was yellow like the sunlight, and there was the smell of ripe cornfields about him. She could feel herself being carried swiftly off across the fields. After such a reverie she would rise hastily, angry with herself, and go down to the bathhouse that was partitioned off the kitchen shed. There she would stand in a tin tub and prosecute her bath with vigor, finishing it by pouring buckets of cold well-water over her gleaming white body which no man on the Divide could have carried very far. (pp. 205–6)

Cather's critics have been intrigued by this enigmatic passage and not surprisingly have offered varied interpretations, seeing the male figure as a vegetation god, the animus, the Genius of the Divide. If we fix his identity in this way, however, we run into one problem: Cather does not portray the land as male. In the earlier passage describing Alexandra's response to the "Genius of the Divide" with "love and yearning," Cather is careful to refer to the spirit as "it," leaving gender indeterminate; and her actual descriptions of the soil create a strong, procreative, maternal presence. Moreover, trying to establish the dream-figure's identity is difficult since he changes as the dream changes over time, suggesting that his "meaning" can be found only in relationship to his author, the dreamer Alexandra.

This version of Alexandra's fantasy—for which she wants to "prosecute" her body with cold water—occurs most often during her "girlhood" (p. 205). But as she grows older and takes on more responsibilities, the fantasy changes, occurring at night when she is tired rather than in the morning when she is "fresh and strong." Sometimes after a day of work on the farm, her body "aching with fatigue," just before she falls asleep, Alexandra will feel "the old sensation of being lifted and carried by a strong being who took from her all her bodily weariness" (p. 207). Here the "strong being" does not possess the same erotic power as does the godlike man "yellow like the sunlight"; not given a specific gender, this "being" is a parental figure who soothes and comforts the child's weary body. Later in the novel the dream changes once more. Depressed, lonely, and hopeless after the deaths of Emil and Marie, Alexandra again experiences ''the old illusion of her girlhood, of being lifted and carried lightly by some one very strong.'' But this is, we discover, a death wish: Alexandra, ''tired of life,'' longing to escape consciousness, imagines herself being carried off by the ''mightiest of all lovers'' to a place where she will be ''free from pain'' (pp. 282–83).

Although Alexandra feels that death is the true identity of the male figure (whom she can at last see ''clearly'' [p. 282]), we have to question her interpretation. He has no ''clear'' coherent identity; constructed by Alexandra, he is a character whose manifestation and meaning depend on her psychological and emotional state. She creates the erotic god when she is lying in bed ''luxuriously idle,'' the hooded death-figure when she longs to be ''free from her own body'' and the memories of her loss. Like an author creating two characters or a dreamer splitting herself among the various actors in a dream, Alexandra projects herself both into the figure who lifts and the one who is lifted.

Although the male figure's meaning and function change over time, the dream always involves the woman's yielding and submission. In this fantasy Alexandra thus expresses feelings she enacts only with the land but never, in reality, with another person. On the contrary, she is the strong parent-figure for everyone in the novel, the mother/father who either cares for or controls others: in addition to managing the farm, she protects Ivar, watches out for Mrs. Lee, helps Emil to escape the cornfields, advises Marie, organizes her brothers' work. No one takes care of her. Carl Linstrum is the only character who does not place her in a parental role, but he must leave her and the Divide to prove himself a man. Only in the dream does she create a figure stronger than herself, one who can lift her up, remove bodily weariness, and release her from pain.

The repressed part of the self that Alexandra expresses in the dream is thus not the male figure upon which most critical attention has been focused, but the version of herself who is lifted and borne away. Alexandra both fears and wants to release the yielding, passive, regressive aspects of the self, those she at first expresses only with the land. Once the land is tamed and her artistic project completed, however, she does not have that form of release, and so the fantasy becomes more threatening and self-destructive.

Yet Alexandra does not yield to the final version of her fantasy; when Carl Linstrum returns she begins to express her suppressed needs for support, companionship, and tenderness with him. Carl is not a titanic male figure who will lift her away from all human trials, but he offers her human-scale support that Cather regards positively: ''She leaned heavily on his shoulder. 'I am tired,' she murmured. 'I have been very lonely, Carl.' '' (p. 309). A childhood friend rather than an imaginary lover, Carl will live on Alexandra's land instead of carrying her away from it; his love will support but not engulf her. ''Lean[ing] heavily'' on someone else rather than finding relief for fatigue and loneliness only in her daydreams, Alexandra finds in Carl a human version of her mythic bond with the Divide.

•

By the end of the novel, Alexandra has thus established two strong and enduring relationships: with Carl and with the land. In forming bonds with both a male and a female presence, Alexandra reflects a larger pattern in the novel, for Cather wants to attribute her character's success as farmer and artist to her receiving a patrilineal and a matrilineal inheritance. Simultaneously Cather connects her own emergence as a writer to the male and female literary traditions.

Interweaving these two inheritances, both Alexandra and Cather can write a new story, a parallel Cather suggests by the many similarities she draws between Alexandra's role as daughter in the Bergson family and her own role as daughter in a literary family headed by Walt Whitman and Sarah Orne Jewett. Whereas the male Adamic self must be "happily bereft of ancestry, untouched and undefiled by the usual inheritances of family" to achieve autonomous existence and original expression, it is precisely because these two daughters receive inheritances that they can become originators.

At first Cather emphasizes Alexandra's continuity with her male "strain of blood" (*MME,* p. 82). She is said to resemble both her father and grandfather in her intellegence, directness, and strength of will. Although John Bergson would have preferred to see these family likenesses in one of his sons, he knows "it was not a question of choice" and is grateful that in his daughter he has an heir "to whom he could entrust the future of his family and the possibilities of his hard-won land" (p. 24). And so he wills his daughter the land, thereby passing on his will to her just as he hopes that the strength that has vanished from his hands will flow into Alexandra's "strong ones" (p. 25). Cather thus opens her novel with a traditional scene: the dying patriarch choosing an heir. But she reverses the patrilineal inheritance of land, the pattern in her own family as in the larger society, by having the father give power and authority to the daughter. Bergson gives his daughter not a family estate but the injunction to continue the struggle to create one. His work is not finished, and so this is a liberating gift: she will have to complete the work he has begun, thus making it her own.

And Alexandra does make the land her own, cultivating it with "love and yearning" rather than her father's will to dominate. Yet the inheritance from her father remains important; in choosing her as his heir, Bergson gives her the public legitimacy and authority the daughter needs in a patriarchal society. So when Lou and Oscar challenge their sister's right to the land, asserting the patrilineal rights of inheritance ("The property of a family really belongs to the men of the family, no matter about the title" [p. 169]), Alexandra does not have to rely on her spiritual and emotional affinity with the Divide to refute them. Thanks to her father, she has patriarchal law on her side, represented by her legal title to the land: she is entitled to it. "You are talking nonsense," she tells her brothers impatiently. "Go to the county clerk and ask him who owns my land, and whether my titles are good" (p. 168).

So titles do matter: they grant a woman publicly recognized authority and ownership. Cather claimed a similar authority for herself—the authority of authorship—when she chose the title for her book, thus declaring that her subject belonged to her. Simultaneously she was declaring herself Whitman's heir, an even bolder act than Alexandra's accepting the gift of her father's land: the literary daughter was appropriating her father's literary mission, to write the story of America—in effect staking a claim to his land.

If we look at the poem that gave Cather her title, her literary relationship to the paternal figure becomes even more complex. In contrast to her *April Twilights* poems and stories like "The Willing Muse," in which she was being a "bond slave" to A. E. Housman and Henry James, in *O Pioneers!* Cather was being rebellious as well as respectful in alluding to Whitman. Her novel radically revises the vision of pioneering and settlement Whitman advances in "Pioneers! O Pioneers!," a jingoistic hymn to progress, manifest destiny, and the Westward Movement.

Whitman imagines his pioneers as male, referring to them as "Colorado men" and "Western youths" filled with "manly pride and friendship." Surrounding them with images of violence, war, and penetration, he describes the course of empire as a masculine triumph over a feminine land:

> Come my tan-faced children,
> Follow well in order, get your weapons ready,
> Have you your pistols? Have you your sharp-edged axes?
> Pioneers! O Pioneers!

The connection between such weaponry and the violation of nature's body emerges more fully later in the poem:

> We primeval forest felling,
> We the rivers stemming, vexing we and piercing deep the mines within,
> We the surface broad surveying, we the virgin soil upheaving,
> Pioneers! O Pioneers![14]

Whitman thus celebrates what Cather condemns, the masculine urge to subdue and to violate: his land is to be tamed by pioneers brandishing the guns that Cather, like Ivar, would like to exclude from her territory. And so in her novel she offers another vision of the taming of the land, one erasing the polarities and hierarchies in the Whitman poem: male/female, culture/nature, subject/object.

But "Pioneers! O Pioneers!" expresses only one aspect of Whitman, the strident, heartily patriotic poet enraptured with America's progress and power. There was another side to Whitman that is quite consistent with Cather's vision in *O Pioneers!*—the mystic, yielding, erotic speaker of "Song of Myself" who integrates body and soul, male and female, self and other and becomes an artist fueled by this integration of opposites.[15] The likelihood that Cather had "Song of Myself" in mind as well as "Pioneers! O Pioneers!" is suggested by the parallels between the endings of "Song" and *O Pioneers!* Whitman ends his poem envisioning himself reincarnated in the grass, both nature's grass and his own *Leaves of Grass* ("I bequeath myself to the dirt to grow from the grass I love,/ If you want me again look for me under your boot-soles"). Similarly, in the last paragraph of her novel Cather imagines Alexandra becoming part of a continuous cycle of life, one day returning to the "bosom" of the Divide, to be reborn in the "yellow wheat, in the rustling corn, in the shining eyes of youth"—eyes that will read Alexandra's story in Nebraska's "leaves of grass," the wheat and corn she helped the land to produce (p. 309).[16]

While Alexandra and her creator both rewrite the inheritances they receive from their fathers, absorbing what they can use and altering what they cannot, they draw more directly on their mothers' gifts. "Thinking back through their mothers," to paraphrase Virginia Woolf, they imagine themselves.[17] Alexandra's inheritance from her mother is private and unofficial. Mrs. Bergson cannot give her the authority to cultivate the land, but she shows her daughter how to farm it by the model of creativity she exemplifies. "Alexandra often said that if her mother were cast upon a desert island, she would thank God for her deliverance, make a garden, and find something to preserve." By making a garden in the desert, Mrs. Bergson both adapts to the land and expresses creativity and ingenuity as she tries to make the preserves she loves with new materials:

> Preserving was almost a mania with Mrs. Bergson. . . . She made a yellow jam of the insipid ground-cherries that grew on the prairie, flavoring it with lemon peel; and she made a sticky dark conserve of garden tomatoes. She had experimented even with the rank buffalo-pea. . . . When there was nothing more to preserve, she began to pickle. (p. 29)

Alexandra's mother preserves on a symbolic level as well, continuing traditions and customs from the Old World that give her family a sense of "household order" and ritual in the new (p. 20).

Even before the daughter takes over the family land, then, the mother has made her mark on the soil: evidence of "human striving" is in her log house, her garden, her glass jars of preserves. Although she does not make the mother–daughter relationship a prominent feature in the book, the link between Mrs. Bergson and Alexandra reveals Cather's new recognition of artistry among the rituals of domesticity. When Alexandra extends her mother's efforts and transforms the wilderness into a fruitful, orderly garden, she gains the public recognition her mother lacked. But since the daughter is in a sense carrying on and expanding the work her mother began, the two women are collaborators, not competitors: the seeds of the daughter's achievement are in the mother's garden.

Although Alexandra never pays direct tribute to her mother's example—the bond between the two women is only briefly suggested—Cather publicly acknowledged the influence of Sarah Orne Jewett on her novel when she dedicated *O Pioneers!* to her mentor. Jewett's literary advice and supportive friendship had been important to Cather, but the dedication calls our attention in particular to the links between Jewett's fiction and Cather's novel.

The American pastoral was not completely a male-dominated genre, and Cather may have wanted to associate her own pastoral with Jewett's *The Country of the Pointed Firs,* which also portrays rural people living harmoniously within their environment. Somewhat concerned that the two parts of her pastoral might not have been perfectly integrated, Cather may also have found consolation in remembering that *The Country of the Pointed Firs* also intertwined separate stories which, taken together, painted the portrait of a community. Moreover, the mother–daughter bond between Mrs. Todd and the narrator in *Country* and that between Mrs. Bergson and Alexandra share an important characteristic—in both texts daughter/artists receive a creative inheritance from mothers whose arts are domestic.

Even stronger parallels exist, however, between *O Pioneers!* and "A White Heron," one of the short stories Cather included in her edition of Jewett's fiction. In Jewett's story a young girl's peaceful union with nature is disrupted by an intrusive male who brings sexuality and violence to her secluded rural world. In asking her to reveal the white heron's hidden nest, the hunter threatens not only Sylvia's bond with nature but also her autonomy, her very self. Even though this girl's "woman's heart" is "vaguely thrilled by a dream of love" inspired by the handsome intruder, the "piteous sighs of thrushes and sparrows dropping silent to the ground, their songs hushed and their pretty feathers stained and wet with blood" warn her to protect the heron, nature, and herself, and she refuses to "tell the heron's secret and give its life away" (*CPF*, pp. 161–171). The imagery recalls the scene where Emil shoots the wild ducks, in which Cather is similarly condemning male violence to nature (and symbolically to women).

Sylvia's respectful bond with the natural world suggests Alexandra's. The "happiest day of her life" occurs when she, like Sylvia, discovers a bird's secret hiding place: one day she and Emil come upon an inlet protected by overhanging willows where a wild duck "was swimming and diving and preening her feathers, disporting herself very happily in the flickering light and shade." The two watch the bird for a long time, and years later Alexandra thinks of the duck "as still there, swimming and diving all by herself in the sunlight, a kind of enchanted bird that did not know age or change" (pp. 204–5). She assumes that Emil shares her appreciation, but as we learn he does not: Emil wants to kill the ducks to possess them, whereas—like the artist—Alexandra possesses the duck in memory, where she grants it enduring life. The references to "A White Heron" thus signify both Cather's and Jewett's joint effort to revise the male-authored story of woman and nature as well as the younger writer's willingness to acknowledge her debt to her precursor.

Cather may have found it easy to honor Jewett in *O Pioneers!* because, like Alexandra, she also extends her mother's range. Going beyond Jewett's sketches to write her first successful "long story," giving her novel a title that does not limit it to a region, associating herself with Whitman's epic mode, Cather announces her intention to represent a central pattern in American history and culture. And in intertwining "The White Mulberry Tree" with *O Pioneers!* she includes disturbing subjects that generally did not find their way into Jewett's Maine pastorals: passion, violence, death.

•

The story of Alexandra's triumph occupies only about the first quarter of *O Pioneers!* What is so impressive about the novel is that—even though Cather projects her own creative success into Alexandra's triumph—she does not write a self-congratulatory hymn to the woman artist's glories, as she would in the concluding section of *The Song of the Lark*. Once the land is settled Cather moves to other concerns, giving increasing attention to the lovers' story (which Alexandra does not understand or perceive) and to Alexandra's breakdown after the lovers' murder.

Some of Cather's readers have criticized her for writing an episodic novel in which the lovers' subplot is imperfectly integrated into the whole, but she had good reasons for intertwining her two stories. Passion, Cather knew, was the artist's "open secret," the source of creative energy; and in her carefully contrasted stories, she exposes the insufficiency—even the danger—of passion channeled into romantic love and the grandeur of passion directed toward "something complete and great" (*MA*, p. 18). These concerns she would explore throughout her career, returning to the dangers of romantic love in *A Lost Lady, My Mortal Enemy,* and *Lucy Gayheart,* and exploring the human desire for self-transcendence in *The Song of the Lark, My Ántonia, One of Ours, The Professor's House,* and *Death Comes for the Archbishop.*[18]

In telling the story of Emil's and Marie's passion and death, Cather was retelling one of the "two or three" plots that recur in Western literature: the *Liebestod,* the story of adulterous or unlawful lovers whose desire cannot be contained by social structures and conventions. She was also writing a variation of the "heroine's text" of descent and death: the plot of *Clarissa, The Mill on the Floss,* and *The Awakening* as well as of the male-authored tales of adulterous love she associated with Chopin's novel, *Madame Bovary* and *Anna Karenina.*[19] Cather was revising this plot not by rewriting it but by integrating it into Alexandra's story, which begins and ends her novel, just as Alexandra was born before and outlives Marie. Although she tells the lovers' story with empathy and compassion, Cather wants to subordinate their narrative to the other narrative she was writing: not only that of Alexandra's artistic triumph but also her unconventional relationship with Carl, the friend she will someday marry.

Cather admires Marie's spontaneity, vitality, and warmth, but she is critical of the ways in which her imagination is structured by social and narrative conventions. Like the "self-limited" Edna Pontellier, Marie is one of the "spoils of the poets," unable to envision a life-story for a woman outside of the romantic plot. Cather's description of Marie as a child hints that she will live out a common female destiny, although at first we assume marriage rather than death will be the end of her story. More traditionally feminine than the young Alexandra, who breaks gender and dress codes, Marie has "brown curly hair, like a brunette doll's," and dresses in the "Kate Greenaway" manner, already a follower of fashion.

Marie's choice for a desirable future also conforms to feminine fashions. We first see her flirting with the men in the general store who tell her that "she must choose one of them for a sweetheart" and who reward her with "bribes"—"candy, and little pigs, and spotted calves." Accepting their bribes, Marie looks around the group "archly" and then chooses her uncle by saying "Here is my sweetheart." When she grows up Marie still limits her dreams to the selection of sweethearts: first Frank, then Emil. As the heroine of a love story, her only choice is to keep saying "Here is my sweetheart," using the vocabulary given to her by men early in the novel (pp. 11–12). Cather demonstrates Marie's continuing imaginative preoccupation with romantic narrative in the fortune-telling scene. Dressed in a Bohemian costume at the church fair, Marie entertains the crowd by telling them the stories they want to hear, all tales of love and marriage—much like the popular women writers Cather had once wanted to disown.

Marie thrives on romantic separation, her passion intensifying when she can shape the absent lover according to the requirements of her own imagination. She becomes infatuated with Frank during her years in the convent when she dreams over the photographs he sends her; similarly her thoughts and desires turn to Emil when he is in Mexico, sending her letters that "enlist her imagination in his behalf" (p. 199). When Emil tells her that he must finally leave the Divide, she is strangely exhilarated, anticipating the "perfect love" she can experience when he is gone (p. 249). When Emil rushes to Marie in the orchard, where she has "lived a day of her new life of perfect love" in a deathlike trance, she gives him a conditional greeting: "I was dreaming this . . . don't take my dream away!" She will accept him as long as he corresponds to the "perfect" dream-lover she has created.

Cather demonstrates that a self-absorbed romantic love is not solely a female preoccupation. Emil also experiences trancelike states of lassitude and reverie in which he releases the romantic rather than the creative imagination, projecting his own desires into an image of Marie rather than into another identity as the writer does in creating a character. The desire prompting the romantic lover's imagination is narcissistic, Cather suggests; when Emil finally finds "the thing he looked for" in Marie's eyes, he glimpses "his own face" (p. 259).

Cather wants to distinguish this sort of dreaming—which resembles Bartley Alexander's solipsistic fantasies of Hilda—from Alexandra's visionary glimpses of the future and the artist's creative imagination, the dreaming Cather was doing in descending to the "bottom of . . . consciousness" in writing *O Pioneers!* Because the lovers' dreams construct an idealized other in terms of the self—like infantile fantasies of a perfect mother—they are regressive rather than creative. So the lovers' deaths are not merely the punishments required by society and the narrative for unlawful passion; like Bartley's drowning in the St. Lawrence, they signify the extent to which the lovers have already drowned in "self-limited" fantasies.

When Ivar discovers the murdered lovers in the orchard, the narrator tells us that the "story of what had happened was written plainly on the orchard grass, and on the white mulberries that had fallen in the night and were covered with dark stain" (p. 268). But since, unlike Alexandra, the lovers do not find an imaginative and enriching bond with something beyond the self, their story is not only one that has already been written, the love/death plot: unlike Alexandra's, it is also impermanent, a temporary mark on the land.

•

Although Cather turns her attention to the lovers' plot after she gives Alexandra her early triumph, she continues to develop her heroine's unconventional story when she portrays her sorrow and despair after she loses her brother and her friend. During her years of success Alexandra has very little "realization of herself" (p. 203); she puts her personality into her enterprises, and as a result does not express her own emotional needs or understand those of others. Opening her eyes to the Divide, she closes them to Emil and Marie, never sensing their intensifying passion; her imagination fails her when it comes to human relationships.

Yet the depression into which she is plunged after their deaths is ultimately a good thing, Cather suggests: this tragedy is so great that Alexandra cannot repress her grief and loss, and this emotional emergence allows her finally to acknowledge her love for Carl, whom she has kept at arm's length throughout the novel. Carl can now overcome his own fear of being judged a "little" man by conventional standards and returns because he knows he is needed. So Cather ends her novel with a calmer undoing of repression than she had portrayed in "The Bohemian Girl" or in the lovers' story: Alexandra is able to feel and to express her needs for intimacy, for dependence, for companionship.

Although a man and a woman acknowledge their love and plan their marriage in the concluding pages of *O Pioneers!,* this is not the traditional ending of nineteenth-century women's fiction. Nor is it simply a token marriage, Cather's half-hearted attempt to follow narrative conventions and please her readers. Neither lover conforms to the expected gender role, and they will marry after rediscovering each other in mid-life, drawn together, as are the married couple in "The Sentimentality of William Tavener," by their shared memories and their common history. And although Cather has impressed us with Alexandra's strength early in the novel, at the end she suggests that Carl has powers that her heroine lacks. Sensitive to the emotional and psychological nuances that Alexandra misses, he has to interpret the lovers' story to her. When she asks him why—if he had sensed their attraction—he had done nothing about it, he replies, "My dear, it was something one felt in the air, as you feel the spring coming, or a storm in summer. I didn't *see* anything. . . . After I got away, it was all too delicate, too intangible, to write about" (p. 305). Gifted with the supposedly female power of intuition, Carl can read human relationships the way Alexandra can read the land, an ability that she—and the author—respects.

"When friends marry, they are safe," Alexandra tells Carl at the end of the novel. "I think we shall be very happy" (p. 308). Although some readers might lament the exclusion of passion from their future, because Cather associates sexual desire both with female subordination and with romantic narcissism, she wants to use friendship as the model for a revised male–female bond. Alexandra and Carl will be "safe" not only because they will be protected from the dangers of passion but also from the romantic lovers' denial to each other of an independent selfhood. We can see the difference between the two ways of loving in Carl's childhood memory of his bond with Alexandra. Sitting at the place where the "Bergson pasture joined the one that had belonged to his father," he remembers how "just there" he and Alexandra "used to do their milking together, he on his side of the fence, she on theirs" (p. 126). The setting—the point where the fields both meet and are distinguished by the fence—represents their own union, one that balances separation and connection. Unlike Emil and Marie, in joining together Alexandra and Carl do not lose themselves.

Alexandra's projected marriage to Carl thus is not Cather's gesture to the marriage plot, her attempt to placate her readers by giving an untraditional narrative a traditional conclusion. Instead, it exemplifies Cather's revision of another inherited genre, the nineteenth-century woman's *Künstlerroman* in which love and work, femininity and creativity, can never be combined, assumptions Cather shared herself during the 1890s and early 1900s. In nineteenth-century works like *The Story of Avis* or *The Awakening,* the lover or husband is invariably a menace to the heroine's desire for creative expression. As Rachel Blau DuPlessis argues, when twentieth-century women writers revise this plot they generally dispense with the patriarchal husband/lover and attribute the heroine's empowerment to her discovery of a maternal heritage.[20] In *O Pioneers!,* however, Cather both describes the heroine's discovery of female creative power in her bond with the land and rewrites the heterosexual love story to give Alexandra a tender friend and lover who appreciates rather than threatens her achievements. To be sure, the story of Alexandra's work and love is written and achieved sequentially rather than simultaneously. But the structure of the novel demonstrates Cather's desire to reject the oppositions between private and public, gender and vocation, affection and ambition that had circumscribed nineteenth-century novels of gifted women as well as her own *Troll Garden* stories.[21]

•

Even though Cather felt confident in her newly released creative powers, because she had drawn deeply on the self in writing *O Pioneers!* she feared that she might have committed the unforgivable female literary sin of overwriting. Perhaps her portrait of the land was too lyrical, her writing too emotional? Had she been able to control and to shape the passion inspiring the novel? She begged Elizabeth Sergeant to tell her the "unvarnished truth," to come down on her "hard" if she found such flaws (*WC:AM,* p. 92). But Sergeant liked the book, and so did Ferris Greenslet, who told his Houghton Mifflin colleagues that *O Pioneers!* would establish Cather as a "novelist of the first rank."[22]

Once the reviews began to come out, Cather must have been even more encouraged to follow in her new literary direction. Not only were most of them positive, but some reviewers even sensed that this was an unusual story for a woman writer to tell and applauded Cather's unorthodoxy in creating a strong, creative heroine. *The Bookman* found *O Pioneers!* to be merely an ordinary local-color novel and professed to have anticipated all the turns of plot ("it requires no keen guesswork to forsee that the young neighbour of her youth will return"), but the *New York Times* found a much more innovative novel. The anonymous reviewer praised Cather for creating a "new mythology" in her novel by replacing the traditional American hero—a man who leaves the farm for the city—with a female "goddess of fertility" who possesses a "deep instinct for the land" and a respect for lives lived close to the soil. Cather had actually substituted three heroines for one hero—Alexandra, Marie, and the land—and some might even term the novel "feminist," the reviewer acknowledged, but Cather could have had nothing so "inartistic" in mind. Nevertheless, the reviewer concluded, this was a "feminine" story because it expressed an "expansion of the very essence of femininity"; and at the same time *O Pioneers!* was also "American in the best sense of the word."[23]

Cather must have been pleased by the *Times*' assessment, but she may have been even happier to read the glowing notice in the Lincoln *Sunday State Journal:* the debacle of "A Wagner Matinee" was not going to be repeated, and her work was going to be appreciated in her home state. Although reviewer Celia Harris wanted *O Pioneers!* to be a realistic and comprehensive portrayal of Nebraska life and customs rather than a consciously shaped fiction—and so lamented the absence of native-born American characters—she praised Cather's "extraordinary" and "beautiful" book. Harris particularly applauded Cather's transition from an exotic early style to a more simple, unaffected, and "American" prose.[24]

In drawing on her own memories, affections, fears, and experiences, Cather had indeed taken more command of the language, leaving behind the convoluted, stilted style of her Jamesian stories for a simpler and more resonant prose. Writing of her own country and people, she knew what words to use. When a Bryn Mawr teacher of Elizabeth Sergeant's criticized Cather's use of "globule" in *O Pioneers!*, suggesting that "drop" would have been the better word to describe dew, Cather was not intimidated. She defended her choice "stoutly," Sergeant remembered, saying that dewdrops "could be of several shapes," but only "globule" described the "firm round drop that formed on prairie grass" (*WC:AM,* p. 119).

•

This simple comment reveals that Cather had moved from apprentice to accomplished writer: she now knew what stories to tell and how to tell them. More comfortable with the language, having heard her own voice emerge in *O Pioneers!*, Cather was ready to see resemblances between herself and Olive Fremstad, the Wagnerian soprano she met soon after completing *O Pioneers!* In the course of writing an article on women opera singers for *McClure's,* she interviewed Fremstad and began an important friendship. The Swedish–American singer made a powerful impression on Cather, possibly because she saw an analogue to her own recent transition from the short story to the novel in Fremstad's courageous decision to extend her range from contralto to soprano. Singing teachers, music critics, and opera-goers had all agreed that Fremstad's strength was in her lower tones, but she believed that the "Swedish voice is always long." Because she did not apply traditional musical categories to herself, the contralto defied conventional wisdom by extending her upper scale "tone by tone, without much encouragement." "I do not claim this or that for my voice," Fremstad told Cather. "I do not sing contralto or soprano. I sing Isolde. What voice is necessary for the part I undertake, I will produce."[25]

As the author of *O Pioneers!* who had also produced the voice she needed for the part, Cather delighted in the correspondences she saw between herself and Fremstad and decided to make her next artist/heroine an opera singer. In *The Song of the Lark* we see the stories of Cather and Fremstad combined in Thea Kronborg, the singer who discovers the power of her voice after a liberating sojourn in the Southwest. In writing this veiled autobiographical novel—her portrait of the artist—Cather was inspired in part by her regeneration in the Southwest and her meeting with Fremstad. But the strongest biographical source for this novel was the emergence of her writer's voice in *O Pioneers!*: *The Song of the Lark* could not have been written if Cather were not already sure of her own creative abilities.

The encouragement she derived from her success in *O Pioneers!* can be seen in the letters Cather wrote to editor Ferris Greenslet about *The Song of the Lark*.

Unlike her hesitant letters to Sergeant about *O Pioneers!,* these exude excitement and self-confidence. Writing Greenslet on December 13, 1914, she told him that her new novel would be much more interesting and colorful than *O Pioneers!,* and she was sure he would like her heroine. A few days later Cather informed Greenslet that she knew this was a good novel; she was writing at a record pace and enjoying it more than she ever had before. Finally in March 1915 she confessed that she thought so well of her book she had better try not to give her opinion—but she would say that he would not publish a story like it every day.[26]

There were many reasons for Cather's self-assured delight in her new story: her exhilarating friendship with Fremstad, her success with *O Pioneers!,* her unleashing of pent-up creative energies, her sense that she, like the opera singers she admired, possessed a distinctive voice. Underlying these was her new freedom to tell, if she wished, a woman's story. She knew she was doing this in *The Song of the Lark:* in addition to the internal evidence of the novel, we have her statement to Greenslet that Houghton Mifflin should advertise *Song* in women's colleges. Students at Smith, Barnard, and Radcliffe would like Thea, she knew: they also wanted that defiant, unsentimental kind of success.[27]

With this comment, Willa Cather shows us how much her creative breakthrough owed not just to Jewett but to the vital, imaginative continuity she had established with the women in her own past—the storytellers of Willow Shade and the Divide. Once they had told stories to her; now Cather was the storyteller herself, addressing a receptive female audience. And the stories she told in *O Pioneers!* and *The Song of the Lark* demonstrate the alliance between womanhood and creativity that Cather's infatuation with male values had obscured for so many years. The younger writer's literary emergence would come, Sarah Orne Jewett had thought, from "recognition," from knowing again. And so it did: from Cather's return to what she had once known as child, but had forgotten—the narrative power of women's voices.

And finally Cather loved writing *The Song of the Lark* because her novel enclosed and preserved the imaginative energy she had poured into it. The book had so much of the West in it, she told Greenslet, that even when she was old and could no longer run about the desert she knew that the landscape she loved would always be in the book for her. She would only have to lift the lid.[28] To lift the lid, to open a box, to pull out a drawer—after *O Pioneers!,* the creative process for Willa Cather was at last the disclosure and unveiling of a woman writer's power.

NOTES

1. Rachel Blau DuPlessis, *Writing Beyond the Ending: Narrative Strategies of Twentieth-Century Women Writers* (Bloomington: University of Indiana Press, 1985), p. 4. These alternatives correspond to the "euphoric" and "dysphoric" poles Nancy Miller identifies as structuring the "heroine's text" in the eighteenth-century novel—marriage and social in-

tegration on the one hand, death and isolation on the other (Nancy K. Miller, *The Heroine's Text: Readings in the French and English Novel, 1722–1782* [New York: Columbia University Press, 1980], p. xi).

2. Marianne Hirsch, "Spiritual *Bildung:* The Beautiful Soul as Paradigm," *The Voyage In: Fictions of Female Development,* ed. Elizabeth Abel, Marianne Hirsch, and Elizabeth Langland (Hanover, N.H.: University Press of New England, 1983), p. 28. As she observes, in nineteenth-century women's fiction the novel of development is frequently the "story of the potential artist who fails to make it" (p. 28).

3. Sandra M. Gilbert and Susan Gubar, *The Madwoman in the Attic: The Woman Writer and the Nineteenth-Century Literary Imagination* (New Haven: Yale University Press, 1979), p. 67.

4. Although both genres seem to reflect and shape American experience, the pastoral has been a more enduring part of the American literary imagination. As Leo Marx observes, the "pastoral ideal has been used to define the meaning of America since the age of discovery" because the genre's story of withdrawal from a corrupt civilization to a virginal landscape seemed attainable on the new continent (*The Machine in the Garden: Technology and the Pastoral Ideal in America* [New York: Oxford University Press, 1964; rpt., 1969], p. 3).

5. R. W. B. Lewis, *The American Adam: Innocence, Tragedy and Tradition in the Nineteenth Century* (Chicago: University of Chicago Press, 1955), p. 5. The ways in which the Adamic myth both stereotypes and excludes women have been explored by Annette Kolodny in *The Lay of the Land: Metaphor as Experience and History in American Life and Letters* (Chapel Hill: University of North Carolina Press, 1975), and by Nina Baym in "Melodramas of Beset Manhood: How Theories of American Fiction Exclude Women Writers," *American Quarterly* 33, no. 2 (Summer 1981): 123–39.

6. Baym, pp. 132–33. As Baym observes, women can rewrite the male-centered American myth in many ways, and she suggests that Cather does so in *O Pioneers!* by casting the land as male, thus reversing the traditional gender roles. But although Alexandra's dream-figure is male, the land is described as feminine and maternal—and hence Cather is rewriting the traditional story in a more complicated way, suggesting that the land—like her heroine—cannot be easily categorized by applying socially derived notions of gender. (Baym, p. 136.)

7. Paul A. Olson explores Cather's use of epic conventions to describe the settling of the Midwest in "The Epic and the Great Plains Literature: Rolvaag, Cather, and Neihardt," *Prairie Schooner* 55 (Spring/Summer 1981): 263–85. As he observes, Cather draws primarily on the *Aeneid.* Olson argues, however, that *My Ántonia* depends most heavily on Virgil's story of the founding of a civilization; in my view *O Pioneers!* is Cather's attempt to write a pastoral epic, *My Ántonia* her attempt to write a myth.

8. In part Cather used the parallel with Cassandra to exalt Rossetti's art when she described her as "never yielding to accepted forms" and writing with the "mystic, enraptured faith of Cassandra, which is a sort of spiritual ecstasy, and which is to the soul what passion is to the heart." Yet she also hinted that Rossetti had not been greatly favored by Apollo: given only a slight "spark" of the "divine fire," she had a heart where the "laurel had not taken root kindly" (*KA*, pp. 347–49).

9. Early in the novel Carl tells Alexandra that he wants to "paint some slides" for his magic lantern "out of the Hans Andersen book" (p. 17).

10. See Kolodny, pp. 3–9.

11. Sherry Ortner, "Is Female to Male as Nature Is to Culture?" in *Women, Culture and Society,* ed. Michelle Zimbalist Rosaldo and Louise Lamphere (Stanford: Stanford University Press, 1974), pp. 67–87.

12. Henry David Thoreau, *Walden,* ed. Sherman Paul (Boston: Houghton Mifflin [Riverside ed.], 1959, p. 114.

13. Frank's killing of his wife is linked with his urge to dominate her: his idea of love is based on the memories of their first days of marriage, when she had been his "slave" (p. 222).

14. Walt Whitman, *Leaves of Grass,* ed. Harold W. Blodgett and Sculley Bradley (New York: Norton, 1965), pp. 229–32.

15. As Paul Zweig observes, before he became a "poet in full possession of his powers," Walt Whitman would integrate opposites within the self—active and passive, masculine and feminine. Whitman represents this union symbolically in the marriage described in the opening stanzas of "Song of Myself," his great poem which records his own creative emergence: another parallel between "Song and Myself" and *O Pioneers!,* which also records the process of its own composition (*Walt Whitman: The Making of a Poet* [New York: Basic Books, 1984], p. 87). The connections Zweig makes between Whitman's experience of the creative process and its representation in "Song of Myself" have several parallels in Cather's literary evolution. See in particular Chapter 7, "Song of Myself."

16. John J. Murphy develops several parallels between *O Pioneers!* and *Leaves of Grass* in "A Comprehensive View of Cather's *O Pioneers!*" in *Critical Essays on Willa Cather,* ed. John J. Murphy (Boston: G. K. Hall, 1984), pp. 113–27, esp. pp. 124–26.

17. Virginia Woolf, *A Room of One's Own* (New York: Harcourt, Brace, 1929), p. 79.

18. E. K. Brown refers to the novel's "happy looseness of structure" while David Daiches thought this "unevenly patterned" novel contained "disparate elements which are never wholly resolved into a unity" (E. K. Brown, *Willa Cather: A Critical Biography,* completed by Leon Edel [New York: Knopf, 1953], p. 179; David Daiches, *Willa Cather* [Ithaca: Cornell University Press, 1951], pp. 28–29). More recent readers have been convinced of the novel's unity: see David Stouck, *Willa Cather's Imagination* (Lincoln: University of Nebraska Press, 1975), p. 32; Sharon O'Brien, "The Unity of Willa Cather's 'Two-Part Pastoral': Passion in *O Pioneers!,*" *Studies in American Fiction* (Fall 1978): 157–71; and Bruce P. Baker II, *O Pioneers!:* The Problem of Structure," *Great Plains Quarterly* 2, no. 4 (Fall 1982): 218–23.

19. Miller, p. xi. As Miller observes, the "heroine's text is the text of an ideology that codes femininity in paradigms of sexual vulnerability" (p. xi). Miller uses her terminology to analyze eighteenth-century French and English fiction, but she argues that in the nineteenth-century novel "the ideological underpinnings of the old plot" are not seriously threatened: "experience for women characters is still primarily tied to the erotic and the familial," and the "sexual faux pas is still a fatal step" (p. 157).

20. Du Plessis, pp. 91, 93–94.

21. It is possible that as a lesbian writer Cather found it easier to revise male–female relationships than did her contemporaries Edith Wharton and Ellen Glasgow. Not invested in the heterosexual love plot herself, never having experienced an oppressive or unsatisfactory marriage, Cather may have been able more easily to envision alternatives. And her portrayal of Carl and Alexandra may also owe something to her own marriage-like friendship with Edith Lewis.

22. Quoted in Brown, p. 179.

23. Frederick Taber Cooper, "Big Moments in Fiction and Some Recent Novels," *The Bookman* 37 (August 1913): 666–67 (reprinted in Murphy, pp. 112–13); "A Novel Without a Hero," *New York Times,* September 14, 1913.

24. Celia Harris [Review of *O Pioneers!*], Lincoln *Sunday State Journal,* August 3,

1913, p. A-7; reprinted in *Willa Cather Pioneer Memorial and Educational Foundation Newsletter* 16, no. 2 (Spring 1972), n.p.

25. Willa Cather, "Three American Singers," *McClure's* 42, no. 2 (December 1913): 33–48, esp. 45–46.

26. Willa Cather to Ferris Greenslet, December 13, [1914], December 31, 1914, and March 28, 1915, Houghton Library, Harvard University, Cambridge, Mass.

27. Cather to Greenslet, September 26, [1915].

28. Cather to Greenslet, March 28, 1915.

Index